# Topics in Current Genetics



Series Editor: *Stefan Hohmann*

Available online at
SpringerLink.com

Lilia Alberghina · Hans V. Westerhoff  (Eds.)

# Systems Biology:
## Definitions and Perspectives

With 88 Figures, 10 in Color; and 22 Tables

 Springer

Professor Dr. Lilia Alberghina
Dept. of Biotechnology and Biosciences
University of Milano-Bicocca
Piazza della Scienza 2
20126 Milano
Italy

Professor Dr. Hans V. Westerhoff
BioCentrum Amsterdam
Mathematical Biochemistry
University of Amsterdam
and
Molecular Cell Physiology
Faculty of Earth and Life Sciences
Free University
De Boelelaan 1087
1081 HV Amsterdam
The Netherlands

The cover illustration depicts pseudohyphal filaments of the ascomycete *Saccharomyces cerevisiae* that enable this organism to forage for nutrients. Pseudohyphal filaments were induced here in a wild-type haploid MATa Σ1278b strain by an unknown readily diffusible factor provided by growth in confrontation with an isogenic petite yeast strain in a sealed petri dish for two weeks and photographed at 100X magnification (provided by Xuewen Pan and Joseph Heitman).

ISSN 1610-2096
ISBN 978-3-540-74269-2        e-ISBN 978-3-540-31453-0

Library of Congress Control Number: 2007934276



Springer-Verlag is a part of Springer Science + Business Media
springer.com



Printed on acid-free paper – 39/3152-YK – 5 4 3 2 1 0

# Table of contents

## METABOLISM

**Modeling the *E. coli* cell: The need for computing, cooperation, and consortia ...... 163**



**Metabolic flux analysis: A key methodology for systems biology of metabolism ...... 191**

# List of contributors

Aguilera-Vázquez, Luciano
Institute for Biochemical Engineering, Allmandring 31, 70569 Stuttgart, Germany

Alberghina, Lilia
Dept. of Biotechnology and Biosciences, University of Milano-Bicocca, Piazza della Scienza 2, 20126 Milano, Italy
lilia.alberghina@unimib.it

Bentele, Martin
Division Theoretical Bioinformatics, German Cancer Research Center (DKFZ), 69120 Heidelberg, Germany

Berg, Otto
Department of Molecular Evolution, Uppsala University, Norbyvägen 18 C S-752 36 Uppsala, Sweden

Bettenbrock, Katja
Systems biology group, Max-Planck-Institut für Dynamik, komplexer technischer Systeme, Sandtorstr. 1, 39106 Magdeburg, Germany

Bruggeman, Frank J.
Molecular Cell Physiology & Integrative Bioinformatics, Biocentrum & Faculty of Earth and Life Sciences, De Boelelaan 1085, 1081 HV, Vrije Universiteit, Amsterdam, The Netherlands

Chen, Katherine C.
Department of Biology, Virginia Polytechnic Institute and State University, Blacksburg, Virginia, VA 24061, USA

Demin, Oleg V.
A.N. Belozersky Institute of Physico-Chemical Biology, Moscow State University, Moscow, Russia

Ehrenberg, Måns
Department of Cell and Molecular Biology, Uppsala University, Box 596, Uppsala, Sweden
ehrenberg@xray.bmc.uu.se

Eils, Roland
Division Theoretical Bioinformatics, German Cancer Research Center (DKFZ), 69120 Heidelberg, Germany
r.eils@dkfz.de

Elf, Johan
Department of Cell and Molecular Biology, Uppsala University, Box 596, Uppsala, Sweden
johan.elf@icm.uu.se

Fell, David A.
School of Biological & Molecular Sciences, Oxford Brookes University, Headington, Oxford OX3 0BP, UK
dfell@brookes.ac.uk

Finney, Andrew
Science and Technology Research Institute, University of Hertfordshire, Hatfield, AL10 9AB, UK

Fischer, Sophia
Systems biology group, Max-Planck-Institut für Dynamik, komplexer technischer Systeme, Sandtorstr. 1, 39106 Magdeburg, Germany

Gilles, Ernst D.
Systems biology group, Max-Planck-Institut für Dynamik, komplexer technischer Systeme, Sandtorstr. 1, 39106 Magdeburg, Germany
gilles@mpi-magdeburg.mpg.de

Goryanin, Igor I.
GlaxoSmithKline, Scientific Computing & Mathematical Modeling, Stevenage, SG2 8PU, UK
goryanin@inf.ed.ac.uk

Heinrich, Reinhart
Humboldt University Berlin, Institute of Biology, Department of Theoretical Biophysics, Invalidenstraße 42, 10115 Berlin, Germany
reinhart.heinrich@biologie.hu-berlin.de

Hofmeyr, Jan-Hendrik S.
Dept. of Biochemistry, University of Stellenbosch, Private Bag X1, Matieland 7602, Stellenbosch, South Africa

Hohmann, Stefan
Department of Cell and Molecular Biology, Göteborg University, Box 462, S-40530 Göteborg, Sweden
hohmann@gmm.gu.se

Hucka, Michael
Control and Dynamical Systems, California Institute of Technology, Pasadena, CA 91125-8100 USA

Kholodenko, Boris N.
Department of Pathology, Anatomy and Cell Biology, Thomas Jefferson University, 1020 Locust St., Philadelphia, PA 19107, USA
Boris.Kholodenko@jefferson.edu

Kitano, Hiroaki
Sony Computer Science Laboratories, Inc, 3-14-13 Higashi-Gotanda, Shinagawa, Tokyo 141-0022 Japan, and, The Systems Biology Institute, Suite 6A, M31 6-31-15 Jingumae, Shibuya, Tokyo 150-0001 Japan
kitano@symbio.jst.go.jp

Klipp, Edda
Berlin Center for Genome Based Bioinformatics (BCB), Max-Planck Institute for Molecular Genetics, Dept. Vertebrate Genomics, Ihnestr. 73, 14195 Berlin, Germany

Kofahl, Bente
Berlin Center for Genome Based Bioinformatics (BCB), Max-Planck Institute for Molecular Genetics, Dept. Vertebrate Genomics, Ihnestr. 73, 14195 Berlin, Germany

Kolupaev, Alex G.
A.N. Belozersky Institute of Physico-Chemical Biology, Moscow State University, Moscow, Russia

Kremling, Andreas
Systems biology group, Max-Planck-Institut für Dynamik, komplexer technischer Systeme, Sandtorstr. 1, 39106 Magdeburg, Germany

Kummer, Ursula
EML Research, Schloss-Wolfsbrunnenweg 33, 69118 Heidelberg, Germany
ursula.kummer@eml-r.villa-bosch.de

Lebedeva, Galina V.
A.N. Belozersky Institute of Physico-Chemical Biology, Moscow State University, Moscow, Russia

Mauch, Klaus
Institute for Biochemical Engineering, Allmandring 31, 70569 Stuttgart, Germany

Metelkin, Eugeniy A.
A.N. Belozersky Institute of Physico-Chemical Biology, Moscow State University, Moscow, Russia

Müller, Dirk
  Institute for Biochemical Engineering, Allmandring 31, 70569 Stuttgart, Germany
  Mueller@ibvt.uni-stuttgart.de

Nordlander, Bodil
  Department of Cell and Molecular Biology, Göteborg University, Box 462, S-40530 Göteborg, Sweden

Novák, Béla
  Molecular Network Dynamics Research Group, Budapest University of Technology and Economics and Hungarian Academy of Sciences,
  1111 Budapest, Gellért tér 4, Hungary
  bnovak@mail.bme.hu

Olsen, Lars Folke
  Celcom, Department of Biochemistry, University of Southern Denmark, Campusvej 55, 5230 Odense M, Denmark

Paulsson, Johan
  Department of Applied Mathematics and Theoretical Physics, Centre for Mathematical Sciences, Wilberforce Road, University of Cambridge, Cambridge CB3 0WA, UK

Plyusnina, Tatyana Y.
  Biophysics Department, faculty of Biology, Moscow State University, Moscow, Russia

Porro, Danilo
  University of Milano – Bicocca, Department of Biotechnology and Biosciences, Piazza della Scienza 2, 20126 Milano, Italy

Reuss, Matthias
  Institute for Biochemical Engineering, Allmandring 31, 70569 Stuttgart, Germany

Rossi, Riccardo L.
  University of Milano – Bicocca, Department of Biotechnology and Biosciences, Piazza della Scienza 2, 20126 Milano, Italy

Sauer, Uwe
  Institute of Molecular Systems Biology, ETH Zürich, CH-8093 Zürich, Switzerland
  sauer@biotech.biol.ethz.ch

Sauro, Herbert M.
    Keck Graduate Institute, 535 Watson Drive, Claremont, CA 91106, USA, and
    Control and Dynamical Systems, California Institute of Technology, Pasadena,
    CA 91125, USA

Snoep , Jacky L.
    Department of Biochemistry, University of Stellenbosch, Private Bag X1,
    Matieland, 7602, South Africa
    Department of Molecular Cell Physiology, Vrije Universiteit, De Boelelaan
    1087, NL-1081 HV Amsterdam, The Netherlands
    jls@sun.ac.za

Stelling, Jörg
    Institut für Computational Science, ETH Zentrum HRS H 28, Hirschengraben
    84, 8092 Zürich, Switzerland

Tobin, Frank
    GlaxoSmithKline, Scientific Computing & Mathematical Modeling, Upper
    Merrion, USA

Tyson, John J.
    Department of Biology, Virginia Polytechnic Institute and State University,
    Blacksburg, Virginia, VA 24061, USA

Vanoni, Marco
    University of Milano – Bicocca, Department of Biotechnology and Biosci-
    ences, Piazza della Scienza 2, 20126 Milano, Italy

Wanner, Barry L.
    Department of Biological Sciences, Purdue University, West Lafayette, IN
    47907-2054 USA
    blwanner@purdue.edu

Westerhoff, Hans V.
    BioCentrum Amsterdam, Mathematical Biochemistry, University of Amster-
    dam and Molecular Cell Physiology, Faculty of Earth and Life Sciences, Free
    University, De Boelelaan 1087, NL-1081 HV Amsterdam, The Netherlands,
    and Manchester Interdisciplinary Biocentre, Department of Systems Biology
    hans.westerhoff@falw.vu.nl

Zobova, Ekaterina A.
    A.N. Belozersky Institute of Physico-Chemical Biology, Moscow State Uni-
    versity, Moscow, Russia

# INTRODUCTION

# Systems Biology: Did we know it all along?

Hans V. Westerhoff and Lilia Alberghina

## Abstract

It is often suggested that Systems Biology is nothing new, or that it is irrelevant. Its central paradigm, i.e. that much of biological function arises from the interactions of macromolecules, is not generally appreciated. We here contend that much like molecular biology in its past, Systems Biology is new and old at the same time. It looks in a new way and with new and improved reincarnations of existing and new technologies at scientific issues that the existing disciplines describe but do not resolve. Its main focus is to understand in quantitative, predictable ways the regulation of complex cellular pathways and of intercellular communication so as to shed light on complex biological functions (e.g. metabolism, cell signaling, cell cycle, apoptosis, differentiation, and transformation). It is for the lack of achieving this understanding of living systems that the existing paradigms for biomedical research fail for the majority of diseases on the Northern hemisphere. Systems Biology appears appropriate for these complex and multifactorial diseases.

But of course it is not for us to define what Systems Biology is or should be. Yet, it is important that there be an end to suggestions that Systems Biology is vague or can be anything. As is molecular biology, Systems Biology is rich, wide, and diverse, not vague: most aspects of biology and of mathematical modeling are not part of Systems Biology. This book serves to define this rich and heterogeneous Systems Biology 'bottom up', i.e., by having systems biologists themselves define it.

## 1 Is Systems Biology something new?

All too often one of our colleagues 'confesses' to not knowing what Systems Biology is, or that Systems Biology is 'nothing but an existing science with a new touch'. Systems Biology 'should be nothing but good old physiology', Systems Biology 'is molecular biology claiming additional money', or 'Systems Biology is the explanation by engineers of how biology works'.

Of course, *this* is nothing new. When *molecular* biology began, it was considered a branch of biochemistry and biophysics. Its first major success was a combination of crystallography, theoretical biology, and chemistry. Later, it was made synonymous with molecular genetics. This view was right and wrong at the same time: on the one hand, molecular biology used concepts that came from the surrounding sciences. Its people were trained biochemists and physicists. On the

Topics in Current Genetics, Vol. 13
L. Alberghina, H.V. Westerhoff (Eds.): Systems Biology
DOI 10.1007/b137744 / Published online: 13 May 2005

other hand, molecular biology became a discipline of its own, which outgrew its founding sciences. Its success resulted from its focus on a type of molecule that led to a vast number of new and powerful concepts and techniques, each of which consisted of new combinations of existing chemistry and physics. Functional genomics is the contemporary and revolutionary result of its line of work.

How would we define molecular biology? Well, it is the natural science that deals with the individual macromolecules of living organisms, with an emphasis on information-rich molecules. As such it should use chemistry, physics, and mathematics whenever necessary. As molecular biology expanded, it emancipated from the dominance of physical chemists to which it was subject to early on. Biologists with a dislike of physics and mathematics took their place and ultimately molecular biologists even refused to express their concepts with the use of formal mathematical techniques. The cartoon became the highly efficient expression of their models.

Systems Biology is similarly new and not new at the same time. It does use classical physics, chemistry, molecular biology, and mathematics. However, it thrives on the *integration* of these and other sciences, and that *is* relatively new; recent molecular biology made some use of mathematics but only minimally and until recently many molecular biology journals would discriminate against papers with mathematical equations. Systems Biology (or the part thereof that we focus on in this book) has the living cell as object of study, has as its predecessor cell biology. Yet, it is much more than cell biology ever was. Systems Biology is after the mechanisms by which macromolecules through dynamic interactions produce the functional properties of living cells. Systems Biology does not just observe and describe functions in and of living systems, such as physiology does. Systems Biology adds the mechanistic interest of biochemistry and physics to physiology, and of course the analysis tools of mathematics. Indeed, Systems Biology is a science in that it is after principles and generalities rather than special cases. In order to discover those principles, it uses whatever science or technology is available. It is mathematics (in the sense of deducing principles from *a priori*'s) and biology (in the sense of addressing functional issue related to Life) at the same time. And then perhaps most importantly, Systems Biology is also biology in that it is after the principles of Life, the principles that are specific to living systems. These principles are the result of an evolutionary optimization that led to a local maximum in fitness for some habitat. The principles are also confined by the hysteresis of evolution, and by the feature that new life has always been an extension of existing Life. Here Systems Biology combines principles of physics and chemistry with principles of microbiology.

From the above it may be clear that Systems Biology is nothing new, yet highly new at the same time: it is in the *combination* of previous disciplines and in a *new focus* that Systems Biology distinguishes itself from other sciences.

# 2 Is it important?

In 1995 and 2001 mankind witnessed two of its greatest scientific achievements, i.e. the elucidation of the complete code for a living organism, and for a human being, respectively. Soon, a plethora of new techniques made it possible also to measure the expression of this code, at the level of mRNA, protein and in quite a few cases function (metabolome, fluxome). Thus, in one sense we are now able to determine what Life is, in terms of the concentrations of virtually all its molecular constituents. If we know precisely the contents of living organisms, for sure we must know Life?

As functional genomics data flood the scientific literature, its reader is increasingly confronted with a paradox: one may 'know' everything without understanding it. With every new publication on p53 we seem to understand less, rather than more of how life functions (Lazebnik 2002). What is the problem?

Unlike digital computers, the human mind is indeed much better at understanding a few things than at understanding many. We are confused by larger numbers of data and by many degrees of freedom. Human understanding boils down to the ability to order observations along the lines of relatively few patterns which we then call (empirical) 'laws'. Understanding is even better if we can deduce the one empirical law from the other, or from a small set of underlying principles. Human understanding is fundamentally qualitative. If two factors stimulate process $A$ little and another factor inhibits it much more, then we have no way to intuit what the total effect will be. It becomes even more difficult with nonlinear and recursive interactions.

With genomics came the definitive appreciation of the minimum size of Life, i.e. some 300 processes (or at least a number of processes specified by some 300 genes) (Hutchison et al. 1999). The simultaneous action of 300 processes is way above the action of the five that we might be able to understand. Moreover, it is not clear that the principles or laws that govern the behavior of molecules in living cells are as simple as those in physics or chemistry. They may well be based on strongly nonlinear principles that engage tens of degrees of freedom at the same time. Many of the concepts that exist in biology are formulated qualitatively rather than quantitatively and in terms of interactions between already complex objects ('a bird sees a fish and therefore tries to capture it'). Much of Systems Biology may be too complex for the human mind, unless the latter is aided by some kind of information technology. Even in hypothesis driven research, the hypotheses may need to be generated by computers (King et al. 2004). Like in the days of empiricism when physics came about, empirical science, now called data driven hypothesis generation, may become important again. Clearly, with a complexity that is substantial yet bounded by what is needed to sustain Life, Systems Biology is an enormous challenge for science itself, of the same grandeur perhaps as relativistic and quantum mechanics have been before.

Systems Biology is important for science, but beyond that, is it important for society? The point is made often but perhaps not often enough: progress has been appallingly slow in the medical biology that should lead to cures for the diseases that remain a threat to health in developed societies. Bacterial infectious disease is what we are good at, thanks to antibiotics, but even there the parasite strikes back by being selected for resistance. Viral infections, cancer, heart disease, arthritis, diabetes are all major diseases of this society. Although biology has pinpointed numerous factors that affect the etiology of these diseases, cures are mostly empirical, and often ineffective. Cancer research started off trying to identify the single molecular or other factor responsible for this disease. It was found that in many tumors, glycolysis is increased and biologists went after glycolytic factors as causes of this disease. When this did not work, assays were developed for determining factors that when deleted or expressed ectopically promoted tumorigenesis. Many such oncogenes were scored. Likewise many factors have been found to affect type II diabetes, none of them determining the disease completely. Molecular cell biology research continues to identify single molecules that correlate with these diseases, then studies their direct mechanisms of action, and is funded for it. Functional genomics programs continue to be directed towards determining the patterns of change of all the genes that correlate with the disease phenotype. This effort may work for diagnostics, but will it work for understanding multifactorial disease and for rational development of new drug targets and therapies?

It is time for us to recognize that the biochemistry and molecular biology that we were raised with, is not a good paradigm for the many diseases that have not been eradicated from the wealthy societies of our world. We need something else, yet something that is equally rational and scientific. Many of the diseases have been found to be multifactorial and strongly nonlinear, i.e. the effect of the one factor being determined by the strength of the other (cf. the chapter by Hofmeyr & Westerhoff). Many of these diseases reflect the system's nature of the human being. We propose that the new scientific paradigm that is needed is precisely Systems Biology.

We realize that recognizing the failure of biochemistry and molecular biology and suggesting that systems biology may be required for the battles against cancer, type-II diabetes, arthritis, and heart disease, may be considered iconoclastic. However, society has long been directing its medical biology research towards molecular biology, and there is appreciable conservatism *vis-à-vis* funding the new Systems Biology. The slow change of the funding agenda in some countries and continents is not only costing society money, it also retards the development of cures and drugs, and it slows down the new economic development that should emerge from a better manageable biotechnology.

## 3 What is it?

Is a definition important? A definition can help to identify a new area of science where there is much potential for progress. It can also help direct research effort to where it should be rather than continuing to be spent on the same topics but under a new name. Scientists wishing to continue doing their own thing under a new funding flag often proclaim that they do not know what Systems Biology is and that Systems Biology is vague, or that it is just the same emperor in new cloths. For the same reasons it has proven important that molecular biology was defined. It led to scientific organizations that promoted the area almost exclusively (e.g. EMBO) and even to scientific institutes (e.g. EMBL).

Is it important that the definition is precise? Yes, it is, because otherwise old things compete with the new topics for funding and human capital. Is it important that the definition is uniform or homogeneous? Paradoxically perhaps: No. Many excellent scientific disciplines are heterogeneous. Chemistry and molecular biology are just two examples.

How does one then define Systems Biology in such a heterogeneous way? Well, by examples, i.e. by challenging Systems biologists to explain what they find is Systems Biology. This is what this book is about: to define Systems Biology by examples. Reading through the chapters, the reader will find that Systems Biology indeed consists of a number of related, well-defined topics, based in physics, chemistry, biology, *and* mathematics, all focusing on the mechanisms behind the emergence of functionality. Yes, we reckon that the heterogeneous definition given in this book should take precedence over the more homogeneous and limited definition that we use (cf. below). We emphasize that the definition is heterogeneous and dynamic, not vague. By this definition, the majority of present day cell biology, molecular biology, biophysics, and mathematics is excluded: Systems Biology is a new science.

The following is the definition that *we* use for Systems Biology: The science that discovers the principles underlying the emergence of the functional properties of living organisms from interactions between macromolecules. What is it not then? Well, Systems Biology is not the biology of systems, nor is it the chemistry/physics/molecular genetics of molecules *in* biological systems. It is the difference between the two.

This is the definition that we use and actually also the definition we communicated to the authors of the chapters of this book before they started writing. We then asked them to challenge this definition and to add what they found, i.e. the more important aspects of Systems Biology, and then, in order not to get stranded in words, show by example what they mean.

Indeed, most chapters integrate conceptual (theoretical) and experimental (factual, molecular) aspects of cell function. In each case the essence is to demonstrate what Systems Biology is. The examples describe the author's approach and show that the system had properties that were not in the individual molecules, but only arise when the molecules are together and active, and are important for biological function.

At the same time, this book is much more than a book that defines Systems Biology. It contains examples of the most exciting Systems Biology of our times. It is thereby full of suggestions for the new systems biologists. And if the reader understands all the chapters, (s)he is ready to go!

## 4 Did we know it all along?

Did we know all along that functional properties are not in the molecules? Yes, of course, and many physiologists will state that this is the very essence of physiology and that, therefore, Systems Biology is nothing but physiology. On the other hand, Systems Biology demands the understanding of functional properties in terms of how they emerge in or persist through the many nonlinear interactions of those molecules. The understanding should be in terms of time, chemistry, space, and history. And, this should be accomplished with complete reference to experimental reality, i.e. by assaying the molecules and their interactions. This is not what physiology has done, but much more what molecular biology has been doing. Accordingly Systems Biology is not new with respect to molecular biology either, or...... it is, as the nonlinear synthesis of molecular biology and physiology.

## 5 Will it work?

Of course, the next bit of criticism to a development that appears to be new and important, is that it will not work. This is a point of criticism that we actually like most, because it constitutes a scientific challenge. Perhaps we should not give away the secret that it has worked, for us. One of us has been able to formulate a new mechanism for cell size control during the cell cycle, which is, at last, testable experimentally. She has also been able to get the experiments in gear with the theory and modeling, and to motivate the measurement of the dynamics of cell cycling quantitatively. While the other one of us has co-discovered new drug targets, understood why trypanosomes have glycosomes and why phosphatases are sometimes more important than kinases for signal transduction, figured out why phosphofructokinase is not the rate-limiting step, and found a way to determine experimentally rather than assume intuitively how much of regulation is actually transcriptional. Below, you will find chapters perhaps even more convincing examples of how Systems Biology has worked.

However, let us take it as the challenge for science and technology, i.e. the 'Grand Challenge', that Systems Biology will enable us now to understand the molecular basis for human health, disease, and therapy. This book may help get this Grand Challenge organized, by defining a little bit better what we need to focus on.

*Quousque tandem......*

# References


Hutchison CA, Peterson SN, Gill SR, Cline RT, White O, Fraser CM, Smith HO, Venter JC (1999) Global transposon mutagenesis and a minimal Mycoplasma genome. Science 286:2165-2159

King RD, Whelan KE, Jones FM, Reiser PG, Bryant CH, Muggleton SH, Kell DB, Oliver SG (2004) Functional genomic hypothesis generation and experimentation by a robot scientist. Nature 427:247-252

Lazebnik Y (2002) Can a biologist fix a radio?-Or what I learned while studying apoptosis. Cancer Cell 2:179-182

Westerhoff HV, Palsson BO (2004) The evolution of molecular biology into systems biology. Nature Biotechnol 22:1249-1252



Alberghina, Lilia
Dept. of Biotechnology and Biosciences, University of Milano-Bicocca, Piazza della Scienza 2, 20126 Milano, Italy
lilia.alberghina@unimib.it

Westerhoff, Hans V.
BioCentrum Amsterdam, Mathematical Biochemistry, University of Amsterdam and Molecular Cell Physiology, Faculty of Earth and Life Sciences, Free University, De Boelelaan 1087, NL-1081 HV Amsterdam, The Netherlands, and Manchester Interdisciplinary Biocentre, Department of Systems Biology
hans.westerhoff@falw.vu.nl

# METHODS

# From isolation to integration, a systems biology approach for building the Silicon Cell

Jacky L. Snoep and Hans V. Westerhoff

## Abstract

In the last decade, the field now commonly referred to as systems biology has developed rapidly. With the sequencing of whole genomes and the development of analysis methods to measure many of the cellular components, we have now entered the realm of complete descriptions at a cellular level. Although we have been seeing that larger and larger systems were being described, making a description complete is much more important than just adding additional components. The possibility of making complete descriptions will cause a paradigm shift in our approaches, on a theoretical, as well as a modeling and an experimental level. We will here present our view on systems biology and specifically focus on modeling strategies to build cellular models on the basis of detailed enzyme kinetic information: an approach advocated in the Silicon Cell project (http://www.siliconcell.net) making use of the JWS Online database of kinetic models (http://jjj.biochem.sun.ac.za).

## 1 Systems biology

Any system consists of components that interact in such a way that they form a functional unit. Systems engineered by us usually have a clear function and one could state that the function defines the system. Although not necessarily engineered by us, we can still use a functional definition to delineate biological systems. For instance, an enzyme can be considered as a biological system consisting of amino acids that interact in a specific way such that the enzyme can catalyze a reaction. Another example of a system could be a set of enzymes, such as the glycolytic enzymes, which together catalyze the conversion of glucose to pyruvate and whose function would be the production of free energy and reducing power in the form of ATP and NADH, respectively. Clearly, systems can be defined at different hierarchical levels and this will determine the extent of detail at which the components of a system should be described usefully. For instance when considering an enzyme as a system one could look at the amino acid level while someone studying glycolysis would typically look at the enzyme level.

Understanding how a system functions with 'understanding' defined operationally as the ability to describe the system on the basis of the characteristics of its

Topics in Current Genetics, Vol. 13
L. Alberghina, H.V. Westerhoff (Eds.): Systems Biology
DOI 10.1007/b106456 / Published online: 11 January 2005

components, is our fundamental aim in doing systems biology. Such an understanding of for instance glycolysis would also enable us to predict the glycolytic flux and intermediate concentrations, on the basis of the characteristics of the glycolytic enzymes. In this chapter, we will largely focus on systems at the cellular level with an aim to understand systems behavior at the level of its catalytic units, the enzymes.

It is important to give the systems description at the enzyme level for several reasons: 1) enzymes are the catalysts in the systems, which can be isolated and characterized separately from the rest of cellular metabolism; 2) gene expression modulation affects the concentration of enzymes in the cell and to take this into account one needs a representation of the different enzymes in the systems description.

## 2 What makes systems biology different from other systems approaches?

A unique characteristic of biological systems is the ability to reproduce, i.e. given suitable conditions and enough time; a cell will grow, divide, and lead to two cells. Such growth is most strongly displayed by uni-cellular organisms like bacteria and yeasts, but in principle most cells contain the information to reproduce themselves. This ability to grow has as the important corollary that biological systems are variable in their make-up. Such variability can be quantitative, making more of the same, but also qualitative, inducing different components. Such qualitative changes are often made as adaptations to variable external conditions, again most strongly expressed in uni-cellular organisms experiencing such variations more directly than cells that are part of a homeostatically controlled organism such as a mammal. The ability to adapt to changes that can in part be due to the cells own activities, can lead to circular control loops in which cause and effect are not easily distinguishable. Robust analysis methods, which have been developed for biological systems such as Metabolic Control Analysis (Kacser and Burns 1973; Heinrich and Rapoport 1974), and extensions thereof, for instance, Hierarchical Control Analysis (Hofmeyr and Westerhoff 2001), are important tools to disentangle such systems.

In addition to changes in expression level of existing genes, the cellular blueprint itself, the DNA molecule, can undergo changes via mutations in the base sequence, on an evolutionary time scale. Thus, the biological cell has truly remarkable capabilities, 1) it catalyses in a coordinated fashion several thousands of reactions in a very small space; 2) these reactions include copying the whole cellular machinery to make another cell. In addition the cell has a variable make-up enabling it 3) to adapt to varying conditions as encoded in the DNA and lastly 4) the base sequence in the DNA is subject to mutations which change the cellular blueprint, and thereby, the cellular composition. These different aspects of the cells' reactivity and ability to change have different time scales, going from fast to slow one can distinguish between diffusion of molecules, enzyme catalysis, gene expression, cell division, and evolution.

Systems biology deals with living systems and functions of (parts of) the system should be put into this functional context; for instance, free energy transduction, or generation of building blocks for biosynthesis. Biological systems have evolved such that these functions can be fulfilled under a variety of conditions; cells are robust, exactly because of their ability to adapt themselves, via regulation of activity, gene expression or even evolutionary changes. Ultimately, cellular function and its ability to adapt are dependent on the interactions between the systems components. This is the focus point for our Systems Biology work, to understand the systems behavior on basis of the interactions and characteristics of the cellular components. Of course, other systems are dependent on interactions as well, but specifically in Systems Biology these interactions are 1) non-linear and 2) are highly variable due to the ability of the cell to adapt.

Our approach follows from a minimal definition of a Systems Biology study as a study that investigates how interactions between biological components lead to the systems functionality. Within this definition there is no restrictions to the size of the system but the emphasis should be on components interactions and functionality, typically the components interactions would be related to characteristics of the components and the functionality would be a systemic property.


## 3 Isolation and characterization

Recent developments in the –omics fields have led to a dramatic increase in the level of detail of cellular descriptions. However, such descriptions, if merely on the level of identification or quantification of components, do not lead to a functional understanding of the system. Although identifying the cellular components is important to know the players of the game, their characteristics and interactions will determine the systems behavior. Our test of understanding is the ability to predict systems behavior on the basis of characteristics of the underlying components. We focus on the enzymes, the main catalysts in biological systems, and would need to determine their relevant characteristics and test if we can describe the system as a function of these characteristics.

Enzymology is a well-established discipline and Systems Biology can benefit greatly from this field when enzymes need to be characterized. However an important difference between the two disciplines should be kept in mind; whereas for enzymologists the enzyme is the system, for systems biologists, enzymes are part of the system. This might not seem a big issue but it has important implications. An enzymologist will tend to work under optimal conditions for the enzyme, even if these are not physiological, for instance high pH values for certain dehydrogenases. This has as a consequence that many of the available kinetic parameter values might not be valid under the physiological conditions for which the system biologist needs them. Although not all kinetic parameters might be influenced strongly by the assay conditions and, for instance, the kinetic mechanism might remain the same, the difference between the enzymologists' and the systems biologists' approaches might make some of the published kinetic values not useable for systems biology approaches.

Since a system biologist would like to simulate systems behavior on the basis of enzyme kinetic information, it is important that the enzyme characterization is performed under physiological conditions, i.e. conditions similar to those under which the enzyme is active in the complete system. Such a characterization is rather difficult since it is not trivial to isolate the enzyme in its cellular environment and make perturbations in substrates, products, and allosteric effectors. In principle, a sensitive NMR machine would allow for such an *in vivo* enzyme characterization if it were possible to make independent perturbations in substrates and products.

In practice, however, most enzymes are characterized in (partly) purified form, thus, isolated from the rest of the system. In such *in vitro* characterizations an effort should be made to try and mimic the cellular environment as much as possible. In principle, it should be possible to analyze all enzymes in the same reaction buffer, ion composition, osmotic strength, and pH. All could be made similar to the cytosolic condition. The enzyme activity should be analyzed such that a description can be given of the activity as a function of its effectors concentrations. It is not essential to know the enzyme kinetic mechanism to fit an equation to the experimental data set. However, if such a mechanism is known a mechanistic interpretation can be given for the enzyme kinetic parameters. If the enzyme kinetic mechanism is not known a generic reversible rate equation can be used in which a thermodynamic (with an equilibrium constant and a mass action ratio) and kinetic part (with binding constants and Vmax value) can be distinguished. Typically rate equations suited for systems biology modeling approaches should be reversible (although the equilibrium constant might be high) with a product binding component (although the binding constant might be high). This is again a difference from many enzyme characterizations made in enzymology studies where typically initial rate measurements are made and product effects ignored. Clearly in metabolic pathways where the product of the one enzyme is the substrate of the next, product concentrations will not be zero and enzymes' sensitivity towards their products must be taken into account.

In summary, when characterizing the enzymes in isolated form it is important to mimic the physiological conditions under which the enzyme is active as much as possible and to investigate every enzyme as a reversible enzyme or at least take product sensitivity into account.

## 4 A modular approach

The sheer multitude of components and kinetic parameters, and the non-linear interactions between the components make the construction of a cellular model a daunting task. A project aiming on isolating all the enzymes in a cell, characterizing them and trying to integrate all the reactions in a model has a small chance of success. Then, how should one go about making a detailed kinetic model of a cell? In our viewpoint, a modular approach would stand a better chance of success. In such an approach, the cell would be divided up in modules for each of which a kinetic model is made and validated. Subsequently, these models are joined, one at a time with validation steps in between. Crucial in this approach is how to divide up the cell. Here the key should lie in the validation. A module should to some extent be isolated from the rest of metabolism, such that it can be analyzed on its own, making an independent validation possible. Such isolation can be literal in the sense that few links exist between the module and the rest of metabolism but it can also be a time separation. If a module has a much faster response time than other parts of metabolism, this could be a way to distinguish between the two (de la Fuente et al. 2002). Also, if the extent of flux through the module is orders of magnitude smaller than other parts of metabolism this could be used in analysis methods (e.g. Snoep et al. 2004).

Important recent developments in structural (or stoichiometric) approaches could be very useful in helping decide on defining modules (e.g. Schuster et al. 2000). In line with the validation criterion, a minimal validation test would be to measure a steady state flux through the module that is tested. Typical measurements that can be made on whole cells are substrate consumption and product formation rates. A simple stoichiometry analysis of a metabolic system will indicate how many independent fluxes must be measured in order to calculate all steady state fluxes in a system. However, many a system will be underdetermined, i.e. it will not be possible to calculate all internal fluxes on the basis of external fluxes only and then other methods must be used. Isotope distribution methods can be used as an independent additional method to calculate all fluxes in a system and this method has been optimized extensively over the last years and is now regularly used in systems biology studies (e.g. Wiechert 2001).

In addition, more elaborate stoichiometric analysis methods can give information on parts of metabolism that can function in isolation of the rest of metabolism and lead to a steady state. Examples of such methods involve the determination of elementary modes (Schuster and Hilgetag 1994), extreme pathways (Schilling et al. 2000) or extreme currents (Clarke 1981). Although these parts of metabolism can lead to steady state on their own, in isolation of the rest of metabolism, this does not mean that in the real system they will also function in isolation of the rest of metabolism. Thus, not every elementary mode is necessarily a good choice as a module but each module that is selected should be able to lead to a steady state on its own.

An advantage of these structural models is that their input, the stoichiometry of the chemical reactions, is known for many reactions and is independent of the experimental conditions and can to some extent be derived from the DNA sequence. As a result, much larger systems have been modeled by structural models than by kinetic models. Of course these models are limited in the sense that they can make statements on possible flux distributions leading to a steady state but cannot make predictions about actual flux distributions since they do not contain kinetic information.

## 5 Integration

Once a suitable module has been selected and the reactions and their kinetics characterized, this information can be integrated using computer models. Typically, a model would be cast in the form of a set of ordinary differential equations, which can be analyzed numerically. Many different computer programs are available for such an analysis ranging from general programming environments such as Pascal and Python for which specific packages for biochemical pathway analysis have been written, e.g. PysCeS (Olivier 2002), to mathematical programs such as Mathematica (http://www.wolfram.com), Maple (http://www.maplesoft.com), MatLab (http://mathworks.com), to dedicated programs developed for the analysis of biochemical pathways such as Gepasi (Mendes 1997), Scamp (Sauro 1993), Jarnac (Sauro 2000). Good numerical solving routines are available for all of these programs and usually the available software does not limit the analysis of ODE's. Solvers such as LSODA can readily analyze even so-called stiff systems, and large time separation between different reactions does not generally cause problems.

We should mention here that although ODE's can be solved readily this does not hold for systems that are analyzed using partial differential equations. Even systems of moderate size can become already very computer intensive and software packages to analyze such systems tend to be much more technical.

## 6 Validation

An important aspect of building the Silicon Cell will be model validation. Validation of a model should be independent of the model construction, i.e. model parameters should not be fitted on the same data set that is used for validation of the model. In our view, kinetic parameters should be determined in experiments on the isolated enzyme, this is also the strategy followed in the Silicon Cell project (http://www.siliconcell.net). Such isolation can be a physical separation from the rest of metabolism but it can also be done intracellular if the effectors of the enzyme and its activity can be measured in the cell. Validation will typically involve measurements on the whole system, for instance, measurement of steady state fluxes or metabolite concentrations, which are dependent on the complete system. After validation experiments have been performed, model predictions (fluxes and metabolite concentrations) can be compared with the experimental measurements. Differences between the model prediction and the experimental measurements can be addressed in several ways. First, it should be analyzed whether the differences between the values can be related to errors in experimentally determined values. In other words, can the experimental data be described accurately within, say, a 5% error margin of the kinetic parameters in the kinetic model? If the answer to this question were negative, a second approach would be to analyze the model step by step. This is possible on the basis of the steady state metabolite concentrations and the steady state flux. The check is to take the rate equations one at a time and analyze the enzyme activity if the measured metabolite concentrations are substituted into the rate equation. If the resulting activity is different from the experimentally measured flux through the enzyme then that specific rate equation is not correct. A further characterization of this enzyme would be necessary to check on the kinetic parameters and investigate the existence of other regulators of the enzyme not incorporated in the current model.

In the Silicon Cell project, we advocate the use of experimentally determined kinetic parameter values for the construction of detailed kinetic models. Only by such a characterization of the individual components and subsequent integration can we hope to come to a quantitative understanding of the system. As before, the criteria to assess understanding is the ability to describe the system on the basis of the characteristics of the isolated components. This is an objective different than making the most accurate model for description of systems behavior. For such a purpose it would be much easier and better to fit the model parameters on the systems behavior. However, although such a model would be more accurate in its description of the system it would not help in *understanding* the system since one would not be able to relate the model parameters to physical entities of the system. For the same reason, we think it is important to try to make a duplicate of the biological system *in silico*. Different from the traditional approach where a model would always be reduced to the most simple form to address a specific question, we do not aim at simplification of the system: We would like to copy the complete and detailed system and use the model as a tool to come to an understanding of the system.

Once a model has been validated, it can be stored in a model repository (see below) and it can be linked to other existing models with which it interacts. Upon the linking of models it is important to again make a validation, i.e. to ask whether the combined models describe the experimental data set accurately.

## 7 Yeast glycolysis as an example

In our group, we posed the question whether our current knowledge of the best-studied pathway, yeast glycolysis, is good enough to describe the systems behavior quantitatively on the basis of its enzyme characteristics (Teusink et al. 2000). Thus, we made an extensive literature search for each of the enzymes in yeast glycolysis and collected kinetic information mostly with respect to enzyme kinetic mechanisms. Subsequently, we made a large number of *in vitro* enzyme kinetic measurements in a standard assay buffer used for all the enzymes with a composition aimed at mimicking the intracellular environment. We fitted the known kinetic mechanisms (or assumed a random order equilibrium binding mechanism) on these *in vitro* measurements to obtain the kinetic parameter values. An example of such kinetic measurements for the phosphofructokinase of yeast is given in Figure 1, with the rate equation of the Monod Wymann Changeux mechanism fitted through the data points. Subsequently, these kinetic rate equations were analyzed and solved for steady state. It became quickly clear that our initial system

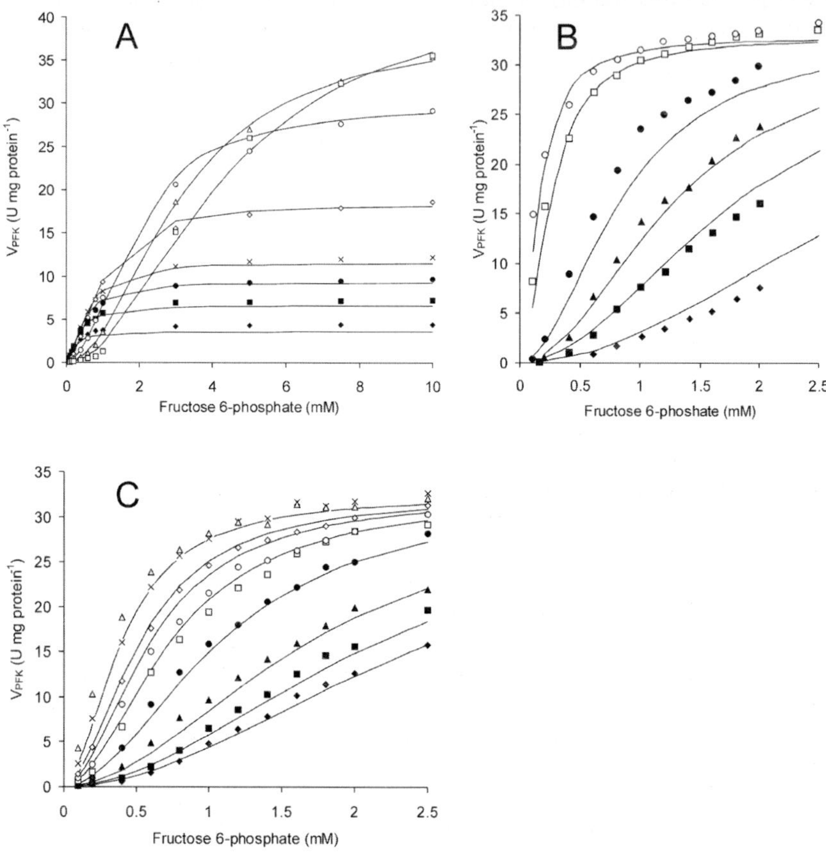

**Fig. 1.** Kinetics of PFK as a function of its substrate F6P and the effects of ATP, F2,6bP and AMP. A shows the effect of different ATP concentrations (ranging from 0.0125 (♦) to 0.5 mM (□)), B the effect of F26bP (ranging from 0 (♦) to 0.02 mM (o)), C the effect of AMP (ranging from 0.025 (♦) to 5 mM (x)) on the saturation of PFK with F6P.

definition was too narrow. No steady state was obtained when we limited our system to a linear system with ethanol as the sole product. Experimentally, it was observed that in addition to ethanol, also glycerol, succinate, trehalose, and glycogen were formed and upon incorporating these branches in the model, a steady state was obtained.

The experimental set up to measure the *in vivo* fluxes and metabolite levels was under non-growing, anaerobic conditions (the model contains only fermentative routes and no biosynthesis). Yeast cells were incubated in the buffer and external fluxes and internal metabolite concentrations were measured over time. After about 15 minutes, the system relaxed to a steady state and the model was validated on its ability to describe this steady state.

The model accurately described the steady state flux but some of the metabolite concentrations were more than a factor 5 off. However, upon testing each of the kinetic equations and allowing for a 5 % variation in the kinetic parameter values an exact description of the steady state metabolite concentrations and fluxes was possible.

Here, we should emphasize that we modeled only one specific steady state and analyzed the system under non-growing conditions to simplify the system in as much as possible. The branches added later onto the model (glycerol, trehalose, glycogen, and succinate formation) were modeled with simple kinetics since no detailed kinetic information was available. We could make such simplifications, as our aim was to test whether we could simulate the steady state behavior under one specific experimental condition, i.e. one specific steady state. Of course these simplifications in the model limited its versatility vis-à-vis describing different steady states.

In addition to validations using steady state experimental data, a more stringent validation test can be made on dynamic behavior of the model. Experimentally yeast has been observed to show limit cycle oscillations in the glycolytic pathway. The model as published describes a stable steady state only, however, realizing the specific requirements that need to be made experimentally to obtain limit cycle oscillations, i.e. glucose starvation and incubation with cyanide (Richard et al. 1993, 1994, 1996), it was not surprising that a few adjustments needed to be made to the model to obtain the dynamic behavior. We made small changes to the transport kinetic step (it is known that yeast alters its glucose transporters upon glucose starvation) and to the ATP hydrolysis step. Furthermore, we modeled the equilibrium blocks in the Teusink model explicitly in the dynamic model. With these small changes the model showed limit cycle oscillations. However, the description is not in close agreement with the experimental results; the frequency of the oscillations is 0.5 min$^{-1}$ while the experimental frequency is 1.5 min$^{-1}$ and the amplitude of the oscillations of the metabolites is too low (Reijenga 2002). We have not analyzed the sensitivity of the amplitude values for parameter values.

To be able to observe the limit cycles experimentally the oscillations in different cells must be synchronized. Without synchronization, small phase differences in the yeast cells would level out the oscillations of the population. Our hypothesis is that synchronization works through acetaldehyde, a volatile compound that diffuses rapidly through the membrane and can thus act as a communicating molecule (Richard et al. 1996). Cyanide complexes with acetaldehyde and oscillations are only observed in a narrow range of cyanide concentrations. To accommodate the synchronization issue we needed to incorporate an acetaldehyde efflux reaction and a reaction between extra cellular acetaldehyde and cyanide. With these additional reactions added to the model we could show active synchronization between different yeast cells if the biomass concentration was high enough. At low biomass concentration the cells desynchronized, in agreement with experimental observations (manuscript in preparation).

With this modeling project on yeast glycolysis we have shown that detailed *in vitro* kinetic information can be used to make quantitative descriptions of steady state metabolite levels and fluxes. In addition, we have shown that qualitative descriptions of dynamic behavior could also be made with the same model. Currently, we are trying to extend the model such that it can describe different steady states.

## 8 The Silicon Cell

Constructing the detailed model of yeast glycolysis as described above involved many man-hours of work, both for the experimental and the modeling work. Still, the model only contains detailed kinetic information for 12 reactions. Making a kinetic model for several thousands of reactions would be an impossible task for a single laboratory and would necessitate splitting up the system in manageable modules. Such modules would allow different laboratories to work on building kinetic models simultaneously and also make it possible to have validation steps for models with a limited number of kinetic parameters. We envision that if we are serious in trying to make a Silicon Cell we need to do this in a collaborative effort where a large number of research groups divide the work up. Clearly, a number of agreements will have to be made to make such a project work, such as concerning, what strain(s) are to be used, who works on what parts of metabolism, what assay conditions will be used, how the strains will be grown. An initial focus would probably be on metabolism of one steady state but eventually this will have to be expanded to different conditions making variable gene expression and signal transduction pathways necessary extensions to the model.

The International Workgroup for Yeast Systems Biology forms an ideal forum to coordinate such an effort for building a Silicon Yeast Cell. Clearly such a project needs input from both experimentalists and modelers, ideally working in the same group but otherwise working closely together. Some examples of other initiatives for large scale cellular models are the hepatocyte project initiated in Germany (http://www.systembiologie.de), the *E. coli* alliance (IECA, http://www.uni-giessen.de/~gx1052/IECA/ieca.html) and the E-cell project (http://www.e-cell.org).

One might ask what the use of such a detailed kinetic model would be or how one would analyze such a model, which will tend to become as complex as the real system. Clearly one will need good analysis tools to work with models as big as we envision to be making, consisting of several thousands of ODE's. We propose to use the framework of Metabolic Control Analysis (MCA) as such an analysis tool. MCA allows to relate steady state systemic behavior (using so-called control coefficients) to the characteristics of the isolated components (using elasticity coefficients). In addition, MCA quantifies the importance of each of the catalytic steps for any steady state variable. It would be a trivial task to select, given a detailed kinetic model, from a list of three thousand enzymes, the ten enzymes that are most important in limiting the oxygen consumption rate. In our view, MCA or its extension Hierarchical Control Analysis (HCA) would be used as a meta-language to analyze the model, still being able to relate systemic behavior to enzyme characteristics.

In addition to being a tool for understanding, a Silicon Cell model can also be used to test whether theoretical concepts that have been developed in the field of theoretical biology, can be applied to a specific biological system. A detailed kinetic model aiming to be a replica of the real system is much more accessible for experimentation (via simulation) than the real system, and any (testable) hypothesis can be investigated *in silico* without experimental limitations.

## 9 JWS - Online Cellular Systems Modelling

To make Silicon Cell models will involve a good collaboration between many different research groups from fields as far apart as experimental biology and computer science. Making detailed kinetic models of systems consisting of several thousands of enzymes will require a large number of groups to divide up the work and make models of parts of metabolism, to be combined later. These models need to be stored in a central database accessible for the other modeling groups and for the experimentalists to use in validation experiments and in setting up new experiments.

Important for such a database is that models can be downloaded in a format accessible to all users. Currently no standard format for model descriptions is available although SBML (http://sbml.org) seems to become the accepted standard. Equally important to a generally accepted format is model curation and accessibility. The models contained in the database should be testable for the users, independent of specific software with which the model was constructed. Lastly, the database should be up to date and not only contain models made by a specific workgroup but be a reflection of all published kinetic models.

Having started to make a repository of kinetic models in 2000 by coding them from literature, we quickly realized the importance of the model curation. More than 90% of the models we attempted to code were incompletely (or wrongly) described in the published manuscripts and upon contacting the authors many of the models could not be reproduced any more. This strengthened our conviction that we needed to conserve existing models in a repository, preferably widely accessible to the public. In 2001, we had started a web site with a collection of model that could be interrogated by everybody with Internet access. A list of the models currently present in the JWS database is given in Table 1. The JWS Online Cellular Modeling site (http://jjj.biochem.sun.ac.za) uses a client-server set up with all calculations being done on the server site with minimal computer requirements on the client side (Oliver and Snoep 2004). In addition, the graphical interface is very user friendly and allows for a wide range of simulations. JWS runs on two mirror sites, in the Netherlands at the Free University of Amsterdam (http://www.jjj.bio.vu.nl) and in the US at the Virginia Bioinformatics Institute in Blacksburg, VA (http://jjj.vbi.vt.edu).

**Table 1.** Detailed kinetic models in the JWS database. The table is a screenshot from the JWS website taken on 5 July 2004.

| Model | Reference |
| --- | --- |
| Detailed glycolytic model in *Lactococcus lactis* http://jjj.biochem.sun.ac.za/database/test/index.html | Hoefnagel et al. 2002 |
| Glycolysis in *Trypanosoma brucei* http://jjj.biochem.sun.ac.za/database/bakker/index.html | Helfert et al. 2001 |
| A computational model for glycogenolysis in skeletal muscle http://jjj.biochem.sun.ac.za/database/lambeth/index.html | Lambeth et al. 2002 |
| Pyruvate branches in *Lactococcus lactis* http://jjj.biochem.sun.ac.za/database/snoep1/index.html | Hoefnagel et al. 2002 |
| Glycolysis in *Saccharomyces cerevisiae* http://jjj.biochem.sun.ac.za/database/teus1/index.html | Teusink et al. 2000 |
| Sucrose accumulation in sugarcane http://jjj.biochem.sun.ac.za/database/rohwer1/index.html | Rohwer et al. 2001 |
| Bacterial phosphotransferase system http://jjj.biochem.sun.ac.za/database/pts/index.html | Rohwer et al. 2001 |
| Threonine synthesis pathway in *E. coli* | Chassagnole et al. 2001 |
| Kinetics of histone gene expression | Koster et al. 1988 |
| Glycolysis in *Saccharomyces cerevisiae*, 6 variables http://jjj.biochem.sun.ac.za/database/galazzo1/index.html | Galazzo et al. 1990 |
| Full scale model of glycolysis in *Saccharomyces cerevisiae* http://jjj.biochem.sun.ac.za/database/hynne/index.html | Hynne et al. 2001 |
| Quantification of short term signaling by the epidermal GFR http://jjj.biochem.sun.ac.za/database/kholodenko/index.html | Kholodenko et al. 1999 |
| Red blood cell model | Mulquiney et al. 1999 |
| Mechanism of protection of peroxidase activity by oscillatory dynamics http://jjj.biochem.sun.ac.za/database/olsen/index.html | Olsen et al. 2003 |
| Dynamic model of *Escherichia coli* tryptophan operon http://jjj.biochem.sun.ac.za/database/bhartiya/index.html | Bhartiya et al. 2003 |
| MCA of glycerol synthesis in *Saccharomyces cerevisiae* http://jjj.biochem.sun.ac.za/database/cronwright/index.html | Cronwright et al. 2003 |
| Mathematical modelling of the urea cycle http://jjj.biochem.sun.ac.za/database/maher/index.html | Maher et al. 2003 |
| A kinetic model of the branch-point between the methioninehttp://jjj.biochem.sun.ac.za/database/curien/index.html | Curien et al. 2003 |
| Modelling photosynthesis and its control http://jjj.biochem.sun.ac.za/database/poolman/index.html | Poolman et al. 2000 |
| cell cycle model http://jjj.biochem.sun.ac.za/database/tyson2001/index.html | Tyson et al. 2001 |
| *In situ* kinetic analysis of glyoxalase I and glyoxalase II in *Saccharomyces* http://jjj.biochem.sun.ac.za/database/martins/index.html | Martins et al. 2001 |
| Kinetic model of human erythrocytes http://jjj.biochem.sun.ac.za/database/holzhutter/index.html | Holzhutter et al. 2004 |

**Table 2.** Experimental data and model predictions for glycolysis in *Saccharomyces cerevisiae*. Experimental data and model prediction are published in Teusink et al. 2000 and can be tested on http//jjj.biochem.sun.ac.za/yeast.

| | Experiment | Model Teusink 2000 | Model Extended |
|---|---|---|---|
| Fluxes | (mmol/min/L-cyt) | (mmol/min/L-cyt) | (mmol/min/L-cyt) |
| Glucose | 108 | 88 | 89 |
| Ethanol | 135 | 129 | 152 |
| $CO_2$ | 154 | - | - |
| Glycogen | 6.0 | 6.0* | 6.0* |
| Trehalose | 4.8 | 4.8* | 4.8* |
| Glycerol | 18.2 | 18.2* | 4.1 |
| Succinate | 2.9 | 3.6* | 0.81* |
| Pyruvate | 2.2 | - | - |
| Acetate | 0.5 | - | - |
| - | - | - | - |
| Metabolites | (mmol/L-cyt) | (mmol/L-cyt) | (mmol/L-cyt) |
| G6P | 2.45 | 1.07 | 3.42 |
| F6P | 0.62 | 0.11 | 0.54 |
| F1,6bP2 | 5.51 | 0.60 | 4.14 |
| DHAP | 0.81 | 0.74 | 3.62 |
| Gly3P | 0.15 | - | 0.47 |
| 3PGA | 0.90 | 0.36 | 0.70 |
| 2PGA | 0.12 | 0.04 | 0.09 |
| PEP | 0.07 | 0.07 | 0.21 |
| Pyr | 1.85 | 8.52 | 12.3 |
| Acetaldehyde | 0.17 | 0.17 | 0.04 |
| ATP | 2.52 | 2.51* | 3.28* |
| ADP | 1.32 | 1.29* | 0.74* |
| AMP | 0.25 | 0.30* | 0.08* |
| NAD | 1.20 | 1.55 | 1.36 |
| NADH | 0.39 | 0.04 | 0.23 |

The large number of errors or incomplete descriptions of kinetic models in the literature indicated to us that a need exists to improve reviewing of manuscripts containing kinetic models. Very few reviewers will set out to code a kinetic model to check simulation results if such a model contains more than say five ODE's. In an attempt to improve reviewing of manuscripts containing kinetic models and to keep our database up to date we started collaborating with scientific journals, notably the *European Journal of Biochemistry, Systems Biology* and *Microbiology*.

Authors that submit a manuscript containing a kinetic model provide us with a model description; we code the model and allow access for reviewers to a secure site for them to interrogate the model. Thus, we are secure of new models and the journals have better reviewing tools for their manuscripts.

Clearly this web-based database of kinetic models can play an important role in Systems Biology. Groups working together to model a specific system could post all their models in the database making it accessible to the other members of the workgroup. Such members could be other modeling groups but also experimental groups that could use the model without having to learn how to use a specific modeling package using the friendly web interface. In addition, the more hard-core modelers could download the model, in SBML format, or other supported download formats and work in their favorite modeling package.

The ambitious project to group the different models together in a growing Silicon Cell model could also be coordinated via the JWS Online site. Subsequent postings of models of parts of metabolism can be linked to a growing model. Let yeast again serve as an example. We have currently collected seven models for this organism and started putting some of these models together in a "growing" Silicon Yeast Cell model (http://jjj.biochem.sun.ac.za/yeast). Thus, we changed the simple kinetics in the glycerol branch with detailed kinetics (Cronwright et al. 2003) and we attached a methylglyoxal pathway to the model (Martins 2001). Addition of these reactions to the existing glycolytic model without making any additional changes led to a slightly improved prediction of metabolite concentrations in the top part of glycolysis and did not affect the steady state flux predictions (Table 2). Of course, these are small additions but it shows the principle of the modular approach.

## 10 How far are we, and what needs to be done?

In our view, the tools are available to try to start building kinetic models of complete biological systems at the cellular level. It would be best to start with a number of assumptions that make the task a bit easier and which might be relieved in a later stage, such as assuming the cell to be a well stirred reactor and work with ordinary differential equations. In a later stage, when it becomes evident that such a simplification does not hold for all metabolites, we might be able to mix ODE and PDE types of models to get a more accurate description of the system. Another simplification would be to start with building a model for a single steady state and relieving the complication of variable gene expression. Once a good working model has been described for the one steady state one can start incorporating variable gene expression to allow the cell to adapt to different environmental conditions. The mathematical and computational tools seem to be sufficient for building and analyzing detailed kinetic models at the cellular level if ODE's are used. Obtaining experimentally determined kinetic parameters for these differential equations is a much bigger task.

In the last decade remarkable progress has been made in the development of experimental tools for systems biology. After whole genomes had been sequenced complete description of all cellular components became possible in principle at least in some sense. Using this knowledge it has been possible to construct for a number of model organisms a nearly complete stoichiometric network map. However, values for kinetic constants cannot be derived from the DNA sequence and need to be determined at the bench. With at least four parameters for each rate equation it will be a big task to measure all of these constants, certainly for enzymes that have a low expression level and for which no simple assay methods are available.

Again as an initial modeling approach, we will investigate the possibility to start with a detailed structural model, containing the complete stoichiometry of the system and filling in detailed kinetics for the components as this information becomes available. One could try to use very simple kinetics in the structural model based on the measured fluxes. Of course such kinetics need to be replaced ultimately with experimentally determined reaction kinetics, in line with the Silicon Cell approach but such simple kinetics might fulfill a role in initial stages of the project. In this respect, the constraint based modeling approach of the Palsson group should be mentioned (e.g. Price 2003). Starting from the network structure, similar to the structural approaches mentioned above, this approach adds additional constraints on top of the network structure, such as maximal flux values, and has been successful, also for bigger systems. Although we see advantages to such approaches in the initial modeling stages, we adhere to experimentally determined kinetic parameters as input for the final model. Again, we think such an input is essential to *understand* the system as a function of its components. After all, the constraints based modeling approach assumes fluxes to have a maximal value. It does not elucidate how the molecules in the cell enable the cell to make the fluxes as high as they are. And, for microbiology, it is well known that neither fluxes nor efficiency are as high as they could be theoretically (cf. Westerhoff and Van Dam 1987).

In addition to being important for model construction, experiments at the system level will also be important for model validation. Here the measurement of a large number of internal metabolites might be difficult although remarkable progress has been made to measure mRNA and protein levels. Also, high throughput methods for measurement of intermediates are developing rapidly, using a wide variety of mass spectroscopy techniques.

In our view, it will be important to combine efforts from a large number of modeling as well as experimental laboratories to collect and interpret the necessary data for large scale modeling projects such as building a detailed model of a biological cell. Clearly this will involve good management and coordination but we think such a project is doable, albeit a lot of work.

Will the result, the detailed kinetic model be worth all this effort? This question can only be answered with certainty after the model has been constructed, but the value of such models will be very great for fundamental as well as for applied studies. For instance in medical applications, detailed kinetic models will be a great asset in drug target identification. Detailed kinetic models of e.g. a parasite and his host would enable to identify enzymes that have a high control in the parasite and a small control in the host. Such enzymes would be good drug targets as a maximal effect of the drug on the parasite with minimal side effects on the host is expected (Bakker et al. 2002).

## References


Bakker BM, Assmus, HE, Bruggeman F, Haanstra JR, Klipp E, Westerhoff HV (2002) Network-based selectivity of antiparasitic inhibitors. Mol Biol Rep 29:1-52

Bhartiya S, Rawool S, Venkatesh, KV (2003) Dynamic model of *Escherichia coli* tryptophan operon shows an optimal structural design. Eur J Biochem 270:2644–2651

Chassagnole C, Fell DA, Rais B, Kudla B, Mazat J-P (2001) Control of the threonine-synthesis pathway in *Escherichia coli*: a theoretical and experimental approach. Biochem J 356:415-423

Clarke BL (1981) Complete set of steady states for the general stoichiometric dynamical system. J Chem Phys 75:4970-4979

Cronwright GR, Rohwer JM, Prior BA (2003) Metabolic control analysis of glycerol synthesis in *Saccharomyces cerevisiae*. Appl Environ Microbiol 68:4448-4456

Curien G, Ravanel S, Dumas R (2003) A kinetic model of the branch-point between the methionine and threonine biosynthesis pathways in *Arabidopsis thaliana*. Eur J Biochem 270:1–13

De la Fuente A, Snoep JL, Westerhoff HV, Mendes P (2002) Metabolic control in integrated biochemical systems. Eur J Biochem 269:4399-4408

Galazzo JL, Bailey JE (1990) Fermentation pathway kinetics and metabolic flux control in suspended and immobilized *Saccharomyces cerevisiae*. Enz Microb Technol. 12:162-172

Heinrich R, Rapoport TA (1974) A linear steady-state treatment of enzymatic chains. Eur J Biochem 42:89-95

Helfert S, Estevez AM, Bakker B, Michels P, Clayton C (2001) Roles of triosephosphate isomerase and aerobic metabolism in *Trypanosoma brucei*. Biochem J 357: 117-125

Hofmeyr JHS, Westerhoff HV (2001) Building the cellular puzzle. J Theor Biol 208:261-285

Hoefnagel MHN, Starrenburg MJC, Martens DE, Hugenholtz J, Kleerebezem M, Van Swam II, Bongers R, Westerhoff HV, Snoep JL (2002) Metabolic engineering of lactic acid bacteria, the combined approach: kinetic modelling, metabolic control and experimental analysis. Microbiology 148: 1003-1013.

Hoefnagel MHN, Van Der Burgt A, Martens DE, Hugenholtz J, Snoep JL (2002) Time dependent responses of glycolytic intermediates in a detailed glycolytic model of Lactococcus lactis during glucose run-out experiments. Mol Biol Rep 29: 157-161

Holzhütter H-G (2004) The principle of flux minimization and its application to estimate stationary fluxes in metabolic networks. Eur J Biochem 271:2905-2922

Hynne F, Dano S, Sorensen PG (2001) Full-scale model of glycolysis in *Saccharomyces cerevisiae*. Biophys Chem 94:121-163

Kacser H, Burns JA (1973) The control of flux. In: Davies DD (ed) Rate control of biological processes. Cambridge University Press, London, pp 65-104

Kholodenko BN, Demin OV, Moehren G, Hoek JB (1999) Quantification of Short Term Signaling by the Epidermal Growth Factor Receptor J Biol Chem 274:30169–30181

Koster JG, Destrée OHJ, Westerhoff HV (1988) Kinetics of Histone Gene Expression during Early Development of *Xenopus laevis*. J Theor Biol 135:139-167

Lambeth MJ, Kushmerick MJ, (2002) A Computational Model for Glycogenolysis in Skeletal Muscle Ann Biomed Eng 30: 808–827

Maher AD, Kuchel PW, Ortega F, de Atauri P, Centelles J, Cascante M (2003) Mathematical modelling of the urea cycle. Eur J Biochem 270, 3953–3961

Martins AM, Mendes P, Cordeiro C, Freire AP (2001) *In situ* kinetic analysis of glyoxalase I and glyoxalase II in *Saccharomyces cerevisiae*. Eur J Biochem 268:3930-3936

Mendes P (1997) Biochemistry by numbers: simulation of biochemical pathways with Gepasi 3. TIBS 22:361-363

Mulquiney PJ, Kuchel PW, (1999) Model of 2,3-bisphosphoglycerate metabolism in the human erythrocyte based on detailed enzyme kinetic equations: computer simulation and metabolic control analysis. Biochem J 342: 597–604

Poolman MG, Fell DA, Thomas S (2000) Modelling photosynthesis and its control. J Exp Bot 51:319-328

Olivier BG and Snoep JL( 2004) Web-based kinetic modelling using JWS Online. Bioinformatics 20:2143-2144

Olivier BG, Rohwer JM, Hofmeyr JHS (2002) Modelling cellular processes with Python and Scipy. Mol Biol Rep 29:249-254

Olsen LF, Hauser MJB, Kummer U, (2003) Mechanism of protection of peroxidase activity by oscillatory dynamics. Eur J Biochem 270:2796–2804

Price ND, Papin JA, Schilling CH, Palsson BO (2003) Genome-scale microbial *in silico* models: the constraints-based approach. TRENDS Biotechnol 21:162-169

Richard P, Bakker BM, Teusink B, Westerhoff HV, Van Dam K (1993) Synchronisation of glycolytic oscillations in intact yeast cells. In: Schuster S, Rigoulet M, Ouhabi R, Mazat JP (Eds) Modern trends in Biothermokinetics. Plenum Press, London, pp: 413-416

Richard P, Teusink B, Westerhoff HV, Van Dam K (1994) Around the growth phase transition *S. cerevisiae's* make-up favours sustained oscillations of intracellular metabolites. FEBS Lett 318:80-82

Richard P, Teusink B, Van Dam K, Westerhoff HV (1996) Acetaldehyde mediates the synchronization of sustained glycolytic oscillations in yeast-cell populations. Eur J Biochem 235:238-241

Reijenga K (2002) Dynamic control of yeast glycolysis. PhD thesis, Vrije Universiteit Amsterdam.

Rohwer JM, Meadow ND, Roseman S, Westerhoff HV and Postma PW (2000) Understanding glucose transport by the bacterial phosphoenolpyruvate:glycose phosphotransferase system on the basis of kinetic measurements *in vitro*. J Biol Chem 275:34909-34921

Rohwer JM, Botha FC (2001) Analysis of sucrose accumulation in the sugar cane culm on the basis of *in vitro* kinetic data. Biochem J 358:437-445

Sauro HM (1991) SCAMP: a general-purpose simulator and metabolic control analysis program. CABIOS 9:441-450

Sauro HM (2000) Jarnac: a system for interactive metabolic analysis. In: Hofmeyr JHSH, Rohwer JM, Snoep JL (eds) Animating the cellular map: Proceedings of the 9th international meeting on biothermokinetics. Stellenbosch University Press, Stellenbosch, pp: 221-228

Schilling CH, Letscher D, Palsson BO (2000) Theory for the systemic definition of metabolic pathways and their use in interpreting metabolic function from a pathway-oriented perspective. J Theor Biol 203:229-248

Schuster S, Fell DA, Dandekar T (2000) A general definition of metabolic pathways useful for systematic organization and analysis of complex metabolic networks. Nature Biotech 18:326-332

Schuster S, Hilgetag C (1994) On elementary flux modes in biochemical reaction systems in steady state. J Biol Syst 2:165-182

Snoep JL, Hoefnagel MHN, Westerhoff HV (2004) Metabolic engineering of branched systems: redirecting the main pathway flux. In: Westerhoff HV, Kholodenko B (eds) Metabolic engineering in the post-genomic era. Horizon Scientific Press, Norwich, UK pp 357-377

Teusink B, Passarge J, Reijenga CA, Esgalhado E, Van der Weijden CC, Schepper M, Walsh MC, Bakker BM, Van Dam K, Westerhoff HV, Snoep JL (2000) Can yeast glycolysis be understood in terms of in vitro kinetics of the constituent enzymes? Testing biochemistry. Eur J Biochem 267:5313-5329

Tyson JJ, Novak B (2001) Regulation of the eukaryotic cell cycle: Molecular antagonism, hysteresis, and irreversible transitions. J Theor Biol 210:249-263

Westerhoff HV, Van Dam (1987) Thermodynamics and control in biological free-energy transduction. Elsevier, Amsterdam, The Netherlands.

Wiechert W (2001) 13C metabolic flux analysis. Metab Eng 3:95–206

Snoep , Jacky L.
Department of Biochemistry, University of Stellenbosch, Private Bag X1, Matieland, 7602, South Africa
Department of Molecular Cell Physiology, Vrije Universiteit, De Boelelaan 1087, NL-1081 HV Amsterdam, The Netherlands
jls@sun.ac.za

Westerhoff , Hans V.
Department of Molecular Cell Physiology, Vrije Universiteit, De Boelelaan 1087, NL-1081 HV Amsterdam, The Netherlands
hw@bio.vu.nl

# Kinetic modelling of the *E. coli* metabolism

Oleg V. Demin, Tatyana Y. Plyusnina, Galina V. Lebedeva, Ekaterina A. Zobova, Eugeniy A. Metelkin, Alex G. Kolupaev, Igor I. Goryanin, and Frank Tobin

## Abstract

We describe a general strategy that enables us to develop kinetic models of large-scale metabolic systems by collecting and using available metabolic and gene regulation experimental data. The approach could be used to explore the local and global regulatory properties of metabolic pathways, and to predict how cell genome modifications can meet specific biotechnological and biomedical criteria. We have successfully applied the strategy for the development and applications of detailed kinetic models of catabolic and anabolic pathways of *E. coli*.

## 1 Introduction

The last several years have seen substantial progress in molecular biology and genetic research of *E.coli* (Perna et al. 2001; Kolisnichenko et al. 2002). Sequence information on genomes of more than 100 of different organisms has stimulated the emergence of functional genomics, a discipline that sets out to understand the meaning of sequenced data using high throughput small molecules, genes, and proteins expression data. Life scientists have transformed old style protein chemistry to proteomics, traditional biochemistry to metabolomics. These new fields provide essential clues to the underlying metabolic, gene regulation, and signalling networks that operate in cells, tissues, and organisms under different conditions.

Cellular metabolism, the integrated interconversion of thousands of metabolic substrates through enzyme-catalyzed biochemical reactions, is the most investigated system of intracellular molecular interactions. When one has knowledge of most, or all, of the major biological entities and stoichiometry of their interactions an illusion could appear that this voluminous knowledge will enable us to predict behaviours of the whole cell for the purpose of mechanistic understanding and bioengineering control. Indeed, in some cases, it is possible to make plausible predictions based on "static" information without relying upon kinetic data (Edwards et al. 2001). Unfortunately, this is not generally the case. Overall, cellular behaviour is determined not only by available biological entities, but mainly by their dynamic interactions and individual properties. Activities of most if not all of the enzymes involved in cellular metabolism are regulated by end products and intermediates of corresponding pathways. This complex network of positive and negative feedback as well as genetic regulation of expression level provide flexible ad-

Topics in Current Genetics, Vol 13
L. Alberghina, H.V. Westerhoff (Eds.): Systems Biology
DOI 10.1007/4735_85 / Published online: 21 June 2005
© Springer-Verlag Berlin Heidelberg 2005

aptation of metabolic network to fast and low changes in external environment correspondingly. It is the overall dynamic nature of the cell that determines its present properties, but its future ones as well. The cellular regulatory system is responsible for maintenance of homeostasis and for transitions between different physiological states. So, it is extremely important to include regulatory properties (effects) for metabolic pathways for cell models. These could be achieved by using kinetic modelling approach presented here.

The approach for constructing large-scale kinetic network of the cellular metabolism is described in this contribution. *E.coli* histidine pathway is used as an example. A novel way to collect, mine large-scale experimental data, and use them to build and validate kinetic models is discussed. We also show how to apply the approach to practical problems of biotechnology and bioengineering. The kinetic model of branched amino acid biosynthesis pathway has been used to optimize isoleucine and valine production.

## 2 Basic principles of kinetic model construction

The term "kinetic model" is often used dually. In the biological sense, it is a network of interactions between biological entities. The entities are always changing, even if the resulted fluxes could become zero. Generally, any two entities are connected, if, for any time, there is a flux connecting them. All temporal genotypic and phenotypic changes could be accounted by kinetic models. In the mathematical sense, "kinetic" refers to a system of mechanistic ordinary differential equations that determine the temporal state of the corresponding system of biochemical reactions. Conservation laws usually apply in the production and consumption in the form:

$$dX/dt = v_{production} - v_{consumption},$$

Where $v_{production}$ and $v_{consumption}$ are the respective rates of production and consumption of biological entity X.

The development of kinetic models of the biological networks is accomplished in several steps. Each stage of the process with examples is described.

### 2.1 Development of system of ordinary differential equations (ODEs) describing dynamics of selected biochemical system

The first step is to develop a static model of the biological system, i.e., to find out all cellular players, intermediates, enzymes, small molecules, co-factors, and all non-enzymatic processes in the cellular network. The resulting network (i.e. a directed bond graph) comprises all known interactions connecting all the entities. Ideally, the whole cellular network should contain **interconnected** entities, so each biological entity should have a source and sink, or at least participate in one the reactions. Disconnected fragments, resulting from incomplete knowledge, could, optionally, be considered as part of the one whole cellular network. Fragments

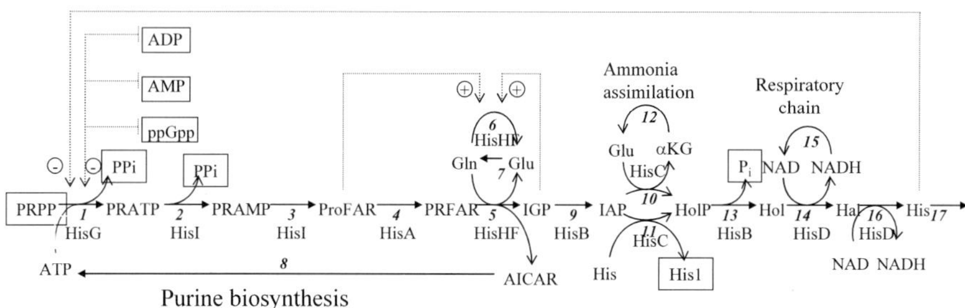



**Fig. 1.** Scheme of histidine biosynthesis pathway in *Escherichia coli*.

will be treated as separate networks or isolated pathways for all practical purposes. The pathway of histidine biosynthesis in *Escherichia coli* (Umbarger 1996) comprises 17 reactions and 17 entities (Fig. 1). Taking into account intermediates (Table 1, 2) and reactions the static model in the form of stoichiometric matrix could be created:

| $v_1$ | $v_2$ | $v_3$ | $v_4$ | $v_5$ | $v_6$ | $v_7$ | $v_8$ | $v_9$ | $v_{10}$ | $v_{11}$ | $v_{12}$ | $v_{13}$ | $v_{14}$ | $v_{15}$ | $v_{16}$ | $v_{17}$ | |
|---|---|---|---|---|---|---|---|---|---|---|---|---|---|---|---|---|---|
| −1 | 0 | 0 | 0 | 0 | 0 | 0 | 1 | 0 | 0 | 0 | 0 | 0 | 0 | 0 | 0 | 0 | *ATP* |
| 1 | −1 | 0 | 0 | 0 | 0 | 0 | 0 | 0 | 0 | 0 | 0 | 0 | 0 | 0 | 0 | 0 | *PRATP* |
| 0 | 1 | −1 | 0 | 0 | 0 | 0 | 0 | 0 | 0 | 0 | 0 | 0 | 0 | 0 | 0 | 0 | *PRAMP* |
| 0 | 0 | 1 | −1 | 0 | 0 | 0 | 0 | 0 | 0 | 0 | 0 | 0 | 0 | 0 | 0 | 0 | Pr*oFAR* |
| 0 | 0 | 0 | 1 | −1 | 0 | 0 | 0 | 0 | 0 | 0 | 0 | 0 | 0 | 0 | 0 | 0 | *PRFAR* |
| 0 | 0 | 0 | 0 | −1 | −1 | 1 | 0 | 0 | 0 | 0 | 0 | 0 | 0 | 0 | 0 | 0 | *G*ln |
| 0 | 0 | 0 | 0 | 1 | 1 | −1 | 0 | 0 | −1 | 0 | 1 | 0 | 0 | 0 | 0 | 0 | *Glu* |
| 0 | 0 | 0 | 0 | 1 | 0 | 0 | −1 | 0 | 0 | 0 | 0 | 0 | 0 | 0 | 0 | 0 | *IGP* |
| 0 | 0 | 0 | 0 | 1 | 0 | 0 | −1 | 0 | 0 | 0 | 0 | 0 | 0 | 0 | 0 | 0 | *AICAR* |
| 0 | 0 | 0 | 0 | 0 | 0 | 0 | 0 | 1 | −1 | −1 | 0 | 0 | 0 | 0 | 0 | 0 | *IAP* |
| 0 | 0 | 0 | 0 | 0 | 0 | 0 | 0 | 0 | 0 | −1 | 0 | 0 | 0 | 0 | 1 | −1 | *His* |
| 0 | 0 | 0 | 0 | 0 | 0 | 0 | 0 | 0 | 1 | 1 | 0 | −1 | 0 | 0 | 0 | 0 | *HolP* |
| 0 | 0 | 0 | 0 | 0 | 0 | 0 | 0 | 0 | 1 | 0 | −1 | 0 | 0 | 0 | 0 | 0 | *αKG* |
| 0 | 0 | 0 | 0 | 0 | 0 | 0 | 0 | 0 | 0 | 0 | 0 | 1 | −1 | 0 | 0 | 0 | *Hol* |
| 0 | 0 | 0 | 0 | 0 | 0 | 0 | 0 | 0 | 0 | 0 | 0 | 0 | −1 | 1 | −1 | 0 | *NAD* |
| 0 | 0 | 0 | 0 | 0 | 0 | 0 | 0 | 0 | 0 | 0 | 0 | 0 | 1 | −1 | 1 | 0 | *NADH* |
| 0 | 0 | 0 | 0 | 0 | 0 | 0 | 0 | 0 | 0 | 0 | 0 | 0 | 1 | 0 | −1 | 0 | *Hal* |

Columns of the matrix correspond to reactions of the pathway and rows correspond to metabolites.

The next stage of the kinetic model development is to write the system of differential equations describing dynamics of network or pathway:

$$\frac{d\mathbf{x}}{dt} = \mathbf{N} \cdot \mathbf{v}, \quad \mathbf{x}(0) = \mathbf{x_0} \tag{2.1}$$

Here, $\mathbf{x} = [x_1,\dots,x_m]^T$ is vector of intermediates concentrations, $\mathbf{x_0} = [x_{10},\dots,x_{m0}]^T$ is vector of initial concentrations of intermediates, $\mathbf{v} = [v_1,\dots,v_n]^T$ is vector of reaction rates, $\mathbf{N}$ is stoichiometric matrix which has $N$ columns and $M$

**Table 1.** Values of kinetic parameters of histidinol dehydrogenase.

| Parameter | Values resulted from model identification against experimental data published in (Loper et al. 1965; Adams 1955) | Parameter | Values resulted from model identification against experimental data published in (Loper et al. 1965; Adams 1955) |
|---|---|---|---|
| $k_1$ | 1073 min$^{-1}$ | $K_{NAD}$ | 47.2 mM |
| $k_{-1}$ | 645 min$^{-1}$ | $K_{Hol}$ | 0.208 mM |
| $k_2$ | 1073 min$^{-1}$ | $K_{NAD,Hol}$ | $6.7 \cdot 10^{-3}$ mM |
| $K_0^1$ | $3.8 \cdot 10^{-7}$ mM | $K_{Hal}$ | 1.8 mM |
| $K_0^2$ | $8.5 \cdot 10^{-7}$ mM | $K_{Hal,NAD}$ | 0.21 mM |
| $K_{Hol}^1$ | $10^{-7}$ mM | $K_{NADH}$ | 0.63 mM |
| $K_{Hol}^2$ | $4.2 \cdot 10^{-6}$ mM | $K_{His}$ | 0.2 mM |
| $K_{Hal,His}^1$ | $2.2 \cdot 10^{-7}$ mM | $K_{Hal,NADH}$ | $2.8 \cdot 10^{-4}$ mM |
| $K_{Hal,His}^2$ | $2.8 \cdot 10^{-3}$ mM | $K_{NADH,His}$ | 2.1 mM |

**Table 2.** Intermediates of histidine biosynthesis pathway in *Escherichia coli.*

| Intermediate designation | Chemical name of intermediate |
|---|---|
| PRATP | N1-(5'-phosphoribosyl)-ATP |
| PRAMP | N1-(5'-phosphoribosyl)-AMP |
| ProFAR | Pro-phosphoribosyl-formimino-5-aminoimidazole-4-carboxamide ribonucleotide |
| PRFAR | Phosphoribosyl-formimino-5-aminoimidazole-4-carboxamide ribonucleotide |
| IGP | Imidazoleglycerol phosphate |
| AICAR | 5-aminoimidazole-4-carboxamide ribonucleotide |
| IAP | Imidazoleacetol phosphate |
| HolP | L-Histidinol phosphate |
| Hol | L-Histidinol |
| Hal | L-Histidinal |
| His | L-Histidine |
| ATP | Adenosine triphosphate |
| Gln | Glutamine |
| αKg | αketoglutarate |
| Glu | Glutamate |
| NAD | Nicotineamide adenine dinucleotide phosphate oxidized |
| NADH | Nicotineamide-adenine-dinucleotide phosphate reduced |

rows. In the case of pathway of histidine biosynthesis both $M$ and $N$ are equal. Vectors of intermediate concentrations and initial conditions are:

$\mathbf{x} = $ [ATP, PRATP, PRAMP, ProFAR, PRFAR, Gln, Glu, IGP, AICAR, IAP, His, HolP, αKG, Hol, NAD, NADH, Hal]$^{T}$,

$\mathbf{x_0}$= [$ATP_0$, $PRATP_0$, $PRAMP_0$, $ProFAR_0$, $PRFAR_0$, $Gln_0$, $Glu_0$, $IGP_0$, $AICAR_0$, $IAP_0$, $His_0$, $HolP_0$, $\alpha KG_0$, $Hol_0$, $NAD_0$, $NADH_0$, $Hal_0$]$^T$.

Methods of rate equation derivation will be addressed later in detail. First, we will focus on static modelling. The static model is important for an understanding of the cellular machinery. It enables us to derive relationships between steady state fluxes and to find out conservation laws both for concentrations and for fluxes (re-action rates). By solving the system of linear algebraic equations

$$\mathbf{N \cdot v = 0} \tag{2.2}$$

one could find that any steady state reaction rate (steady state flux), vi, i=1,...$N$, can be expressed as linear combination of $NF$ independent steady state rates. The $NF$ equals the dimension of kernel of matrix $\mathbf{N}$, and coefficients are fully deter-mined by stoichiometric matrix. As an example, we consider one for possible rela-tionships between steady state rates of histidine biosynthesis pathway (see Fig. 1):

$$v_{16} = v_{14} = v_{13} = v_9 = v_8 = v_5 = v_4 = v_3 = v_2 = v_1,$$
$$v_7 = v_1 + v_6,$$
$$v_{10} = v_{17} = v_{12},$$
$$v_{11} = v_1 - v_{12},$$
$$v_{15} = 2 \cdot v_1$$

From these relationships it follows that any steady state rate can be expressed in terms of the three independent rates v1, v6, and v12, i.e., *s* is equal to 3.

The relationship between fluxes could be used to predict multiple solutions and which cell is likely to use. Different techniques exist, including extreme pathway analysis (Schilling 2000), elementary mode analysis (Schuster 1999), flux balance analysis (FBA) (Bonarius 1997), and minimization of metabolic adjustments (Segre 2002).

Metabolite conservation laws are first, linear integrals of the system of differen-tial equations (2.1). As a simplest example of the conservation law valid for the pathway of histidine biosynthesis, we could consider the following algebraic ex-pression:

$$NAD + NADH = const_1. \tag{2.3}$$

So, the sum of concentrations of NAD and NADH does not change with time. The number of concentration conservation laws $NC$ of the model of $M$ intermediates connected with $N$ reactions is given by following formula:

$$NC = M - N + NF. \tag{2.4}$$

In the case of histidine biosynthesis pathway, both *m* and *n* are equal and *s* is equal to 3. Consequently (2.4), the number of conservation laws for histidine biosynthe-sis pathway shown in Figure 1 is equal to 3. Relationship (2.3) is one of the three conservation laws.

$$PRATP + PRAMP + ProFAR + PRFAR + AICAR + ATP = const_2$$
$$Glu + Gln + \alpha KG = const_3 \tag{2.5}$$

The values of parameters consti, i=1,2,3, are completely determined by initial conditions, since equations (2.3) and (2.5) are always valid:

$$const_1 = NAD_0 + NADH_0,$$
$$const_2 = PRATP_0 + PRAMP_0 + ProFAR_0 + PRFAR_0 + AICAR_0 + ATP_0$$
$$const_3 = Glu_0 + Gln_0 + \alpha KG_0$$

## 2.2 Basic principles of kinetic description of enzymatic reactions using *in vitro* experimental data

Once an appropriate static network has been chosen and analyzed, the second stage is to generate rate equations, and to describe the dependence of each reaction rate on concentrations of intermediates involved in the selected pathway. To make the models scalable and comparable with different kinds of experimental data we have developed both detailed and reduced descriptions for every biochemical process in the model. The *detailed* reaction description includes the exact molecular mechanism of the bio-molecular reaction (i.e. enzyme catalytic cycle) and takes into account all possible states of the protein, including possible non-active states (i.e. phosphorylated) or dead-end inhibitor complexes. Usually, the detailed description comprises a set of algebraic differential equations. Ordinary differential equations for fluxes are combined with and non-linear algebraic equations, if flux or conservation laws have been taken into account.

The *reduced* description represents the reaction rates as an explicit analytic function of the substrates and products. The catalytic cycle could be identified from the literature or hypothesized on 3D structure and other relative biological information for each active protein involved in the model (i.e. protein with catalytic function).

To derive the rate equations from the catalytic cycle, quasi-steady state, and rapid equilibrium approaches are being used. The quasi-steady state of the system (if possible) is calculated as a function of substrates, products, inhibitors, activators, total protein concentrations and all kinetic constants. The rate law for every process is derived as a flux from the catalytic cycle for this quasi-steady state. The rate law depends on the total concentration of the protein, concentrations of the effectors (activators, inhibitors, agonists, and antagonists), substrates, products, and the values of the kinetic parameters (Km, Ki, Kd, and elementary rate constants). While derived using a quasi-steady state assumption, rate laws are being used in simulations without any such simplifications made.

The level of the detailed elaboration of catalytic cycle and the subsequent derivation of rate equation are fully determined by experimental data available on structural and functional organization of the enzyme.

Rate equations could be derived directly if the catalytic cycle of the enzyme has been established and proven experimentally. The "minimal" catalytic cycle model is considered if enzyme mechanism is unknown. The "minimal" catalytic cycle model should

1) Satisfy all structural and stoichiometric data available from literature
2) Allow deriving the rate equation. This equation should explain all available kinetic experimental data
3) Represent the simplest catalytic cycle mechanism and comply with first two clauses

Kinetic experimental *in vitro* data published in the literature have usually been measured under diverse conditions (pH, temperature). Catalytic cycle mechanisms and corresponding rate law should be derived to satisfy all available data dependences on pH, temperature and other experimental conditions. The "minimal" catalytic cycle mode should also describe dependence of reaction rate on pH and temperature.

Parameter estimation is the third stage of model development. To estimate the kinetic parameter values the following sources could be used

1)   Literature data on values of $K_m$, $K_i$, $K_d$ , rate constants, pH optimum, etc;

2)   Electronic databases; only a few databases with specific kinetic content are available at the moment, in particular EMP (Selkov et al. 1996) and BRENDA (Schomburg et al. 2002).

3)   Experimentally measured dependencies of the initial reaction rates on concentrations of substrates, products, inhibitors, and activators.

4)   Time series data for enzyme kinetics and whole pathways.

As we have found, many processes (i.e. enzyme reactions) have not been studied kinetically. Many kinetic parameters cannot be estimated from the literature or databases due to a lack of available experimental data. One remedy is to express these unknown or "free" parameters in terms of other measured kinetic parameters. The result is the establishment of functional relationships between "free" parameters and measured kinetic parameters. Each parameter value, of course, is constrained by physico-chemical properties and any other information available from related organisms or related processes. The more constraints are available, the more dimensionality reduction can occur.

The modified version of the software package DBSolve (Goryanin et al. 1999) has been used to develop and to analyze kinetic models.

## 2.3 Derivation of rate equation of histidinol dehydrogenase of *Escherichia coli* and estimation of its kinetic parameters using *in vitro* experimental data

To illustrate basic principles of the catalytic cycle construction and derivation of rate equation we consider enzyme catalyzing two successive reactions of histidine biosynthesis pathway of *Escherichia coli* (see Fig. 1). This enzyme (EC 1.1.1.23), encoded by gene *hisD*, is able to catalyze two separate reactions (Umbarger 1996): oxidation of histidinol to histidinal (histidinoldehydrogenase activity)

Hol + NAD = Hal + NADH   (2.6)

and oxidation of histidinal to histidine (histidinaldehydrogenase activity)

Hal + NAD = His + NADH   (2.7)

*Availability of the experimental data.* To construct catalytic cycle of histidinol dehydrogenase the following data on structural and functional features of the enzyme have been used:

i.   When histidinol is used as a substrate of the enzyme, one histidine molecule is formed and two NAD molecules are reduced per one histidinol molecule consumed, but histidinal accumulation is not experimentally detected (Lopel et al. 1965; Adams 1955)

ii.  When histidinal is used as a substrate of the enzyme, one histidine molecule is formed and one NAD molecule is reduced per one histidinal molecule consumed (Loper et al. 1965)

iii.  The enzyme has one catalytic site (Loper et al. 1965; Adams 1955), in other words, different substrate of histidinol dehydrogenase and histidinal dehydrogenase reactions compete to each other

iv.  The binding of substrates as well as dissociation of products proceed in random order (Umbarger 1996; Loper et al. 1965; Adams 1955)

v.  pH optimum of histidinoldehydrogenase reaction differs from that of histidinaldehydrogenase reaction by two units of pH (Loper et al. 1965)

Kinetic properties of the enzyme operating as histidinoldehydrogenase, i.e., catalyzing reaction (2.6) only, has been partly studied in a paper (Loper et al. 1965) where turnover number and Michaelis constants for Hol and NAD have been estimated at different pH values:

$$K_{m,Hol}^{pH=7.7} = 0.029 \ mM, \ pH = 7.7; \quad K_{m,Hol}^{pH=9.3} = 0.012 \ mM, \ pH = 9.3$$

$$K_{m,NAD}^{Hol} = 1.53 \ mM, \ pH = 7.5; \quad K_{m,NAD}^{Hol} = 1.26 \ mM, \ pH = 9.3$$

$$k_{cat,Hol} = 1073 \ \text{min}^{-1}; \quad pH = 9.4 \tag{2.8}$$

From equations (2.8) it follows that an increase in pH from 7.7 to 9.3 decreases the Michaelis constant for histidinol by almost three times but changes the Michaelis constant for NAD ( $K_{m,NAD}^{Hol}$ ) by less than 20 percentages, i.e., in the range of experimental error. From this one could conclude that the Michaelis constant for NAD does not depend on pH.

Kinetic properties of the enzyme operating as histidinaldehydrogenase, i.e., catalyzing reaction (2.7) only, has been partly studied in the paper (Loper et al. 1965) where turnover number and Michaelis constants for Hal and NAD were estimated at different pH values:

$$K_{m,Hal} = 0.0078 \ mM, \ pH = 7.7;$$

$$K_{m,NAD}^{Hal} = 0.21 \ mM, \ pH = 7.5;$$

$$k_{cat,Hal} = 1073 \ \text{min}^{-1}; \quad pH = 7.6 . \tag{2.9}$$

Besides, dependencies of maximal activity of histidinoldehydrogenase and histidinaldehydrogenase on pH have been measured in (Loper et al. 1965) and the time dependence of NADH accumulation at presence of 2.7 mM of histidinol dehydrogenase and pH equal to 8.9 has been followed in (Adams 1955).

*Construction of the catalytic cycle.* To construct the catalytic cycle of histidinol dehydrogenase (see Fig. 2A), we have used experimental data described in clauses (i) – (iv) of the previous section. We have assumed that the enzyme operated following the Random Bi Bi mechanism of Cleland's classification (Cleland 1963). It has been conformed by statement (iv) of the previous section. The suggested mechanism has four ternary complexes Hol°E°NAD, Hal°E°NADH, Hal°E°NAD, His°E°NADH. The transition of ternary enzyme-substrate complex Hol°E°NAD

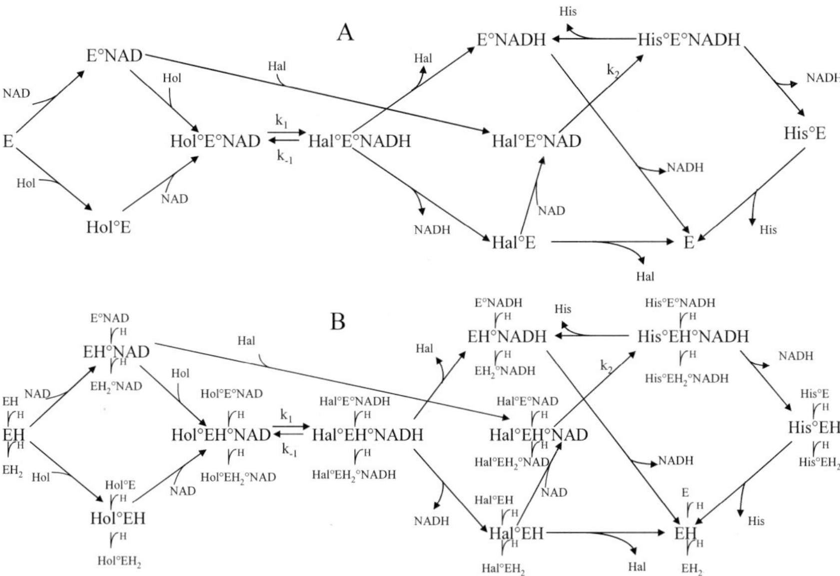

**Fig. 2.** Catalytic cycle of histidinol dehydrogenase without (A) and with (B) mechanism assigning pH dependence.

to the complex of enzyme with products Hal°E°NADH corresponds to histidinoldehydrogenase activity of the enzyme but the irreversible (i.e. far-from-equilibrium) transition (Loper et al. 1965) of ternary complex Hal°E°NAD to complex His°E°NADH corresponds to histidinaldehydrogenase activity of the enzyme.

In accordance with statement (iii) histidinol should compete with histidinal for the catalytic site of the enzyme. To take this fact into account we have introduced additional reaction of binding of histidinal to enzyme state with NAD bound in catalytic site (E°NAD→ Hal°E°NAD).

If histidinol and NAD are used as initial substrates then in accordance with the suggested catalytic cycle one molecule of histidine is formed and two molecules of NAD are reduced per one histidinol molecule consumed. This stoichiometry corresponds to successive transitions via the following stages of catalytic cycle: E→E NAD→Hol E NAD → Hal E NADH→Hal E→ Hal E NAD→His E NADH→His E→E. Note that histidinal molecule, being intermediate on the way of histidine formation from histidinol, is not released during cycling through these stages of the catalytic cycle but remains in the catalytic site of the enzyme and is oxidized to histidine. This means that it is not possible to detect free histidinal experimentally if the enzyme mainly operates via the selected cycle of stages of catalytic cycle. This corresponds to statement (i) of the previous section. If histidinal and NAD are used as initial substrates then one molecule of histidine is formed and one molecule of NAD is reduced per one histidinal molecule consumed. This stoichiometry corresponds to successive transitions via the following stages of catalytic cycle: E→E°NAD→ Hal°E°NAD→His°E°NADH→His°E→E and is in agreement with statement (ii) of the previous section.

The next stage of catalytic cycle construction consists in taking into account possibility of proton binding to different states of enzyme. The catalytic cycle (Fig. 2A) should be modified and a rate equation should be derived according following experimental observations/constraints on pH:

  A.  pH optima of histidinoldehydrogenase activity differs from that of histidinaldehydrogenase one (that is in agreement with clause (v) of previous section)
  B.  The Michaelis constant with respect to histidinol depends on pH
  C.  The Michaelis constant with respect to NAD of histidinoldehydrogenase reaction does not depend on pH.

In order to introduce the pH dependence of enzyme operation and to fit the requirements formulated above, the approach described in Cornish-Bowden (2001) was used.

The enzyme (or, in other words, one or several amino acid residues of catalytic cycle directly participating in catalysis) could be deprotonated, once protonated and twice protonated. The catalytic cycle has been modified under the assumption that singly protonated enzyme is catalytically active (Fig. 2B).

*Derivation of rate equations.* Using scheme of the catalytic cycle (Fig. 2B) rates of histidinoldehydrogenase and histidinaldehydrogenase reactions can be obtained

$$V_{Hol} = k_1 \cdot Hol°EH°NAD - k_{-1} \cdot Hal°EH°NADH \qquad (2.10)$$
$$V_{Hal} = k_2 \cdot Hal°EH°NAD. \qquad (2.11)$$

In order to derive rate equations describing the dependence of rates of histidinoldehydrogenase and histidinaldehydrogenase reactions on concentrations of substrates, products, and effectors, we assume that reactions of substrate binding, product dissociation and enzyme protonation are much faster than the catalytic reactions designated by numbers 1 and 2 in Figure 2B. Consequently, we consider all fast reactions as quasi-equilibrium and obtain analytical expressions of concentrations of enzyme states (Hol°EH°NAD, Hal°EH°NADH, Hal°EH°NAD) included in right hand sides of equations (2.10) and (2.11). This enables us to derive dependencies of rates of histidinoldehydrogenase and histidinaldehydrogenase reactions on substrates, products and proton concentrations:

$$v_{Hol} = \frac{HisD}{\Delta} \cdot \left( k_1 \cdot \frac{Hol}{K_{NAD,Hol}} \cdot \frac{NAD}{K_{NAD}} - k_{-1} \cdot \frac{NADH}{K_{Hal,NADH}} \cdot \frac{Hal}{K_{Hal}} \right)$$

$$v_{Hal} = \frac{HisD \cdot k_2 \cdot NAD \cdot Hal / K_{Hal,NAD} / K_{Hal}}{\Delta}$$

$$\Delta = h_E + h_{Hol \circ E} \cdot \frac{Hol}{K_{Hol}} + h_{E \circ NAD} \cdot \frac{NAD}{K_{NAD}} + h_{Hol \circ E \circ NAD} \cdot \frac{Hol}{K_{NAD,Hol}} \cdot \frac{NAD}{K_{NAD}} +$$

$$h_{Hal \circ E} \cdot \frac{Hal}{K_{Hal}} + h_{Hal \circ E \circ NAD} \cdot \frac{NAD}{K_{Hal,NAD}} \cdot \frac{Hal}{K_{Hal}} + h_{E \circ NADH} \cdot \frac{NADH}{K_{NADH}} +$$

$$h_{Hal \circ E \circ NADH} \cdot \frac{NADH}{K_{Hak,NADH}} \cdot \frac{Hal}{K_{Hal}} + h_{His \circ E} \cdot \frac{His}{K_{His}} + h_{His \circ E \circ NADH} \cdot \frac{His}{K_{NADH,His}} \cdot \frac{NADH}{K_{NADH}}$$

$$(2.12)$$

Here, the functions $h_X$ define level of once protonated enzyme state X, X $\in$ {E, E°NAD, Hol°E°NAD etc}:

$$h_X = h_X(pH) = 1 + \frac{K_X^1}{H} + \frac{H}{K_X^2} \; . \tag{2.13}$$

Here, H is proton concentration; $K_X^1$ and $K_X^2$ are dissociation constants describing proton dissociation from once and twice protonated enzyme state X, respectively. HisD is the total enzyme concentration: $K_A$ is the dissociation constant of substrate (or product) A from free enzyme; $K_{BA}$ is the dissociation constant of substrate (or product) A from the complex of the enzyme with substrate (or product) B; $k_1$, $k_{-1}$, $k_2$ are rate constants of catalytic steps of catalytic cycle. Dissociation constants of substrates/products and rate constants are shown in Figure 2B near corresponding reactions.

Equations (2.12) involve 32 parameters: 12 parameters are rate and dissociation constants describing kinetic properties of individual steps of catalytic cycle (Fig. 2B), and 20 parameters characterize proton binding to different enzyme states. In accordance with our approach we introduce several assumptions, which do not conflict with experimental data cited above on the one hand and simplify mechanism of pH dependence of enzyme operation on the other hand. This enables us to reduce number of unknown parameters. Let us assume:

1.  10 enzyme states of catalytic cycle, depicted in Figure 2A, can be subdivided into 3 groups
    Group 1: E, E°NAD, E°NADH
    Group 2: Hol°E, Hol°E°NAD
2.  Group 3: Hal°E°NADH, Hal°E°NAD, His°E°NADH, Hal°E, His°E
3.  Protonation of amino acid residues of catalytic site of enzyme states included in one group are described by identical dissociation constants

$$h_E = h_{E \circ NAD} = h_{E \circ NADH} = h_0 = 1 + \frac{K_0^1}{H} + \frac{H}{K_0^2} \tag{2.14}$$

$$h_{Hol \circ E} = h_{Hol \circ E \circ NAD} = h_{Hol} = 1 + \frac{K_{Hol}^1}{H} + \frac{H}{K_{Hol}^2} \tag{2.15}$$

$$h_{Hal\circ E} = h_{Hal\circ E\circ NAD} = h_{Hal\circ E\circ NADH} = h_{His\circ E} = h_{His\circ E\circ NADH} =$$

$$= h_{Hal,His} = 1 + \frac{K^1_{Hal,His}}{H} + \frac{H}{K^2_{Hal,His}} \tag{2.16}$$

These assumptions allow us to reduce number of unknown parameters for pH dependence of enzyme operation from 20 to 6 and to rewrite equations (2.12) as:

$$v_{Hol} = \frac{HisD}{\Delta} \cdot \left( k_1 \cdot \frac{Hol}{K_{NAD,Hol}} \cdot \frac{NAD}{K_{NAD}} - k_{-1} \cdot \frac{NADH}{K_{Hal,NADH}} \cdot \frac{Hal}{K_{Hal}} \right)$$

$$v_{Hal} = \frac{HisD \cdot k_2 \cdot NAD \cdot Hal / K_{Hal,NAD} / K_{Hal}}{\Delta}$$

$$\Delta = h_0 \cdot \left( 1 + \frac{NAD}{K_{NAD}} + \frac{NADH}{K_{NADH}} \right) + h_{Hol} \left( \frac{Hol}{K_{Hol}} + \frac{Hol}{K_{NAD,Hol}} \cdot \frac{NAD}{K_{NAD}} \right) +$$

$$h_{Hal,His} \left( \frac{Hal}{K_{Hal}} + \frac{NAD}{K_{Hal,NAD}} \cdot \frac{Hal}{K_{Hal}} + \frac{NADH}{K_{Hal,NADH}} \cdot \frac{Hal}{K_{Hal}} + \frac{His}{K_{His}} + \frac{His}{K_{NADH,His}} \cdot \frac{NADH}{K_{NADH}} \right)$$

$$\tag{2.17}$$

It is easy to see that this simplified mechanism of pH dependence completely satisfies experimentally established facts (A), (B), and (C) cited in previous section. Moreover, it is not difficult to prove that further simplification of the mechanism of pH dependence (such as subdivision of all enzyme states to two groups instead of three) does not allow us to derive rate equations describing these experimental facts correctly. So, the suggested mechanism of pH dependence is minimal one of all possible mechanisms able to describe experimental facts (A), (B), and (C).

*Estimation of kinetic parameters of the rate equations using in vitro experimental data.* To reduce the number of unknown parameters of equations (2.17) we have used values of kinetic parameters (2.8) and (2.9) measured experimentally. Loper and colleagues have estimated turnover numbers and Michaelis constants with respect to substrates of histidinoldehydrogenase and histidinaldehydrogenase reactions at different pH values (Loper et al. 1965). We have obtained analytical expressions for these kinetic parameters via parameters of the catalytic cycle, i.e., dissociation and rate constants. Then, using these relationships between kinetic parameters and parameters of catalytic cycle, we have obtained an expression for 7 parameters in equation (2.17) via experimentally measured kinetic parameters (2.8) and (2.9) and 11 remaining parameters of the catalytic cycle:

$$k_1 = k_{cat,Hol} \cdot h_{Hol}(pH = 9.4) \ , \ k_2 = k_{cat,Hal} \cdot h_{Hal,His}(pH = 7.6) \ , \ K_{Hal,NAD} = K^{Hal}_{m,NAD}$$

$$K_{NAD} = \frac{K_{Hol} \cdot K^{Hol}_{m,NAD}}{K^{pH=9.3}_{m,Hol}} \cdot \frac{h_0(pH = 9.3)}{h_{Hol}(pH = 9.3)} \ , \ K_{NAD,Hol} = K^{pH=9.3}_{m,Hol} \cdot \frac{h_{Hol}(pH = 9.3)}{h_0(pH = 9.3)}$$

$$K_0^1 = \frac{\dfrac{K_{m,Hol}^{pH=9.3}}{K_{m,Hol}^{pH=7.7}} \cdot \dfrac{h_{Hol}(pH=9.3)}{h_{Hol}(pH=7.7)} \cdot \left(1+\dfrac{10^{-4.7}}{K_0^2}\right) - \left(1+\dfrac{10^{-6.3}}{K_0^2}\right)}{\dfrac{1}{10^{-6.3}} - \dfrac{1}{10^{-4.7}} \cdot \dfrac{K_{m,Hol}^{pH=9.3}}{K_{m,Hol}^{pH=7.7}} \cdot \dfrac{h_{Hol}(pH=9.3)}{h_{Hol}(pH=7.7)}} \quad ,$$

$$K_{Hal} = \frac{K_{Hol} \cdot K_{m,NAD}^{Hol} \cdot K_{m,Hal}}{K_{m,NAD}^{Hal} \cdot K_{m,Hol}^{pH=9.3}} \cdot \frac{h_0(pH=9.3)}{h_{Hol}(pH=9.3)} \cdot \frac{h_{Hal,His}(pH=7.7)}{h_0(pH=7.7)}$$

$$(2.18)$$

where,

$$h_0(pH) = 1 + \frac{K_0^1}{10^{-3+pH}} + \frac{10^{-3+pH}}{K_0^2} \quad , \qquad h_{Hol}(pH) = 1 + \frac{K_{Hol}^1}{10^{-3+pH}} + \frac{10^{-3+pH}}{K_{Hol}^2}$$

$$h_{Hal,His}(pH) = 1 + \frac{K_{Hal,His}^1}{10^{-3+pH}} + \frac{10^{-3+pH}}{K_{Hal,His}^2} \quad . \qquad\qquad (2.19)$$

To estimate values of the remaining 11 parameters experimental data (Loper et al. 1965) on the pH dependence of maximal velocity of histidinoldehydrogenase and histidinaldehydrogenase have been used. Data (Adams 1955) on time dependence of NADH accumulation at presence of 2.7 mM of histidinol dehydrogenase and pH equal to 8.9 has been used too.

pH dependence of maximal velocity of histidinoldehydrogenase activity of the enzyme normalized to its value at optimal pH is determined:

$$V_{max,norm}^{Hol}(pH) = \frac{1 + 2 \cdot \sqrt{\dfrac{K_{Hol}^1}{K_{Hol}^2}}}{h_{Hol}(pH)} \quad . \qquad\qquad (2.20)$$

In a similar way, the pH dependence of maximal velocity of histidinaldehydrogenase activity normalized to its value at optimal pH is determined:

$$V_{max,norm}^{Hal}(pH) = \frac{1 + 2 \cdot \sqrt{\dfrac{K_{Hal,His}^1}{K_{Hal,His}^2}}}{h_{Hal,His}(pH)} \quad . \qquad\qquad (2.21)$$

Values of the four parameters from the equations (2.20) and (2.21) have been chosen so that pH dependencies of maximal velocities of histidinol dehydrogenase have coincided with experiments measured in (Loper et al. 1965). Figure 3 shows pH dependencies of maximal velocities calculated from equations (2.20) and (2.21) (solid lines) and experimentally measured ones (symbols). The values of estimated parameters are listed in Table 3.

The remaining 7 parameters of equation (2.17) have been estimated from experimental data (Adams 1955). The kinetic experiment has been conducted as follows: the reaction was started by adding 2.7 µM of histidinol to a cuvette containing 500 µM of NAD and 2.7 µM of histidinol dehydrogenase at pH 8.9.

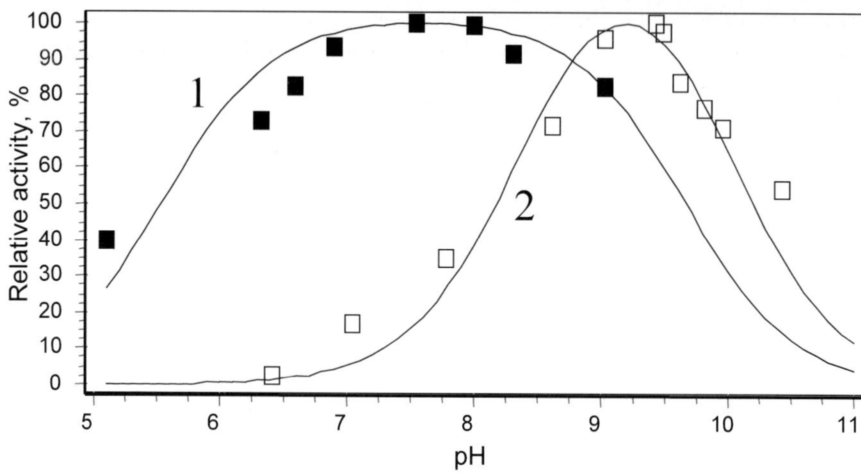

**Fig. 3.** Dependencies of experimentally measured (symbols, Loper et al. 1965) and calculated from model (solid lines) histidinoldehydrogenase (empty squares and line 2) and histidinaldehydrogenase (filled squares and line 1) activities of histidinol dehydrogenase on pH.

$$\text{NAD} \quad \text{NADH} \quad \text{NAD} \quad \text{NADH}$$
$$\text{Hol} \rightleftharpoons \text{Hal} \longrightarrow \text{His}$$

**Fig. 4.** Kinetic scheme of experiment described in (Adams 1955).

Accumulation of NADH was followed in time. The kinetic model has been updated to incorporate this experiment Figure 4:

$$dNADH/dt = v_{Hol} + v_{Hal}$$
$$NAD + NADH = 500 \ \mu M$$
$$Hol + Hal + His = 20 \ \mu M$$
$$NADH + 2{\cdot}Hol + Hal = 40 \ \mu M \tag{2.22}$$

The initial values of model variables corresponded to initial concentrations of substrates in experiment:

$$Hol = 20 \ \mu M,$$
$$NAD = 500 \ \mu M$$
$$NADH = Hal = His = 0 \tag{2.23}$$

Seven unknown parameters for rate equations of histidinodehydrogenase and histidinadehydrogenase reactions have been chosen so that the time dependence of NADH accumulation resulting from numerical solution of system (2.22), (2.23) (solid line in Fig. 5) coincided with the corresponding time dependence measured experimentally (symbols in Fig. 5) in (Adams 1955). Values of estimated parameters are listed in Table 3.

**Table 3.** Reactions of histidine biosynthesis pathway in *Escherichia coli.*

| Reaction number | Chemical equation | Enzyme | Gene |
|---|---|---|---|
| 1 | PRPP + ATP = PRATP + PPi | ATP-phosphoribosyltransferase | HisG |
| 2 | PRATP = PPi + PRAMP | Phosphoribosyl-ATP-pyrophosphohydrolase: Phosphoribosyl-AMP cyclohydrolase | HisI |
| 3 | PRAMP = ProFAR | Phosphoribosyl-ATP-pyrophosphohydrolase: Phosphoribosyl-AMP cyclohydrolase | HisI |
| 4 | ProFAR = PRFAR | Phosphoribosyl-formimino-5-amino-1-phosphoribosyl-4-imidazole-carboxamide isomerase | HisA |
| 5 | PRFAR + Gln = Glu + IGP + AICAR | IGP synthase | HisHF |
| 6 | Gln = Glu | IGP synthase | HisHF |
| 7 | Glu = Gln | Glutamine-synthatase | |
| 8 | AICAR = ATP | Purine biosynthesis | |
| 9 | IGP = IAP | IGP dehydratase | HisB |
| 10 | IAP + Glu = HolP + $\alpha$KG | Histidinol phosphate aminotransferase | HisC |
| 11 | IAP + His = HolP + His1 | Histidinol phosphate aminotransferase | HisC |
| 12 | $\alpha$KG = Glu | Ammonia assimilation | |
| 13 | HolP = Hol + Pi | Histidinol phosphatase | HisB |
| 14 | Hol + NAD = Hal + NADH | Histidinoldehydrogenase | HisD |
| 15 | NADH = NAD | Respiratory chain | |
| 16 | Hal + NAD = His +NADH | Histidinaldehydrogenase | HisD |
| 17 | His $\rightarrow$ | Histidine consumption | |

# 3 Application of the *Escherichia coli* branched-chain amino acid biosynthesis model. Prediction of possible genetic changes that should maximize isoleucine and valine production

Valine and isoleucine are branched-chain amino acids that are widely used in biomedicine and biotechnology. Indeed, valine is used in food industry for flavour additive production. Moreover, this amino acid is one of the sources of cephamycin antibiotic biosynthesis. However, pathways of valine and isoleucine biosynthesis are strongly coupled in *E. coli* because four of five steps in valine biosynthesis pathway are catalyzed by enzymes involved in the isoleucine biosynthesis pathway. This strong coupling between biosynthetic  pathways raises the question

**Table 4.** Intermediates of isoleucine and valine biosynthesis pathway in *Escherichia coli*.

| Intermediate designation | Chemical name of intermediate |
| --- | --- |
| OA | Oxaloacetate |
| PEP | Phosphoenolpyruvate |
| Thr | Threonine |
| Kb | Ketobutirate |
| Pyr | Pyruvate |
| Ahb | Aceto-hydroxybutyrate |
| Al | Acetolactate |
| Dhv | Dihydroxy-methylvaleriate |
| Dhi | Dihydroxy-isovaleriate |
| Ktv | Keto-methylvaleriate |
| Kti | Keto-isolvaleriate |
| Ile | Isoleucine |
| Val | Valine |
| NADP/NADPH | Nicotineamide adenine dinucleotide phosphate |
| Glt | Glutamate |
| $\alpha$Kg | $\alpha$Ketoglutorate |

**Table 5.** Enzymes of isoleucine and valine biosynthesis pathway in *Escherichia coli*.

| Designation | Enzyme |
| --- | --- |
| TDH | Threonine dehydrotase |
| AHAS (I, III) | Acetolactate synthase (I, III) |
| IR | Acetohydroxy acid isomeroreductase |
| DHAD | Dihyroxy-acid dehydrotase |
| BCAT | Branch-chain-amino-acid transaminase |

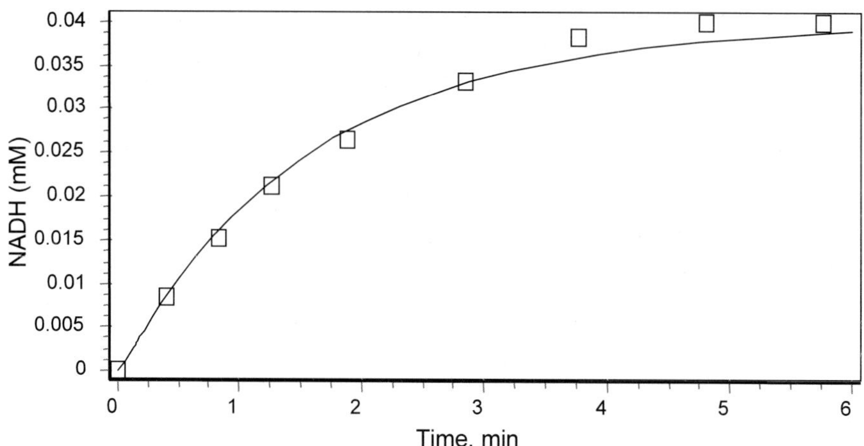

**Fig. 5.** Experimentally measured (empty squares, Adams 1955) and calculated from model (solid line) time dependence of NADH accumulation.

if it is possible to introduce such genetic modifications into some chosen strain that should allow to increase valine production and to decrease production of isoleucine, which possibly contaminates valine production processes in industry. In this section, we develop the kinetic model of the biosynthesis of valine and isoleucine in *E.coli* and validate it against *in vitro* experimental data as described in the previous section. Then we apply this model to predict two strain improvement strategies: the first one enables to maximize valine yield and to decrease isoleucine production and the second strategy predict such genetic modifications which, on the contrary, should provide an increase in isoleucine production and a decrease valine yield. However, it is necessary to stress that both strain improvement strategies suggested in this section are based on heterogeneous experimental data taken from different literature sources and measured for different strains of *E. coli*. That's why these results cannot be applied directly but should require further tuning of the kinetic model in accordance with peculiarities of any *E. coli* strain chosen for actual production. This means that the aim of the section is to merely demonstrate how to apply kinetic modelling to develop a strategy for improving the production of the purposeful metabolite.

## 3.1 Model development

The initial substrates of isoleucine and valine biosynthesis pathways shown in Figure 6 are pyruvate (*Pyr*), which is required for the synthesis of both amino acids and L-Threonine (*Thr*), which is a precursor of L-Isoleucine (*Ile*). To connect our model with the whole cell metabolism we consider the influx to threonine (*Thr*) from oxaloacetate (*OA*) (Step 1, Fig. 6) and influx to Pyruvate (*Pyr*) from phosphoenolpyruvate (*PEP*) (Step 8, Fig. 6).

The reaction of deamination (Step 2, Fig. 6) catalyzed by threonine dehydratase (TDH) transforms L-Threonine to α-Ketobutyrate (*Kb*). Then decarboxylation of pyruvate catalyzed by acetolactate synthase (AHAS) results in α-Aceto-α-hydroxybutyrate (*Ahb*) in the isoleucine pathway (Step 3, Fig. 6) and α-Acetolactate (*Al*) in the valine pathway (Step 9, Fig. 6). Both acetohydroxybutyrate and acetolactate are converted to dihydroxy acids by isomeroreductase (IR), corresponding to dihydroxy-methylvaleriate (*Dhv*) in the isoleucine pathway (Step 4, Fig. 6) and to dihydroxy-isovaleriate (*Dhi*) (Step 10, Fig. 6). These dihydroxy acids are converted to α-keto acids by dihyroxy-acid dehydrotase (DHAD), respectively to ketomethylvaleriate (*Ktv*) in the isoleucine pathway (Step 5, Fig. 6) and to keto-isolvaleriate (*Kti*) in the valine pathway (Step 11, Fig. 6). The final step in the biosynthesis of both amino acids is a transamination reaction catalyzed by branched-chain-amino-acid transaminase (BCAT) with glutamate as an amino donor. Keto-methylvaleriate is converted to isoleucine (*Ile*) (Step 6, Fig. 6) and keto-isolvaleriate is converted to valine (*Val*) (Step 12, Fig. 6). To couple our model to intracellular energy metabolism and to processes consuming isoleucine and valine, we take into account the recycling of NADPH (Step 14, Fig. 6) and the effluxes of isoleucine (Step 7, Fig. 6) and valine (Step 13, Fig. 6).

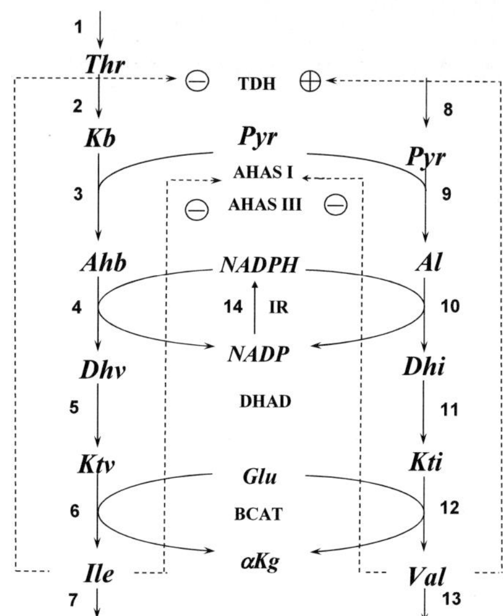

**Fig. 6.** Kinetic scheme of isoleucine and valine biosynthesis pathways. Metabolite notations are in text. Minus sign (-) corresponds to inhibition, plus sign (+) corresponds to activation.

**Table 6.** The system of differential equation describing time behaviour of kinetic model. Indices 1 refer to isoleucine pathway, indices 2 refer to isoleucine pathway.

| Isoleucine pathway | Valine pathway |
|---|---|
| $\dfrac{dThr}{dt} = v_{influx1} - v_{TDH}$ | $\dfrac{dPyr}{dt} = v_{influx2} - 2(v_{AHAS\,I,2} + v_{AHAS\,III,2}) - (v_{AHAS\,I,1} + v_{AHAS\,III,1}$ |
| $\dfrac{dKb}{dt} = v_{TDH} - (v_{AHAS\,I,1} + v_{AHAS\,III,1})$ | |
| $\dfrac{dAhb}{dt} = v_{AHAS\,I,1} + v_{AHAS\,III,1} - v_{IR1}$ | $\dfrac{dAl}{dt} = v_{AHAS\,I,2} + v_{AHAS\,III,2} - v_{IR2}$ |
| $\dfrac{dDhv}{dt} = v_{IR1} - v_{DHAD1}$ | $\dfrac{dDhi}{dt} = v_{IR2} - v_{DHAD2}$ |
| $\dfrac{dKtv}{dt} = v_{DHAD1} - v_{BCAT1}$ | $\dfrac{dKti}{dt} = v_{DHAD2} - v_{BCAT2}$ |
| $\dfrac{dIle}{dt} = v_{BCAT1} - v_{efflux1}$ | $\dfrac{dVal}{dt} = v_{BCAT2} - v_{efflux2}$ |

$$\frac{dNADP}{dt} = v_{IR1} + v_{IR2} - v_{NADPH\ recycling}$$

$$\frac{dNADPH}{dt} = v_{NADPH\ recycling} - v_{IR1} - v_{IR2}$$

$$NADP + NADPH = const$$

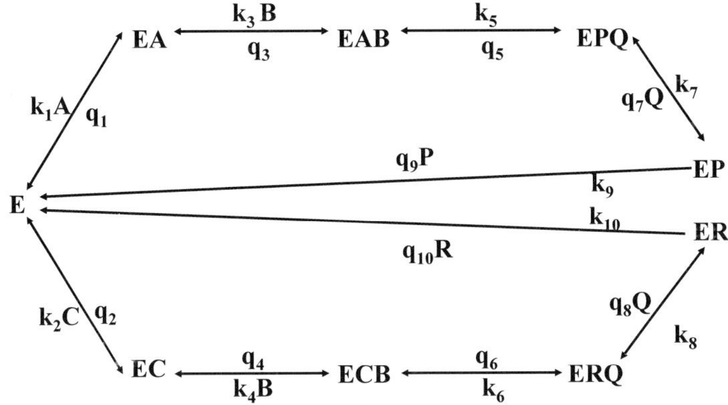


**Fig. 7.** Catalytic cycle of the enzyme catalyzing two competing reactions. A, B and C are substrates, P, Q and R are products. E is enzyme in free form. $k_i$ and $q_i$ are the constants of elementary stages.

The kinetic model of the branched-chain amino acids biosynthesis consists of mass balance equations describing time behaviour of intracellular metabolites, which take the form of (2.2). Table 6 lists the complete set of these balance equations.

The pathway of isoleucine and valine biosynthesis has several features that are worthy of notice:

 i. The pathway includes four enzymes (AHAS, IR, DHAD, and BCAT), each is able to catalyze two different reactions: one belonging to the isoleucine biosynthesis pathway, other belonging to the pathway of valine biosynthesis. From this, it follows that intermediates of isoleucine biosynthesis pathway compete with intermediates of valine biosynthesis pathway for these four enzymes.

 ii. The reactions of deamination and decarboxylation (steps 2 and 3 in Fig. 6) are regulated by isoleucine and valine.

 iii. There are several isoezymes of acetolactate synthase that provides differences in regulation by the end products of the pathways.

### 3.1.1 Derivation of the rate equations

Since most of the enzymes involved in branched-chain amino acid formation are able to catalyze two reactions (one belonging to isoleucine biosynthesis pathway and another one belonging to the pathway of valine biosynthesis), they should be described so as to take into account competition between substrates of the reactions for the catalytic site of the enzyme. As a first step we have constructed a catalytic cycle describing this feature. Let's consider such a catalytic cycle for the reaction with two substrates and two products depicted in Figure 7. Substrate *A* interacts with substrate *B* giving products *Q* and *P*. At the same time substrate *C* can interact with substrate *B* giving products *Q* and *R*. Substrates *A* and *C* compete with each other for the binding site. Applying the quasi steady states approach we derived rate equations corresponding to both reactions, which in simplified form have a following view:

$$v_1 = \frac{\dfrac{V_{m1}}{F_1 K_{m1}^B K_m^A}\left(A\cdot B - \dfrac{P\cdot Q}{K_{eq}}\right)}{1+\left(\dfrac{P}{K_{m1}^P}+\dfrac{Q}{K_m^Q}\right)\left(1+\dfrac{A}{F_1 K_m^A}\dfrac{B}{K_{m1}^B}\right)+\left(\dfrac{P}{K_{m2}^P}+\dfrac{R}{K_m^R}\right)\left(1+\dfrac{C}{F_2 K_m^C}\dfrac{B}{K_{m2}^B}\right)} \qquad (3.1\ 24)$$

$$v_2 = \frac{\dfrac{V_{m2}}{F_2 K_{m2}^B K_m^C}\left(C\cdot B - \dfrac{P\cdot R}{K_{eq}}\right)}{1+\left(\dfrac{P}{K_{m1}^P}+\dfrac{Q}{K_m^Q}\right)\left(1+\dfrac{A}{F_1 K_m^A}\dfrac{B}{K_{m1}^B}\right)+\left(\dfrac{P}{K_{m2}^P}+\dfrac{R}{K_m^R}\right)\left(1+\dfrac{C}{F_2 K_m^C}\dfrac{B}{K_{m2}^B}\right)} \qquad (3.2\ 25)$$

$$F_1 = 1+\frac{A}{K_m^A}+\frac{B}{K_{m1}^B}\ ,\quad F_2 = 1+\frac{C}{K_m^C}+\frac{B}{K_{m2}^B}\ .$$

Each of them corresponded to an ordered Bi-Bi mechanism (in Cleland terms), where $V_m$, $K_m$ and $K_{eq}$ are maximal velocity, Michaelis constants and equilibrium constant, respectively. The same approach can be applied to derive rate equations for competing reactions with two substrates and one product (Ordered Bi-Uni according to Cleland classification)

$$v_1 = \frac{\dfrac{V_{m1}}{F_1 K_{m1}^B K_m^A}\left(A\cdot B - \dfrac{Q}{K_{eq}}\right)}{1+\dfrac{Q}{K_m^Q}\left(1+\dfrac{A}{F_1 K_m^A}\dfrac{B}{K_{m1}^B}\right)+\dfrac{R}{K_m^R}\left(1+\dfrac{C}{F_2 K_m^C}\dfrac{B}{K_{m2}^B}\right)}\ , \qquad (3.3\ 26)$$

$$v_2 = \frac{\dfrac{V_{m2}}{F_2 K_{m2}^B K_m^C}\left(C\cdot B - \dfrac{R}{K_{eq}}\right)}{1+\dfrac{Q}{K_m^Q}\left(1+\dfrac{A}{F_1 K_m^A}\dfrac{B}{K_{m1}^B}\right)+\dfrac{R}{K_m^R}\left(1+\dfrac{C}{F_2 K_m^C}\dfrac{B}{K_{m2}^B}\right)}\ , \qquad (3.4\ 27)$$

as well as for competing reactions with one substrate and one product (Uni-Uni according to Cleland classification)

$$v_1 = \frac{\dfrac{V_{m1}}{K_m^A}\left(A - \dfrac{Q}{K_{eq}}\right)}{1+\dfrac{Q}{K_m^Q}+\dfrac{A}{K_m^A}+\dfrac{R}{K_m^R}+\dfrac{C}{K_m^C}}\ , \qquad (3.5\ 28)$$

$$v_2 = \frac{\dfrac{V_{m2}}{K_m^C}\left(C - \dfrac{R}{K_{eq}}\right)}{1+\dfrac{Q}{K_m^Q}+\dfrac{A}{K_m^A}+\dfrac{R}{K_m^R}+\dfrac{C}{K_m^C}}\ . \qquad (3.6\ 29)$$

### 3.1.2 Detailed description of pathway steps

For estimation of kinetic parameters of rate equations describing individual enzymes of the pathway we have fitted rate equations against *in vitro* experimental data taken from different literature sources. When experimental curves were not available we have used values of kinetic parameters available from electronic databases Brenda and EMP, and other published information on the individual enzyme. In the cases of lack of any experimental data we put "free" values (notation f.p in Table 7) of parameters.

*Influxes.* We describe the reactions supplying initial substrates to the branch-chain amino acid biosynthesis in the simplest form according to the mass action law. Threonine, the precursor of isoleucine, is derived from oxaloacetate (*OA*). Pyruvate, which takes part in the both pathways, is derived from phosphoenolpyruvate:

$$v_{influx1} = V_{influx1}\left(OA - \frac{Thr}{K_{eq}^{in1}}\right) \;,\; v_{influx2} = V_{influx2}\left(PEP - \frac{Pyr}{K_{eq}^{in2}}\right) \;.$$

Parameters of $V_m$ and $K_{eq}$ were evaluated, described below.

*Threonine dehydratase (TDH).* Threonine dehydratase catalyzes the conversion of threonine to ketobutyrate; it is inhibited by isoleucine and is activated by valine. According to results published in (Wessel et al. 2000; Umbarger et al. 1996) isoleucine inhibits TDH via enhancement of the cooperativity of substrate binding while valine abolishes cooperativity and allows the enzyme to bind threonine with normal Michaelis-Menten kinetics. To describe reversible kinetics of TDH and to take into account allosteric properties of the enzyme we used the following rate equation taken from (Ivanitzky et al. 1978):

$$v_{TDH} = \frac{\dfrac{V_m^{TDH}}{K_m^{Thr}}\left(Thr - \dfrac{Kb}{K_{eq}^{TDH}}\right)}{\left(1 + \dfrac{Thr}{K_m^{Thr}} + \dfrac{Kb}{K_m^{Kb}}\right)} \frac{1 + LF^{n-1}H^n}{1 + LF^n H^n} \;.$$

Where

$$F = \frac{1 + \dfrac{Thr}{K_m^{Thr}}\dfrac{K_r^{Thr}}{K_t^{Thr}} + \dfrac{Kb}{K_m^{Kb}}\dfrac{K_r^{Kb}}{K_t^{Kb}}}{1 + \dfrac{Thr}{K_m^{Thr}} + \dfrac{Kb}{K_m^{Kb}}} \;,\; H = \frac{\left(1 + \dfrac{Ile}{K_t^{Ile}}\right)\left(1 + \dfrac{Val}{K_t^{Val}}\right)}{\left(1 + \dfrac{Ile}{K_r^{Ile}}\right)\left(1 + \dfrac{Val}{K_r^{Val}}\right)} \;.$$

To estimate the parameters of the equation, we have fitted it against experimental curves from (Wessel et al. 2000) (Fig. 8). We have fitted all experimental curves simultaneously and obtained a unique set of kinetic parameters listed in Table 7. The values of *Vm* and *Keq* were estimated as described below.

*Acetolactate synthase (AHAS).* Acetolactate synthase catalyzes the conversion of ketobutyrate to aceto-hydroxybutyrate in the pathway of isoleucine biosynthesis and the conversion of pyruvate to acetolactate in the valine biosynthesis pathway. Three different enzymes catalyzing the formation of acetohydroxy acids have been

**Table 7.** Values of kinetic parameters of isoleucine and valine biosynthesis pathway in *Escherichia coli*.

| Enzyme | Parameter values | Source or estimation |
| --- | --- | --- |
| TDH | $K_r^{Thr}=654.99$ | Fitting of literature data (Wessel et al. 2000) |
| | $K_t^{Thr}=2.62$ | |
| | $K_r^{Kb}=1.0$ f.p | |
| | $K_t^{Kb}=1.0$ f.p | |
| | $K_r^{Ile}=0.0058$ | |
| | $K_t^{Ile}=7.98$ | |
| | $K_r^{Val}=180.15$ | |
| | $K_t^{Val}=0.05$  $L=0.09$ | |
| | $n=4$ | |
| AHAS | $K_m^{Kb}=0.053$ | Fitting of literature data (Engel et al. 2000; Vyazmensky et al. 1996; Bar-Ilan et al. 2001; Hill et al. 1998; Hill et al. 1997; Eoyang et al. 1984; Barak et al. 1987) |
| | $K_{m1}^{Pyr}=0.145$ | |
| | $K_m^{Ahb}=1.0$ f.p | |
| | $K_{m2}^{Pyr}=0.00098$ | |
| | $K_m^{Al}=1.0$ f.p | |
| | $K_i^{Ile}=1.416$ | |
| | $K_i^{Val}=0.033$ | |
| IR | $K_{m1}^{NADPH}=0.073$ | Brenda and literature data (Madhavi et al. 1997; Aulabaugh et al. 1990; Chunduru et al. 1998) |
| | $K_m^{Ahb}=0.002$ | |
| | $K_{m1}^{NADP}=0.0042$ | |
| | $K_m^{Dhv}=1$ f.p | |
| | $K_{m2}^{NADPH}=0.073$ | |
| | $K_m^{Al}=0.014$ | |
| | $K_{m2}^{NADP}=0.206$ | |
| | $K_m^{Dhi}=1$ f.p | |
| DHAD | $K_m^{Dhv}=0.13$ | Fitting of literature data (Limberg et al. 1995; Myers 1961) |
| | $K_m^{Dhi}=0.1$ | |
| | $K_m^{Ktv}=0.1$ f.p | |
| | $K_m^{Kti}=0.1$ f.p | |

| Enzyme | Parameter values | Source or estimation |
| --- | --- | --- |
| BCAT | $K_m^{Ktv}=0.2$ | Brenda and literature data (Hall et al. 1993; Lee-Peng et al. 1979; Inoue et al. 1988) |
|  | $K_{m1}^{Glt}=1$ f.p |  |
|  | $K_m^{Ile}=1.1$ |  |
|  | $K_{m1}^{Ktg}=0.6$ |  |
|  | $K_m^{Kti}=0.2$ |  |
|  | $K_{m2}^{Glt}=1$ f.p |  |
|  | $K_m^{Val}=5$ |  |
|  | $K_{m2}^{Ktg}=3$ |  |

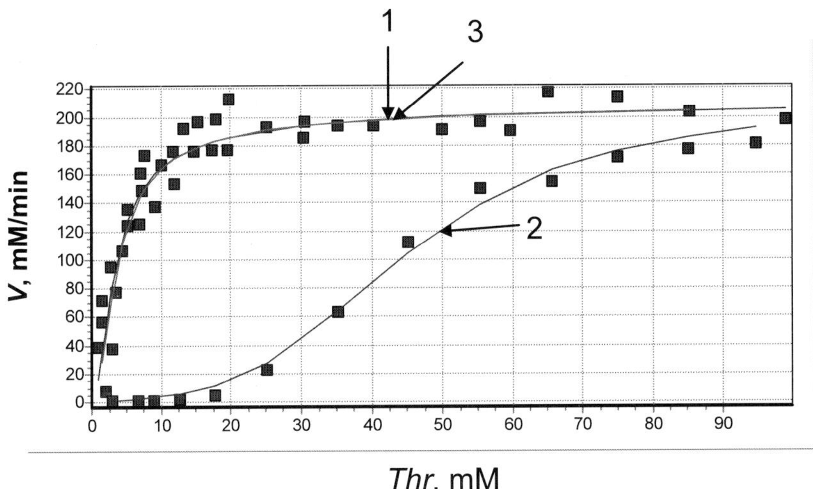



**Fig. 8.** Steady-state kinetics of TDH. Dependence of the initial rate of product formation on threonine concentration. Symbols corresponds to experimental data taken from (Wessel et al. 2000), lines designates model curves. 1 - without effectors, *Ile*=0 mM, *Val*=0 mM. 2 - with isoleucine as inhibitor, *Ile*=0.05 mM, *Val*=0 mM. 3 - with both effectors, *Ile*=0.05 mM *Val*=0.5 mM.

found (Umbarger 1996; Barak et al. 1987). Each isozyme is sensitive to the inhibition by isoleucine and valine, but at a different level. To simplify the model, we have not considered all three isozymes and have taken into account AHAS I and AHAS III only. These isozymes show maximal differences in their sensitivity to the inhibitors and enable to explore the possible differences in regulation in framework of whole model. The rate equations of reactions 3 and 9 (see Fig. 6) catalyzed by AHAS have been taken in the form of (3.3) and (3.4), respectively. The rate equations for AHAS I and AHAS III are the same in form but differ in values of kinetic parameters only:

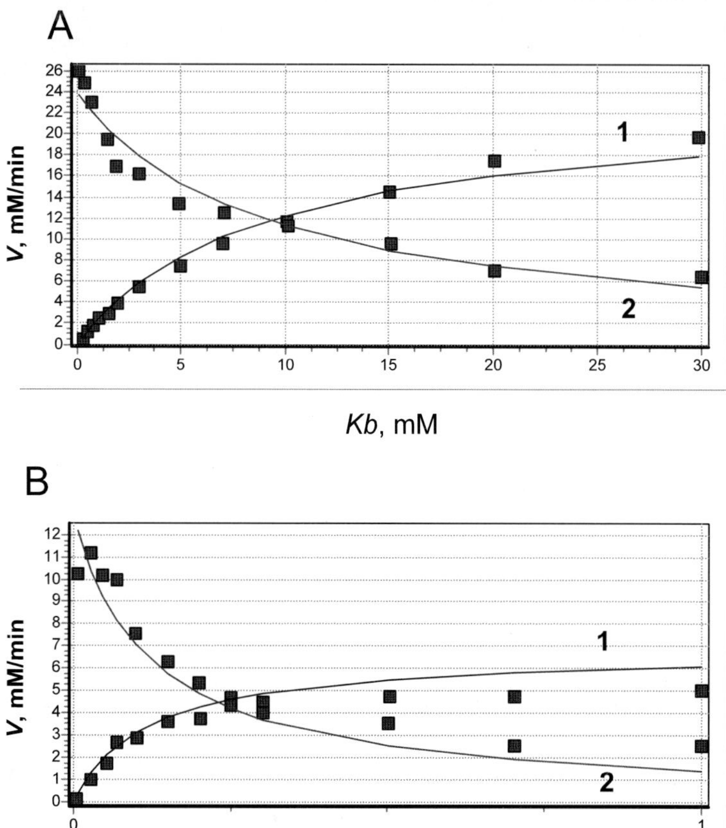

**Fig. 9.** Dependence of acetohydroxybutyrate (curves 1) and acetolactate (curves 2) formation catalyzed by AHAS I on ketobutyrate concentration with concentration of pyruvate of 10 mM (A) and with concentration of pyruvate 2 mM (B). Symbols corresponds to experimental data taken from (Barak et al. 1987), lines designates model curves.

$$v_{AHAS1} = \frac{\dfrac{V_{m1}^{AHAS}}{F_1 K_{m1}^{Pyr} K_m^{Kb}}\left(Kb \cdot Pyr - \dfrac{Ahb}{K_{eq}^{AHAS}}\right)}{1 + \dfrac{G_1}{F_1} + \dfrac{G_2}{F_2} + \dfrac{Ile}{K_i^{Ile}} + \dfrac{Val}{K_i^{Val}}} \;,\; v_{AHAS2} = \frac{\dfrac{V_{m2}^{AHAS}}{F_2 K_{m2}^{Pyr} K_m^{Al}}\left(Pyr^2 - \dfrac{Al}{K_{eq}^{AHAS}}\right)}{1 + \dfrac{G_1}{F_1} + \dfrac{G_2}{F_2} + \dfrac{Ile}{K_i^{Ile}} + \dfrac{Val}{K_i^{Val}}} \;.$$

Where

$$F_1 = 1 + \frac{Pyr}{K_{m1}^{Pyr}} \;,\; F_2 = 1 + \frac{Pyr}{K_{m2}^{Pyr}} \;,\; G_1 = \frac{Kb}{K_m^{Kb}} + \frac{Ahb}{K_m^{Ahb}} \;,\; G_2 = \frac{Pyr}{K_{m2}^{Pyr}} + \frac{Al}{K_m^{Al}} \;.$$

To estimate values of parameters of these equations, we have fitted them against experimentally measured curves published in Engel et al. (2000), Vyazmensky et

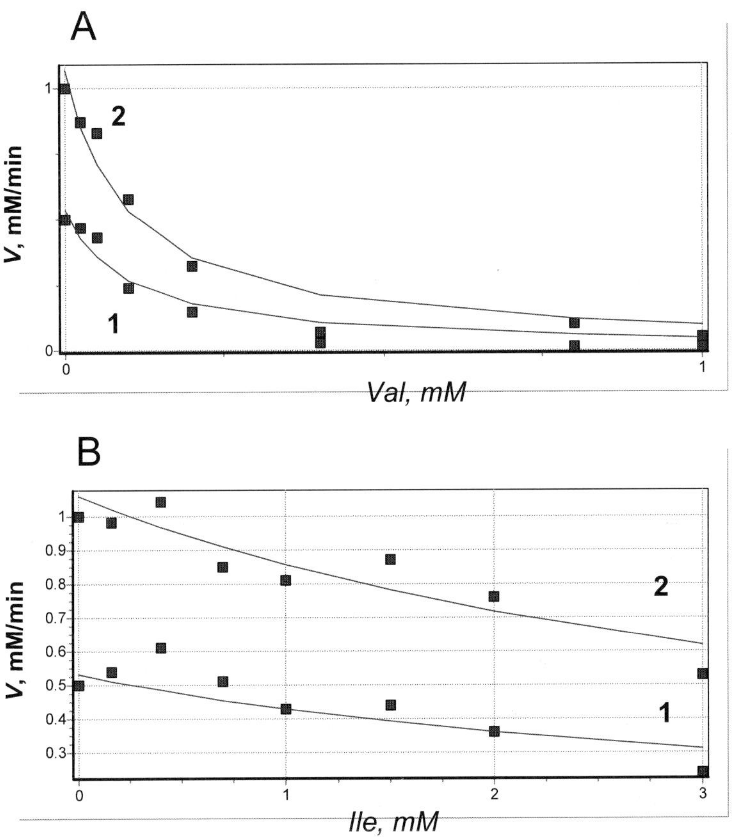



**Fig. 10.** Dependence of acetohydroxybutyrate (curves 1) and acetolactate (curves 2) formation on valine (A) and isoleucine (B) concentration. Experimental data taken from (Barak et al. 1987).

al. (1996), and Bar-Ilan et al. (2001). First, we have evaluated Michaelis constants for AHAS I. We have fitted four experimental curves (Figs. 9A, B) simultaneously and obtained a unique set of kinetic parameters listed in Table 7. Then, we have fitted experimental data enabling us to evaluate the inhibition constants of valine (Fig. 10A) and isoleucine (Fig. 10B) listed in Table 7. By the same way, we have estimated the kinetic parameters of AHAS III (Figs. 11A, B and Figs. 12A, B). The values of the parameters are listed in the Table 7.

*Acetohydroxy acid isomeroreductase (IR).* Acetohydroxy acid isomeroreductase catalyzes the conversion of aceto-hydroxybutyrate to dihydroxy-methylvaleriate in the isoleucine pathway and the conversion of acetolactate (*Al*) to dihydroxy-isovaleriate (*Dhi*) in the valine pathway. These reactions are coupled with *NADPH* reduction. The rate equations of IR catalyzed reactions have been

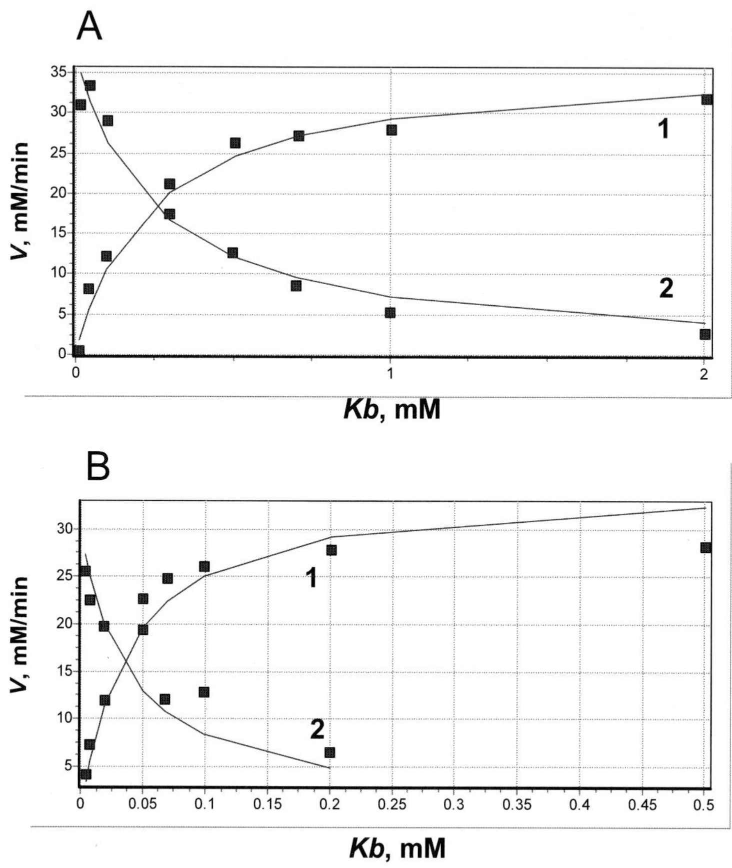



**Fig. 11.** Dependence of acetohydroxybutyrate (curves 1) and acetolactate (curves 2) formation catalyzed by AHAS III on ketobutyrate concentration mM (A) and with concentration of pyruvate 2 mM (B). Experimental data taken from (Barak et al. 1987).

taken in the form of (3.1) for the isoleucine pathway and in the form of (3.2) for the valine pathway:

$$v_{IR1} = \frac{\dfrac{V_{m1}^{IR}}{F_1 K_{m1}^{NADPH} K_m^{Ahb}}\left(NADPH \cdot Ahb - \dfrac{NADP \cdot Dhv}{K_{eq}^{IR}}\right)}{1 + \left(\dfrac{NADPH}{K_{m1}^{NADPH}} + \dfrac{Dhv}{K_m^{Ahb}}\right)\left(1 + \dfrac{Ahb}{F_1 K_m^{Ahb}}\dfrac{NADP}{K_{m1}^{NADP}}\right) + \left(\dfrac{NADPH}{K_{m2}^{NADPH}} + \dfrac{Dhi}{K_m^{Al}}\right)\left(1 + \dfrac{Al}{F_2 K_m^{Al}}\dfrac{NADP}{K_{m2}^{NADP}}\right)}$$

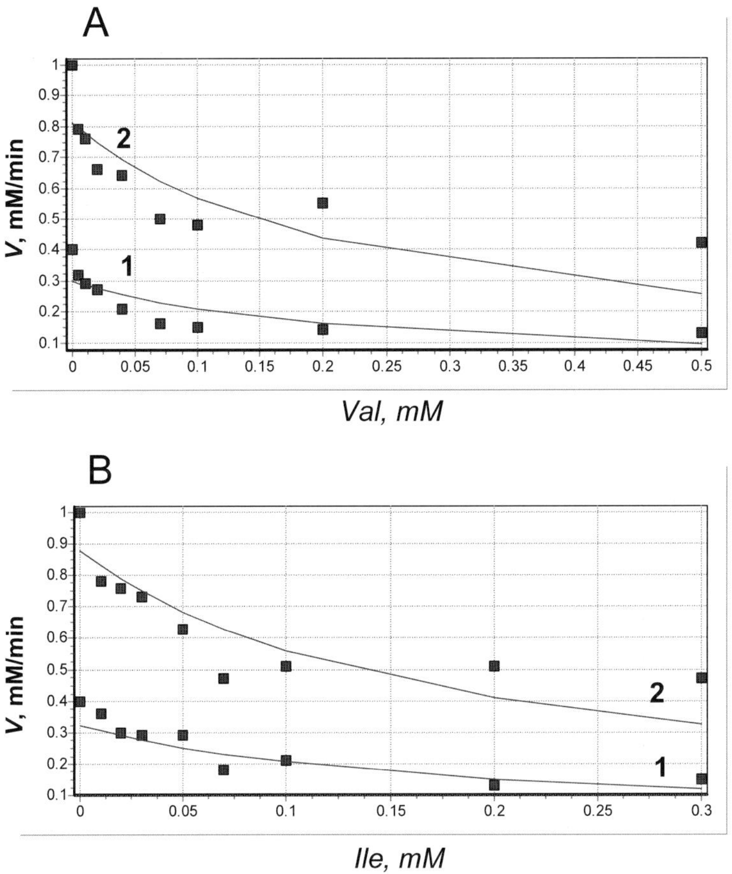

**Fig. 12.** Dependence of acetohydroxybutyrate (curves 1) and acetolactate (curves 2) formation on valine (A) and isoleucine (B) concentration. Experimental data taken from (Barak et al. 1987).

$$v_{IR2} = \frac{\dfrac{V_{m2}^{IR}}{F_2 K_{m2}^{NADPH} K_m^{Al}} \left( NADPH \cdot Al - \dfrac{NADP \cdot Dhi}{K_{eq}^{IR}} \right)}{1 + \left( \dfrac{NADPH}{K_{m1}^{NADPH}} + \dfrac{Dhv}{K_m^{Ahb}} \right) \left( 1 + \dfrac{Ahb}{F_1 K_m^{Ahb}} \dfrac{NADP}{K_{m1}^{NADP}} \right) + \left( \dfrac{NADPH}{K_{m2}^{NADPH}} + \dfrac{Dhi}{K_m^{Al}} \right) \left( 1 + \dfrac{Al}{F_2 K_m^{Al}} \dfrac{NADP}{K_{m2}^{NADP}} \right)}$$

Where

$$F_1 = 1 + \frac{Ahb}{K_m^{Ahb}} + \frac{NADP}{K_{m1}^{NADP}} \ , \quad F_2 = 1 + \frac{Al}{K_m^{Al}} + \frac{NADP}{K_{m2}^{NADP}} \ .$$

We have not found experimental curves describing IR kinetics and have used the values of kinetic constants listed in database Brenda and in papers (Madhavi et al.

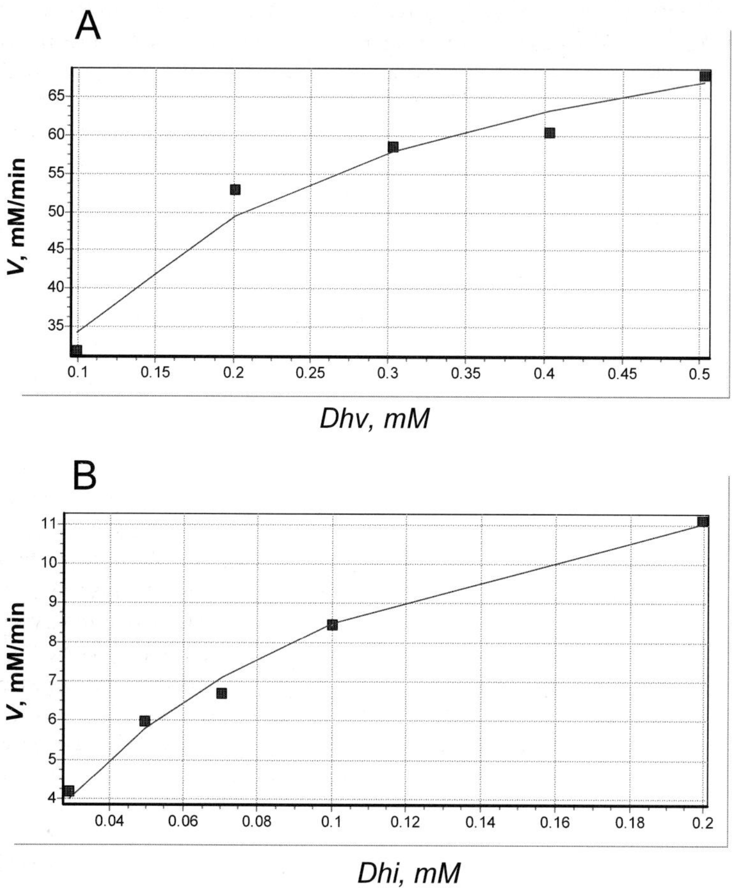



**Fig. 13.** Dependence of the initial rate of product formation catalyzed by DHAD on dihy-droxy-methylvaleriate (*A*) and dihydroxy-isovaleriate (*B*). Experimental data taken from (Myers 1961).

1997; Aulabaugh et al. 1990; Chunduru et al. 1998). The values of the parameters are listed in the Table 7.

*Dihyroxy-acid dehydrotase (DHAD)*. Dihyroxy-acid dehydrotase catalyzes the conversion of dihydroxy-methylvaleriate to keto-methylvaleriate in the isoleucine pathway and the conversion of dihydroxy-isovaleriate to keto-isolvaleriate in the valine pathway. The rate equations of DHAD catalyzed reactions have been taken in the form of (3.5) for the isoleucine pathway and in the form of (3.6) for the valine pathway:

$$v_{DHAD1} = \frac{\dfrac{V_{m1}^{DHAD}}{K_m^{Dhv}}\left(Dhv - \dfrac{Ktv}{K_{eq}^{DHAD}}\right)}{1 + \dfrac{Dhv}{K_m^{Dhv}} + \dfrac{Ktv}{K_m^{Ktv}} + \dfrac{Dhi}{K_m^{Dhi}} + \dfrac{Kti}{K_m^{Kti}}},$$

$$v_{DHAD2} = \frac{\dfrac{V_{m2}^{DHAD}}{K_m^{Dhi}}\left(Dhi - \dfrac{Kti}{K_{eq}^{DHAD}}\right)}{1 + \dfrac{Dhv}{K_m^{Dhv}} + \dfrac{Ktv}{K_m^{Ktv}} + \dfrac{Dhi}{K_m^{Dhi}} + \dfrac{Kti}{K_m^{Kti}}}.$$

To estimate the parameters of the equation, we have fitted (see Fig. 13 for fitting quality) these rate equations against experimental data published in (Limberg et al. 1995; Myers 1961). The values of the parameters are listed in the Table 7.

*Branched-chain-amino-acid transaminase (BCAT)*. Branched-chain-amino-acid transaminase catalyzes the conversion of keto-methylvaleriate to isoleucine and the conversion of keto-isolvaleriate to valine. The reaction is coupled with the conversion of glutamate (*Glt*) to ketolgutorate (*6Kg*). The rate equations of BCAT catalyzed reactions have been taken in the form of (3.1) for the isoleucine pathway and in the form of (3.2) for the valine pathway:

$$v_{BCAT1} = \frac{\dfrac{V_{m1}^{BCAT}}{F_1 K_m^{Ktv} K_m^{Glt}}\left(Ktv \cdot Glt - \dfrac{Ile \cdot Kgl}{K_{eq}^{BCAT}}\right)}{1 + \left(\dfrac{Kgl}{K_{m1}^{Kgl}} + \dfrac{Ktv}{K_m^{Ktv}}\right)\left(1 + \dfrac{Ile}{F_1 K_m^{II}} \dfrac{Glt}{K_{m1}^{Glt}}\right) + \left(\dfrac{Kgl}{K_{m2}^{Kgl}} + \dfrac{Kti}{K_m^{Kti}}\right)\left(1 + \dfrac{Val}{F_2 K_m^{II}} \dfrac{Glt}{K_{m2}^{Glt}}\right)}$$

$$v_{BCAT2} = \frac{\dfrac{V_{m2}^{BCAT}}{F_2 K_m^{Ktv} K_m^{Glt}}\left(Kti \cdot Glt - \dfrac{Val \cdot Kgl}{K_{eq}^{BCAT}}\right)}{1 + \left(\dfrac{Kgl}{K_{m1}^{Kgl}} + \dfrac{Ktv}{K_m^{Ktv}}\right)\left(1 + \dfrac{Ile}{F_1 K_m^{II}} \dfrac{Glt}{K_{m1}^{Glt}}\right) + \left(\dfrac{Kgl}{K_{m2}^{Kgl}} + \dfrac{Kti}{K_m^{Kti}}\right)\left(1 + \dfrac{Val}{F_2 K_m^{II}} \dfrac{Glt}{K_{m2}^{Glt}}\right)}$$

Where

$$F_1 = 1 + \frac{Ktv}{K_m^{Ktv}} + \frac{Kgl}{K_{m1}^{Kgl}} \;,\quad F_2 = 1 + \frac{Kti}{K_m^{Kti}} + \frac{Kgl}{K_{m2}^{Kgl}} \;.$$

We have not found kinetic experimental data for BCAT and have used the values of kinetic constants listed in database Brenda and in papers (Hall et al. 1993; Lee-Peng et al. 1979; Inoue et al. 1988). The values of the parameters are listed in the Table 7.

*NADP-recycling and effluxes*. The rate equations describing kinetics of effluxes and *NADP*-recycling have been written in the simplest form as the product of appropriative constants and concentrations.

$$v_{nadp} = V_m^{NADP} NADP$$

$$v_{efflux1} = V_{efflux1} Ile$$

$$v_{efflux2} = V_{efflux2} Val$$

Parameters of $V_m$ were evaluated by the way described below.

**Table 8.** Values of steady state concentrations, maximal rates and equilibrium constants of isoleucine and valine biosynthesis pathway in *Escherichia coli*.

| Steady state concentrations (mM) | Maximal rates (mM/min) | Equilibrium constants |
|---|---|---|
| OA=1 | $V_{influx1}=1$ | $K_{eq}^{in1}=1.1$ |
| Pep=2.8 (Chassagnole et al. 2002) | $V_{influx2}=1.2$ | $K_{eq}^{in2}=0.9$ |
| Thr=0.1 | | |
| Kb=0.1 | $V_m^{TDH}=10$ | $K_{eq}^{TDH}=10$ |
| Pyr=2.8 (Chassagnole et al. 2002) | $V_{m1}^{AHAS}=0.11$ | $K_{eq1}^{AHASI}=10$ |
| Ahb=0.05 | $V_{m2}^{AHAS}=0.11$ | $K_{eq1}^{AHASIII}=10$ |
| Al=0.05 | | |
| Dhb=0.03 | $V_{m1}^{IR}=0.34$ | $K_{eq2}^{AHASI}=0.1$ |
| Dhi=0.03 | $V_{m2}^{IR}=0.34$ | $K_{eq2}^{AHASIII}=0.1$ |
| Ktv=0.02 | $V_{m1}^{DHAD}=0.03$ | $K_{eq1}^{IR}=100$ |
| Kti=0.02 | $V_{m2}^{DHAD}=0.03$ | $K_{eq2}^{IR}=3$ |
| Ile=0.001 | $V_{m1}^{BCAT}=0.55$ | $K_{eq1}^{IR}=3$ |
| Val=0.001 | $V_{m2}^{BCAT}=0.55$ | |
| NADP=0.19 (Chassagnole et al. 2002) | | $K_{eq2}^{DHAD}=3$ |
| NADPH=0.06 (Chassagnole et al. 2002) | $V_{efflux2}=1$ | $K_{eq1}^{BCAT}=10$ |
| Glu=0.1 | $V_{efflux2}=1.2$ | $K_{eq2}^{BCAT}=10$ |
| αKg=0.1 | $V_m^{NADP}=0.1$ | |

### 3.1.3 Evaluation of maximal reaction rates

To estimate maximal reaction rates we used an approach described in (Chassagnole et al. 2002). According to this approach at steady state each rate equation corresponding to reaction catalyzed by enzyme $i$ is given by

$$\tilde{v}_i = V_i^{\max} f_i(\tilde{C}_i, \tilde{P}_i) , \tag{3.7}$$

where $\tilde{P}_i$ is the parameter vector and $\tilde{C}_i$ is the steady state concentration vector of the metabolites involved in reaction. From the equation (3.7) it follows that the maximal rate can be calculated in following manner:

$$V_i^{\max} = \frac{\tilde{v}_i}{f_i(\tilde{C}_i, \tilde{P}_i)} .$$

The stationary rates are estimated by the values of stationary fluxes taken from (Holms 1986). The value of stationary rate was calculated by the following expression

$$\tilde{v}_i = 5.56 f_i$$

where $f_i$ – the value of stationary flux.

The steady state concentrations have been chosen according to the *in vivo* concentration of metabolites in the *E.coli* cell. The equilibrium constants have been evaluated to keep up the real steady state concentrations of metabolites. Steady state concentrations, maximal rates and equilibrium constants are collected in the Table 8.


## 3.2 Application of kinetic model to optimize production of isoleucine and valine

We suppose that this and similar models could be used for the prediction of the effect of genetic changes when new bacterial strains are constructed. Actually modern biotechnology possesses powerful methods to accomplish different genetic modifications of bacterial genome (mutations, knockouts, amplifications of genes, as well as addition of genes from other organisms to the host genotype). Genetic changes of this type allow us to modify significantly the metabolic system of the cell by means of changes in certain enzymes activity. Engineering of new bacterial strains is usually directed at increased synthesis of some products by bacterial cells. However, until recently, the effect of performed genetic changes sometimes could not be predicted. Thus, for example, an incorporated genetic change can lead to lower improvement then the expected one or even have negative effects on the production rate of the "target" metabolite. Certain genetic changes seem positive when tested individually, but frequently, when incorporated with other improvements, they are no longer positive. The difficulties that are experienced in the prediction of effects of genetic changes most likely arise from the high complexity of regulation of the corresponding metabolic pathway. In these cases, it seems appropriate to use mathematical models of corresponding metabolic pathways and to make a computational simulation of the effect of genetic change.

In particular, the pathway of branched-chain amino acid biosynthesis is one of such complicated pathways, which are subject to rather complex system regulation including positive and negative feedback. The activity of the pathway also depends on the intermediates of linked metabolic pathways such as glycolysis, Krebs cycle, biosynthesis of certain amino acids, etc. We have tried to use the developed kinetic model of branched-chain amino acid biosynthesis to simulate the effect of possible genetic changes on the production of one of the pathway metabolites. We chose valine and isoleucine as "target" metabolites and tried to optimize their production in *E.coli* cells.

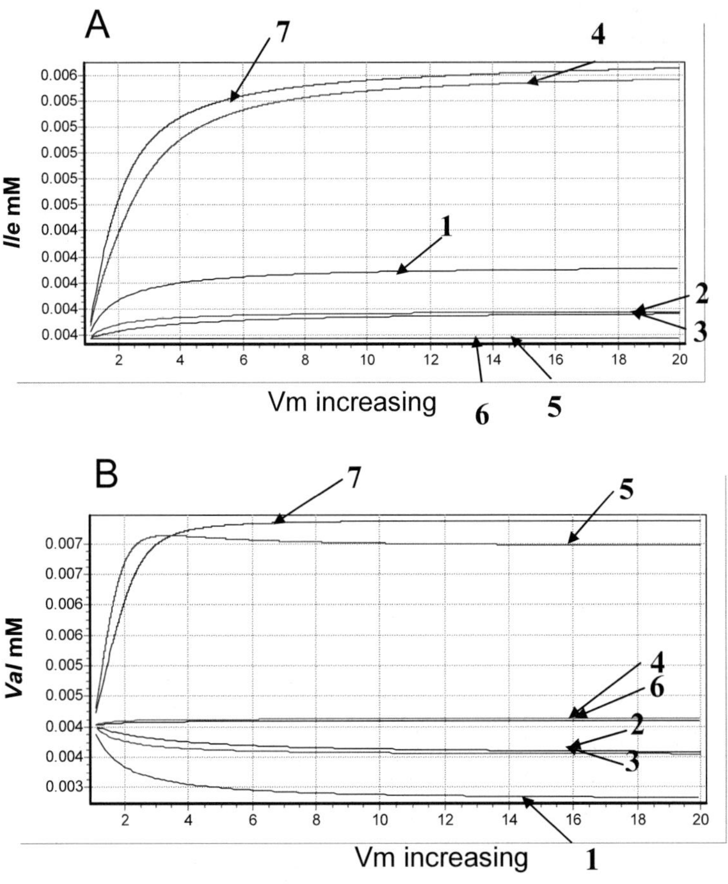


**Fig. 14.** Dependence of steady state concentrations of valine (A) and isoleucine (B) on maximal rates of TDH (line 1), AHAS I (line 2), AHAS III (line 3), IR (line 4), DHAD (line 5), BCAT (line 6) and NADP recycling (line 7).

To simulate the effect of various genetic changes on the specific amino acid production we assume that:

  i.   Knockout of a gene means that corresponding enzyme concentration is zero (or Vm=0)
  ii.  Mutation in the promoter of a gene means a change (a decrease) in concentration (or
  iii. Vm) of corresponding enzyme
  iv.  Gene amplification means an increase of enzyme concentration (or increase of Vm of corresponding reaction).

One of the main aims of the model development was to find out two strain improvement strategies: the first one enables to maximize valine yield and to decrease isoleucine production and the second strategy should predict such genetic

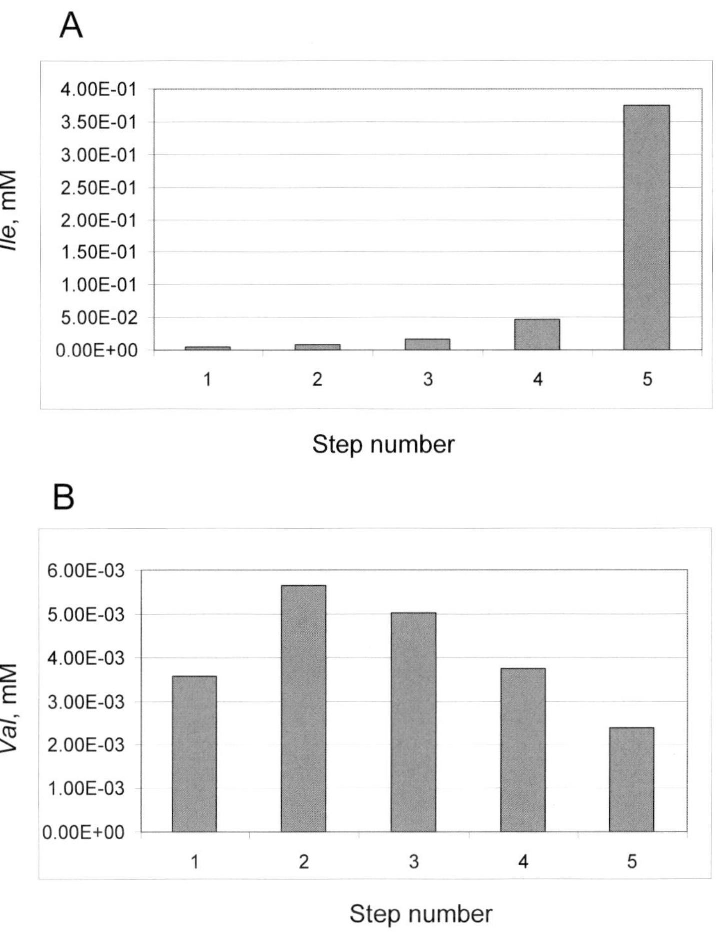



**Fig. 15.** Optimization of valine production. Changes in valine (A) and isoleucine (B) steady state concentrations resulted from different level of enzymes amplification: 1 - the base level; 2 - NADPH recycling; 3 – DHAD; 4 – BCAT; 5 – IR.

modifications that, on the contrary, provide increase in isoleucine production and decrease valine yield.

The first step was to study the changes in steady state concentrations of these amino acids in dependence of pathway enzyme concentrations (parameters of $Vm$ in the model). We assumed that every enzyme could be amplified no more than 20 times. So, to simulate the enzyme amplification, we increased the parameters of maximal rates of every enzyme involved in amino acids formation by 20 times (see Fig. 14). Using this approach, we could choose the most sensitive stage, which for both amino acids was the stage of NADPH recycling (Fig. 14A, B).

We have found that the first step in strain improvement should be amplification of enzymes catalyzing the formation of NADPH involved, for example, in the

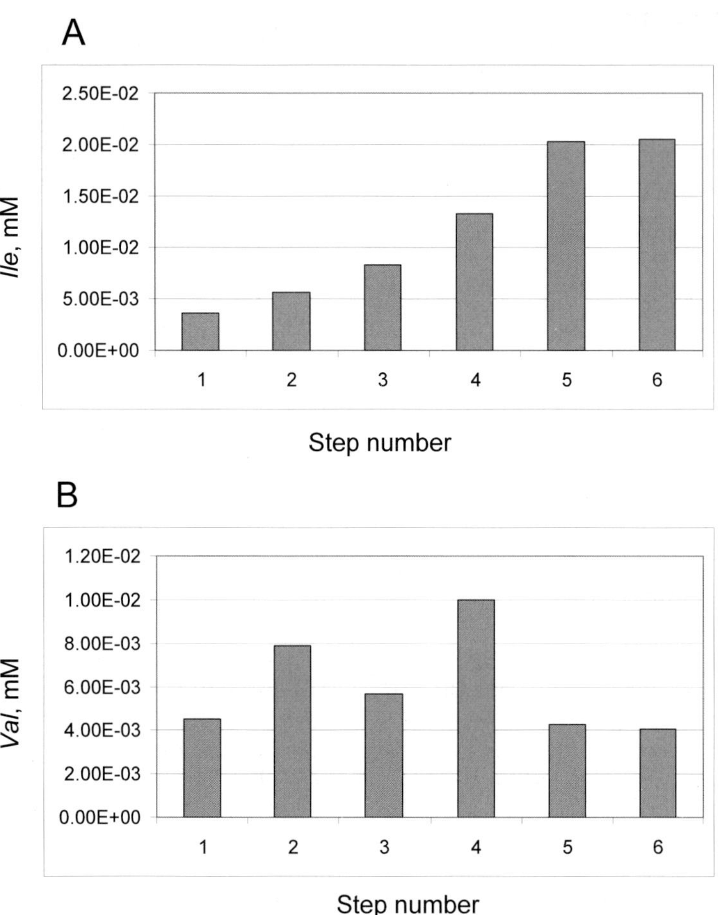

**Fig. 16.** Optimization of isoleucine production. Changes in isoleucine (A) and valine (B) steady state concentrations resulted from different level of enzymes amplification: 1 - the base level; 2 - NADPH recycling; 3 – TDH; 4 – DHAD; 5 – IR; 6 - AHAS III.

Krebs cycle. The NADPH recycling amplification results in a new steady state level, which is sensitive to the changes of the other enzymes concentrations. We studied step-by-step the changes in isoleucine and valine steady state concentrations resulting from amplifications of the other enzymes and found that amplifications of NADPH recycling enzymes (step 2 in Fig. 15), DHAD (step 3 in Fig. 15), BCAT (step 4 in Fig. 15), and IR (step 5 in Fig. 15) led to maximization of valine yield and decrease in isoleucine production. On the other hand, the kinetic model predicts that amplification of NADPH recycling enzymes (step 2 in Fig. 16), TDH (step 3 in Fig. 16), DHAD (step 4 in Fig. 16), IR (step 5 in Fig. 16), and AHAS III (step 6 in Fig. 16) results in increase in isoleucine production and decrease valine yield.

# 4 Discussion

Currently, there are a lot of attempts to make definition of systems biology broader, and even to avoid defining it. While systems biology could deal with a wide range of problems in biological and medical sciences, it is essential for the systems biology approach to use theory of dynamic systems for explaining and analyzing biological phenomena. This approach considers biological objects as SYSTEMS. Using it, one could identify and quantify all elements of biological systems, understand their interactions as a whole by dynamic networks analysis. In this chapter, we have shown that pathway of branched-chain amino acid biosynthesis is a subject to complex system regulation including positive and negative feedback. Activity of the pathway depends on the intermediates of linked metabolic pathways such as glycolysis, Krebs cycle, and biosynthesis of amino acids. We have shown that developing of kinetic models is very important. Integrating them into the whole cell kinetic model will raise the investigation of dynamic biological systems to another level of complexity.

# References

Aulabaugh A, Schloss JV (1990) Oxalyl hydroxamates as reaction-intermediate analogs for ketol-acid reductoisomerase. Biochemistry 29:2824-2830

Barak Z, Chipman DM, Gollop N (1987) Physiological implications of the specificity of acetohydroxy acid synthase isozymes of enteric bacteria. J Bacteriol 169:3750-3756

Bar-Ilan A, Balan V, Tittmann K, Golbik R, Vyazmensky M, Hubner G, Barak Z, Chipman DM (2001) Binding and activation of thiamin diphosphate in acetohydroxyacid synthase. Biochemistry 40:11946-11954

Chassagnole C, Noisommit-Rizzi N, Schmid JW, Klaus Mauch K, Reuss M (2002) Dynamic modeling of the central carbon metabolism of *Escherichia coli*. Biotechnol Bioeng 79:53-73

Chunduru SK, Mrachko GT, Calvo KC (1998) Mechanism of ketol acid reductoisomerase. Steady-state analysis and metal ion requirement. Biochemistry 28:486-493

Cleland WW The kinetics of enzyme-catalysed reactions with two or more substrates or products. Biochim Biophys Acta 67:104-137

Cornish-Bouden A (2001) Fundamentals of enzyme kinetic. Portland Press, Cambridge

Edwards JS, Ibarra RU, Palsson BO (2001) *In silico* predictions of *Escherichia coli* metabolic capabilities are consistent with experimental data. Nat Biotechnol 19:125-130

Elijah Adams (1955) I-histidinal, a biosynthetic precursor of histidine. J Biol Chem 217:325-344

Engel S, Vyazmensky M, Barak Z, Chipman DM, Merchuk JC (2000) Determination of the dissociation constant of valine from acetohydroxy acid synthase by equilibrium partition in an aqueous two-phase system. J Chromatogr B 743:225-229

Eoyang L, Silverman PM (1984) Purification and subunit composition of acetohydroxyacid synthase I from *Escherichia coli* K-12. J Bacteriol 157:184-189

Goryanin I, Hodgman TC, Selkov E (1999) Mathematical simulation and analysis of cellular metabolism and regulation. Bioinformatics 15:749-758

Hall TR, Wallin R, Reinhart GD, Hutson SM (1993) Branched chain amino transferase isoenzymes. Purification and characterization of the rat brain isoenzyme. J Biol Chem 268:3092-3098

Hill CM, Duggleby RG (1998) *Escherichia coli* acetohydroxyacid synthase II mutants. Biochem J 335:653-661

Hill CM, Pang SS, Duggleby RG (1997) *Escherichia coli* acetohydroxyacid synthase II. Biochem J 327:891-898

Holms WH (1986) The central metabolic pathways of *Escherichia coli*: relationship between flux and control at a branch point, efficiency of conversion to biomass, and excretion of acetate. Curr Top Cell Regul 28:69-104

Inoue K, Kuramitsu S, Aki K, Watanabe Y, Takagi T, Nishigai M, Ikai A, Kagamiyama H (1988) Branched-chain amino acid amino transferase of *Escherichia coli*: overproduction and properties. J Biochem 104:777-784

Ivanitzky GR, Krinsky VI, Selkov EE (1978) Mathematical biophysics of the cell. Nauka, Moscow

Lee-Peng FC, Hermodson MA, Kohlhaw GB (1979) Transaminase B from *Escherichia coli*: quaternary structure, amino-terminal sequence, substrate specificity, and absence of a separate valine-α-ketoglutarate activity. J Bacteriol 139(2):339-345

Limberg G, Klaffke W, Thiem J (1995) Conversion of aldonic acids to their corresponding 2-keto-3-deoxy-analogs by the non-carbohydrate enzyme dihydroxy acid dehydratase (DHAD). Bioorg Med Chem 3:487-494

Loper J, Adams E (1965) Purification and properties of histidinol dehydrogenase from *Salmonella typhimurium*. J Biol Chem 240:788-795

Myers JW (1961) Dihydroxy acid dehydrase: an enzyme involved in the biosynthesis of isoleucine and valine. J Biol Chem 236:1414-1418

Perna NT, Plunkett G 3rd, Burland V, Mau B, Glasner JD, Rose DJ, Mayhew GF, Evans PS, et al. (2001) Genome sequence of enterohaemorrhagic *Escherichia coli* O157:H7. Nature 409:529-533

Rane MJ, Calvo KC (1997) Reversal of the nucleotide specificity of ketol acid reductoisomerase by site-directed mutagenesis identifies the NADPH Binding Site1. Arch Biochem Biophys 338:83-89

Selkov E, Basmanova S, Gaasterland T, Goryanin I, Gretchkin Y, Maltsev N, Nenashev V, Overbeek R, Panyushkina E, Pronevitch L, Yunis I (1996) The metabolic pathway collection from EMP: the enzymes and metabolic pathways database. Nucleic Acids Res 24:26-28

Shomburg I, Chang A, Shomburg D (2002) BRENDA, enzyme data and metabolic information. Nucleic Acids Res 30:47-49

Umbarger HE (1996) *Escherichia coli* and *Salmonella*: cellular and molecular biology: ASM Press, Washington DC:442-458

Vyazmensky M, Sella C, Barak Z, Chipman DM (1996) Isolation and characterization of subunits of acetohydroxy acid synthase isozyme III and reconstitution of the holoenzyme. Biochemistry 35:10339-10346

Wessel PM, Graciet E, Douce R, Dumas R (2000) Evidence for two distinct effector-binding sites in threonine deaminase by site-directed mutagenesis, kinetic, and binding experiments. Biochemistry 39:15136-151143

Demin, Oleg V.
A.N. Belozersky Institute of Physico-Chemical Biology, Moscow State University, Moscow, Russia

Goryanin, Igor I.
GlaxoSmithKline, Scientific Computing & Mathematical Modeling, Stevenage, SG2 8PU, UK
goryanin@inf.ed.ac.uk

Kolupaev, Alex G.
A.N. Belozersky Institute of Physico-Chemical Biology, Moscow State University, Moscow, Russia

Lebedeva, Galina V.
A.N. Belozersky Institute of Physico-Chemical Biology, Moscow State University, Moscow, Russia

Metelkin, Eugeniy A.
A.N. Belozersky Institute of Physico-Chemical Biology, Moscow State University, Moscow, Russia

Plyusnina, Tatyana Y.
Biophysics Department, faculty of Biology, Moscow State University, Moscow, Russia

Tobin, Frank
GlaxoSmithKline, Scientific Computing & Mathematical Modeling, Upper Merrion, USA

Zobova, Ekaterina A.
A.N. Belozersky Institute of Physico-Chemical Biology, Moscow State University, Moscow, Russia

# Metabolic Control Analysis

David A. Fell

## Abstract

Metabolic Control Analysis (MCA) is a theoretical framework for investigating and understanding control and regulation of metabolism. In particular, it relates the properties of metabolic systems to the kinetic characteristics of the component enzymes. However, not all of the properties of enzymes strongly influence the behaviour of metabolic systems, some of which is generic and is reviewed here. It is argued that MCA is an important component of systems biology that still has much to offer in the development of predictive and integrative biology and the linking of genome to phenotype.

## 1 Introduction

Metabolic Control Analysis (MCA) was developed in order to provide a mathematical and quantitative framework for understanding the control and regulation of either the whole of metabolism, or, more usually, of definable subsystems (Kacser and Burns 1973; Heinrich and Rapoport 1974). Other approaches, such as Biochemical Systems Analysis (Savageau 1976) and Flux-Oriented Theory (Crabtree and Newsholme 1985) were also developed that shared these aims, but it is not my purpose in this chapter to review their similarities and differences. One of the key features of MCA is the exploration of the sensitivity of metabolic systems to changes in the amount or activity of a single enzyme (or other step such as a transporter). At the time MCA was developed, the prevailing view was that the steady state metabolic rate of any pathway (or *flux* as it is called in MCA terminology) was set by a key rate-limiting enzyme, which could be identified in the main by the characteristics of the enzyme itself (see e.g. Newsholme and Start 1973). Other enzymes were classed as non-rate limiting, again with many of the criteria related to their individual properties. MCA challenged this binary classification, and replaced it with the concept that the degree of influence of an enzyme on a particular metabolic flux had to be measured on a continuous scale in terms of a sensitivity coefficient, in this instance called a *flux control coefficient*. Very approximately, an enzyme's flux control coefficient is the percentage change in a specified steady state metabolic flux that is caused by a one percent change in the activity of the enzyme. This definition and other aspects of MCA are explained in further detail in my book (Fell 1997). The flux control coefficient can be shown to represent the enzyme's potential to affect the metabolic flux if that enzyme alone

Topics in Current Genetics, Vol. 13
L. Alberghina, H.V. Westerhoff (Eds.): Systems Biology
DOI 10.1007/b137745 / Published online: 12 May 2005

has its activity altered by an effector such as a signal metabolite, by covalent modification, or by induction of synthesis or degradation.

More than just defining a measure of degree of rate limitation, MCA established the relationships between the values of the flux control coefficients and the kinetic properties of the enzymes. This was first laid down in the initial publications (Kacser and Burns 1973; Heinrich and Rapoport 1974) and then developed further by others (e.g. Fell and Sauro 1985: Reder 1988). For this purpose, the kinetics of the enzymes are represented by *elasticities*, which are the sensitivities that an enzyme rate would show to a change in the concentration of each metabolite in turn whilst all other metabolites that influence the enzyme are held constant at their steady state concentrations in the metabolic system. (Again, the elasticity can be regarded as the percentage change in rate produced by a one percent change in the selected metabolite concentration, but the relationships are described in more detail in Fell 1997.) Two important conclusions flow from the relationships developed within MCA between flux control coefficients and elasticities. Firstly, the expression for a flux control coefficient of any particular enzyme contains elasticity terms for all the other enzymes connected to this metabolic flux, showing that the flux control coefficient can never be deduced solely from the kinetic properties of one enzyme; it is a system property that can in principle depend on *all* the enzymes in the system. Secondly, the elasticities that enter these expressions contain no information about how the enzyme implements this kinetic response at a molecular or mechanistic level. For example, the elasticity of an enzyme to one of its inhibitors will be negative, but it will make no difference to the values of the flux control coefficients whether the inhibition is competitive, uncompetitive or mixed; it is only the magnitude that counts. Therefore the molecular details of enzyme structure and mechanism that implement the elasticity are not strongly connected to the overall behaviour of the metabolic system, at least as represented by the control coefficients.

Similar conclusions can be reached about the relationships between enzymes and the steady state concentrations of metabolites; that is, the effect on concentrations of varying the activity of one enzyme is a function of the elasticities of all the enzymes in the system.

At this point I would like to address two issues. Firstly, is the study of metabolism part of systems biology? Secondly if it, and by extension MCA, is, does it show the general characteristics claimed for systems biology by the editors of this book? That is: "*Systems biology is not the biology of systems, nor is it the chemistry/physics/molecular genetics of molecules in biological systems. It is the difference between the two. It studies how new properties that are functionally important for life arise in interactions.*"

I do not think it will be contentious to affirm, in answer to the first question, that the study of metabolism can be part of systems biology. A significant aspect of an organism's phenotype is represented by the capabilities and adaptability of its metabolic network. The network is coded in the genome, but which parts of it are available depend on which enzyme and transporter genes are transcribed, which transcripts are translated, and what post-translational modifications occur. The metabolite concentrations and fluxes that are achieved through all this constitute part of the phenotype. Connecting genotype to phenotype may therefore be pursued through transcriptomics, proteomics and metabolomics, but ultimately the connection will be incomplete unless we can relate these to the fluxes in the metabolic network. Hence, sooner or later we must come to the type of questions addressed by MCA: how much does a change in the active amount of a specified enzyme affect fluxes in metabolism? Furthermore, if systems biology aims to be predictive rather than descriptive, then MCA provides conceptual and quantitative tools to aid understanding of how the metabolic phenotype derives from the genome sequence. However, although I am arguing that systems biology must necessarily encompass metabolism, I could not claim that all studies of metabolism are intrinsically an aspect of systems biology, especially as much classical biochemistry was blind to the way the behaviour of enzymes was dependent on their context and their interactions, mediated by metabolites, with other enzymes of the network (as indicated above in the discussion on rate-limiting enzymes).

Given this, MCA does then exemplify the editors' definition of systems biology. The development of MCA was a reaction against the concept that control of metabolic systems could be inferred from the kinetics of certain enzymes taken in isolation. I have outlined above how the elasticities of all the enzymes in the system determine the amount of influence that any given enzyme has on a metabolic flux; in other words the distribution of flux control arises as an interaction between enzymes through metabolites. I have also referred to the fact that it is not important for these conclusions how any particular enzyme generates an elasticity at the molecular level. So although there are many good reasons for studying the amino acid sequence, molecular structure, and reaction mechanisms of enzymes in detail, the case that these are an essential part of systems biology merely because metabolism is a system of interconnected enzymes looks weak. On the other hand, the experimental basis of systems biology cannot be complete without addition to metabolomics and the other –omics technologies of the study of enzyme kinetics and activities (with the aim of establishing an enzyme's response to metabolites under intracellular conditions, not for elucidating molecular mechanisms).

In the rest of this chapter, I will expand this argument by showing that MCA can explain a number of other important biological characteristics that arise from the interactions of enzymes and metabolites.


## 2 Relating system variables to enzyme kinetics

The relationship between the kinetics of single enzymes and the behaviour of metabolic systems needs a little more exploration. If we possess an equation that represents the full kinetic response of an enzyme to all its substrates, products, and effectors, then we can calculate the elasticity of the enzyme to any metabolite, given that we know the concentrations of all the metabolites in the enzyme's environment at steady state in the cell; examples are given in Fell (1997). But however complicated the elasticity function may be, it generates, under given conditions, a single value for the elasticity that could, in principle, have arisen from any number of different enzyme kinetic equations. Thus, in the case of the response of a metabolic pathway flux to an inhibitor, that response is determined by the product of the flux control coefficient of the inhibited enzyme and its elasticity with respect to the inhibitor (Kacser and Burns 1973). As stated above, this means that the immediate response to the inhibitor does not depend on the type of the inhibitor, only its strength as represented by the magnitude of the elasticity.

This analysis does underplay two factors. Firstly, the values of all the elasticities depend on the steady state values of the metabolite concentrations, and these, as with all other steady state variables, do depend on the full kinetic equations of the enzymes in the system. That is, different enzyme mechanisms do give rise to different steady states. An important illustration of this is the fact that predicted steady states are markedly different if product inhibition of enzymes is neglected (Cornish-Bowden and Cárdenas 2001). MCA, as an approximation to behaviour around the steady state, is not a tool that can predict the steady state itself, which must be determined experimentally or calculated numerically from the enzyme rate functions.

Secondly, MCA is only strictly valid for small changes. When large alterations are made in an enzyme activity or an inhibitor concentration, metabolite concentrations start to change so that both the flux control coefficient of the enzyme and its elasticity with respect to an inhibitor changes, the latter in a manner that depends on the inhibition type. For example, if the inhibitor is of the competitive type, adding a large amount of it will generally cause the steady state concentration of the competing substrate to rise, counteracting the effect of the inhibition. In the same circumstances, the rise in substrate concentration will potentiate the effect of an uncompetitive inhibitor. Eisenthal and Cornish-Bowden (1998) pointed out that it would be preferable to search for drug candidates that were uncompetitive inhibitors, whereas structure-based design inevitably leads to competitive inhibitors that then need to be given at much higher concentrations to be effective. This is one of the factors that cause the $IC_{50}$ (the concentration required for 50% inhibition of a function – here assumed to be metabolic flux) of a drug always to be significantly greater than its $K_i$ as an inhibitor. The other factor is that 50% inhibition of an enzyme is not usually sufficient to impact on metabolic flux (see below). Although the reasons for this are well understood from an MCA perspective (Groen et al. 1982), it is not clear that the message has yet been absorbed by the pharmaceutical industry (Cascante et al. 2002).


# 3 Generic properties of metabolic systems


MCA has led to a number of generalizations about the properties of metabolic systems that support its claim to be a part of systems biology. One of the most famous of these is the flux summation theorem, first derived by Kacser and Burns (1973). The flux summation theorem states that for any chosen flux in a metabolic system, the control coefficients on that flux of every enzyme in the system add up to one. The metabolic system is not restricted; in principle it can include every enzyme and transport reaction in the cell that is connected by mass flow to the flux under question. In the case of a simple linear pathway, given that its enzymes do not show such relatively rare kinetic characteristics as substrate inhibition or product activation, all the flux control coefficients can be shown to be zero or positive. For such a pathway, if any one enzyme had a flux control coefficient of one (making it an undisputed 'rate-limiting enzyme'), all the other enzymes would necessarily have flux control coefficients of zero. However, consideration of the relationships between control coefficients and elasticities makes it more likely that the control will be shared, albeit unevenly, between all of the enzymes in the pathway. The more enzymes there are in the system, the smaller the average flux control coefficient is likely to be (Kacser and Burns 1979). From the point of view of controlling metabolic fluxes, this concept of limited and distributed control seems a negative and surprising conclusion: how can metabolic fluxes be controlled when no one enzyme has a strong influence on the flux? More recently, emphasizing the positive aspect has become more popular, and the principle is cast as the robustness of metabolic systems – their ability to maintain fluxes relatively unchanged in the face of perturbations in the activities or other kinetic parameters of the enzymes.

The concept of limited and distributed control has been the subject of recurrent debate. One reason for this is that any metabolic network that contains branches and cycles will have negative flux control coefficients, and in such cases the summation to a total of one does not necessarily limit the magnitude of control coefficients. Whilst this possibility theoretically exists, there does not seem to be any good evidence of its widespread occurrence; measurements are generally consistent with limited, distributed control. The importance of this question arises because Kacser and Burns (1981) explicitly linked a biochemical explanation of the genetic phenomenon of dominance of wild type over mutant phenotypes to the summation theorem and the associated expectation that flux control coefficients would be small on average. They argued that the most important aspect of metabolic phenotype was metabolic fluxes, and that the majority of enzymes would have small flux control coefficients. In the case of diploids, heterozygotes with one wild type allele and one loss-of-function mutant allele would contain approximately 50% of the wild type content of enzyme, but a 50% reduction in enzyme content of an enzyme with a small flux control coefficient has an experimentally negligible effect on metabolic flux, so the metabolic phenotype would still appear to be wild type. Compared to other explanations of dominance, which assume it to be an evolved property of diploid organisms, this one suggests that dominance is an inevitable by-product of the general characteristics of natural metabolic networks. This has received experimental support from Orr (1991) who showed that a strictly haploid organism displayed dominance in artificial, partial diploids.

The other side of this argument is that, at the wild type levels of enzymes in cells, the response to over-expression of any one enzyme is even more likely to be small, except where the flux control coefficient approaches one. MCA itself does not allow accurate calculation of the response to large degrees of over-expression, but the likely effect can be approximated (Small and Kacser 1993).

Even though MCA has not very frequently been applied to large numbers of components, this does not mean that it cannot be applied to large systems. There is always the option of defining a small number of interacting components (or blocks), each of which is composed of several, even many, enzymes. This has been the strategy used in top-down control analysis (Brand et al. 1988), which has been applied at several scales from mitochondrial oxidative phosphorylation to the analysis of ATP production and consumption in thymocytes (Buttgereit and Brand 1995). In the latter case, the blocks can include substantive cellular processes such as protein synthesis, and the analysis provides quantitative estimates of how much each energy-requiring block is able to control ATP production, and the ATP consumption of other blocks. Since top-down analysis aims to derive insight into system function from the fewest possible number of measurements, it demonstrates that not all approaches to systems biology require massively-parallel data sets.

Hofmeyr and Cornish-Bowden's supply-demand analysis (1991, 2000) can be regarded as a specific instance of top-down analysis that reveals some of the generic characteristics of metabolic systems, which can usually be conceptually divided by choosing some specific metabolite, thereby giving a block of supply reactions that lead to this metabolite, with the rest of metabolism being the demand system that removes it. Even for such a simplified system, widely applicable results are obtained. For example, better metabolite homeostasis can be obtained if the control of flux is exerted largely by the demand block (Thomas and Fell 1996; Fell 1997). One of the ways that control is transferred to the demand steps is by feedback control loops from the intermediary metabolite between the blocks to enzymes in the supply block. Hofmeyr and Cornish-Bowden (1991) showed that cooperativity of feedback interactions benefited metabolite homeostasis more than flux control itself. In this way, MCA again provides a framework to understand generic characteristics of the features in metabolic pathways without needing to know extensive details of their molecular implementation.

Another general conclusion from MCA that applies to large metabolic systems is the means by which cells are able to alter metabolic fluxes. It has been mentioned above that up to 50% inhibition of an enzyme generally has little effect on metabolic flux, and that likewise activation or over-expression also makes little impact on flux. In fact, even in cases where there is an enzyme with a significant flux control coefficient, activation of that enzyme alone causes its flux control coefficient to fall in most cases, so its ability to increase the flux is self—limiting. How then can cells alter metabolic fluxes? If all enzymes in a metabolic network are activated by exactly the same factor, then all the fluxes will increase by the same factor and the metabolite concentrations will be unchanged. (This is, in fact, another way of expressing the summation theorems for fluxes and concentrations.) In order to achieve a more selective activation, a possible solution is to activate, by the required factor, all the enzymes in the same linear segment that carries the target flux (the 'Universal Method': Kacser and Acerenza 1993). The segment of the network from which the target flux branched will also need activation, but probably by a lesser amount proportionately because it feeds other metabolic branches as well. With Simon Thomas (Fell and Thomas 1995), I proposed that the principles of Kacser and Acerenza's 'Universal Method' of flux increases was discernible in the control mechanisms of metabolic pathways. Although many researchers had emphasized the importance of one or other site in a pathway where control could be exerted, a striking fact is that most pathways contained several enzymes that are susceptible to control. If these multiple sites acted in concert, by 'multisite modulation' (Fell and Thomas 1995), then metabolic flux change would be possible. At the same time, Korzeniewski et al. (1995) similarly proposed that 'proportional activation' of both supply and demand in supply-demand systems would allow large flux changes without perturbation of metabolite concentrations. Some examples of this phenomenon have been analysed (Schafer et al. 2004; Korzeniewski 2003; Vogt et al. 2002) that demonstrate that it is the most likely explanation of the observed quantitative changes in fluxes and concentrations. The role of MCA in these investigations has been to establish quantitative expectations for the relative changes in fluxes and concentrations that would accompany different control mechanisms and to compare these different expectations to the observations. It also underlines how it is the system context that determines the effect of activating an enzyme; if only one enzyme is activated then its flux control generally falls, whereas the same degree of activation in the context of perfect multisite modulation or proportional activation freezes the control distribution, so that even after activation, further activation is still just as able to modify the metabolic flux. In addition to the examples referred to above, other cases where control mechanisms are known to operate at multiple sites are discussed in Fell (1997), though the evidence that proportional activation/multisite modulation must be invoked as an explanation of the systems behaviour is circumstantial in most of these. Thorough investigation would require simultaneous measurements of enzyme activities, metabolite concentrations, and fluxes throughout a metabolic system in at least two distinct states. The development of metabolomics technology holds the promise that such system-wide measurements should eventually be easier, though issues of precision and quantitation still have to be solved.

The analyses just referred to are concerned with rapid adjustments of metabolism. Longer-term adjustments in metabolism are generally accompanied by changes in the amounts of enzymes, but again there is evidence to support the necessity for changing many or most of the enzymes in a pathway. I have reviewed a number of published cases where the enzymes in a pathway have all shown near-proportional changes in their activity in response to a physiological or environmental signal (Fell 2000). In some cases, these changes are known to involve changes in the balance of synthesis and degradation of enzyme protein, but most of the studies involved classical, low-throughput biochemical assays. An analysis of flux and enzyme activity changes in *T. vaginalis* and other parasitic protists (ter Kuile 1996; ter Kuile and Westerhoff 2001) showed, for different enzymes in the same pathway, variable proportions of the contributions from 'hierarchical control' (that is, from genome through to the enzyme level) and from control at the metabolic level through mechanisms modulating activity via metabolic effectors. As part of the post-genomics agenda, it would be useful to be able to determine the extent to which such co-ordinate changes can be traced quantitatively from changes in gene transcription to changes in enzyme activity to effects on metabolites. One confounding factor is that for individual enzymes, the relationship between changes in transcript levels and changes in steady state protein levels shows a high degree of variability (Ideker et al. 2001). There is, however, a significant underlying trend (Fell 2001), and therefore, reason to expect that enzymes carrying the same metabolic flux should show correlated patterns of transcription as their pathway is up- or down-regulated. Schuster et al. (2002) observed such correlations between members of 'enzyme subsets', which are the smallest functionally associated units in metabolic networks, in a small set of transcript data for *Saccharomyces cerevisiae*. More recently, an extensive analysis on a very large set of *S. cerevisiae* microarray experiments has shown that adjacent enzymes and enzymes classified as part of the same functional classification in the Kegg database (http://www.genome.jp/kegg/) are more likely to show transcriptional correlation (Ihmels et al. 2004). The lessons from these studies are considered in the final section.


## 4 Perspectives for the future


I have argued that MCA is indeed the systems biology of metabolic networks, even though it does not typically deal with individual consideration of all the molecular components comprising the system. Indeed, it would be unwieldy if it did deal with large numbers of components. What then, can MCA offer systems biology in its guise as an experimental biological science that is increasingly permeated by the highly parallel, high-throughput –omics technologies? The agenda of post-genomic science, functional genomics, supposes that we will come to be able to predict how genomes in cells, interacting with the environment, generate a phenotype. But metabolic fluxes and metabolite concentrations constitute a very important part of the phenotype, so eventually we will have to be able to trace the interactions that connect genes through mRNA, protein, and enzyme activities to metabolic activity. Currently, there is a mismatch in the available forms of data: mRNA, and to a lesser extent proteins and metabolites can be measured in large numbers at a time in high-throughput methods, but generally with a coefficient of variation that is large relative to the typical small changes in flux and metabolite concentrations that accompany many physiological changes in metabolic status. Protein and mRNA levels are, moreover, only partially indicative of enzyme activities, which are of prime importance for the behaviour of metabolic systems. As yet, enzyme activities cannot be determined in a parallel high-throughput methodology, but it is worth noting that old low-throughput methods were capable of much greater precision in detecting small changes in enzyme activity. Therefore, the challenge is to be able to predict those changes in transcription that have significant effects at the phenotypic level, and those that do not, and furthermore, those changes that interact with one another to produce a significant effect, when individually they are virtually powerless. MCA will have a role in enabling the creation of predictive biology from –omics science because it provides such concepts as the small and distributed sensitivity of systems to single perturbations, and a toolkit for understanding the functioning of metabolic systems. An example already exists in the FANCY method (Raamsdonk et al. 2001) for inferring the metabolic module to which a gene of unknown function may belong. This draws on the concepts of the co-response coefficient (Cornish-Bowden and Hofmeyr 1994) and modular MCA (Kahn and Westerhoff 1991; Schuster et al. 1993).

Additional potential for exploiting concepts from MCA comes in the search for correlations in mRNA and protein expression data; this remains problematic because of the extremely large number of potential comparisons coupled with the noisiness of the data. However, from the principles of the universal method and of enzyme subsets (see above), it is possible to predict which enzyme activities must change in parallel to facilitate a metabolic change, and therefore to assign higher prior probability to a correlation between the expression levels of members of the set. (These sets are more tightly drawn than the broad pathway designations in the database ontologies.)

Other areas for development include broadening the scope of MCA. Already, hierarchical control analysis provides a framework for analysing systems where there are interactions between gene transcription, translation, post-translational modification and metabolism (Kahn and Westerhoff 1991). Admittedly, it is still not easy to apply these theoretical concepts in an experimental setting, though some applications exist (e.g. Jensen et al. 1999). A variation on top-down MCA has been used to quantify the contributions of different signal transduction routes to transmission of a signal to metabolism (Krauss and Brand 2000), and some aspects of MCA are transferable to signal transduction (Peletier et al. 2003). All these emphasize that the MCA-like approaches are not confined to cellular metabolism. Indeed, Brown (1994) proposed that the analysis could be extended upwards to the contributions of different organs to the overall metabolism of multicellular organisms. Although this proposal has not been developed, it has been approached from a different angle. Weibel and Taylor (Weibel et al. 1991, 2000) proposed the concept of *symmorphosis* at the physiological level. Their argument is that, for a specific function such as running in mammals, the limitations to maximum performance are distributed throughout the contributing anatomical components and physiological subsystems, such that any increase would require adaptations throughout. They have illustrated this with detailed analysis at the ultrastructural and biochemical level, from the lungs, through the circulatory system to the muscles. Although the theory of symmorphosis is different in detail from MCA, there is no mistaking the common concept of distributed limitations to performance necessitating multiple, distributed points of action to change the overall capacity to perform. The reason is almost certainly that, at both the biochemical and physiological level, the systems have evolved to achieve a competitive level of performance with limited investment in resources. This argument has been explored in MCA (Kacser and Beeby 1984; Klipp and Heinrich 1999), and the results seem to instantiate the principle of maximum utility from economics.

In conclusion, MCA provides a theoretical bridge from the kinetic properties of enzymes to the systems properties of metabolism. As a component of systems biology, it can be extended by Hierarchical Control Analysis so that it spans the range from genome to metabolome. Furthermore, it exhibits no discontinuity at the cell boundary; it is compatible with cognate approaches to the control of physiological systems of multicellular organisms. For practitioners of MCA, a major challenge is now to embrace the –omics measurement technologies, which offer more detailed insights into the functioning of the cell than has ever been available before, but together these approaches can lead to a greater understanding of biological systems than either can achieve alone.

# References


Brand MD, Hafner RP, Brown GC (1988) Control of respiration in non-phosphorylating mitochondria is not shared between the proton leak and the respiratory chain. Biochem J 255:535-539

Brown GC (1994) Control analysis applied to the whole body: Control by body organs over plasma concentrations and organ fluxes of substances in the blood. Biochem J 297:115-122

Buttgereit F, Brand MD (1995) A hierarchy of ATP-consuming processes in mammalian cells. Biochem J 312:163-167

Cascante M, Boros LG, Comin-Anduix B, de Atauri P, Centelles JJ, Lee PWN (2002) Metabolic control analysis in drug discovery and disease. Nat Biotechnol 20:243-249

Cornish-Bowden A, Cárdenas ML (2001) Information transfer in metabolic pathways. Effects of irreversible steps in computer models. Eur J Biochem 268:6616-6624

Cornish-Bowden A, Hofmeyr JHS (1994) Determination of control coefficients in intact metabolic systems. Biochem J 298(Mar):367-375

Crabtree B, Newsholme EA (1985) A quantitative approach to metabolic control. Curr Top Cell Regul 25:21-76

Eisenthal R, Cornish-Bowden A (1998) Prospects for antiparasitic drugs: The case of *Trypanosoma brucei*, the causative agent of African sleeping sickness. J Biol Chem 273:5500-5505

Fell DA (1997) Understanding the Control of Metabolism. Portland Press, London

Fell DA (2000) Signal transduction and the control of expression of enzyme activity. Advan Enzym Regul 40:35-46

Fell DA (2001) Beyond genomics. Trends Genet 17:680-682

Fell DA, Sauro HM (1985) Metabolic Control Analysis: Additional relationships between elasticities and control coefficients. Eur J Biochem 148:555-561

Fell DA, Thomas S (1995) Physiological control of flux: the requirement for multisite modulation. Biochem J 311:35-39

Groen AK, van der Meer R, Westerhoff HV, Wanders RJA, Akerboom TPM, Tager JM (1982) Control of metabolic fluxes. Metabolic Compartmentation, ed. Sies H, Academic Press, London, pp. 9-37

Heinrich R, Rapoport TA (1974) A linear steady-state treatment of enzymatic chains; general properties, control and effector strength. Eur J Biochem 42:89-95

Hofmeyr JHS, Cornish-Bowden A (1991) Quantitative assessment of regulation in metabolic systems. Eur J Biochem 200:223-236

Hofmeyr JHS, Cornish-Bowden A (2000) Regulating the cellular economy of supply and demand. FEBS Lett 476:47-51

Ideker T, Thorsson V, Ranish JA, Christmas R, Buhler J, Eng JK, Bumgarner R, Goodlett DR, Aebersold R, Hood L (2001) Integrated genomic and proteomic analyses of a systematically perturbed metabolic network. Science 292:929-934

Ihmels J, Levy R, Barkai N (2004) Principles of transcriptional control in the metabolic network of *Saccharomyces cerevisiae*. Nat Biotechnol 22:86-92

Jensen PR, Van der Weijden CC, Jensen LB, Westerhoff HV, Snoep JL (1999) Extensive regulation compromises the extent to which DNA gyrase controls DNA supercoiling and growth rate of *Escherichia coli*. Eur J Biochem 266:865-877

Kacser H, Acerenza L (1993) A universal method for achieving increases in metabolite production. Eur J Biochem 216:361-367

Kacser H, Beeby R (1984) Evolution of catalytic proteins or on the origin of enzyme species by means of natural selection. J Mol Evol 20:38-51

Kacser H, Burns JA (1973) The control of flux. Symp Soc Exp Biol 27:65-104. Reprinted in Biochem Soc Trans (1995) 23:341-366

Kacser H, Burns JA (1979) Molecular democracy: who shares the controls? Biochem Soc Trans 7:1149-1160

Kacser H, Burns JA (1981) The molecular basis of dominance. Genetics 97:639-666

Kahn D, Westerhoff HV (1991) Control theory of regulatory cascades. J Theor Biol 153:255-285

Klipp E, Heinrich R (1999) Competition for enzymes in metabolic pathways: Implications for optimal distributions of enzyme concentrations and for the distribution of flux control. Biosystems 54:1-14

Korzeniewski B (2003) Regulation of oxidative phosphorylation in different muscles and various experimental conditions. Biochem J 375:799-804

Korzeniewski B, Harper ME, Brand MD (1995) Proportional activation coefficients during stimulation of oxidative phosphorylation by lactate and pyruvate or vasopressin. Biochim Biophys Acta 1229:315-322

Krauss S, Brand MD (2000) Quantitation of signal transduction. FASEB J 14:2581-2588

Newsholme EA, Start C (1973) Regulation in Metabolism. Wiley and Sons, London

Orr HA (1991) A test of Fisher's theory of dominance. Proc Natl Acad Sci USA 88:11413-11415

Peletier MA, Westerhoff HV, Kholodenko BN (2003) Control of spatially heterogeneous and time-varying cellular reaction networks: A new summation law. J Theor Biol 225:477-487

Raamsdonk LM, Teusink B, Broadhurst D, Zhang NS, Hayes A, Walsh MC, Berden JA, Brindle KM, Kell DB, Rowland JJ, Westerhoff HV, Dam KV, Oliver SG (2001) A functional genomics strategy that uses metabolome data to reveal the phenotype of silent mutations. Nat Biotechnol 19:45-50

Reder C (1988) Metabolic control theory: a structural approach. J Theor Biol 135:175-201

Savageau MA (1976) Biochemical Systems Analysis: a Study of Function and Design in Molecular Biology. Addison-Wesley, Reading, Mass

Schafer JRA, Fell DA, Rothman D, Shulman RG (2004) Protein phosphorylation can regulate metabolite concentrations rather than control flux: The example of glycogen synthase. Proc Natl Acad Sci USA 101:1485-1490

Schuster S, Kahn D, Westerhoff HV (1993) Modular analysis of the control of complex metabolic pathways. Biophys Chem 48(1):1-17

Schuster S, Klamt S, Weckwerth W, Moldenhauer F, Pfieffer T (2002) Use of network analysis of metabolic systems in bioengineering. Bioprocess Biosyst Eng 24:363-372

Small JR, Kacser H (1993) Responses of metabolic systems to large changes in enzyme activities and effectors. 1. the linear treatment of unbranched chains. Eur J Biochem 213:613-624

ter Kuile BH (1996) Metabolic adaptation of *Trichomonas vaginalis* to growth rate and glucose availability. Microbiology 142:3337-3345

ter Kuile BH, Westerhoff HV (2001) Transcriptome meets metabolome: Hierarchical and metabolic regulation of the glycolytic pathway. FEBS Lett 500:169-171

Thomas, S, Fell, DA (1996) Design of metabolic control for large flux changes. J. Theor Biol 182: 285-298

Vogt AM, Poolman M, Ackermann C, Yildiz M, Schoels W, Fell DA, Kubler W (2002) Regulation of glycolytic flux in ischemic preconditioning - A study employing metabolic control analysis. J Biol Chem 277:24411-24419

Weibel ER (2000) Symmorphosis. On Form and Function in Shaping Life. Harvard University Press, Cambridge

Weibel ER, Taylor CR, Hoppeler H (1991) The concept of symmorphosis: a testable hypothesis of structure-function relationship. Proc Natl Acad Sci USA 88:10357-10361

Fell, David A.
School of Biological & Molecular Sciences, Oxford Brookes University, Headington, Oxford OX3 0BP, UK
dfell@brookes.ac.uk

# No music without melody: How to understand biochemical systems by understanding their dynamics

Ursula Kummer and Lars Folke Olsen

## Abstract

The dynamics of the concentration of biochemical species is a systems property that arises through the interaction of metabolites and other molecules. Thanks to improving experimental technology it becomes increasingly possible to tightly follow these concentrations in the cell measuring the dynamics of concentration changes. Due to this development, it is now apparent that dynamics, e.g. nonlinear dynamics like oscillations carry important functionality and should not be neglected in the search for a full understanding of cellular processes. Examples are discussed in this contribution.

## 1 Introduction: Dynamics is a systems property essential for systems biology

Systems biology is the study of how new properties, that are functionally important to life, arise in interactions. As we will point out in the following, this implies that a new dimension, which has been badly neglected in the past, has to be added to cellular biology, namely time. This means that one has to study the dynamics of biochemical systems in detail. Dynamics is a true systems property, and therefore, should be one of the foci in systems biology. An isolated molecule can obviously not display any dynamics in this sense (since it does not have a concentration) and there is also no way to compute or estimate potential dynamical behaviour arising in a sample of molecules without knowing how these molecules and their reactions interact with each other. The knowledge of the dynamics of biochemical systems in the cell is limited up to this moment and the view of this systems property has therefore been simplified in the past. However, it should be pointed out that in the few well-studied cases this has been a field where the integration of experimental and theoretical approaches by us and by others has been extremely fruitful. We think that the goal mentioned above, namely the understanding of the systems properties in the cell can only be reached by following this combined approach. Moreover, as we will describe below, it is of crucial importance that experimental investigations are done as non-invasively as possible to minimize experimental ar-

Topics in Current Genetics, Vol. 13
L. Alberghina, H.V. Westerhoff (Eds.): Systems Biology
DOI 10.1007/b138740 / Published online: 25 May 2005

tefacts and maximize time resolution. Otherwise, our understanding of the dynamic nature of biochemical processes will likely remain as limited as it is now.

According to a typical textbook view, the dynamics of biochemical systems is restricted to a steady state, which is reached as quickly as possible and maintained unless something dramatic happens. Similar to the reductionist approach in studying isolated proteins and genes, this simple view has led to many useful insights into cell biochemistry, which to some extent justifies its use. Moreover, limitations in experimental techniques have often prevented the direct observation of dynamical behaviours of metabolite concentrations in the cell. Usually, times series of concentrations have been obtained by quenching the cells and extracting their cell material every few minutes or even hours and measuring the concentrations in the extract. This tedious approach offers only very limited insight into the systems dynamics, since the time-scale is usually not adequate compared to the time-scale in which events happen in the cell and the method is error prone. This is still true for modern experimental techniques from genomics and proteomics, e.g. microarrays, where the number of samples is often even smaller due to the high costs, and measurements are typically restricted to before and after a certain parameter change (e.g. stress) is applied.

However, trying to understand what is going on in the cell will require the whole picture including the true dynamics of biochemical processes and simplifications (which of course are unavoidable at times) have to be made in such a way that important information needed for the elucidation of cell function is not forgotten. It is a fact that the interactions of reactions and other molecular events (e.g. transport) give rise to all kinds of dynamics, where steady state behaviour is only one out of many possible dynamic behaviours (Goldbeter 1996). Steady state behaviour offers a continuous and steady level of concentrations of metabolites in the cell. This steadiness is of course in many cases advantageous. By reaching a stable steady state and maintaining this by employing certain feedback mechanisms the organism/cell ensures that none of the involved concentrations of metabolites in the cell builds up or is completely broken down.

Improving experimental technology, however, has led to a completely different view on the dynamics of biochemical species in the cell. Even though, there are still only a few examples where it is possible to observe concentrations in the living cell in a continuous way, almost all of these examples have shown that steady state behaviour is not necessarily the typical behaviour of concentrations in the cell. One finds a whole zoo of dynamic behaviours present in biochemistry. Among these are oscillations of diverse complexity and even something as exotic as chaos, all of them being representatives of nonlinear dynamic behaviour (Goldbeter 1996; Scheeline et al. 1997). Apart from being just intrinsically natural and therefore abundant all over biology these dynamic states enable the cell to do a lot more than simply hovering over a specific concentration. As we will describe below, employing nonlinear dynamics enables the cell e.g. to encode information in signal transduction systems, to stabilize enzymes, and to protect the cell from harmful processes.

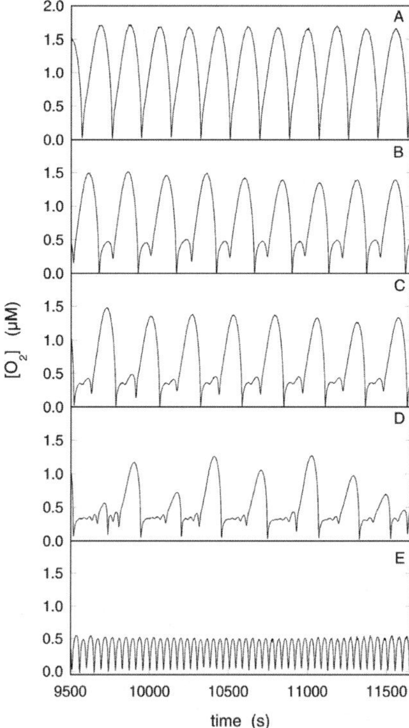

**Fig. 1.** Examples of dynamic behaviour displayed by the PO reaction when supplied continuously with the substrates NADH and oxygen. The rate of NADH supply was increased from A-E and the oxygen concentration measured. Experimental setup and parameters as in Moller and Olsen (1999).

Describing the plus in information that one gains by understanding the dynamics of biochemical processes, we want to restrict ourselves to examples from our own work rather than giving a complete overview over the field. Our examples represent a single-enzyme reaction, signalling pathways, and metabolic pathways. This categorisation is a little bit arbitrary with certainly fuzzy borderlines, but it is common and therefore we stick to it.

## 2 Nonlinear dynamics displayed and used by single-enzyme reactions

Very early on in the history of nonlinear chemistry a group headed by Isao Yamazaki presented the first observations of oscillatory dynamics in a single-enzyme reaction, now known as the peroxidase-oxidase (PO) reaction. First, they observed only damped oscillations (Yamazaki et al. 1965), but later also sustained oscillations were obtained (Nakamura et al. 1969). These experiments showed that even relatively simple biochemical reactions far from equilibrium have nonlinear properties that allow them to display complex dynamic behaviours rather than simple steady state behaviour. Later it was shown that the reaction is also capable of displaying complex oscillatory behaviour and chaos (Olsen and Degn 1977, 1978). Fig. 1 shows some examples of experimentally observed dynamic behaviours.

The overall stoichiometry of the PO reaction is:

$$2NADH + O_2 + 2H^+ \rightarrow 2NAD^+ + 2H_2O$$

The reaction is catalyzed by peroxidases from many different sources (Kummer et al. 1996). Oscillations and chaos are observed when oxygen and NADH are continuously supplied to a reaction mixture containing peroxidase and a suitable aromatic cofactor (Scheeline et al. 1997). It is interesting that many of these cofactors are naturally occurring (Kummer et al. 1997), the most prominent one being melatonin (Olsen et al. 2001). The PO reaction is a branched chain reaction (Scheeline et al. 1997) and it has several reactive oxygen species, e.g. superoxide and hydrogen peroxide, as intermediates. The mechanism and the parameters are known in great detail, which allows for the quantitative computational modelling of this biochemical system (e.g. Olsen et al. 2003b).

Of course, the question remains whether this system is a good example for dynamic behaviour in biochemistry in the context of systems biology since most of the references deal with the isolated system *in vitro*. However, there are three reasons to include it in this discussion:

- it very probably plays at least one important role *in vivo*, namely during the activation of leukocytes (see below)
- it shows how versatile the dynamics of even a single-enzyme reaction can be
- the dynamics protect the enzyme from inactivation

Now, the latter fact is a nice example for implications that the dynamics of such a system can have. It was rather a coincidental observation doing experiments with the PO reaction that one of us noticed that the enzyme is much more stable while in an oscillatory state compared to a steady state. This somewhat surprising result was obtained, since the enzyme also shows bistability meaning that two different dynamic states - a steady state and an oscillatory state - are observable for the very same parameter set (Aguda et al. 1990). Therefore, the average flux through the system is roughly the same in both states. However, when the reaction is in the oscillatory state the enzyme is decomposed at a much lower rate than when the reaction is in the steady state (Fig. 2) (Hauser et al. 2001). Computer simulations of the system revealed that even though the overall flux through the system is the same in both states there are significant differences in the concentrations of intermediates between the two dynamic states. During oscillations oxygen radicals are produced to a much lesser degree. Since these radicals can harm the enzyme, it is clear that this protects the enzyme and explains the difference in stability (Olsen et al. 2003c). Therefore, evolution in this system should have led to a pronounced and thus likely oscillatory regime in order to result in a stable enzyme, which is usable for a longer period of time. An intrinsically stable enzyme enables

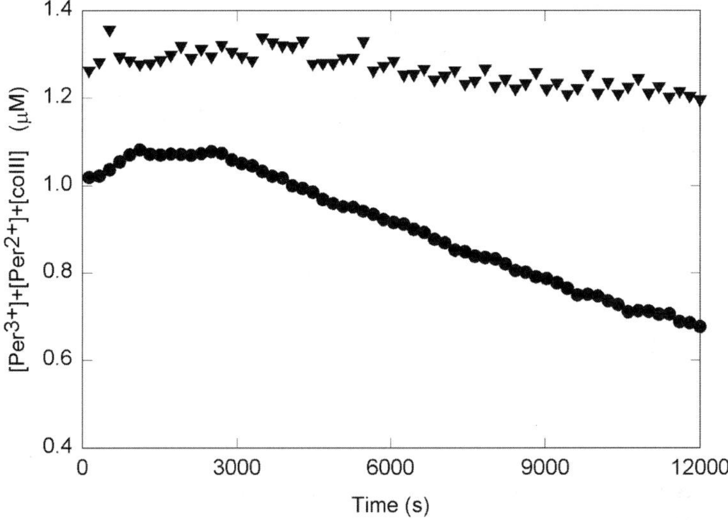

**Fig. 2.** Average concentration of the peroxidase intermediates $Per^{3+}$, $Per^{2+}$, and compound III in the course of the PO reaction fed continuously with the substrates. Circles denote the enzyme concentration during steady state while triangles show the enzyme concentration during oscillations. Experimental setup and parameters described in Hauser et al. (2001).

the system to control its activity by the usual diverse factors like degrading enzymes, inhibitors, and activators, etc. all under the influence of the biochemical network. The activity of an enzyme that is very much under the influence of destruction by oxygen radicals is much harder to control. It is needless to say that oxygen radicals at least at high concentrations would also be harmful for other components in the cell.

Another more speculative implication of this dynamics concerns the possibility that intermediates of this reaction, the above mentioned oxygen radicals are second messengers in the cell (Sundaresan et al. 1995; Crawford et al. 1996). Similar to the information processing in the case of calcium signal transduction (see below) the cell could use the dynamic profile (e.g. amplitude and frequency) of the concentration of these messengers to encode information. However, this remains to be shown experimentally.

## 3 Nonlinear dynamics displayed and used by metabolic pathways

The earliest evidence for nonlinear dynamics in biochemistry found in a living organism have been the oscillations displayed by a major metabolic pathway namely glycolysis (Chance et al. 1964). The fact that the concentrations of metabolites participating in glycolysis are able to oscillate under certain conditions has been a matter of studies around the world. However, even though there is some speculation about the involvement of glycolytic oscillations in the triggering of insulin oscillations in mammals (e.g. Westermark and Lansner 2003), it is not yet known, if these oscillations offer any beneficial aspects.

The early observation of glycolytic oscillations were possible because of the fluorescent properties of one of the metabolites, namely NADH. This property also enabled researchers to observe oscillations of NAD(P)H in human neutrophilic leukocytes (Amit et al. 1999). Additionally, other metabolites like reactive oxygen species, protons, and calcium ions were shown to oscillate in these cells. All compounds seem to oscillate with the same frequency.

The oscillations are accompanied by travelling waves of NADPH, protons, and oxygen species that serve to direct a localized pulsating release of antibactericidal oxidants (Kindzelskii and Petty 2002). This should help preventing damage to the neutrophils themselves and to other neighbouring tissue. It has been suggested that the PO reaction described above is involved in these oscillations and a mathematical model capable of reproducing the experimental observations has been constructed (Olsen et al. 2003a). This model is a two-compartment model involving the hexose monophosphate shunt and the enzymes myeloperoxidase and NADPH oxidase. Experimental results were predicted on the basis of this model (see Fig. 3 for an example) and later on verified. Much of the dynamics of this system can be approximated by the one-enzyme reaction involving only peroxidase (Olsen et al. 2003b). Furthermore, it is interesting to note that myeloperoxidase apparently has structural properties which favour that it participates in an oscillating reaction (Gabdoulline et al. 2003).

Again, the dynamics of the system seems to serve at least one important function. Here, it enables the neutrophil to aim at the target (the invading pathogen or a tumour cell) with a directed pulse of highly concentrated oxygen radicals. If the dynamics were absent, the oxygen radical concentration would increase homogeneously in the whole cell and subsequently in its surroundings causing unnecessary damage to healthy tissue. It is interesting that individuals lacking the functional myeloperoxidase seem to have more occurrences of cancer of the connective tissue (Lanza 1998). Since neutrophils also kill tumour cells an explanation could be that impaired dynamics might lead to a loss of tumour killing.

## 4 Nonlinear dynamics displayed and used by signal transduction systems

Signal processing in human technology inevitably involves the dynamics of the signal. Information is transferred as an oscillating signal with the message encoded in e.g. the amplitude and frequency. This principle is also employed by nature and everybody is aware that nerve signals are a major example for this. However, in the context of signal transduction systems in which small molecules play a role as so-called second messengers, this was long not recognized.

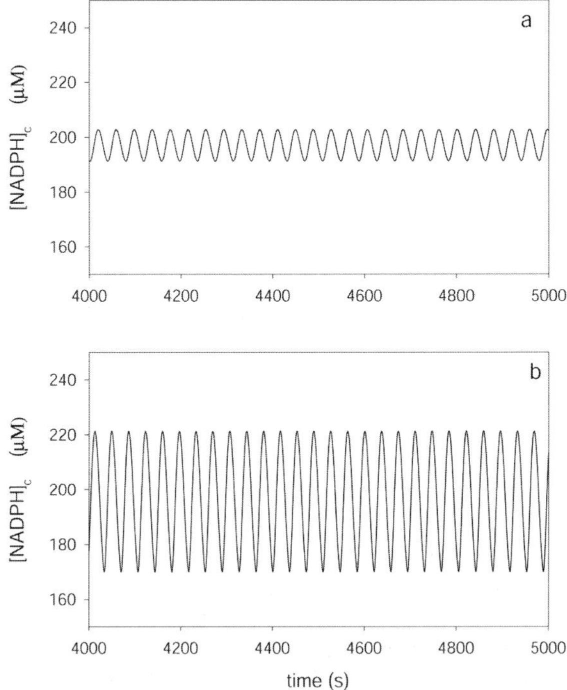

Fig. 3. Simulated concentration of NADPH in activated neutrophils in the presence of low (top) and high (bottom) melatonine concentrations. The predicted effect of amplitude increase was later on verified experimentally. Model described in Olsen et al. (2003a).

Again, the reason for this lack of knowledge is mainly to be sought in the difficulty of measuring metabolite concentrations in the cell over very short time intervals. However, as soon as chromophores that can be introduced in the cell and bind specifically to certain substances became available, it was also discovered that the concentration of second messengers often oscillate and information is encoded in the oscillatory dynamics of the signal.

The most widely studied example for this principle is the signal transduction via calcium ions. Actually, without the finding that information is hidden in the dynamics of the concentration changes, it would be extremely puzzling how the information can be processed at all. At the membrane surface of e.g. hepatocytes there are many different hormone receptors which all activate the calcium signal transduction machinery in the cell (for a review see Clapham 1995). Of course, almost all of these hormones have a different task to fulfil. So merely activating the liberation of calcium into the cytosol and thereby raising its concentration does not really make sense in the context of hormone induced signals, since one and the same messenger would not be able to carry information about a diversity of different hormones. The finding that the dynamics of the calcium concentration is involved in the signal transduction process at least offers a partial solution to this

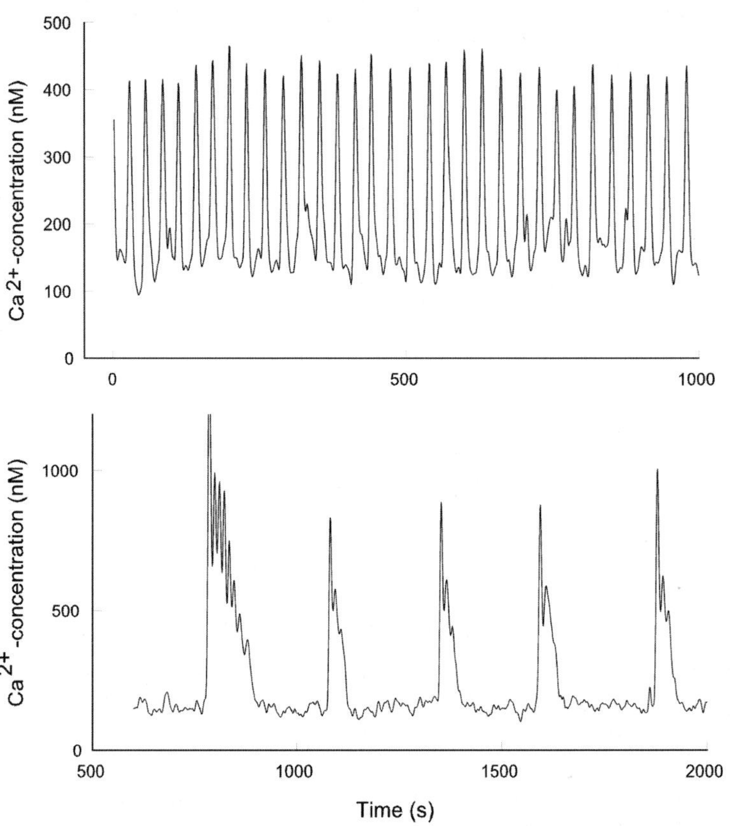

**Fig. 4.** Concentration of calcium in hepatocytes during stimulation with vasopressin (top) and ATP (bottom). Data kindly supplied by Anne Green and Jane Dixon.

problem. However, there doubtless is still much to encounter with respect to e.g. the signals resulting from stimulation with multiple combinations of different hormones.

Diverse types of oscillations have been found in response to different hormones e.g. in hepatocytes. Simple (spiking) oscillations as well as complex (bursting) oscillations are examples (Woods et al. 1986; Dixon et al. 1990) (Fig. 4). For hepatocytes it has been shown that the more hormone is applied the higher the frequency of the signal becomes (Woods et al. 1986). Thus, the kind of hormone can be encoded in the type of oscillation whereas the amount is encoded in the frequency.

While the encoding process is still relatively simple to observe (isolated hepatocytes are stimulated with a certain type and amount of hormone and the response is recorded), the decoding is much harder to see. For this purpose, in principle, one would have to subject an isolated enzyme or gene to a certain calcium signal and

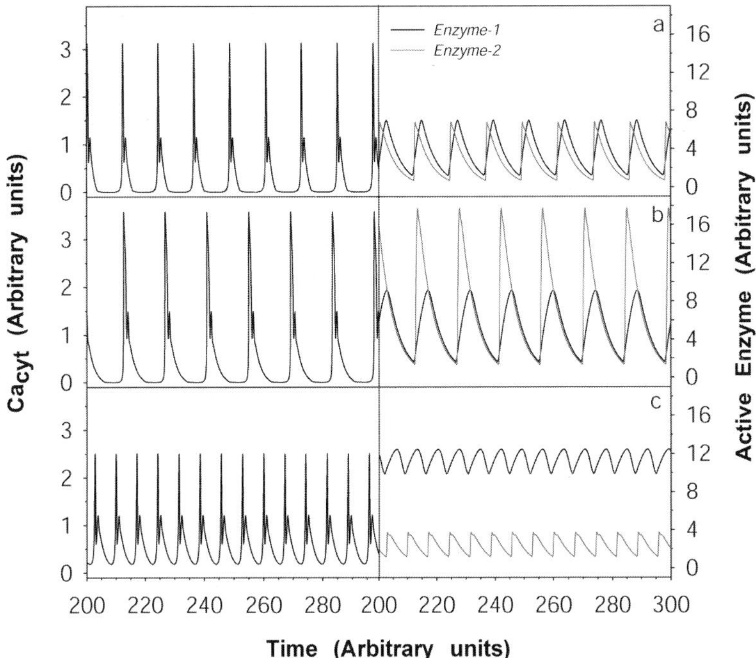



**Fig. 5.** Example of decoding of a calcium signal (left panels) by two different enzymes obeying the same kinetics but different binding parameters for calcium. Model described in Larsen et al. (2004).

measure the effect. Of course, it is extremely difficult to generate calcium concentration signals like those observed in living cells without employing the respective cellular machinery. Therefore, only a few examples with simple calcium signals have been published (for a review see Larsen and Kummer 2003). These examples show that enzyme activities and gene expression are capable of decoding specific frequency information. However, exactly how this is achieved is still a matter of speculation.

Theoretical work by us (Larsen et al. 2004) and by others (for a review see Larsen and Kummer 2003) has shed some light into the decoding mechanisms. One of the features of calcium binding enzymes and other target proteins is that they have multiple calcium binding sites and bind calcium in a cooperative manner. This cooperativity could be the key to the decoding mechanisms since it allows for very different answers to a calcium signal if the binding parameters vary as they do in reality (Larsen et al. 2004). Fig. 5 shows an example how two different enzymes with different binding parameters decode one and the same calcium signal.

As experimental technology advances more and more examples of messenger molecules displaying nonlinear dynamics become known. Diverse hormones (for a review on e.g. insulin see Simon and Brandenberger 2002), cAMP (for a review see Nanjundiah 1998), oxygen radicals (e.g. Amit et al. 1999) (also see above) are prominent examples here.

## 5 Recent developments, summary, and outlook

All of the above examples show that it is crucial for the observation of dynamic properties of biochemical systems to be able to measure diverse metabolites without disruption of the system and as continuously as possible. Such measurements have not been possible except in a very few cases until recently. Fast and high-resolution fluorescent microscopy as well as other spectroscopic developments enables us to follow metabolic concentrations quite closely. The resulting observations show that indeed much of the intrinsic dynamics of biochemical networks happens on a very short timescale. This is underlined by the fact that due to advanced technologies and in the course of the last months more oscillatory biochemical processes in the cell have been discovered. Thus, the NF-B signalling pathway has been shown to oscillate and thereby influence gene expression (Nelson et al. 2004), as well as the concentration of tumour suppressor p53 in individual cells after DNA damage (Lahav et al. 2004).

However, this development is still in the beginning and fluorescent probes that specifically bind to metabolites and thus form the basis for the measurement are still rare. A recent development to build almost arbitrary probes on a protein coupled basis (Fehr et al. 2002) or to encapsulate fluorescent probes in polymer nanoparticles (Clark et al. 1999; Xu et al. 2001) might be the starting point for efforts to solve this problem.

The examples from our own work described above show that it is clearly not sufficient to average over concentrations in the cell assuming that an averaged steady state level is all that is needed to understand what is going on. There is important functionality in the dynamics of biochemical systems, which cannot be elucidated without actually seeing it and without modelling it to understand the mechanistic basis behind it. These functionalities discussed above encompass information processing, stability, and protection from harmful radicals in the context of immune response. It is not an overstatement to say that most organisms would not be able to sustain themselves without such functionalities.

Moreover, the examples described above certainly cover only the range of functionalities that we are working with. There are many more examples already known and given the fact that the experimental techniques enabling us to observe these phenomena are only starting to take off many more discoveries are awaiting us.

Apart from the need for experimental developments as described above, it is also clear that computational methodologies have to be adjusted in order to facilitate the studying of the dynamics of biochemical systems. On the one hand, many theoretical frameworks in the context of systems biology so far rely on the steady state as a basis and are therefore not applicable to all systems with nonlinear dynamics. On the other hand, theoretical methods from nonlinear dynamics (which are mainly applied in physics) often need adjustment to the specific needs of biological problems.

All in all, we see that the understanding of biochemical systems in terms of their dynamics is currently discovering new and exciting functionalities in biochemistry. This research is supported by the general awareness that it is systems properties that need more attention, which is the main goal of systems biology. We are certain that melodious times lie ahead.

## Acknowledgments

Ursula Kummer wishes to thank the Klaus Tschira Foundation while Lars Folke Olsen thanks the Danish Natural Science and Medical Research Councils for funding. We also would like to acknowledge our collaborators Jane Dixon, Anne Green, Marcus Hauser, Ann Zahle Larsen, and Howard Petty for their support.

## References

Aguda BD, Frisch LLH, Olsen LF (1990) Experimental evidence for the coexistence of oscillatory and steady states in the peroxidase-oxidase reaction. J Am Chem Soc 112:6652-6656

Amit A, Kindzelskii AL, Zanoni J, Jarvis JN, Petty HR (1999) Complement deposition on immune complexes reduces the frequencies of metabolic, proteolytic, and superoxide oscillations of migrating neutrophils. Cell Immunol 194:47-53

Chance B, Schoener B, Elaesser S (1964) Control of the waveform of oscillations of the reduced pyridine nucleotide level in a cell-free extract. Proc Natl Acad Sci USA 52:337-341

Clapham DE (1995) Calcium signalling. Cell 80:259-268

Clark HA, Kopelman R, Tjalkens R, Philibert MA (1999) Optical nanosensors for chemical analysis inside single living cells. 2. Sensors for pH and calcium and the intracellular application of PEBBLE sensors. Anal Chem 71:4837-4843

Crawford LE, Milliken EE, Irani K, Zweier JL, Becker LC, Johnson TM, Eissa NT, Crystal RG, Finkel T, Goldschmidt-Clermont PJ (1996) Superoxide-mediated actin response in post-hypoxic endothelial cells. J Biol Chem 271:26863-26867

Dixon CJ, Woods NM, Cuthbertson KSR, Cobbold PH (1990) Evidence for two Ca2+-mobilizing purinoreceptors on rat hepatocytes. Biochem J 269:499-502

Fehr M, Frommer WB, Lalonde S (2002) Visualization of maltose uptake in living yeast cells by fluorescent nanosensors. Proc Natl Acad Sci 99:9846-9851

Gabdoulline RR, Kummer U, Olsen LF, Wade RC (2003) Concerted simulations reveal how peroxidase compound III foremation results in cellular oscillations. Biophys J 85:1421-1428

Goldbeter A (1996) Biochemical oscillations and cellular rhythms. Cambridge University Press, Cambridge

Hauser MJB, Kummer U, Larsen AZ, Olsen LF (2001) Oscillatory dynamics protect enzymes and possibly cells against toxic substances. Faraday Discuss 120:215-227

Kindzelskii A, Petty HR (2002) Apparent role of traveling metabolic waves in oxidant release by living neutrophils. Proc Natl Acad Sci USA 99:9207-9212

Kummer U, Valeur KR, Baier G, Wegmann K, Olsen LF (1996) Oscillations in the peroxidase-oxidase reaction: a comparison of different peroxidases. Biochim Biophys Acta 1289:397-403

Kummer U, Hauser MJB, Wegmann K, Olsen LF, Baier G (1997) Oscillations and complex dynamics in the peroxidase-oxidase reaction induced by naturally occurring aromatic substrates. J Am Chem Soc 119:2084-2087

Lahav G, Rosenfeld N, Sigal A, Geva-Zatorsky N, Levine AJ, Elowitz MB, Alon U (2004) Dynamics of the p53-Mdm2 feedback loop in individual cells. Nat Gen 36:147-150

Lanza F (1998) Clinical manifestation of myeloperoxidase deficiency. J Mol Med 76:676-681

Larsen AZ, Kummer U (2003) Information processing in calcium signal transduction. In Understanding Calcium Dynamics (Falcke M, Malchow D, eds.) Springer, Heidelberg, Germany

Larsen AZ, Olsen LF, Kummer U (2004) On the encoding and decoding of calcium signals in hepatocytes. Biophys Chem 107:83-99

Moller AC, Olsen LF (1999) Effect of magnetic fields on an oscillating enzyme reaction. J Am Chem Soc 121:6351-6354

Nakamura S, Yokota K, Yamazaki I (1969) Sustained oscillations in a lactoperoxidase, NADH and O2 system. Nature 222:794

Nanjundiah V (1998) Cyclic AMP oscillations in *Dictyostelium discoideum*: models and observations. Biophys Chem 5:1-8

Nelson DE, Ihekwaba AEC, Elliott M, Johnson JR, Gibney CA, Foreman BE, Nelson G, See V, Horton CA, Spiller DG, Edwards SW, McDowell HP, Unitt JF, Sullivan E, Grimley R, Benson N, Broomhead D, Kell DB, White MRH (2004) Oscillations in NF-B signaling control the dynamics of gene expression. Science 306:704-708

Olsen LF, Degn H (1977) Chaos in an enzyme reaction. Nature 267:177-178

Olsen LF, Degn H (1978) Oscillatory kinetics of the peroxidase-oxidase reaction in an open system -- experimental and theoretical studies. Biochim Biophys Acta 523:321-334

Olsen LF, Lunding A, Lauritsen FR, Allegra M (2001) Melatonin activates the peroxidase-oxidase reaction and promotes oscillations. Biochem Biophys Res Commun 284:1071-1076

Olsen LF, Kummer U, Kindzelskii AL, Petty HR (2003a) A model of the oscillatory metabolism of activated neutrophils. Biophys J 84:69-81

Olsen LF, Lunding A, Kummer U (2003b) Mechanism of melatonin-induced oscillations in the peroxidase-oxidase reaction. Arch Biochem Biophys 410:287-295

Olsen LF, Hauser MJB, Kummer U (2003c) Mechanism of protection of peroxidase activity by oscillatory dynamics. Eur J Biochem 270:2796-2804

Scheeline A, Olson DL, Williksen EP, Horras GA, Klein ML, Larter R (1997) The peroxidase-oxidase oscillator and its constituent chemistries. Chem Rev 97: 739-756

Simon C, Brandenberger G (2002) Ultradian oscillations of insulin secretion in humans. Diabetes 51:S258-s261

Sundaresan M, Zu-Xi Y, Ferrans VJ, Irani K, Finkel T (1995) Requirement for generation of H2O2 doe platelet-derived growth factor signal transduction. Science 270:296-299

Westermark PO, Lansner A (2003) A model of phosphofructokinase and glycolytic oscillations in the pancreatic beta-cell. Biophys J 85:126-39

Woods NM, Cuthbertson KSR, Cobbold PH (1986) Repetitive transient rises in cytoplasmic free calcium in hormone-stimulated hepatocytes. Nature 319:600-602
Yamazaki I, Yokota K, Nakajima R (1965) Oscillatory oxidations of reduced pyridine nucleotide by peroxidase. Biochem Biophys Res Commun 21:582-586
Xu H, Aylott JW, Kopelman R, Miller TJ, Philbert MA (2001) A real-time ratiometric method for the determination of molecular oxygen inside living cells using sol-gel-based spherical optical nanosensors with applications to rat C6 glioma. Anal Chem 73:4124-4133

Kummer, Ursula
EML Research, Schloss-Wolfsbrunnenweg 33, 69118 Heidelberg, Germany
ursula.kummer@eml-r.villa-bosch.de

Olsen, Lars Folke
Celcom, Department of Biochemistry, University of Southern Denmark, Campusvej 55, 5230 Odense M, Denmark

# Mesoscopic kinetics and its applications in protein synthesis

Johan Elf, Johan Paulsson, Otto Berg, and Måns Ehrenberg

## Abstract

Molecular biology emerged through unification of genetics and nucleic acid chemistry that took place with the discovery of the double helix (Watson and Crick 1953). Accordingly, molecular biology could be defined as the sum of all techniques used to perform genetic experiments by manipulating DNA. One consequence of the development of these techniques is large-scale sequencing of genomes from an ever increasing number of organisms. However, it became clear from this development that genetic information *per se* is not enough to grasp the most interesting functional and evolutionary aspects of cells and multi-cellular organisms. In fact, understanding how genotype leads to phenotype depends on concepts and techniques from areas that so far have been largely alien to molecular biological research, like physics, mathematics, and engineering. From the bits and pieces from these and other scientific fields new tools must be generated to make possible an understanding of the dynamic, adapting, and developing living systems that somehow take shape from the instructions given by their genomes. The growing total of these tools and their integration in experimental and theoretical approaches to understand complex biological processes in ways previously out of reach could be a way to define systems biology, in analogy with the above definition of molecular biology.

## 1 Introduction

The time evolution of an intracellular chemical system is often described by ordinary differential equations for average concentrations. This macroscopic approach has been very successful, but attempts at faithful descriptions of intracellular processes must consider the fact that all chemical reactions are discrete and stochastic (Singer 1953; van Kampen 1992): discrete in the sense that there is a finite number of molecules that changes in steps of integers, and stochastic in the sense that the time points at which the reactions occur are inherently unpredictable.

In chemical systems at equilibrium, fluctuations in numbers of molecules obey near Poisson statistics, so that the standard deviation of a copy number normalized to its mean value is equal to the inverse of the square root of the mean (Gardiner 1985). This important result from equilibrium statistical mechanics has led to the

Topics in Current Genetics, Vol. 13
L. Alberghina, H.V. Westerhoff (Eds.): Systems Biology
DOI 10.1007/4735_86 / Published online: 21 June  2005

idea that fluctuations can significantly affect intracellular processes carried out by molecules present at low copy numbers, and that intracellular noise can be neglected when molecule copy numbers are high (Schrödinger 1944). But chemical reactions in the living cell often occur very far from thermodynamic equilibrium, so that the relative size of fluctuations and their importance for cell physiology only can be assessed by careful analysis of the stochastic properties of each individual system (Delbruck 1940; Novick and Weiner 1957; Rigney and Schieve 1977; Berg 1978). Some molecules may be present in a few copies without detectable relative fluctuations, while others can be present in thousands of copies and still show spontaneous fluctuations that are large compared to the mean. When the fluctuations are big, the macroscopic rate equations will not correctly describe the average behavior of the system, because the reaction rates usually depend nonlinearly on the molecule concentrations (Renyi 1954).

In this chapter, we will summarize some useful methods for the characterization of stochastic properties of intracellular kinetics. These approaches (Gardiner 1985; Keizer 1987; Érdi and Tóth 1989; van Kampen 1992) will be illustrated by a few examples of how the sensitivity amplification (Kacser and Burns 1973; Savageau 1976; Goldbeter and Koshland 1982) of metabolite pools to perturbations in their in- or out-flows is related to the stochastic properties of the sizes of these pools. At the end of the chapter, the stochastic analysis will be used to shed light on some proposed properties of aminoacyl-tRNA pools: under conditions of amino acid limitation in the cell, hyper-fluctuations in aminoacyl-tRNA concentrations will emerge with significant consequences for protein synthesis.

## 2 Chemical reactions in the living cell

Consider a set of biochemical reactions with $N$ different chemical components (species) homogeneously distributed in a living cell (system) with volume $\Omega$. The state of such a homogenous system is defined by the number of molecules of each species as summarized by the copy number vector $\mathbf{n} = \left[ n_1 \cdots n_N \right]^T$, with $n_i$ as the current number of molecules of species $i$. Different species are not necessarily different molecules, they may also be specific conformations or states of the same a molecule, i.e. phosphorylated, bound to DNA, etc.

A state change takes place by any one of $R$ "elementary" reactions. When reaction $j$ occurs the chemical component number $i$ changes from $n_i$ to $n_i + S_{ij}$ molecules. The integers $S_{ij}$, $i=1,2...N$; $j=1,2,...,R$; are the elements of the $N \times R$ stoichiometric matrix $\mathbf{S}$ of the reaction network. When reaction $j$ is elementary, the probability that it occurs in a small time interval $\delta t$ is given by $\Omega \tilde{f}_j (\mathbf{n},\Omega)\delta t$ , where $\tilde{f}_j (\mathbf{n},\Omega)$ is the transition rate for reaction $j$ or, equivalently, the probability of reaction per unit time and unit volume. The transition rates may also depend on many variables, such as pH, temperature, and pressure, which we will assume are constant, and therefore, can be implicit in the state description. $\Omega$ is kept out of the transition rate, since rate laws commonly are given in intensive units (molar reactions per unit time $[Ms^{-1}]$ instead of number of reactions per unit time $[s^{-1}]$).

For example, a zero order event $\varnothing \xrightarrow{k_j} 2X_1$ with rate constant $k_j$ $[Ms^{-1}]$ has the transition rate $\tilde{f}_j(n,\Omega) = k_j$ and the stoichiometry $S_{1j} = 2$. The first order reaction $X_1 \xrightarrow{k_j} X_2$ with rate constant $k_j$ $[s^{-1}]$ has the transition rate $\tilde{f}_j(n_1,\Omega) = k_j n_1 \Omega^{-1}$ and stoichiometries $S_{\cdot j} = \begin{bmatrix} -1 & 1 \end{bmatrix}^T$. Here, $n_1\Omega^{-1}$ is the concentration of $X_1$ molecules. The transition rate for the second order reaction $X_1 + X_1 \xrightarrow{k} X_2$ with association rate constant $k$ $[M^{-1}s^{-1}]$ is $\tilde{f}(n_1,\Omega) = kn_1(n_1-1)/\Omega^2$. The expression follows from the fact that each of the $n_1$ molecules can react with each of the $n_1$-1 available partners with the rate $k/\Omega^2$. When $n_1 \gg 1$, the mesoscopic transition rate is approximated by its macroscopic counterpart $f(x_1) = \lim_{\Omega \to \infty} \tilde{f}(n_1 = \Omega x_1,\Omega) = \lim_{\Omega \to \infty} k\Omega x_1(\Omega x_1-1)/\Omega^2 = kx_1^2$, where $x_1$ is the average concentration of X molecules.

The transition rates, $\tilde{f}_j(\mathbf{n},\Omega)$, of elementary reactions are assumed to be constant as long as the state $\mathbf{n}$ or the volume $\Omega$ do not change. This suggests that the system has no memory, i.e., that it is Markovian in the specified variables. Because the state description is always contracted from a much larger state description, e.g. including internal energy levels of the molecules, their spatial position and their momentum, this is always an approximation, and assumes that all variables that are not included in the state description reach stationary distributions on a faster time scale. Sometimes, a complex reaction with many chemical intermediates can be approximated as elementary (Keizer 1987). The requirement for this is again that the intermediates attain their steady state distribution on a much shorter time scale than the state changes of the system. The complex reaction can then be represented by its average rate, conditional on the current state of the system. One must, however, take into account that molecules distributed in intermediate states of complex reactions can not participate in other reactions. Such state contractions can greatly simplify mesoscopic descriptions, making it advantageous to always use the smallest state description that keeps the system Markovian.

In this chapter we only discuss homogenous systems, in which the spatial distribution of all molecules is established on a much shorter time scale than the chemical reactions. When this condition is not met, local concentration variations may have to be taken into account (Baras and Mansour 1997; Elf and Ehrenberg 2004).

# 3 Mesoscopic kinetics for homogenous systems

## 3.1 The master equation

The probability $P(\mathbf{n}_m,t+dt)$ that a system of chemical reactions is in a state $\mathbf{n}_m$ (see Section 2 above) at time $t+dt$ will depend on the probability $P(\mathbf{n}_m,t)$ that it was in state $\mathbf{n}_m$ at time $t$ as well on the probability that the system will either reach

$\mathbf{n}_m$ from or leave $\mathbf{n}_m$ to any other state in the system during the short time interval $dt$:

$$P(\mathbf{n}_m, t+dt) \approx P(\mathbf{n}_m, t) + dt \cdot \sum_{k \neq m} W(\mathbf{n}_k, \mathbf{n}_m) P(\mathbf{n}_k, t) - dt \cdot \sum_{k \neq m} W(\mathbf{n}_m, \mathbf{n}_k) P(\mathbf{n}_m, t) \qquad (1)$$

$W(\mathbf{n}_k, \mathbf{n}_m) \cdot dt$ is the probability of a transition from state $\mathbf{n}_k$ to state $\mathbf{n}_m$ in any time interval of length $dt$. Accordingly, the first sum in Eq. (1) is the total probability that the system reaches the state $\mathbf{n}_m$ and the second sum is the total probability that the system leaves $\mathbf{n}_m$ in the time interval $(t, t+dt)$. The state transition rates $W(\mathbf{n}_i, \mathbf{n}_j)$ only depend on the states $\mathbf{n}_i$ and $\mathbf{n}_j$ at time t and not before, meaning that the system lacks memory. The state description can in some cases include the time $t$ as an additional state variable, e.g. for systems contained in increasing volumes (Paulsson and Ehrenberg 2001). Moving $P(\mathbf{n}_m, t)$ in Eq. (1) to the left hand side, dividing by $dt$ and taking the limit $dt \to 0$ lead to the master equation (van Kampen 1992):

$$\frac{dP(\mathbf{n}_m, t)}{dt} = \sum_{k \neq m} W(\mathbf{n}_k, \mathbf{n}_m) P(\mathbf{n}_k, t) - \sum_{k \neq m} W(\mathbf{n}_m, \mathbf{n}_k) P(\mathbf{n}_m, t) \qquad (2)$$

The chemical reaction system discussed in Section 2 is characterized by the transition rates $\Omega \tilde{f}_j(\mathbf{n}, \Omega)$, which can be used to formulate its corresponding master equation with the help of the stoichiometric matrix $\mathbf{S}$ introduced in Section 2. To do this, identify for each reaction $j$ the rate $\Omega \tilde{f}_j(\mathbf{n}, \Omega)$ for the system's transition from state $\mathbf{n}$ to state $\mathbf{n} + S_{\cdot j}$ [1] with $W(\mathbf{n}, \mathbf{n} + S_{\cdot j})$ in Eq. (2) and the rate $\Omega \tilde{f}_j(\mathbf{n} - S_{\cdot j}, \Omega)$ for the system's transition from state $\mathbf{n} - S_{\cdot j}$ to state $\mathbf{n}$ with $W(\mathbf{n} - S_{\cdot j}, \mathbf{n})$. No other transitions to or from $\mathbf{n}$ in a single reaction can occur. This gives the time evolution of the probability for any state $\mathbf{n}$ as

$$\frac{dP(\mathbf{n}, t)}{dt} = \Omega \sum_{j=1}^{R} \left( \mathbb{E}_i^{-S_{\cdot j}} - 1 \right) \tilde{f}_j(\mathbf{n}, \Omega) P(\mathbf{n}, t). \qquad (3)$$

$\mathbb{E}_i^{-S_{\cdot j}}$ is a step operator, defined from $\mathbb{E}_i^{-S_{\cdot j}} g(\mathbf{n}) = g(\mathbf{n} - S_{\cdot j})$, where $g$ is an arbitrary function of the state that due to the step operator is evaluated in a state $\mathbf{n}$ decreased by $S_{\cdot j}$. Eq. (3) fully describes the time evolution of the homogeneous chemical systems defined in section 2, given an initial probability distribution $P(\mathbf{n}, 0)$. Simple examples are given below. Although master equations with more than one state variable usually defy analytical solutions, exact expressions for the moments of copy number distributions can sometimes be obtained from the moment generating function (van Kampen 1992) (see Appendix).

Direct numerical solution of the master equation is usually prevented by the vastness of the state space. Evaluation, therefore, requires approximations, which will be described next. Early references and some of the historical background to the use of the master equation for chemical reactions are given in (McQuarrie 1967).

---

[1] $S_{\cdot j} = \begin{bmatrix} S_{1j} \cdots S_{Nj} \end{bmatrix}^{\mathrm{T}}$

## 3.2 Monte Carlo simulations of system trajectories

One conceptually simple and very useful way to estimate the properties of a master equation is to simulate realizations of the corresponding Markov process using Monte Carlo (MC) methods. These are conveniently based on the sampling of two random events. The first determines the time $(t+\tau)$ at which the next reaction occurs in the system, and the second selects the reaction $(j)$. That is, for a system in state $\mathbf{n}$ at time $t$, the probability $\delta\tau\, p\left(\tau, j \mid \mathbf{n}, t\right)$ that the next reaction event occurs in the small time interval $(t+\tau, t+\tau+\delta\tau)$ and is of type $j$ is given by

$$p\left(\tau, j \mid \mathbf{n}, t\right) = \underbrace{a(\mathbf{n})e^{-a(\mathbf{n})\tau}}_{I}\ \underbrace{\Omega\tilde{f}_{j}\left(\mathbf{n},\Omega\right)/a(\mathbf{n})}_{II} = \Omega\tilde{f}_{j}\left(\mathbf{n},\Omega\right)e^{-a(\mathbf{n})\tau}, \tag{4}$$

where $a\left(\mathbf{n}\right) = \Omega\sum_{j=1}^{R}\tilde{f}_{j}\left(\mathbf{n},\Omega\right)$. The factor $I$ is the probability density for the time $t+\tau$ of *any* next reaction, given that the system was in state $\mathbf{n}$ at time t. The factor $II$ is the probability that the reaction is of type $j$.

This general procedure to implement the discrete Markov process in continuous time was suggested for the simulation of chemical reactions by Gillespie in 1976 (Gillespie 1976) and is in this context known as the Direct Method. See also the related procedure by Bortz, Kalos, and Lebowitz (Bortz et al. 1975). In 2000, Gibson and Bruck presented the Next Reaction Method (Gibson and Bruck 2000), which is more efficient than the Direct Method in sparse networks, where state changes only affect a small number of transition rates. A related strategy is also used by Elf and Ehrenberg (2004) to sample the spatially dependent Markov process defined by the reaction diffusion master equation (Baras and Mansour 1997).

## 3.3 The Fokker-Planck Approximation

When the step operator $\mathbb{E}_{i}^{-S_{.j}}$ in the master equation Eq. (3) operates on a function like $g_{j}(\mathbf{n}) = \tilde{f}_{j}(\mathbf{n},\Omega)P(\mathbf{n},t)$, the function is evaluated in a state, which is shifted from $\mathbf{n}$ to $\mathbf{n}-S_{.j}$. Now, if the displacement is small and the function varies smoothly, the displaced function may be approximated by a Taylor expansion around the state $\mathbf{n}$. This is the idea behind the Kramers-Moyal approximation of the master equation (see Gardiner 1985; Érdi and Tóth 1989). For a general function $g(\mathbf{n})$ of the state vector $\mathbf{n}$, such a Taylor expansion looks like

$$\mathbb{E}^{-S_{.j}} g\left(\mathbf{n}\right) = g\left(\mathbf{n} - S_{.j}\right) \approx \left[1 - \sum_{i} S_{ij}\frac{\partial}{\partial n_{i}} + \frac{1}{2}\sum_{i,k} S_{ij}S_{kj}\frac{\partial^{2}}{\partial n_{i}\partial n_{k}} + \dots\right]g(\mathbf{n}) \tag{5}$$

Insertion of the differential approximation of the step-operator, Eq. (5), in the master equation, Eq. (3), followed by truncation after the second order term leads to the Fokker-Planck (FP) approximation of the master equation (Risken 1984):

$$\frac{dP(\mathbf{n},t)}{dt} = \Omega\sum_{j=1}^{R}\left(-\sum_{i} S_{ij}\frac{\partial\tilde{f}_{j}(\mathbf{n},\Omega)P(\mathbf{n},t)}{\partial n_{i}} + \frac{1}{2}\sum_{i,k} S_{ij}S_{kj}\frac{\partial^{2}\tilde{f}_{j}(\mathbf{n},\Omega)P(\mathbf{n},t)}{\partial n_{i}\partial n_{k}}\right) \tag{6}$$

The FP-equation Eq. (6) is a Partial Differential Equation (PDE) for the N-dimensional probability density $P(n,t)$, which in this approximation is a continuous function of the state. This is unproblematic as long as it is smoothly varying at length scales corresponding to the small jumps in state space given by single reactions. The FP-equation uses the same initial condition as the master equation and additional boundary conditions to preserve the total probability.

For complicated reaction schemes, the FP-equation is almost as complicated to work with as the original master equation. The major advantage of the FP-equation is that it allows for numerical solutions, also in cases when the state space of the full master equation is too large for that approach. Numerical solution of the FP-PDE requires that the continuous state space is discretized on an artificial computational grid. When the probability varies smoothly over the state-space, the computational grid can be sparser than the state space discretization of the original master equation, which is based on molecule copy numbers (Ferm et al. 2004).

## 3.4 The Linear Noise Approximation

The Fokker-Planck approximation would be exact if the jumps in state space were infinitely small. However, the sizes of the jumps in state space caused by chemical reactions are fixed in size, and there is no direct way to take the FP-approximation to a limit where it is exact. One could include more terms from the Kramers-Moyal expansion, but there is no systematic way to know when to truncate to get an approximation with sufficient accuracy. A solution to this technical problem is offered by the $\Omega$ expansion method (van Kampen 1961, 1992). This is a Taylor expansion in powers of $\Omega^{-1/2}$, where $\Omega$ is the volume of the spatially homogenous system with constant intensive rate constants. Truncation after the $\Omega^0$-term leads to the Linear Noise Approximation (LNA) of the master equation (van Kampen 1961, 1992; Elf and Ehrenberg 2003).

Successful application of the LNA requires that transition rates can be accurately approximated by linear functions of the state variables over the whole region of the fluctuations. This is true in the limit of large systems situated far from critical points. Here, the internal fluctuations, i.e. noise originating in the chemical reactions themselves and not supplied from external sources, are small compared to the average value. However, in intracellular systems with small volumes and finite concentrations the fluctuations can be large compared to the average values and the LNA does not have more *a priory* validity than the Fokker-Planck approximation in Eq. (6), unless the reaction rates actually are linear in the state variables. The LNA has, however, proved to be very useful for describing noise in intracellular systems, as it gives simple analytic approximations for the sizes and correlations of fluctuations (Paulsson and Ehrenberg 2001; Elf and Ehrenberg 2003; Elf et al. 2003b; Paulsson 2004; Tomioka et al. 2004). We will recapitulate some properties of the LNA, and for more details we refer to (Keizer 1987; van Kampen 1992; Elf and Ehrenberg 2003).

In the size expansion, a new stochastic variable, $\xi_i$, is defined from the relation $n_i \equiv \Omega x_i + \Omega^{1/2}\xi_i$, where $n_i$ is the copy number of component $i$, $x_i$ is a deterministic function of time and $\Omega$ is the system volume. The properties of $x_i$ and $\xi_i$ are both derived from the master equation using an expansion in powers of $\Omega^{-1/2}$.

It is assumed in the LNA that the concentration vector $\mathbf{x} = \begin{bmatrix} x_1 \cdots x_N \end{bmatrix}^T$ follows from the macroscopic rate equations. These describe the kinetics of the system in the limit of an infinitely large, well stirred, volume. Here, stochastic fluctuations in the state vector $\mathbf{n}$ are negligible, so that the state can be approximated by macroscopic average concentrations $\mathbf{x} = \begin{bmatrix} x_1 \cdots x_N \end{bmatrix}^T$ and each $\tilde{f}_j(\mathbf{n},\Omega)$ by its macroscopic rate law counterpart $f_j(\mathbf{x})$.

The time evolution of the macroscopic concentration vector $\mathbf{x}$ is governed by the deterministic rate equation

$$\dot{\mathbf{x}} = \mathbf{S}\mathbf{f}(\mathbf{x}) , \tag{7}$$

where $\mathbf{S}$ is the stoichiometric matrix (see Eq. (3)) and $\mathbf{f}(\mathbf{x}) = [f_1(\mathbf{x}) \cdots f_R(\mathbf{x})]^T$.

The next order of the expansion gives the linear noise approximation of the fluctuations, $\Omega^{1/2}\xi_i$, around the macroscopic trajectory $\Omega x_i$. The fluctuations are in this approximation characterized by the linear Fokker-Planck equation for the probability density function $\Pi(\xi,t)$:

$$\frac{\partial \Pi(\xi,t)}{\partial t} = -\sum_{i,k} A_{ik} \frac{\partial(\xi_k \Pi)}{\partial \xi_i} + \frac{1}{2}\sum_{i,k} D_{ik} \frac{\partial^2 \Pi}{\partial \xi_i \partial \xi_k} \tag{8}$$

where the $\mathbf{A}$ is the Jacobian and $\mathbf{D}$ is the diffusion matrix evaluated in the state $\mathbf{x}(t)$ as determined by Eq. (7). The elements of the matrices are

$$A_{ik} = \sum_{j=1}^{R} S_{ij} \frac{\partial f_j}{\partial x_k} \quad \text{and} \quad D_{ik} = \sum_{j=1}^{R} S_{ij} S_{kj} f_j(\mathbf{x}) . \tag{9}$$

The stationary solution of the linear Fokker-Planck equation is a multivariate Gaussian with zero average values, i.e. $\langle \mathbf{n} \rangle \approx \Omega \overline{\mathbf{x}}$ as given by $\mathbf{S}\mathbf{f}(\overline{\mathbf{x}}) = 0$. The covariance matrix, $\mathbf{C}$, of the fluctuations around the average is given by the Lyapunov equation

$$\mathbf{AC} + \mathbf{CA}^T + \Omega\mathbf{D} = 0 , \tag{10}$$

where $\mathbf{C}$ is defined as

$$\mathbf{C} = \Omega\left\langle \delta\xi\delta\xi^T \right\rangle = \left\langle (\mathbf{n} - \langle\mathbf{n}\rangle)(\mathbf{n} - \langle\mathbf{n}\rangle)^T \right\rangle . \tag{11}$$

This relation for non-equilibrium stationary fluctuations is sometimes referred to as the Fluctuation-Dissipation Theorem (Keizer 1987), because it summarizes how fluctuations depend on the size of the random events ($\Omega\mathbf{D}$) and the dynamic response to perturbations. The solution reveals that the fluctuations ($\mathbf{C}$) are large if (i) the flows ($\mathbf{D}$) are large and (ii) the eigenvalues of the Jacobian $\mathbf{A}$ are small. Large flows means that many random events occur per time unit and small eigenvalues of $\mathbf{A}$ means that the forces that bring the system back to steady state are weak. A simple analogy is diffusion of a particle in a harmonic potential. A large diffusion constant corresponds to the large flows, and a flat potential with little restoring force when the Brownian particle moves away from the center of the potential corresponds to the small eigenvalues of A.

A simple special case is obtained when molecules of type X are synthesized and consumed, one at the time, in two different reactions with arbitrary rate laws $f_1(x)$ and $f_2(x)$. At the steady state, the macroscopic concentration of X molecules is $\bar{x}$ and the flow through the X pool is $f_1(\bar{x}) = f_2(\bar{x}) = f$. In this case, Eq. (10) is scalar and the LNAs of the variance ($\sigma_x^2$) and the Fano-factor, which is the variance normalized to the mean $<n_x>$, evaluates to

$$\sigma_X^2 \approx -\left.\frac{\Omega f}{f_1'(x) - f_2'(x)}\right|_{x=\bar{x}} \quad \text{and} \quad \frac{\sigma_X^2}{\langle n_X \rangle} \approx -\left.\frac{f}{\bar{x}\left(f_1' - f_2'\right)}\right|_{x=\bar{x}} = \frac{f/\bar{x}}{-\lambda} \tag{12}$$

The new variable $\lambda = f_1'(\bar{x}) - f_2'(\bar{x})$ is the rate of relaxation back to the steady state $(\bar{x})$ after a small perturbation. What determines the size of the fluctuations is therefore the ratio between the rate of turnover of the pool, $f/\bar{x}$, and the macroscopic rate of relaxation back to steady state, $-\lambda$. A typical example of systems with slow rate of relaxation and high pool-turnover are those which display zero-order ultra sensitivity (Goldbeter and Koshland 1981). The mesoscopic properties of such systems were analyzed by Berg et al. (2000). Another example is when a chemical species is synthesized in a reaction with weak product inhibition and is consumed by a nearly saturated enzyme (see below).

If $f_1$ is proportional to a parameter $k$, i.e. $kf_{1k}'(x,k) = f_1(x,k)$ the local sensitivity amplification in $x$ for a change in $k$ can be evaluated by implicit derivation of the steady state expression $f_1(\bar{x},k) = f_2(\bar{x}) = f$. It evaluates to

$$\left.\frac{dx}{dk}\frac{k}{x}\right|_{x=\bar{x}} = \frac{-f_{1k}'}{f_{1x}' - f_{2x}'}\frac{k}{x} = \frac{f/\bar{x}}{-\lambda} \; . \tag{13}$$

From this, it is seen that the sensitivity amplification evaluated in Eq. (13) is equal to the LNA of the Fano-factor in Eq. (12). This close relation between the sensitivity amplification and the Fano factor was described by Paulsson (2000) and Paulsson and Ehrenberg (2001) and is discussed in more detail in (Elf et al. 2003b).


## 4 A master equation with an analytical solution

Consider a reaction scheme where a molecule X is synthesized and consumed in two independent reactions with rates that are functions of the number $n$ of copies of X.

$$\varnothing \xrightarrow{\tilde{f}_1(n,\Omega)} X \xrightarrow{\tilde{f}_2(n,\Omega)} \varnothing \tag{14}$$

In this one-dimensional system the state vector is the scalar $n$. We will now assume that there is product inhibition of an enzyme dependent rate of synthesis of X and that removal of X depends on an enzyme reaction of Michaelis-Menten type. The elementary transition rates $\tilde{f}_1(n,\Omega)$ and $\tilde{f}_2(n,\Omega)$ are then given by

$$\tilde{f}_1(n) = \frac{k}{1 + n/(K_I\Omega)} \qquad \tilde{f}_2(n) = \frac{v_{max}}{1 + K_m\Omega/n} \tag{15}$$

These complex reactions can be approximated as elementary when the binding of X to each of the enzymes equilibrates on a much faster time scale than changes in the state $n$. In the denominators of the two expressions in Eq. (15) we have, furthermore, equated the total with the free number of X molecules in the system. This approximation is valid when the fraction of X molecules bound to both enzymes is much smaller than one.

The master equation for the probability $P(n,t)$ that there are $n$ molecules in the system at time $t$ can be written down directly from the scheme for the state transitions

$$\{n-1\} \underset{\tilde{f}_2(n)}{\overset{\tilde{f}_1(n-1)}{\rightleftharpoons}} \{n\} \underset{\tilde{f}_2(n+1)}{\overset{\tilde{f}_1(n)}{\rightleftharpoons}} \{n+1\} \tag{16}$$

Alternatively, it can be obtained by insertion of the expressions in Eq. (15) together with the stoichiometry matrix $\mathbf{S}=[1\ -1]$ into Eq.(3). In either case, the master equation is

$$\frac{dP(n,t)}{dt} = \Omega \tilde{f}_1(n-1)P(n-1,t) + \Omega \tilde{f}_2(n+1)P(n+1,t)$$
$$-\Omega\left(\tilde{f}_1(n) + \tilde{f}_2(n)\right)P(n,t) \tag{17}$$

The steady state solution $P(n)$ can be obtained iteratively from the equation system that arises when all time derivatives in Eq. (17) are zero:

$$P(n) = P(0)\prod_{n=1}^{n} \frac{\tilde{f}_1(n-1)}{\tilde{f}_2(n)}. \tag{18}$$

Inserting $\tilde{f}_1(n)$ and $\tilde{f}_2(n)$ from Eq. (18) leads to the following closed expression for the stationary distribution $P(n)$:

$$P(n) = P(0)\underbrace{\left(\frac{k}{v_{max}}\right)^n \frac{\Gamma(\Omega K_m + n + 1)}{n!\,\Gamma(\Omega K_m + 1)}}_{\text{negative binomial}} \times \underbrace{\frac{\Gamma(\Omega K_I)}{\Gamma(\Omega K_I + n)}\Omega^n K_I^n}_{\text{displaced Poissonian}}. \tag{19}$$

The normalization constant P(0) follows from the condition that the sum of all probabilities in Eq. (18) must equal one.

In Elf et al. (2003b), this system is described in more detail. It is shown there that there is very good agreement between the exact analytical solution and the LNA estimates for the variance of the fluctuations. In Eq. (13) it is shown that the LNA of the Fano-factor in this case is equivalent to the local sensitivity amplification (local response coefficient) $R_{xk} = (dx/dk)(k/x)$. The sensitivity amplification becomes very large ($R_{xk} \approx \sqrt{K_I/4K_m}$ ; Elf et al. 2003b) when $k = v_{max}$ provided that product inhibition remains small when the enzyme that consumes X operates close to saturation, i.e. when $K_I \gg K_m$ (Fig. 1).

# 5 Stoichiometrically coupled flows

A common motif among the anabolic reactions of the cell is when a product molecule, C, is created by binding of two substrate molecules, X and Y, with a rate constant $k_2$ ($s^{-1}M^{-1}$). The reaction is normally catalyzed by an enzyme and to simplify we will assume that the enzyme is both unsaturated and in rapid equilibrium with both its substrates. We will also assume that X and Y are synthesized with the rate constants $k_X$ and $k_Y$ ($s^{-1}M$), respectively, and that they are diluted or degraded with rate constant $\mu$ ($s^{-1}$). The more detailed analysis including feed-back inhibited synthesis and saturation in the bi-substrate reaction has been described elsewhere (Elf et al. 2003b). The simplified system is defined by the following reactions

$$\varnothing \xrightarrow{k_x} X \qquad \varnothing \xrightarrow{k_y} Y$$
$$X \xrightarrow{\mu} \varnothing \quad Y \xrightarrow{\mu} \varnothing \quad X+Y \xrightarrow{k_2} C \tag{20}$$

The reactions in Scheme (20) are considered to be irreversible; a characteristic that often approximates the behavior of free energy-driven intracellular biochemical pathways. In the macroscopic limit, the concentrations $x$ and $y$ of X and Y obey the following macroscopic rate equations

$$dx/dt = k_X - k_2 xy - \mu x \qquad dy/dt = k_Y - k_2 xy - \mu y \tag{21}$$

This type of system can give rise to very large fluctuations in relation to the mean of the components X and Y (Paulsson 2000; Paulsson and Ehrenberg 2001; Elf et al. 2003b). To see how this behavior emerges, we first consider an extreme case where the rates of synthesis of X and Y are balanced, i.e. $k_x=k_y=k$, and neglect their removal from the system by dilution or degradation ($\mu=0$). Under those conditions, the system is degenerate and the stationary state for the concentrations is a curve $\bar{x} = k/(k_2 \bar{y})$. The random nature of chemical reactions will make the components diffuse freely along this curve. In this idealized system, the variance of the fluctuations increases in an unrestrained manner. When $k_x>k_y$ and $\mu=0$, there is no steady state: $y$ approaches zero and $x$ approaches infinity. Accordingly, the system is infinitely sensitive to a change in either $k_x$ or $k_y$ at the balance point $k_x=k_y$.

When $\mu$ is larger than zero, but small in relation to the rate of disappearance of X and Y in their joining reaction, the system has a single asymptotically stable attractor. However, in the small volumes of living cells, such systems will have properties reminiscent of the completely degenerate case with $\mu=0$. The sensitivity of the concentrations of X and Y to a variation in $k_x$ or $k_y$ and the random fluctuations of their concentrations will still be very large, but finite, at the balance point.

The condition for such near-degenerate system behavior is

$$\mu\bar{x} \ll k \Leftrightarrow \mu \ll \sqrt{kk_2} \; . \tag{22}$$

When this inequality is satisfied, the fluctuations are so large that an accurate description of the system requires a mesoscopic perspective, using the master equation

$$dP/dt = k_X \Omega(\mathbb{E}_X^{-1}-1)P + k_Y \Omega(\mathbb{E}_Y^{-1}-1)P + \left(k_2/\Omega\right)(\mathbb{E}_X^1 \mathbb{E}_Y^1 -1)n_X n_Y P$$
$$+\mu(\mathbb{E}_X^1 -1)n_X P + \mu(\mathbb{E}_Y^1 -1)n_Y P \tag{23}$$

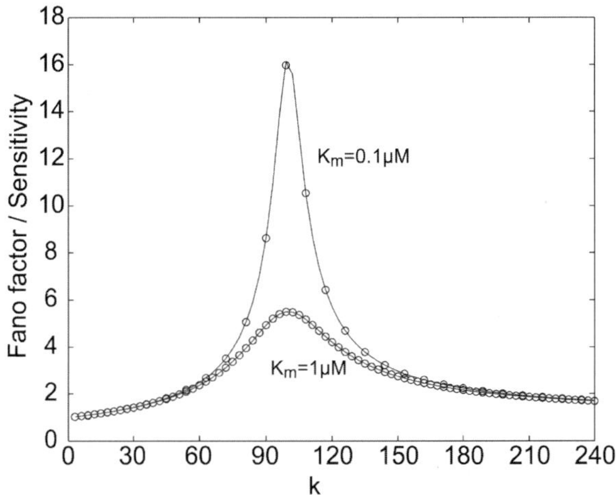


**Fig. 1.** The sensitivity amplification $R_{xk} = (dx/dk)(k/x)$ (lines) are compared to the exact Fano-factor factor ($\sigma_X^2/\langle n_X \rangle$, circles) for two different values of $K_m$. The other parameters are $v_{max}=100s^{-1}$, $K_I=100\mu M$ and $\Omega=10^{-15}$ liter. The sensitivity amplification peaks at $R_{xk} \approx \sqrt{K_I/4K_m}$ when $k = v_{max}$ (Elf et al. 2003b).

P is the probability $P(n_X, n_Y, t)$ that there are $n_X$ molecules of type X and $n_Y$ molecules of type Y in the system at time t. $\Omega$ is the system (cell) volume and $\mathbb{E}$ is a step operator defined from $\mathbb{E}_X^i f(n_X) = f(n_X + i)$. Exact solutions can not be obtained, but evaluation of the moment generating function (see Appendix) gives an exact expression for the variance $\sigma_w^2$ of $n_W=n_X-n_Y$, when $k_X = k_Y (\equiv k)$ :

$$\sigma_w^2 = k\Omega/\mu + \langle n_X \rangle \qquad \frac{\sigma_w^2}{\bar{x}\Omega} \approx k/\bar{x}\mu \gg 1 \tag{24}$$

The variance is also in this case approximated by the flow divided by the rate of relaxation back to steady state after a perturbation, as quantified by the size of the smallest eigenvalue ($\mu$) of the Jacobian. A more general expression can be found in (Elf and Ehrenberg 2003). Since $\sigma_x^2$ cannot be smaller than $\sigma_w^2/4$, even if the X and Y fluctuations are totally anti-correlated, it follows that if the modified fano-factor $\sigma_w^2/\bar{x}\Omega$ is much larger than one, then also $\sigma_x^2/\bar{x}\Omega$ must be much larger than one. The large fluctuations in the near-degenerate case with a small value of $\mu$ and a large coupling flow rate are illustrated in Figures 2B and 2C.

The linear noise approximation can be used to generalize the previous result to a much larger class of reaction schemes:

$$\emptyset \xrightarrow{\tilde{f}_1(n_X,\Omega)} X \quad \emptyset \xrightarrow{\tilde{f}_2(n_Y,\Omega)} Y \quad X+Y \xrightarrow{\tilde{f}_3(n_X,n_Y,\Omega)} \emptyset \ . \tag{25}$$

How the rate of synthesis of X and of Y depend on their respective concentrations and how the reaction where X and Y are joined together are left unspecified. The transition rate vector $\tilde{\mathbf{f}}(\mathbf{n},\Omega)$ and the stoichiometric matrix S are in this case

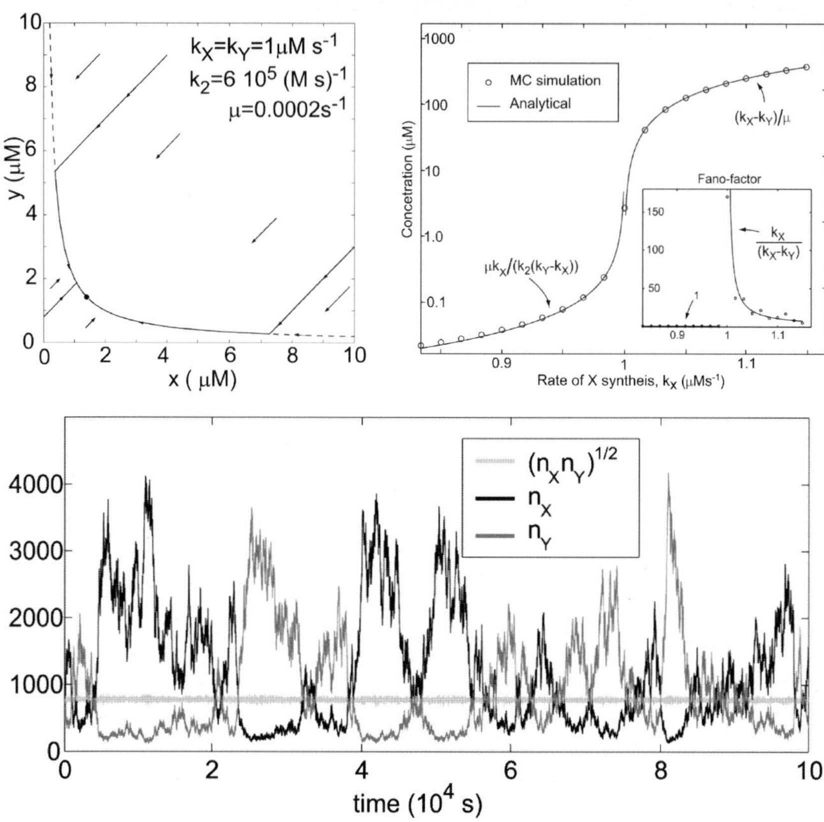


**Fig. 2.** A. Phase-plane of the macroscopic description. Black lines are numerically evaluated trajectories. The dashed line is the curve $k_X = k_Y = k_2 xy$. Arrows indicate the direction of the trajectories. The filled circle represents the macroscopic steady state. $k_X = k_Y = 1\mu M s^{-1}, k_2 = 6 \cdot 10^5 (Ms)^{-1}$ and $\mu = 2 \cdot 10^{-4} s^{-1}$ B. Average stationary concentrations of $x$. The circles are estimated from MC simulation and the solid lines are analytical approximations of the macroscopic steady state value. The parameters are $k_Y = 1\mu M s^{-1}, k_2 = 6 \cdot 10^4 (Ms)^{-1}$ and $\mu = 4 \cdot 10^{-4} s^{-1}$. *Insert*: The Fano-factor $\sigma_{nx}^2 / \langle n_X \rangle$ as estimated by MC-simulation (*dots*) and as approximated by the sensitivity $(dx/k_x)(k_x/x)$ (*line*). C. Near-Critical fluctuations. The system is simulated using Gillespie's direct method. The reaction volume is $\Omega = 10^{-15} \ell$ other parameters are as in A. The fluctuations are correlated such that $n_x n_y$ is nearly constant ($\approx \Omega^2 k/k_2$).

$$\tilde{\mathbf{f}}(\mathbf{x},\Omega) = \left[ \tilde{f}_1(n_X) \quad \tilde{f}_2(n_Y) \quad \tilde{f}_3(n_X, n_Y) \right]^T, \quad \mathbf{S} = \begin{pmatrix} 1 & 0 & -1 \\ 0 & 1 & -1 \end{pmatrix}. \tag{26}$$

The master equation (3) is given by

$$dP(n_X,n_Y,t)/dt = \Omega \tilde{f}_1\left(n_X-1\right)P(n_X-1,n_Y,t)$$

$$+\Omega \tilde{f}_2\left(n_Y-1\right)P(n_X,n_Y-1,t)$$

$$+\Omega \tilde{f}_3\left(n_X+1,n_Y+1\right)P(n_X+1,n_Y+1,t)$$

$$-\Omega\left(\tilde{f}_1\left(n_X\right)+\tilde{f}_2\left(n_Y\right)+\tilde{f}_3\left(n_X,n_Y\right)\right)P(n_X,n_Y,t)$$

$$(27)$$

The macroscopic equation is given by $\dot{\mathbf{x}} = \mathbf{Sf(x)}$ and its steady state, $\bar{\mathbf{x}} = \begin{bmatrix} \bar{x} & \bar{y} \end{bmatrix}$, is given by $\mathbf{Sf(\bar{x})} = 0$. In the LNA, the average stationary copy numbers $\langle \mathbf{n} \rangle$ of the molecules are approximated by their macroscopic steady state values; i.e., $\langle \mathbf{n} \rangle \approx \Omega \bar{\mathbf{x}}$. The covariance matrix $\mathbf{C}$ of the fluctuations around this stationary value is given by Eq. (10) above. The Jacobian matrix $\mathbf{A}$ and the diffusion matrix $\mathbf{D}$ are evaluated at the macoscopic steady state. If it is assumed that the synthesis rates depend on their products in the same way, i.e. that $\tilde{f}_1(n,\Omega) = \tilde{f}_2(n,\Omega)$, then $\bar{x} = \bar{y}$, $f_1(\bar{x}) = f_2(\bar{y}) = f_3(\bar{x},\bar{y})$ and $f'_{1x}(\bar{x}) = f'_{2y}(\bar{y})$. In this special case

$$\mathbf{A} = \begin{bmatrix} f'_{1x} - f'_{3x} & -f'_{3y} \\ -f'_{3x} & f'_{1x} - f'_{3y} \end{bmatrix} \text{ and } \mathbf{D} = f_1 \begin{bmatrix} 2 & 1 \\ 1 & 2 \end{bmatrix}. \tag{28}$$

Here, an explicit algebraic solution can be given for covariance matrix $\mathbf{C}$, although the expressions for $\sigma_X^2, \sigma_Y^2$, and $\sigma_{XY}^2$ are complicated. The variance $\sigma_{X-Y}^2$ for the difference in copy number between X and Y is more informative and is given by

$$\sigma_{X-Y}^2 = \sigma_X^2 + \sigma_Y^2 - 2\sigma_{XY}^2 = \frac{2\Omega f_1\left(f'_{1x} - f'_{3y}\right)}{f'_{1x}\left(2f'_{1x} - f'_{3y} - f'_{3x}\right)}. \tag{29}$$

If $n_X$ and $n_Y$ enter symmetrically in $f_3(n_X,n_Y)$, so that $f'_{3x} = f'_{3y}$, then Eq. (29) reduces to

$$\sigma_{X-Y}^2 \approx \Omega f_1 / f'_1. \tag{30}$$

The sensitivity amplification factor (local response coefficient), $R_{xk} = \left(dx/dk\right)\left(k/x\right)$ is calculated from $f_1(\bar{x},k) = f_2(\bar{y}) = f_3(\bar{x},\bar{y})$ using $f'_{1k}\,dk + f'_{1x}\,dx = f'_{3x}\,dx + f'_{3y}\,dy = f'_{2y}\,dy$. If $f_1$ is proportional to the parameter $k$, i.e. $kf'_1 = f_1$, the sensitivity amplification evaluates to

$$R_{xk} = \left(dx/dk\right)\left(k/x\right) = -\frac{f}{xf'_{1x}}\left(\frac{f'_{1x} - f'_{3x}}{f'_{1x} - 2f'_{3x}}\right) \text{ i.e. } \sigma_{X-Y}^2/2\langle n_X \rangle < R_{kx} < \sigma_{X-Y}^2/\langle n_X \rangle$$

$$(31)$$

Note that the modified Fano factor, $\sigma_{X-Y}^2/\langle n_X \rangle$, is approximately equal to the sensitivity amplification $R_{kx}$. This means that if the fractional change in the concentration of X or Y molecules is numerically much larger than the fractional change in the rate of synthesis of either X or Y, then there will be very large fluctuations in the concentrations of X and Y molecules relative to their mean when their rates of synthesis are balanced.

# 6 Stoichiometrically coupled flows in protein synthesis

In this section we will have a quick look at protein synthesis in *E. coli* with focus on (i) synthesis of the twenty canonical amino acids, (ii) their activation by aminoacyl-tRNA synthetases in reactions that lead to aminoacyl-tRNA molecules which rapidly form ternary complexes with elongation factor Tu and GTP as soon as they have been released from the enzyme and (iii) the consumption of amino acids in protein synthesis as ternary complexes bind to the A site of ribosomes programmed with mRNA codons cognate to the anticodons of the aminoacyl-tRNAs (Fig. 3). The system is described in more detail in (Elf and Ehrenberg 2005). The synthesis and aminoacylation flows of different amino acid are at this level of description independent of each other, while their rates of consumption are stoichiometrically coupled through the frequencies, $f_i$, by which the codons of their tRNA molecules occur on elongating ribosomes in the cell (Fig. 3; Box) (Elf et al. 2003a). In Figure 3, we have simplified the aminoacylation pathways by treating an idealized case with just one aminoacyl-tRNA for each type of amino acid. A more realistic description with the option that several tRNA isoacceptors are cognate to the same amino acid (Björk 1996) leads to very interesting consequences for cells subjected to amino acid limitation (Elf et al. 2003a), and a novel aspect of this situation will be discussed in Section 7 below.

When at least one amino acid is rate limiting for protein synthesis, the model in Figure 3 has similarities with the simple anabolic reaction scheme in Section 5 above, but there are also important differences. These are that (i) there are twenty, rather than two, substrates for protein synthesis, (ii) there are upper bounds to the concentrations [aa-tRNA$_i$] of ternary complexes set by the total concentrations [tRNA$_{0i}$] of tRNA$_i$ in the cell, (iii) the substrates [aa-tRNA$_i$] for protein synthesis do not directly inhibit their own synthesis by aminoacylation, since they are sequestered in ternary complexes with EF-Tu and GTP.

The amino acid with the smallest ratio $s_i = k_{Ei}/\left(f_i[R]k_R\right)$ will be rate limiting for protein synthesis (Elf and Ehrenberg 2005). If, to give an example, the smallest ratio $s_{imin}$ is 0.7 then the total rate of protein synthesis is reduced to 70% of its maximal value. All codons $j \neq i$ are read with maximal rate $k_R$ and codons of type "$i$", responsible for the 30% reduction of the total rate of protein synthesis, are read with rate $f_i/(1/s_{i\min} - 1 + f_i)$ of the maximal (Elf et al. 2001).

The major difference between the response of the anabolic reaction system in Section 4 and protein synthesis to substrate starvation is found in the aminoacylation step which is followed by rapid sequestration of the aminoacylated-tRNA product in ternary complex. This reaction effectively separates the amino acids, mediating feed-back inhibition of their own synthesis (Fig. 3), from the aminoacyl-tRNAs. The absence of direct feed-back inhibition from aminoacylated-tRNA leads to a distinctly switch-like property of the charged levels of tRNAs. When the rate of supply of an amino acid limits protein synthesis, the charged level of the corresponding tRNA is very low and when the rate of supply of this amino acid is in excess, the charged level of its tRNA is near 100% (Elf and Ehrenberg 2005).

To illustrate, we will consider a case when $s_1$ varies from low to high values, $s_2$ is constant and equal to 0.7 and the s-values for all other tRNAs are larger than 1 (Fig. 4A). When $s_1 < 0.7$ the rate of synthesis of aa$_1$ is rate limiting for protein synthesis. When $s_1 = 0.7$, the rates of synthesis of aa$_1$ and aa$_2$ are simultaneously rate limiting and when $s_1 > 0.7$, only aa$_2$ is rate limiting. The figure reveals that the charged level of tRNA$_1$ increases from a very small value to near 100% in an extremely narrow interval of $s_1$ variation around the balance point $s_1 = s_2 = 0.7$. For the parameter set used, the maximal sensitivity amplification of the charged level of tRNA$_1$ in response to the $s_1$ variation is >10 000, which is a remarkably high number.

According to the relation between sensitivity amplification and fluctuations for two coupled pools in Eqs. (30) and (31), one would also expect very large stochastic fluctuations in the charged levels of tRNA$_1$ and tRNA$_2$ at the balance point. This type of behavior is illustrated in Figure 4B, showing Monte Carlo simulations based on the Direct Method (Section 3). Due to product-inhibition, the amino acid concentrations display small random fluctuations at the balance point, while the charged levels of tRNA$_1$ and tRNA$_2$ display very large and anti-correlated fluctuations. An analytical treatment of this interesting problem and some physiological consequences of the zero-order kinetics of aminoacyl-tRNA concentrations when several amino acids are simultaneously rate limiting for protein synthesis can be found in (Elf and Ehrenberg 2005).

# 7 Near-critical fluctuations in the levels of charged tRNA isoacceptors

All amino acids except two (Met, Trp) are encoded by several messenger RNA codons on the ribosome and many amino acids are cognate to several tRNA isoacceptors, that read different codons with partial overlap (Björk 1996). For simplicity, these features of the genetic code and protein synthesis were not included in the scheme in Figure 3 above. It has, however, been shown that the charged levels of isoaccepting tRNA molecules, cognate to the same amino acid and reading different codons, can display selective responses to starvation for that amino acid (Elf et al. 2003a). This type of behavior, with consequences for codon usage in (i) attenuation control mechanisms (Landick and Yanofsky 1987), (ii) the open reading frame of the ribosomal rescue molecule tmRNA, (iii) amino acid synthetic operons (Elf et al. 2003a), was originally anticipated from a simple flow balance argument. The prediction that only a subset of the isoacceptors will lose its charging, irrespective of the severity of the amino acid starvation, has now been confirmed for the Leu, Arg, and Thr-isoacceptor families by microarray and Northern blot experiments on *E. coli* strains that are auxotrophic for Leu (Dittmar et al. 2005) and for two Ser isoacceptors by (Lindsley et al. 2005). If two tRNA isoacceptors that read different codons have identical ratios between their total concentrations and their respective codon frequencies on translating ribosome, then the (macroscopic) theory predicts that their charged levels will decrease in a

**Fig. 3 (overleaf).** Model for amino acid turnover in protein synthesis. Different amino acids (aai) are synthesized with flow rates $J_{Ei}$ by feed-back inhibited enzyme systems ("blocks") with maximal rates $k_{Ei}$ and inhibition constants $K_i$. The amino acids are ester bonded to their cognate tRNAs ($tRNA_i$) with flow rates $J_{Si}$ by aminoacyl-tRNA synthetases, $S_i$, that have maximal rates $k_{si}$, bind tRNAi, with dissociation constants $K_{Si}$ and bind $aa_i$ irreversibly with association rate constants $k_{ai}$. Amino acids, bound to aminoacyl-tRNAs (aa-tRNAi) in ternary complexes with EF-Tu and GTP, are consumed in protein synthesis on active ribosomes, $R$, with flow rates $J_{Ri}$, which are stoichimetrically coupled to the total rate, $J_R$, of protein synthesis in the cell through $J_{Ri}=J_R f_i$, where $f_i$ are the frequencies of codons cognate to $tRNA_i$ on all translating ribosomes in the cell. The maximal rate of translating any codon is $k_R$, here assumed to be uniform. The concentration of ternary complex at which a codon of type "i" is translated with half maximal rate is $K_{Ri}$. Expression of the amino acid

$$J_{Ei}\left([\text{aa}_i]\right) = \frac{k_{Ei}}{1+[\text{aa}_i]/K_i}$$

$$J_{Si} = \frac{[S_i]}{\dfrac{1}{k_{si}}\left(1+\dfrac{K_{Si}}{[\text{tRNA}_i]}\right)+\dfrac{1}{k_{ai}[\text{aa}_i]}}$$

$$J_R = \sum_i J_{Ri} = [R]k_R\left(1+\sum_i \frac{f_i K_{Ri}}{[\text{aa-tRNA}_i]}\right)^{-1}$$

$$J_{Ri} = f_i J_R;\ i=1,2...20$$

biosynthetic operons are in *E. coli* regulated by repressor systems, sensing the amino acid concentrations [$aa_i$], or by transcriptional attenuation systems, sensing the rate of codon translation (Elf et al. 2001). These transcriptional feed-back systems balance the supply capacity of the biosynthetic enzymes to the demand in protein synthesis (Elf 2004). Further information about the scheme and flow equations in Figure 3 is found in (Elf and Ehrenberg 2005). The rate constants used in the simulations described below are: $k_{Si}$=100 s$^{-1}$, $k_R$=20 s$^{-1}$, $k_{ai}$=10$^6$ M$^{-1}$s$^{-1}$, $K_{Si}$=10$^{-6}$ M, $K_I$=10$^{-4}$ M, $K_{Ri}$=10$^{-6}$ M, [tRNA$_{0i}$]=[tRNA$_i$]+[aa-tRNA$_i$]=10$^{-5}$ M, [$S_i$] = 10$^{-6}$ M, [R]=1.67·10$^{-5}$ M, $f_i$=0.05.

uniform fashion as their common amino acid becomes more and more rate limiting for protein synthesis (Elf et al. 2003a). However, at the mesoscopic level this kinetic motif has interesting features which, at least in principle, can be used to implement the principle of stochastic focusing (Paulsson et al. 2000) for improved transcriptional regulation by attenuation mechanisms (Elf 2004).

To see this, consider two different isoacceptors that read two different codons and are aminoacylated with the same kinetic efficiency by a common aminoacyl-tRNA synthetase (RS). The total concentrations of the isoacceptors are $t_{01}$ and $t_{02}$, their charged concentrations are $\alpha_1 t_{01}$ and $\alpha_2 t_{02}$, their uncharged concentrations (1-$\alpha_1$)$t_1$ and (1-$\alpha_2$)$t_2$ and their codon frequencies $f_1$ and $f_2$, respectively. The ratio between the rates of aminoacylation of the two isoacceptors is proportional to the ratio between the concentrations of their deacylated forms and must be equal to the ratio between the frequencies of their codons on ribosomes. That is, for supply to meet demand for both aminoacyl-tRNAs, the following relation must hold

$$\frac{(1-\alpha_1)t_{01}}{(1-\alpha_2)t_{02}} = \frac{f_1}{f_2} \Leftrightarrow \alpha_2 = 1 - \frac{t_{01}f_2}{t_{02}f_1}(1-\alpha_1) = 1 - q(1-\alpha_1) \tag{32}$$

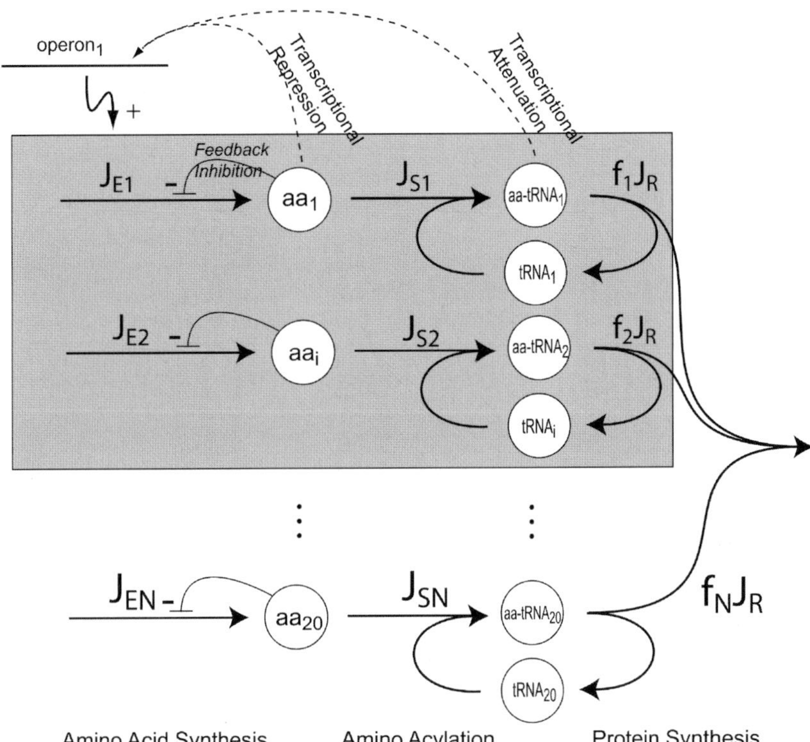


Accordingly the isoacceptor with the lowest $t_0/f$ ratio will lose its charging during amino acid starvation, while the other will keep a charged level higher than 1-q, irrespective of how severe the starvation is. If, however, the ratio between total tRNA and codon frequency is the same for both isoacceptors, then $\alpha_1$ and $\alpha_2$ will be identical according to Eq. (32) and will decrease uniformly with decreasing rate of supply of amino acid. This situation has similarity with the case in the previous section, where two *different* and simultaneously rate limiting amino acids were supplied to protein synthesis in perfect balance with demand. In the present case, there are two tRNAs charged with the *same* rate limiting amino acid by the same aminoacyl-tRNA synthetase, in such a way that the rate of supply of each tRNA meets demand at identical charged levels of the two isoacceptors. It is easily shown that under this condition the aminoacylation reaction will be close to zero order with respect to the charged levels of the two isoacceptors. One reason is that the rates of aminoacylation for the two isoacceptors are determined by the concentrations of their deacylated forms. Since these concentrations are insensitive to large relative variations in the much smaller concentrations of the aminoacylated forms of the isoacceptors, also the rates of aminoacylation of the two isoacceptors will be insensitive to variations in their charged levels. This, in combination with the fact many different combinations of charged levels of the different isoacceptors correspond to the same total rate of  protein synthesis as in Figure 4, gives rise

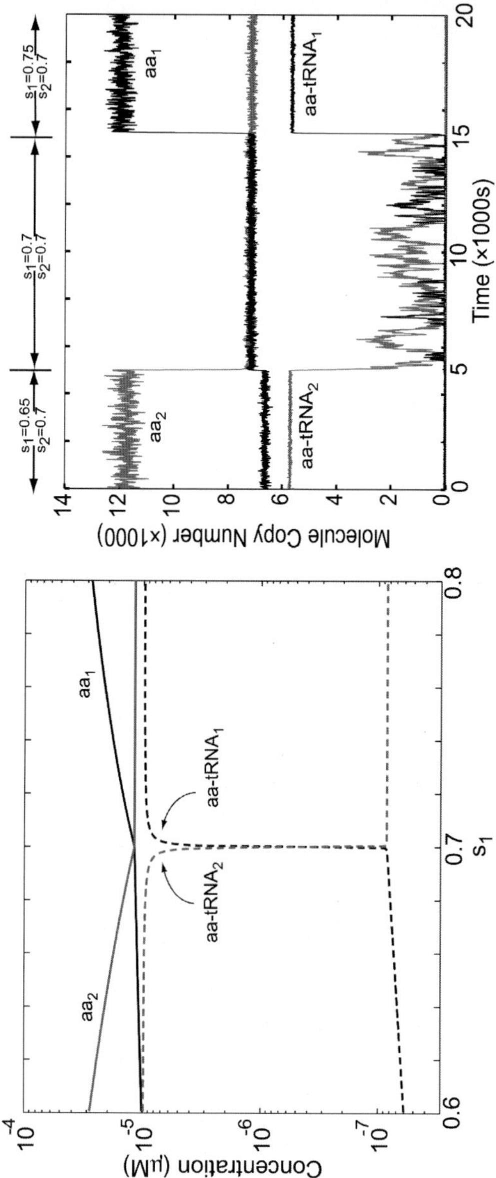



**Fig. 4.** A. Amino acid and aminoacyl-tRNA concentrations for different synthesis capacities ($s_1$). The steady state concentrations for $aa_1$, $aa_2$, aa-tRNA$_1$ and aa-tRNA$_2$ at different synthesis capacities for $aa_1$ ($s_1$) when $s_2$=0.7 and $s_{3\text{-}20}$=1. B. Fluctuations in amino acid and aminoacyl-tRNA copy number. The steady state concentrations for $aa_1$, $aa_2$, aa-tRNA$_1$ and tRNA$_2$ at different synthesis capacities for $aa_1$ ($s_1$) with $s_2$=0.7 and $s_{3\text{-}20}$=1. The figure is reproduced from (Elf and Ehrenberg, 2005) with premission.

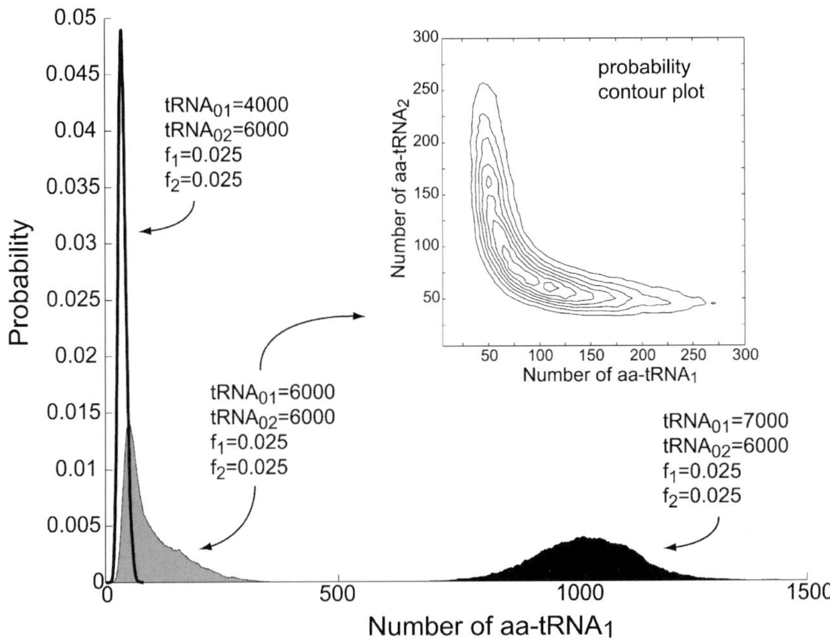



**Fig. 5.** The stationary distribution for of one of two aminoacylated isoacceptors for the limiting amino acid is shown for different total copy numbers of isoacceptor 1 (tRNA1). The two isoacceptors are amino acylated in proportions to the number of deacylated isoacceptors, which a proportionality constant that is chosen to limit the rate of protein synthesis to 50% (vmax/2). The rate of consumption in protein synthesis is $f_i$vmax/$(1+f_i$Km/[aa-tRNA1]$+f_i$Km/[aa-tRNA2]) (Elf et al. 2003a). Km=2μM and the volume is $10^{-15}$ liters. The distributions functions were estimated by Monte Carlo simulations using Gillespie's Direct Method (Gillespie 1976). *insert* The 2D distribution function for the case when the isoacceptor concentrations are perfectly balanced to the usage of their cognate codons.

to the near zero order behavior. The stochastic behavior under amino acid limitation of such a kinetic module containing two isoacceptors, $tRNA_1$ and $tRNA_2$, with total concentrations, $[tRNA_{01}]$ and $[tRNA_{02}]$, nearly or exactly balanced to their respective codon frequencies is illustrated in Figure 5. It is assumed that codons read by $tRNA_1$ and $tRNA_2$ have the same frequencies on translating ribosomes $(f_1=f_2=0.025)$ and that limited supply of the amino acid that is cognate to both isoacceptors reduces the total rate of protein synthesis to 50% of its maximal value. When $[tRNA_{01}]$ is 4000 and $[tRNA_{02}]$ is *6000* molecules per cell, then $[$aa-$tRNA_1]$ is narrowly distributed around about 30 molecules per cell. When $[tRNA_{01}]$ is 7000 and $[tRNA_{02}]$ is 6000 molecules per cell, then $[$aa-$tRNA_1]$ is instead distributed around about 1000 molecules per cell. When, in contrast, $[tRNA_{01}]$ and $[tRNA_{02}]$ both are 6000 molecules per cell, then $[$aa-$tRNA_1]$ and $[$aa-$tRNA_2]$ have the same broad distribution (marginal distribution for $[$aa-$tRNA_1]$ (shown in grey). The charged levels of the two isoacceptors can in this case vary relatively freely on the curve for which the translation rate equals the limiting supply (Fig. 5, insert). This result means that even if codon frequencies match total concentrations of two isoacceptors, the charged levels of the isoacceptors will display very large and anti-correlated stochastic fluctuations (Fig. 5, insert). These fluctuations vary on a time scale that is longer than the time it takes for attenuation mechanisms to "decide" whether transcription should be aborted or continued into the coding parts of an operon (Landick and Yanofsky 1987). This means (Elf 2004) that the response in the rate of codon reading by either one  of these isoacceptors to a variation in the rate of supply of the limiting amino acid can be made more sensitive by stochastic focusing (Paulsson et al. 2000). However, attenuation mechanisms are already hypersensitive due to their multiple step character (Elf et al. 2001) and stochastic focusing is most effective for the improvement of hyperbolic control systems. Therefore, stochastic focusing in attenuation of transcription via large fluctuations in the charged levels of one of two perfectly tuned isoacceptors (Fig. 5, insert) is less likely, but remains an interesting option.

## 8 Conclusions

When the rates of synthesis of two different amino acids are perfectly balanced to the demand from translating ribosomes and at the same time limiting for protein synthesis, then many combinations of charged levels of the corresponding tRNAs lead to the same rate of protein synthesis. This, in combination with the high flow rates of amino acids into nascent peptide chains that characterize bacterial protein synthesis, leads to near-critical behavior of aminoacyl-tRNA concentrations (Fig. 4) due to a phenomenon closely related to zero order ultrasensitivity. This extreme system behavior has immediate consequences for control of gene expression in slowly growing and adapting bacteria (Elf and Ehrenberg 2005).

The principles behind these results have been clarified by applying the statistical tools, explained in Sections 2-6, to a simple global model for amino acid supply, aminoacylation and protein synthesis in growing *E. coli* bacteria (Fig. 3).

## 9 Appendix: The moment generating function

It is possible to derive an exact expression for the stationary variance in $n_A - n_B$ by using moment-generating functions (van Kampen 1992). This gives a point of reference for the numerical and approximate solutions. The moment generating function

$$G(x, y, t) = \sum_{n_X, n_Y = 0}^{\infty} x^{n_X} y^{n_X} P(n_X, n_Y, t)$$

(33)

satisfies the partial differential equation

$$\frac{\partial G}{\partial t} = k_0\Omega(x-1)G + k_0\Omega(y-1)G + \mu(1-x)\frac{\partial G}{\partial x} + \mu(1-y)\frac{\partial G}{\partial y} + \frac{k_2}{\Omega}(1-xy)\frac{\partial^2 G}{\partial x\partial y}$$

$$(34)$$

when $P(n_X,n_Y,t)$ is given by the master equation in Eq. (23). Eq. (34) is obtained by multiplying each term in Eq. (23) by $x^{n_X}y^{n_Y}$ and summing over all $n_X$ and $n_Y$, e.g.

$$\sum_{n_X,n_Y=0}^{\infty} x^{n_X}y^{n_Y}(n_X+1)(n_Y+1)P(n_X+1,n_Y+1,t) =$$

$$(35)$$

$$\sum_{n_X,n_Y=0}^{\infty} x^{n_X-1}y^{n_Y-1}(n_X)(n_Y)P(n_X,n_Y,t) = \frac{\partial^2 G}{\partial x\partial y}$$

Evaluating Eq. (34) in the appropriate limits gives

$$\frac{d\bar{n}_X}{dt} = \frac{\partial}{\partial t}\frac{\partial G}{\partial x}\bigg|_{x,y=1} = k_0\Omega - \mu\bar{n}_X - \frac{k_2}{\Omega}\overline{n_X n_Y}$$

$$(36)$$

$$\frac{d\bar{n}_Y}{dt} = \frac{\partial}{\partial t}\frac{\partial G}{\partial y}\bigg|_{x,y=1} = k_0\Omega - \mu\bar{n}_Y - \frac{k_2}{\Omega}\overline{n_X n_Y}$$

$$(37)$$

$$\frac{d\overline{n_X^2}}{dt} - \frac{d\bar{n}_X}{dt} = \frac{\partial}{\partial t}\frac{\partial^2 G}{\partial x^2}\bigg|_{x,y=1} = 2k_0\Omega\bar{n}_X - 2\mu\left(\overline{n_X^2} - \bar{n}_X\right) - 2\frac{k_2}{\Omega}\left(\overline{n_X^2 n_Y} - \overline{n_X n_Y}\right) \quad (38)$$

$$\frac{d(\overline{n_X n_Y})}{dt} = \frac{\partial}{\partial t}\frac{\partial^2 G}{\partial x\partial y}\bigg|_{x,y=1} = (\bar{n}_X+\bar{n}_Y)k_0\Omega - 2\mu\left(\overline{n_X n_Y}\right) - \frac{k_2}{\Omega}\left(\overline{n_X^2 n_Y} - \overline{n_X n_Y} + \overline{n_Y^2 n_X}\right)$$

$$(39)$$

$$\frac{d\sigma_X^2}{dt} = 2k_0\Omega - \frac{d\bar{n}_X}{dt} - 2\mu\sigma_X^2 - \frac{2k_2}{\Omega}\left(\overline{n_X^2 n_Y} - \bar{n}_X\overline{n_X n_Y}\right)$$

$$(40)$$

$$\frac{d\sigma_Y^2}{dt} = 2k_0\Omega - \frac{d\bar{n}_Y}{dt} - 2\mu\sigma_Y^2 - \frac{2k_2}{\Omega}\left(\overline{n_Y^2 n_X} - \bar{n}_Y\overline{n_X n_Y}\right)$$

$$(41)$$

$$\frac{d\sigma_{XY}^2}{dt} = -2\mu\sigma_{XY}^2 - \frac{k_2}{\Omega}\left(\overline{n_X^2 n_Y} - \bar{n}_X\overline{n_X n_Y} - \overline{n_X n_Y} + \overline{n_Y^2 n_X} - \bar{n}_Y\overline{n_X n_Y}\right)$$

$$(42)$$

where $\sigma_{XY}^2 = \overline{n_X n_Y} - \bar{n}_X\bar{n}_Y$ and bar indicates average. Thus, the variance in $n_X - n_Y$, $\sigma_{X-Y}^2$, can be expressed as

$$\sigma_{X-Y}^2 = \overline{\left((n_X-n_Y)-(\bar{n}_X-\bar{n}_Y)\right)^2} = \left(\sigma_X^2 + \sigma_Y^2 - 2\sigma_{XY}^2\right)$$

$$(43)$$

From Eqs. (40), (41) and (42) we get

$$\frac{d}{dt}\left(\sigma_X^2 + \sigma_Y^2 - 2\sigma_{XY}^2\right) = 2k_0\Omega - 2\mu\left(\sigma_X^2 + \sigma_Y^2 - 2\sigma_{XY}^2\right) + \mu\left(\bar{n}_X + \bar{n}_Y\right).$$

$$(44)$$

At the stationary state, the distribution must be symmetric and $\bar{n}_X = \bar{n}_Y$, $\sigma_X^2 = \sigma_Y^2$, and $\overline{n_X^2 n_Y} = \overline{n_X n_Y^2}$. This gives

$$\sigma_{X-Y}^2 = 2(\sigma_X^2 - \sigma_{XY}^2) = \frac{k_0\Omega}{\mu} + \bar{n}_X.$$

$$(45)$$

# References


Baras F, Mansour MM (1997) Microscopic simulation of chemical instabilities. Adv Chem Phys 100:393-475

Berg OG (1978) A model for the statistical fluctuations of protein numbers in a microbial population. J Theor Biol 71:587-603

Berg OG, Paulsson J, Ehrenberg M (2000) Fluctuations and quality of control in biological cells: zero-order ultrasensitivity reinvestigated. Biophys J 79:1228-1236

Björk G (1996) Stable RNA modification. In: Neidhardt FC (ed) *Escherichia coli* and Salmonella Cellular and Molecular Biology. ASM Press, Washington, D.C., pp 861-886

Bortz A, Kalos M, Lebowitz J (1975) A new algorithm for Monte Carlo simulation of ising spin systems. J Comp Phys 17:10-18

Delbruck M (1940) Statistical fluctuation in autocatalytic reactions. J Chem Phys 8:120-124

Dittmar K, Sørensen M, Elf J, Ehrenberg M, Pan T (2005) Selective charging of tRNA isoacceptors induced by amino acid starvation. EMBO Rep 6:151-157

Ehrenberg M, Elf J, Aurell E, Sandberg R, Tegner J (2003) Systems biology is taking off. Genome Res 13:2377-2380

Elf J (2004) Intracellular flows and fluctuations, Uppsala University, PhD Thesis. http://urn.kb.se/resolve?urn=urn:nbn:se:uu:diva-4291; ISBN: 91-554-5988-9

Elf J, Berg OG, Ehrenberg M (2001) Comparison of repressor and transcriptional attenuator systems for control of amino acid biosynthetic operons. J Mol Biol 313:941-954

Elf J, Ehrenberg M (2003) Fast evaluation of fluctuations in biochemical networks with the linear noise approximation. Genome Res 13:2475–2484

Elf J, Ehrenberg M (2004) Spontaneous separation of bi-stable biochemical systems into spatial domains of opposite phases. Systems Biology 2:230-236

Elf J, Ehrenberg M (2005) Near-critical behavior of aminoacyl-tRNA pools in *E. coli* at rate limiting supply of amino acids. Biophys J 88:132-146

Elf J, Nilsson D, Tenson T, Ehrenberg M (2003a) Selective charging of tRNA isoacceptors explains patterns of codon usage. Science 300:1718-1722

Elf J, Paulsson J, Berg OG, Ehrenberg M (2003b) Near-critical phenomena in intracellular metabolite pools. Biophys J 84:154-170

Érdi P, Tóth J (1989) Mathematical models of chemical reactions. Princeton University Press, Princeton

Ferm L, Lötstedt P, Sjöberg P (2004) Adaptive, Conservative solution of the Fokker-Planck Equation in molecular biology. Technical Report, Department of Information Technology, Uppsala University, Number 2004-054

Gardiner C (1985) Handbook of stochastic methods, second edition edn. Springer-Verlag, Berlin

Gibson M, Bruck J (2000) Efficient exact stochastic simulation of chemical systems with many species and channels. J Phys Chem A 104:1876-1889

Gillespie D (1976) A general method for numerically simulating the stochastic time evolution of coupled chemical reactions. J Comp Phys 22:403-434

Goldbeter A, Koshland DE Jr (1981) An amplified sensitivity arising from covalent modification in biological systems. Proc Natl Acad Sci USA 78:6840-6844

Goldbeter A, Koshland DE Jr (1982) Sensitivity amplification in biochemical systems. Q Rev Biophys 15:555-591

Kacser H, Burns JA (1973) The control of flux. Symp Soc Exp Biol 27:65-104

Keizer J (1987) Statistical Thermodynamics of Nonequlibrium Processes. Springer-Verlag, Berlin

Landick R, Yanofsky C (1987) Transcription attenuation. In: C NF (ed) In *Escherichia coli* and *Salmonella typhimurium*: Cellular and Molecular Biology. ASM Press, Washington D.C., pp 1276-1301

Lindsley D, Bonthuis P, Gallant J, Tofoleanu T, Elf J, Ehrenberg M (2005) Ribosome bypassing at serine codons as a test of the model of selective tRNA charging. EMBO reports 6:147-150

McQuarrie DA (1967) Stochastic approach to chemical kinetics. J Appl Prob 4:413-478

Novick A, Weiner M (1957) Enzyme induction as an all-or-none phenomenon. Proc Natl Acad Sci USA 43:553-565

Paulsson J (2000) The stochastic nature of intracellular control circuits. Acta Universitatis Upsaliensis, Uppsala

Paulsson J (2004) Summing up the noise in gene networks. Nature 427:415-418

Paulsson J, Berg OG, Ehrenberg M (2000) Stochastic focusing: fluctuation-enhanced sensitivity of intracellular regulation. Proc Natl Acad Sci USA 97:7148-7153

Paulsson J, Ehrenberg M (2001) Noise in a minimal regulatory network: plasmid copy number control. Q Rev Biophys 34:1-59

Renyi A (1954) Treating chemical reaction using the theory of stochastic processes. MTA Alk Mat Int Közl 2:93-101

Rigney DR, Schieve WC (1977) Stochastic model of linear, continuous protein synthesis in bacterial populations. J Theor Biol 69:761-766

Risken H (1984) The Fokker-Planck equation. Spinger Verlag, Berlin

Savageau MA (1976) Biochemical systems analysis: a study of function and design in molecular biology. Addison-Wesley, Reading

Schrödinger E (1944) What is life? Cambridge University Press

Singer K (1953) Application of the theory of stochastic processes to the study of irreproducible chemical reactions and nucleation processes. J Roy Statist Soc (B) 15:92-106

Tomioka R, Kimura H, T JK, Aihara K (2004) Multivariate analysis of noise in genetic regulatory networks. J Theor Biol 229:501-521

van Kampen NG (1961) Can J Phys 39:551

van Kampen NG (1992) Stochastic processes in physics and chemistry, second edition, second edn. Elsevier, Amsterdam

Watson JD, Crick FH (1953) A structure for deoxyribose nucleic acid. Nature 171:737-738

Berg, Otto
Department of Molecular Evolution, Uppsala University, Norbyvägen 18 C S-752 36 Uppsala, Sweden

Ehrenberg, Måns
Department of Cell and Molecular Biology, Uppsala University, Box 596, Uppsala, Sweden
ehrenberg@xray.bmc.uu.se

Elf, Johan
 Department of Cell and Molecular Biology, Uppsala University, Box 596, Uppsala, Sweden
 johan.elf@icm.uu.se

Paulsson, Johan
 Department of Applied Mathematics and Theoretical Physics, Centre for Mathematical Sciences, Wilberforce Road, University of Cambridge, Cambridge CB3 0WA, UK

# What is systems biology? From genes to function and back

Hans V. Westerhoff and Jan-Hendrik S. Hofmeyr


## Abstract

The essence of the grand contributions of physiology and molecular biology to biology is discussed in relation to what may be needed to understand living systems. Unanswered is the link between function and molecular behaviour, and emergence of function from the nonlinear interactions, respectively. Systems biology should focus on properties that emerge in nonlinear interactions from the molecular level up, which are crucial for biological function. Pre-genomics approaches such as Metabolic and Hierarchical Control Analysis have already contributed concepts and conclusions to systems biology. Their combination with the genome-wide analyses should now lead to substantial progress in the understanding of life. An aspect of biology at odds with traditional physics and chemistry is the circular causation that occurs in all living systems. By analyzing this phenomenon quantitatively, systems biology can already deal with certain types of circular causation by dissection.


## 1 What came before?

### 1.1. Physiology

The experts differ on whether systems biology has been around for a while or if it is a relatively new science. Both sides may be right, as we shall argue in this chapter. Perhaps biology started in earnest when human beings marvelled over spermatozoa as seen under a microscope, trying to recognize the *homunculus* (little man) in them (*i.e.* supposing that a complete system should be there). Or perhaps biology started with the study of human anatomy, where scientists and artists alike marvelled at the high level of organization in terms of well-defined and fairly autonomous organs, with functions that could almost be understood. Although they were studying biological systems, these disciplines have been called physiology rather than systems biology. They engage in *discourses of nature* where the word 'discourse' is significant in the sense of often being argumentation-driven rather than just data-driven.

Physiology is attractive precisely because it relates directly to function. Indeed, should function be understood, then dysfunction should be understood as well, and avenues towards the treatment of disease should open up. The *discourses* of


Topics in Current Genetics, Vol. 13
L. Alberghina, H.V. Westerhoff (Eds.): Systems Biology
DOI 10.1007/b137122 / Published online: 13 May 2005
© Springer-Verlag Berlin Heidelberg 2005

physiology were often based on rather loose principles however, which appeared to lack generality and were always open to *ad hoc* exceptions. If physiology was at all based on 'laws' of nature, these laws tended to be empirical, such as the law of Haeckel (phylogeny being a recapitulation of ontogeny) and the scale-free relation between respiration rate and biomass. This physiology was often little more than descriptive, useful nonetheless when features would be recurrent. After all, proper diagnosis and then treatment based on experience with previous cases is still one of the most successful qualities of human medicine.

For some time, it seemed that the laws or principles of physiology might be akin to the early formulations of the second law of thermodynamics. This law started off as an empirical law, which appeared to be liable to falsification in any new system under study. The second law of thermodynamics gained enormous standing when it could be derived from the realization that much matter consists of large numbers of rather ill-organized particles, which due to their randomness caused systems to produce entropy, provided that entropy was redefined in terms of an average of molecular properties (*cf.* Westerhoff and Van Dam 1987).

## 1.2 Molecular biology

A remarkable event occurred when Mendel observed regularities when crossing plants, which could be deduced to very simple rules governing the behaviour of quasi particles, later called genes. Apparently, the appearance of a plant system was determined by underlying agents (possibly material). Later, the discovery of the DNA structure with its facile explanation of much of genetics, provided more of a basis to this, however, by then the real biological revolution had already taken place.

The biological revolution had been preceded by the chemical revolution where it had been recognized that the dead world of physics was particle-based. Essentially, this referred to a quantum nature of matter, in the sense that matter comes in different types, all with different properties. These properties are discontinuous, *i.e.* gold differs from silver and there is not necessarily matter with properties halfway in-between. For instance, when hydrogen is made to react with oxygen, the result has completely new properties. The concepts of atoms and molecules revolutionized the ways in which one analyzed the world and for many years the chemical industry delivered many new materials with new and useful properties. Chemistry was perhaps the first clear systems science.

It was soon observed that many of the same molecules that occurred in inorganic matter also occurred in living systems. The new science of chemistry had many ties with biology. No chemical elements were discovered that were unique to living matter. And in fact many chemical molecules for which properties and structure were determined *ex vivo* had a biological origin. Boiling living matter in hydrogen chloride frees a large number of small molecules, including nucleosides and amino acids, all of which inspired organic chemistry into making more similar molecules that could be useful to mankind.

This phase in scientific history suggested that life was perhaps little more than a collection of such molecules. It also inspired *bio*chemistry into a search for the reaction pathways through which those molecules were synthesized. And indeed, rapid progress ensued, up to the elucidation of the molecular basis of life. A major step was the recognition that virtually every chemical reaction in living organisms was catalyzed by a protein and that, therefore, the chemical pathways in life could be delineated by isolating and characterizing the proteins that catalyzed the subsequent steps. Next steps were the discovery of a correspondence between those very same processes and the genes discovered by genetics, and the subsequent identification of genes with parts of the long linear information carrying molecules of DNA. The connection between genetics and biochemistry led to the recognition that life could be studied successfully at the molecular level. This was expressed in the term 'molecular biology', which had an emphasis on the principle that DNA is expressed through RNA into protein, which then catalyzes molecular processes. The primary structure of genes could be elucidated, the corresponding amino acid sequence deduced, the 3-dimensional structure of the corresponding protein could be determined, and action and the mechanism of action of many proteins could be established. Indeed, of every macromolecule, and of every process in living organisms everything could be determined and explained, or so it seemed. This became the triumph of molecular biology and biochemistry combined.

## 1.3 Systems molecular biology?

So, here we are. We will (soon) be able to determine the identity and concentrations of molecules in living organisms. We must surely be close to understanding life and curing its diseases!

While physiology had come close to describing function without really understanding it in solid physical chemical terms, molecular biology now seemed to understand life in precisely those terms, or, at least to do so for the molecules in life. There was an issue with molecular biology, as with physiology, that it might remain an incomplete science and, therefore, it remained a limited scientific discipline in most of the previous century. Living organisms are specified by so many genes and proteins that it seemed that molecular biology could never fully characterize a living system. As a consequence, demonstrated failures of molecular biology to understand living systems could always be attributed to the existence of still unknown factors.

This is where *genomics* caused yet another revolution. With the sequencing of entire genomes and with the ability to study their expression at the level of transcriptome, proteome, and metabolome, every molecular factor became identifiable. And indeed, much activity went into complete identification of transcriptomes, proteomes, and metabolomes. Complete inventories of systems were and are made. Molecular biology became a complete science; no elusive factors remained and scientific explanations by molecular biology became falsifiable, or so it seemed.

Some equated this new molecular biology with systems biology, as it comprised the complete molecular identification of the system of any living organism. With the success of molecular biology in helping us to understand the mechanisms of the individual molecules of life, the extension of molecular biology to the complete set of molecules of life would surely help us comprehend life. Understanding 30,000 times 1/30,000th part of life would surely amount to the understanding of life itself or would it? Is systems biology nothing more than molecular biology of entire systems? Should it be named 'systems molecular biology' (Westerhoff and Palsson 2004)?

## 2 Limits to systems molecular biology

### 2.1 Data floods

A number of caveats with respect to such optimism soon emerged. One of these might prove to be technical only: there is too much data. Data on transcriptomics, proteomics, and metabolomics are now being accumulated at higher rates than that they can be analyzed and structured.

Bioinformatics comes to the rescue here, and so is the ever-increasing power of digital computers. Moreover, the number of types of molecules is limited and for understanding the essence of life, we need perhaps not understand all possible conditions. Furthermore, a better specification of the biological issue one actually wishes to address, and a return to hypothesis-driven research might limit the amount of data needed for analysis significantly. Therefore, molecular biology, genomics plus bioinformatics might still be able to do the job.

### 2.2 Nonlinearity

A second caveat relates to something that is already well known: much of biological function stems from inherent strong nonlinearity. The word nonlinearity is here used in a broad sense and deserves appreciable specification. We shall make this specification by the use of some simple algebra. We ask the less mathematically oriented reader to bear with us; it will be well worth to study this example, because it gives the *crux* of why systems biology is more than the new clothes of the emperor!

We consider molecules of type x and molecules of type y and first assume that a functional property of interest, $f$, depends linearly on both x and y:

$$f = a \cdot x + b \cdot y + c \tag{1}$$

$x$ could be the number of molecules of X, $y$ the number of molecules of y, $a$ and $b$ their respective molecular masses and $c$ the mass of the rest of the cell; $f$ being the mass of the total. The linear dependence of $f$ on both $x$ and $y$ independently has the important property that one can study the dependence of $f$ on X and Y independently and then understand how $f$ behaves by just superimposing the dependencies.

For, denoting changes by delta's, one can first determine experimentally how function changes when only x changes and, thus, determine the value of a.

$$\frac{\Delta f}{\Delta x} = a \tag{2}$$

Similarly one can determine the value of b:

$$\frac{\Delta f}{\Delta y} = b \tag{3}$$

When both $x$ and $y$ change one can then simply calculate the change in function from $a$, $b$ and the actual changes in $x$ and $y$, through:

$$\Delta f = a \cdot \Delta x + b \cdot \Delta y \tag{4}$$

For the mass of a dead cell this might work. But it may not work for many other functions. For instance, it would not work for the mass of a living cell. For, the change in amount of a certain enzyme $x$ would lead to a change in the rate of the reaction it catalyzes and, therefore, to a change in metabolite concentrations and because the cell is an open system a corresponding change in total mass. The latter change could well depend on whether or not the amount of a second enzyme $y$ is also changed at the same time. In the latter case, the total change in mass should not equal the change in mass of enzyme $x$ plus the change in mass of enzyme $y$.

Looking at the living cell from a structural point of view, one might, inadvertently, come to the view that total structure is a linear property: determine the structure of all proteins independently and call these structures x, y, etc., multiply each structure with the number of molecules of the corresponding protein (a, b, etc.). Do this for all proteins and add up all the results so as to obtain the total structure of the cell (one should of course add spatial coordinates for where in the cell the structure sits). The approach seems reasonable, and is valuable perhaps as a first order approach. Yet, one quickly realizes that it may still be incorrect even if one only focuses on the structure and not on the dynamics and functioning of the cell. The approach should fail for instance if two proteins (*e.g.* subunits of a single protein) interact with each and form a complex of a more compact structure, or if chaperons are involved, or if they interact with each other and cause the synthesis of ATP, which causes the phosphorylation of a kinase, the expression of many genes and, therefore, altered levels of many other proteins.

Let us now look at a case where there is no linear relationship between function and the molecules x and y, *e.g.*:

$$f^* = d \cdot x \cdot y^2 \tag{5}$$

Even though this equation is even simpler than the linear one (fewer parameters, *i.e.* only $d$ rather than $a$, $b$, and $c$), it produces many complications. First, the dependence of function ($f^*$) on $x$ is no longer constant but depends on $y$:

$$\frac{\Delta f^*}{\Delta x} = d \cdot y^2 \tag{6}$$

and, hence, cannot be determined once and for all: The functioning of $x$ depends on the activity of $y$.

For the dependence of function on $y$ the situation is even worse: it depends on $x$, on $y$, as well as upon the change in $y$:

$$\frac{\Delta f^*}{\Delta y} = d \cdot x \cdot (y + \Delta y)^2 - d \cdot x \cdot y^2 = 2 \cdot d \cdot x \cdot y + d \cdot x \cdot \Delta y \tag{7}$$

In addition, a change in function due to a change in both $x$ and $y$ cannot be understood as the sum of the change in function due to the change in x and the change in function due to the change in y:

$$\Delta f^* = 2 \cdot d \cdot x \cdot y \cdot \Delta y + d \cdot x \cdot (\Delta y)^2 + d \cdot y^2 \cdot \Delta x + 2 \cdot d \cdot y \cdot \Delta x \cdot \Delta y + d \cdot \Delta x \cdot (\Delta y)^2 \neq$$

$$\frac{\Delta f^*}{\Delta x} \cdot \Delta x + \frac{\Delta f^*}{\Delta y} \cdot \Delta y = d \cdot y^2 \cdot \Delta x + d \cdot x \cdot (2 \cdot y + \Delta y) \cdot \Delta y$$

$$\tag{8}$$

the difference between the two amounting to:

$$SB = \Delta f^* - \frac{\Delta f^*}{\Delta x} \cdot \Delta x - \frac{\Delta f^*}{\Delta y} \cdot \Delta y = 2 \cdot d \cdot y \cdot \Delta x \cdot \Delta y + d \cdot \Delta x \cdot (\Delta y)^2 \tag{9}$$

Only detailed modelling/calculation can relate the functioning of $x$ and $y$ independently to their functioning together.

In the above, $x$ and $y$ were treated as independent molecules. Often they are not independent. And a change in $x$ may also cause $y$ to change. Then already for the linear dependence there is a complication:

$$\frac{\Delta f}{\Delta x} = a + b \cdot \frac{\Delta y}{\Delta x} \tag{10}$$

*i.e.* the functioning of $x$ is not only given by $a$, but also by how $x$ affects function through y and by $b$.

The paradigm of molecular biology is to determine $a$ and $b$. It sees these as universal constants. $a$ and $b$ are the clothes of the emperor. Using the first equation, it then predicts how the molecules determine the functioning of the system. (8 shows that the prediction will be wrong whenever function depends nonlinearly on the molecular behaviour. New clothes for the emperor may not even help: the emperor should engage in a whole new game of redressing himself depending on the active and dynamic conditions he is in).

Again, where are we? Molecular biology determines $x$ and $y$ of the above equations, and perhaps the extents to which they change. Physiology can determine $f^*$ and its changes. The urge is to understand why $f^*$ changes when $x$ and $y$ change. The paradigm that the functional behaviour of the system can be understood from the changes in $x$ and $y$ by just determining $a$ and $b$, is only true in linear systems. Therefore, the issue now is whether biological systems are linear, or more precisely, whether important functional properties of biological systems are linearly related to the properties of their molecules. And the issue is whether the molecular properties are independent of one another.

## 2.3 Nonlinearities and dependencies prevail in real life

There was a time where linear relations were quite popular. Substrate concentrations were assumed to be far below the $K_M$'s of their enzymes such that enzyme kinetics *in vivo* should be linear. Looking at the 'live' database of a number of metabolic pathways (*cf.* www.siliconcell.net ) one readily notes that this assumption is not realistic for the pathways that have been studied to sufficient detail experimentally. Often the substrate concentration is around the $K_M$, so that rates depend non-linearly on concentrations. And, in many cases kinetics is cooperative in terms of substrate concentrations, providing another reason for nonlinearity. And, almost per definition product inhibition, which occurs frequently, is nonlinear.

Indeed, Michaelis-Menten, Monod, and Hill's equations for enzyme and growth kinetics respectively are less than first-order in their dependence on concentration. Quite a few reactions depend on the concentration of more than one compound, *e.g.* when ATP is co-substrate in kinase reactions. Then the rate is bilinear or sub-linear, *i.e.* does not fulfil the linear superposition of two linear equations. Also approximations of kinetic relations in cell function reckon with nonlinear dependencies being the rule rather than the exception. Biochemical Systems Theory (Savageau 1976) uses power laws for this description with the clear intention that the powers need not equal 1. Mosaic non-equilibrium thermodynamics (MNET; Westerhoff and Van Dam 1987) uses linear relationships between the logarithms of concentrations and reaction rates, which translate into nonlinear dependencies of rate on concentrations (*cf.* Wu *et al.* 2004).

How does function depend on gene dosage? For non-redundant essential genes it is clear that function disappears upon a complete knockout. Most such genes (or rather the mutations therein) are recessive, however, this implies that half the gene dosage bestows the organism with full rather than half function: function tends to vary hyperbolically not linearly with gene dosage (Kacser and Burns 1973). The related issue of how pathway flux depends on enzyme activity can be addressed relatively strictly by Metabolic Control Analysis (MCA) and leads to a rather similar answer: pathway flux varies much less than proportionally with enzyme concentration.

There are a number of cases were it is quite obvious that simply adding the behaviour of molecules in isolation will not reproduce their behaviour *in vivo*. One is that the molecules that are involved in the cell-cycle oscillation and the molecules that are involved in glycolytic oscillations would not themselves oscillate in isolation. The oscillation only arises when many molecules of the network are present (see chapter by Novak *et al.*). Yeast glycolysis would not operate as steadily as observed experimentally if the TPS1 regulatory feedback with trehalose phosphate inhibiting hexokinase would not be active (Teusink *et al.* 1998). More generally, reaction rates through metabolic pathways can only attain steady state when they become equal to one another, which they can only do if the enzymes interact with each other, mostly by sensing the concentrations of the metabolites between them. ATP synthesis by the $H^+$-ATPase is only possible when the enzyme occurs in a system with electron–transfer chain linked proton pumps that generate an electrochemical potential difference for protons that is sufficiently high. If the latter are absent, ATP hydrolysis will occur. The difference between ATP synthesis and ATP hydrolysis is crucial for life.

Clearly, the more realistic assumption should be that relationships are nonlinear. Rather than assuming linearity of the dependence of reaction rates on the concentrations of their substrates and products, we use the paradigm that kinetics *in vivo* is massively nonlinear in terms of metabolite concentrations. And, we reckon that just adding knowledge of the individual molecules outside the context of the systems in which they act, will not lead us to understand their functioning in the living organism: outside the system the $y$'s are different, making the prediction of how $x$ affects function, fail. This compromises 'molecular systems biology' and suggests that something is needed in addition to the massive determination of the properties of all the molecules of the living cell, individually.

## 3 Systems biology: Neither the biology of systems nor the biology of all molecules individually

We are able to analyze function of living organisms by physiology without too much reference to molecules. We are also able to determine the properties and concentrations of most molecules that are active in living cells. However, because of the aspect of nonlinearity addressed above, adding all the bits of knowledge about the individual molecules will not lead us to understand the functioning of the living cell. And, on the other hand physiology only understands life in terms of more descriptive properties that are not clearly related to molecular events.

What is needed is a science, which we shall call 'systems biology' that connects physiology to genome-wide molecular biology, *i.e.* a science that addresses specifically how and why the system of macromolecules differs functionally from the sum of the individual behaviours of the molecules that constitute the system. Because of the importance of nonlinearities for function, this science is complex, whether we like it or not. Where physics has profited much from simplification often down to linear first-order approximations, systems biology is inherently more complicated as mostly it cannot engage in such linearization, for fear of approximating away its very essence, *i.e.* the properties that arise precisely due to the nonlinearity of the relations.

The tendency of Non-Equilibrium Thermodynamics to focus on areas where flux-force relations are quasi-linear may seem to make it more appropriate from the physics perspective than from the perspective of systems biology, were it not for the fact that a linear description in a lin-log (Westerhoff and Van Dam 1987; Wu *et al.* 2004) or a log-log world corresponds to a nonlinear world in the true coordinates, and that MNET did away with Onsager's reciprocity relations. Indeed, linear flow-force relations of non-equilibrium thermodynamics are perfectly consistent with the occurrence of oscillations (Cortassa *et al.* 1991).

The feature we noted above, *i.e.* that the dependence of function on a molecular property may depend on the intensity of that property itself and on the intensity of other molecular properties, has the implication that systems biology properties may be much more a conglomerate of special properties that are only valid in a subset of conditions than general properties. In a linear world, many completely general, condition-independent properties should be expected to dominate, but not so in a nonlinear world, although there may still be some such general properties.

And then, one should not forget that systems biology refers to 'biology', *i.e.* to functioning of living systems. This implies that it may not deal with the most general of all nonlinear systems, but mostly with the subset of systems that are found in biology. Accordingly, there is an emphasis on systems that are robust with respect to chemical and even evolutionary fluctuations (*cf.* Carlson and Doyle 2002; Westerhoff and Van Dam 1987). Explosive systems with unbounded reaction rates are unlikely to be common in the living cell.

Is systems biology new then? Well, *No* and *Yes* at the same time. Scientists have long studied cases where new properties arose in molecular interactions (Westerhoff and Palsson 2004). Below, we will discuss show this systems biology *avant la lettre* reached beyond both molecular biology and physiology. We will show how systems biology combines both these disciplines with at least three others, and perhaps even more. We will illustrate how systems biology has already solved issues about biology that many physiologists and molecular biologists were not even able to recognize as issues because of the limitations of their paradigms. And, we will suggest some ways in which parts of systems biology may be developed further.

# 4 Systems biology *avant la lettre*

## 4.1 Self-organization

The molecular biology paradigm sees the cell as a bag of structures kept together by a plasma membrane. The biochemistry paradigm adds that many of the structures correspond to proteins that catalyze or regulate chemical processes. However, macroscopically, and even microscopically, most living organisms do not quite look like amorphous bags of enzymes; rather, they are well structured, the top differing greatly from the bottom and perhaps the left being a mirror image rather than a replica of the right. How do these structures form? Early development of organisms is a case in point, where elaborate spatial structure arises from an apparently spherically symmetrical egg. The apparent breaking of symmetry also occurs in the dimension of time. In a continuous environment, heart cells begin to beat, cell cycles begin to run. The issue connotes with an issue of energetics, *i.e.* how chemical free energy could be converted into and from spatial free energy such as in muscle contraction, in chemicomotion of macromolecules, and in transmembrane electric potentials.

Although preceded by many others, the Brussels school led by Prigogine attracted much attention when addressing these issues (Nicolis and Prigogine 1977). It turned out that small fluctuations, which could themselves break symmetry could be amplified in come cases and lead to a less symmetric final state (both in space and in time) than the initial state. Because in particle physics there should be time invariance and conservation of momentum at the level of individual particles, this still produced a paradox. However, part of this paradox had been resolved before by statistical thermodynamics showing that in systems of many particles some system configurations recognized as single macroscopic states actually corresponded to many more underlying microscopic states than other such macroscopic configurations. Therefore, the former are more likely to be observed. If one were to observe an unlikely macroscopic state first (*e.g.* because it had been pre-prepared by earlier processes), then it should be highly likely that one would see that macroscopic state change to the more likely one. This is what we have learnt to call spontaneous processes. We note that this phenomenon is a property of systems of particles much more than of the particles themselves.

For these transitions between macroscopic states to occur, it is important that the system of particles can move between its microscopic configurations, to then act as if it were searching for the most probable macroscopic state. The movements between the microscopic configurations are called 'fluctuations'. They derive from the continuous bombardment of particles from the environment with energies of all sorts, at least at temperatures above zero Kelvin. The fluctuations cause a progression of a system through its many configurations, which is a random walk in terms of the microscopic states, but a much biased random walk in terms of macroscopic states. In macroscopic terms, it seems that the system exhibits a preference for the most probable macroscopic state, *i.e.* for the macroscopic state that has most microscopic configurations. The latter macroscopic state then is the final 'steady' state and such a state is then stable with respect to fluctuations, *i.e.* after fluctuations the system will return to that macroscopic state. It is this stability against fluctuations (and even of processes themselves) that is at the basis of many systems biology properties, such as robustness and the flux and control properties of dynamic systems including biological ones (*cf.* Westerhoff and Van Dam 1987). For now it is important that this very stability is a property of systems not of their components.

The above statistical thermodynamic argument is often formulated in terms of spontaneous processes producing entropy, where entropy is associated with less ordered states, or chaos. Living organisms are different in this respect as they typically produce order out of chaos, or at least maintain order. Life is, therefore, at the basis of quite a significant extension of thermodynamics where it was made clear that Gibbs (Katchalsky and Curran 1967), or rather metabolic (Westerhoff and Van Dam 1987) free energy has to be destroyed to keep life processes going, whilst some of the input free energy is transduced to free energy in the new biomass. Because this is essentially about life and biology and because it critically depends on the system nature of biology, this has perhaps been the first type of system(s) biology.

Living systems, therefore, need to operate away from equilibrium, yet not so far away that they expend all input free energy and fail to retain some of it for building their own structures. For true symmetry breaking it was shown that systems need to be further away from equilibrium than where the first order Onsager approximation applies (Nicolis and Prigogine 1977; Cortassa *et al.* 1991). Moreover, for such phenomena to occur, more than two components, asymmetrical thermokinetics, and nonlinear kinetics are needed. Such symmetry breaking or 'self-organization' can only happen in systems of molecules not in individual molecules.

## 4.2 Perpetuation

Symmetry breaking of the above, far-from-equilibrium thermodynamic type has the disadvantage that it is non-robust: it depends on the nature of the first fluctuation. If the bifurcation is to lead from spherical to left-handed symmetry for instance, one should expect an equal probability to have an outcome of right-handed symmetry. At a fifty percent error rate, this symmetry breaking would be a highly unreliable mechanism. Because many such symmetry breaking steps need to be made, this should result in little fitness *vis-à-vis* biological evolution. Indeed, experimental results now suggest that there is much less absolute symmetry breaking in biology than was once assumed.

First, most eggs are not quite symmetrical but exist in the context of asymmetrical environment set up by the maternal organism. Probably self-organization mechanisms serve to consolidate symmetry breaking and developmental decisions that have been set in motion by robust asymmetries. The latter are set in place by pre-existing biological matter. Indeed, *perpetuation* is a major characteristic of life as we know it. New cells are not created *de novo* somewhere in the middle of the maternal cell, then to be excreted as a newborn cell. Rather, the mother cell grows in volume and surface area and then either splits into two equal parts, or a small part of the mother cell pinches off and becomes the daughter cell. New proteins are synthesized on old ribosomes, new mRNA is synthesized by old RNA polymerase, and half the DNA of cells already existed in the mother cell. This perpetuation aspect of life makes the issues of symmetry breaking and self-organization much less acute. There is far less self-organization than anticipated earlier on; much of what happens is perpetuation and then division. Self-organization processes serve to maintain and stabilize decisions made through perpetuation.

## 4.3 Chemiosmotic coupling

The maternal organism may convey its own asymmetry to its daughter cell in this mass action way. However, asymmetry can also be conveyed catalytically. Proteins are asymmetrical and are inserted asymmetrically into membranes. Accordingly a cytochrome oxidase catalyzing the macroscopically scalar reaction of oxygen reduction by cytochrome c can couple this reaction to vectorial proton movement by virtue of its asymmetrical 3-D organization. A substantial fraction of the free energy in food is transduced to free energy in new biomass through the electrochemical potential difference for protons across the mitochondrial inner membrane. This involves a transition from chemical free energy, which is a scalar property, through a vectorial property back to a scalar property. Hence, it involves two changes of symmetry, the former of which involves cytochrome c oxidase. For the present chapter, it is important that of necessity, this process requires a system, *i.e.* cannot be carried out by a single macromolecule: it requires a primary proton pump generating the electrochemical potential difference of protons, a closed membrane and a secondary proton pump converting the protonmotive free energy into ATP hydrolytic free energy. Indeed, the chemiosmotic coupling concept by Peter Mitchell (1961) was a second important case of systems biology *avant la lettre*.

## 4.4 Non-equilibrium thermodynamics

Many biological free-energy transduction processes have been described in terms of non-equilibrium thermodynamics. Where thermodynamics was once thought to be devoid of mechanistic detail, and indeed this was proclaimed to be an asset, it was shown that the very essence of processes that are significantly removed from equilibrium is that their thermodynamics does depend on mechanisms (Keizer 1987). MNET was developed precisely to relate systems performance to the mechanism of free-energy transduction, including those leading to incomplete coupling (Westerhoff and Van Dam 1987).

## 4.5 Systems biology *avant la lettre*: Metabolic Control Analysis; laws of systems biology

Non-equilibrium thermodynamics approximates rate equations by linear functions of logarithms of concentrations. This enables analytical solutions at steady state (Katchalsky and Curran 1967). In this approach, however, the parameters are treated as phenomenological and little emphasis is placed on their relationship with parameters that reflect details of mechanisms. Hill's (1977) analysis of complex biological reactions catalyzed by single enzymes showed that although their rate equations obeyed some general format, their parameter values and form also reflected mechanism. For systems of reactions, MNET (Westerhoff and Van Dam 1987) elaborated this and showed how quantitative analysis of biological free-energy transduction systems could lead to conclusions about the functioning of mechanisms such as imperfect coupling and back-pressure.

Biochemical Systems Theory (BST) approximates rate equations with power laws. For linear reaction chains, this again enables analytical solutions of the system equations for steady state (Savageau 1976). There is no emphasis on how the powers relate to underlying mechanisms; the emphasis is on qualitative systems behaviour, *i.e.* on physiology in the above definition. In addition, BST and MNET result in descriptions and sometimes tendencies for systems to behave in certain ways, but not in general principles or 'laws'.

Metabolic Control Analysis (MCA) is both less and more ambitious than BST and MNET. First it does not aim at describing the entire dependence of functional properties on process properties. It focuses on the infinitesimal dependence of functions on those properties. The disadvantage of this is that MCA only considers small changes in the system. The advantage is that in this, MCA is not an approximation but exact, and that as a corollary thereof, MCA has derived some general principles for metabolic systems, *i.e.* some systems biology 'laws'. We shall here illustrate one of these, *i.e.* the summation law for the control of flux through a metabolic pathway by the various reaction steps in that pathway. As in biology almost all reactions are catalyzed by enzymes, the control by a reaction step is related to the control by proteins and to the control by gene expression. The control of the steady-state flux J through a metabolic pathway, such as in Fig. 1, by enzyme i is quantified in terms of the so-called control coefficient of that enzyme *vis-à-vis* that flux:

$$C^J_{e_i} \equiv \left( \frac{d \log|J|}{d \log e_i} \right)_{e_j, j=1,...,n, j \neq i} \tag{11}$$

where *log* stands for logarithm with any base. $e_i$ refers to the catalytic activity of the $i^{th}$ step in the pathway and can in simple cases be replaced by either the $V_{max}$ or the concentration of the enzyme catalyzing that reaction. Technically speaking, the *d* here stands for a partial derivative, with the conditions that the other process activities are held constant and the steady state is re-attained. The logarithmic derivative is taken at the physiological state and can be replaced by the normal derivative provided that the result is then multiplied by the ratio of enzyme activity to flux.

The control coefficient quantifies the importance of the $i^{th}$ step in the pathway for the pathway flux. Taking the molecular biology point of view to the extreme one might wish to determine this flux control coefficient *in vitro* for every enzyme individually and then assume that that control coefficient should be approximately the same *in vivo*. With the enzyme in isolation, the flux through the enzyme is directly proportional to its concentration, which implies that the flux control coefficient equal 1. For a pathway of *n* enzymes, this should imply that all enzymes should be flux-limiting and that the sum of all flux control coefficients should equal *n*. MCA falsifies this conjecture: it has a law that says that the sum of the flux control coefficients for any flux over all processes equal 1:

$$\sum_{i=1}^{n} C^J_{e_i} \equiv 1 \tag{12}$$

Contemplating a two-step metabolic pathway, it is simple to see how this relates to the issue that systems biology deals with nonlinear systems in which processes are dependent:

$$\frac{dJ}{de_2} = \left( 1 - \frac{e_1}{J} \cdot \frac{dJ}{de_1} \right) \cdot \frac{J}{e_2} \tag{13}$$

*i.e.* the functional property flux ($J$) neither depends on the molecular properties $e_1$ and $e_2$ independently, nor on either of them linearly. The more intuitive explanation for this is that when one increases the activity of one enzyme in the pathway to see if it is flux limiting, one simultaneously makes the other enzymes more flux limiting.

Laws of MCA address the sum totals of control coefficients with respect to flux and concentration, as well as relationships between control coefficients and enzyme properties called elasticity coefficients. The latter laws are called connectivity theorems and relate to the stability of the system against fluctuations (*cf.* above and Westerhoff and Van Dam 1987). They are often called theorems but would equally qualify as 'laws' of systems biology, as they indeed address the difference between properties of the systems and properties of the molecules in isolation.

The enzyme properties that are important in the connectivity laws are the elasticity coefficients. They denote the extent to which a reaction step, hence the enzyme that catalyzes it, responds to changes in its metabolic environment. For the elasticity of enzyme 1 of Fig. 1 with respect to the concentration ($z_2$) of metabolite $Z_2$:

$$\varepsilon_{Z_2}^1 \equiv \frac{\partial \log|v_1|}{\partial \log z_2} \tag{14}$$

For the case that enzymes 1 and 2 of the pathway are product insensitive, but enzyme 1 is sensitive to the concentration of metabolite $Z_2$, one finds for the control of the steady state flux through enzyme 1:

$$C_1^J = \frac{\varepsilon_{z_2}^3}{\varepsilon_{z_2}^3 - \varepsilon_{z_2}^1} \tag{15}$$

When enzyme 1 is well regulated by metabolite $Z_2$, the corresponding elasticity $\varepsilon_{z_2}^1$ is quite high in absolute magnitude. Through the above equations, this has the effect that enzyme 1 exerts no control on its own steady state reaction rate; all that control may then reside in enzyme 3 of figure 1. In this way, MCA can illustrate that the *in vitro* control a macromolecule has of its own activity, may be absent in the physiological situation of the intact system. The other factors, residing in different macromolecules, may control its function. This brings home the key issue of systems biology that important aspects of cell function reside in the interaction properties (the elasticity coefficients) rather than in the properties of the individual molecules. It also reinforces the role of MCA as an important theoretical tool in systems biology.

## 4.6 Circular causality and emergence

Biochemistry and molecular biology reinforced the use of the scientific methods of contemporary physics and chemistry in biology, methods which arose from and are still closely tied to the Newtonian view of the world. The Newtonian perspective can be reconciled with three of the four Aristotelian causes, *i.e.*, the state of the system at time $t+\Delta t$ can be explained as the effect of a material cause corresponding to the state of the system at time $t$, an efficient cause corresponding to the mathematical form of the recursive state transition function, and a formal cause corresponding to the initial state at time $t_0$ and other parameter-values. However, as argued by Rosen (see *e.g.* Rosen 1991), there is no place in this view for the fourth Aristotelian causal category, namely that of final cause. The reason

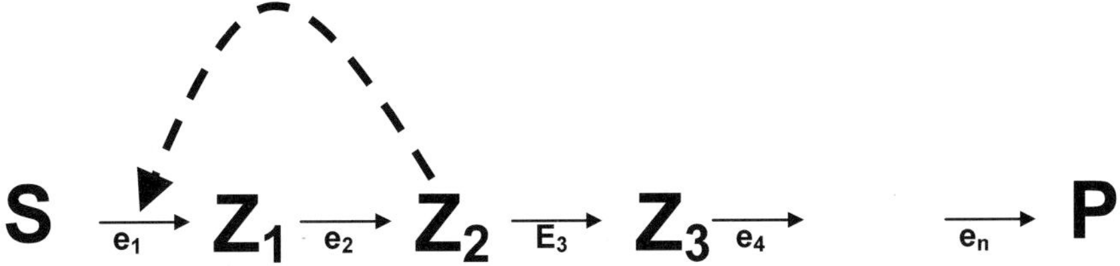


**Fig. 1.** Metabolic pathway, with feedback from the second metabolite on the first reaction. We shall assume all reaction irreversible and not product inhibited.

for this is that whereas material, efficient and formal causes all work in the forward direction, final cause works backwards, which in the Newtonian framework would imply the future affecting the present. From the perspective of physics, chemistry and molecular biology, final cause is, therefore, illegitimate.

Final explanations are closely linked to the concept of function, which is indispensable for many explanations in mainstream biology, and is, despite its uncertain status in formal arguments, often invoked as an inspiration for finding phenomena and even mechanisms. Today many biologists regard evolution and selection of the fittest as an acceptable basis for the use of final causation as part of a scientific explanation. The more frequent type of argument is that a certain mechanism is in place *because* it leads to a higher growth rate or to a higher growth efficiency. The background for this argumentation is often left implicit but is taken to be that because the mechanism improves growth rate, it should have been selected for in evolution, which explains why it is present in the organism under study: ultimately the explanation is then effectively rephrased in terms of formal causation. Common examples of application of final cause include statements such as the occurrence of multidrug resistance proteins at the blood-brain barrier, because it helps to keep toxins out of the brain. Recent examples in systems biology include the flux balance analysis of Palsson and colleagues (Reed *et al.* 2003). Here fluxes are calculated on the basis of earlier experimental results on metabolic pathway genes specifying the metabolic network, and on the assumption of maximal growth rate (which thereby acts as a final cause). The correspondence of the calculated fluxes with experimental fluxes is taken to indicate the correctness of the pathway model.

The first three Aristotelian causes appear to be more solid than final cause. On the other hand, one can see particularly in biology that the implementation of final causes in scientific research can be useful, provided that the final cause (such as the assumed requirement to be optimal in terms of growth rate) is mentioned explicitly. If then, later, it turns out that an organism has not been optimized for growth rate, the argumentation drops for that particular organism but may remain in place for others. After all, other parts of science, with mathematics as a champion, operate on the basis of axioms, which are assumed material causes, and it is accepted that such sciences are useful only for phenomena that obey those axioms.

However, biology may not yet be quite ripe for the acceptance of final cause. First it needs to be certified for every final cause that it has indeed been brought into play in evolutionary selection. Although, it may seem that evolution is able to select for any odd property that could in theory enhance survival potential, this is not so. For instance, lions born with jet engines would be much more successful in hunting, but such lions could certainly not emerge through biological evolution. An understanding of whether natural selection could have indeed selected for a presumed function again requires insight in the functioning of the entire biological system: it requires systems biology. For as long as insufficient systems biology is in place, it is perhaps still better to weed out the implicit use of final (forward) causation that is all too frequent in a biology that pretends to follow rationality only.

The above sketches the way mainstream biology currently attempts to cope with final causation. Contemporary physics and chemistry, on the other hand, are quite strict in eradicating final causes in their scientific arguments. Even more strongly they consider as illegitimate the circular causation implied in explanations in which we say that $A$ causes $B$, $B$ causes $C$ and $C$ causes $A$. However, Rosen (1991) argued persuasively that it is exactly this type of causation that distinguishes living organisms from non-living systems. In brief, Rosen demonstrated that for living systems to be able to fabricate themselves autonomously (*i.e.* be autopoetic in the sense of Maturana and Varela 1980) they need to be organised in a way such that all efficient causes are inside the system. Using the formalism of category theory, he developed a *relational biology* with which this living organisation can be described.

There are other types of circular explanation that have more to do with causation in the Humean sense of one event causing another subsequent event; these also have potential value in biology. Examples are found in induction of gene expression and in the cell cycle: lactose uptake in *E. coli* causes an increase in intracellular allolactose, which binds lac repressor and causes induction of the lac operon which causes enhanced expression of lactose permease, which causes enhanced uptake of lactose, etc. Accordingly, lactose uptake causes (more) lactose uptake. In glycolytic oscillations, activation of phosphofructokinase causes an increase in AMP which causes a further such increase. The ensuing stronger drop in ATP and increase in fructose bisphosphate causes the lower part of glycolysis to make more ATP, which then again stimulates phosphofructokinase which then again decreases ATP and increases fructose bisphosphate which stimulates the lower part of glycolysis. Here phosphofructokinase activation causes phosphofructokinase activation, be it with a time delay; oscillations being the consequence.

Because life is a self-sustaining phenomenon, it has mechanisms in place that cause effects that in turn cause their causes. Although this may not fall within the accepted paradigm of the physical chemical sciences, the phenomenon of circular causation may be so essential to biology that it should be dealt with. As mentioned above, Rosen (1991) has shown how to deal with organisation that leads to circular fabrication. Below we shall show a way that deals with circular event-causality by dissecting it into two or more parts, employing mathematical methodologies.

## 4.7 Networks and hierarchies in life

The identification of most genes encoding the metabolic enzymes of some organism has enabled methodologies for the systematic mapping of metabolic networks. The method first identifies the genes that encode enzymes, then identifies the chemical reactions these enzymes catalyze, and then writes for each enzyme which chemical compounds it produces and consumes, and at which stoichiometry. The stoichiometries for each reaction are then denoted as the column of a huge stoichiometry matrix $N$. The multiplication of $N$ with a vector $v$ of all reaction rates then gives the time dependence of the concentrations of all chemical substances ($dm/dt$) inside the cell:

$$\frac{dm}{dt} = N \cdot v \tag{16}$$

One then tries to determine which combinations of reaction rates should make the right-hand side of the equation equal zero; those are the rates that are compatible with steady states. Technically, these rates are in the Kernel of matrix N, and they may well ones that are biochemically unrealistic for instance by proceeding thermodynamically uphill. In addition, the number of possible reaction rates that lead to steady state is very large. Schuster and colleagues added the requirement that all reactions proceed as allowed by thermodynamics, leading to the so-called elementary reaction modes of the network (Schuster *et al.* 2000). By also taking into account the stoichiometries at which external substrates are utilized and external products are formed, this enabled Schuster and colleagues to examine whether the known network was able to produce certain chemical compounds from certain substrates. Palsson and colleagues added other restrictions such as maximum capacities and later (see above) maximum efficiency or rate of growth (Reed *et al.* 2003). This led them to unique solutions for *v*, which were often close to experimental observations.

These pieces of work are examples of systems biology, in the sense that they depend completely on the connectivity of the network; for an individual reaction such an analysis is impossible. In addition and in contrast to our earlier examples of systems biology, they rely on most, if not all, of the network being known, *i.e.* they are genome-wide.

Both these methods focus on network topology, neither takes kinetics into account. What they, therefore, obtain is *potential* flux patterns, *i.e.* flux patterns that would materialize if indeed kinetics of all the individual reactions were such that metabolite concentrations could evolve to steady state values that are consistent with all those fluxes.

The 'central dogma' of molecular biology states that DNA makes mRNA makes protein makes metabolites. Literally this statement is illustrated in figure 2. Kahn and Westerhoff (1991) observed that what was meant was actually orthogonal to the literal interpretation of the statement: DNA does not make mRNA; rather, RNA-polymerase (efficient cause) makes mRNA (effect) from nucleotide triphosphates (material cause) using DNA as template (formal cause). Likewise,

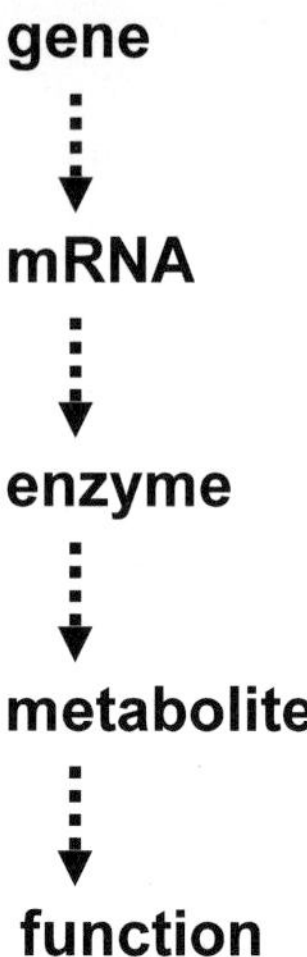


**Fig. 2.** The central dogma of molecular biology: DNA gives mRNA gives protein, gives catalytic activity, gives metabolites and function. 30,000 such diagrams would represent molecular systems biology in humans.

ribosomes, rather than mRNA, make proteins from amino acids using mRNAs as templates. These authors then emphasized that in essence cell function is subdivided in various hierarchical levels, one at the level of mRNA metabolism, one at the level of protein metabolism and one at the level of intermediary metabolism (*cf.* Fig. 3). In essence (though not quite strictly), these levels are *not* converted into each other but regulate conversions inside each other. This means that the stoichiometry matrix $N$ (*cf.* above) is block-diagonal. This feature enabled the development of a new version of MCA called Hierarchical Control Analysis, which led to new laws specific for such hierarchical systems. One of these laws is that:

$$C_{rs}^{J} + C_{rd}^{J} = 0 \qquad (17)$$

acknowledging that transcription of the gene encoding the protein that catalyzes reaction 1 can also control the flux J, and that (at steady state) this control is equally strong as the control by the process that degrades this mRNA. The strength of these controls can readily be 1 and -1 respectively, leading to the effect that the flux through step 2 of the pathway of Fig. 3 can be strongly controlled by a process that is quite remote in the cell's network, *i.e.* the degradation of the mRNA of enzyme 1. This again illustrates that the processes run by macromolecules (such as the enzyme catalyzing step 2 in this example) in living systems, are usually not determined by these macromolecules themselves, but rather by the interactions of all the macromolecules, *i.e.* by what makes the system differ from its components.

Fig. 3 shows how metabolism is not only determined by metabolic control but potentially also by transcription and translation control. It suggests a hierarchy of control, transcription presiding over translation, which would again preside over

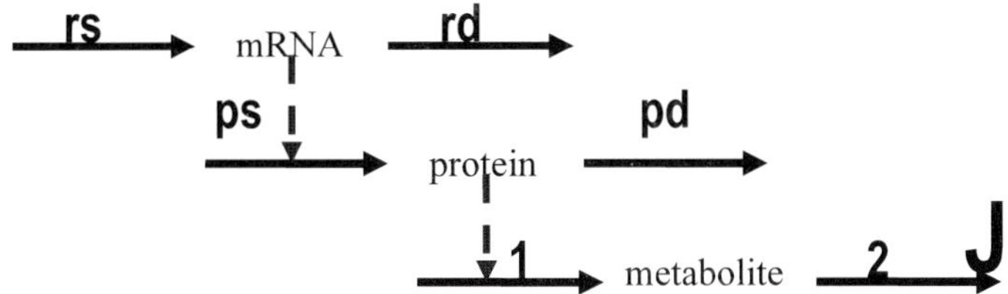


**Fig. 3.** Hierarchical organization of cell function, as simplified for the function flux (J) through a two-step metabolic pathway, leading through reaction 1 from a substrate S, at fixed concentration, not shown, through the metabolite at a variable concentration, through a reaction 2 to a product P, at constant concentration, now shown. Only of step 1 it is shown explicitly that it is catalyzed by a protein, which is synthesized in a 'protein synthesis process 'ps' and degraded in a protein degradation process 'pd'. The protein is not converted to the metabolite, hence, the dashed arrow from protein to reaction 1. The synthesis of this protein occurs in a process that is specified by the corresponding mRNA. This specification, however, corresponds to an influence not to a conversion of mRNA into the protein, hence, the dashed line. The mRNA is synthesized in a process called 'rs' and degraded through a process called 'rd'. Note that feedback from the lower to the upper levels is not taken into account here; this is a dictatorial system.

metabolism. Signal transduction networks also consist of various levels of organization that are not connected by mass flow, just by information flux. Many of the laws Hierarchical Control Analysis also pertain to signal transduction and, many more are still being discovered (Hornberg *et al.* 2005).

The type of control structure in figure 3 has been called a 'dictatorial' control hierarchy. Such dictatorial control systems are highly robust against internal fluctuations, but not adaptive if anything goes wrong structurally. Most biological systems appear to be more sophisticated in that the upper level in the hierarchy (*i.e.* transcription) is not autonomous but adjusts itself to altered requirements at the metabolic, *i.e.* functional level: Transcription regulation often responds to changes at the metabolic level. For instance, allolactose in *E. coli* induces the enzyme that metabolizes it (indirectly, see chapter by Kremling *et al.*) with the effect that if the cell sees lactose (and is able to take it up, see above) it will synthesize more of the enzymes involved in the corresponding catabolic pathway, but only if insufficient such enzymes are present, *i.e.* only if allolactose accumulates to some extent. Biologists have become used to this phenomenon, but perhaps not to the implications that (i) it compromises the central dogma of molecular biology somewhat (*cf.* Fig. 2) in that now metabolism also determines gene expression, and (ii) it provides biology with circular causation, metabolism being the cause of changes in gene expression which in turn causes changes in metabolism.

## 4.8 Systems biology: dealing with the circular causation in biology

Systems with circular causality are shied way from by physics, chemistry, and molecular biology and for good reasons: such systems are much more difficult to handle experimentally and their analysis runs the risk of either becoming futile when the feature of circular regulation is neglected, or becoming inconsistent because cause and effect are interchanged inappropriately.

Again, by taking the feature of circular regulation into account explicitly, systems biology has to deal with one of the features that perhaps distinguish biology most from physics and chemistry. We shall now show how Hierarchical Control Analysis as one of the systems biology approaches *avant la lettre*, is able to deal with the issue of circular regulation. Basically, it does this by cutting away a regulatory link that causes the regulation to be cyclic, by then analyzing the regulation of the remaining non-cyclic parts independently, and by then using mathematics to understand the circularity that occurs in the intact system.

We shall illustrate this for a simplified version of Fig. 3, *i.e.* Fig. 4, which still contains the circular regulation but has neglected the level of translation (protein synthesis and degradation). We shall consider the extent to which enzyme 1 controls its own flux, *i.e.* to what extent the flux through enzyme 1 changes when we activate the $V_{max}$ of that enzyme (through a point mutation, or by reducing the concentration of a non-competitive inhibitor). This increase in its $V_{max}$ will cause an increase in X, which may activate the transcription of gene 1 which will then lead to an increased amount of enzyme 1; activation of step 1 will have been caused by activation of step 1 (we, here, consider the feedback loop to be positive; when the feedback is negative, circular causation is negative, less obvious, but perhaps even more confusing).

We dissect the overall system into two subsystems, one at the level of mRNA metabolism and one at the level of the metabolic pathway. We do this by setting the effect of X on transcription to zero, *i.e.*, by assuming that $\varepsilon_X^t \equiv \dfrac{\partial \log|v_{translation}|}{\partial \log[X]} = 0$. The control coefficients obtained for this dissected system are written as lower-case c's. For the control exerted by enzyme 1 on the metabolic flux this amounts to:

$$c_1^J = \frac{\varepsilon_x^2}{\varepsilon_x^2 - \varepsilon_x^1} \tag{18}$$

For the dissected control of enzyme 1 on the concentration of metabolite X, one finds:

$$c_1^x = \frac{1}{\varepsilon_x^2 - \varepsilon_x^1} \tag{19}$$

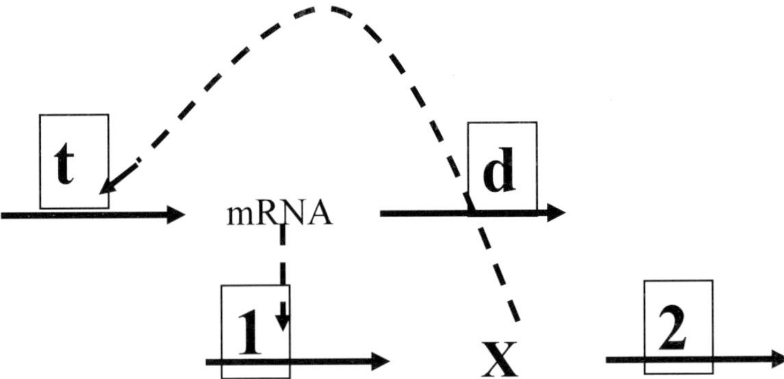


**Fig. 4.** Simplified hierarchical organization of cell function. The transcription and translation levels have been contracted to a single level overlying the metabolic level, in which enzyme 1 synthesizes metabolite X and enzyme 2 degrades it. At the upper level mRNA is synthesized by process 't' and degraded by process 'd'. The dashed arrow from mRNA to enzyme 1 refers to the transcription-translation regulation of the level of enzyme 1. The dashed level from X to 't' refers to transcription regulation by the level of metabolite X. The latter regulation makes the network 'democratic'.

For the control exerted by transcription on the concentration of mRNA for enzyme 1, it leads to:

$$c_t^{mRNA_1} = \frac{1}{\varepsilon_{mRNA_1}^d - \varepsilon_{mRNA_1}^t} \tag{20}$$

For the intact system, Hierarchical Control Analysis has shown that for the control exerted by the $V_{max}$ of enzyme 1 on the flux one finds (Kahn and Westerhoff 1991; Hofmeyr and Westerhoff 2001):

$$C_{V_1}^J = \left(\frac{d\ln J}{d\ln V_1}\right) = \frac{c_1^J}{1-\rho} = \frac{c_1^J}{1 - c_t^{mRNA_1} \cdot \varepsilon_X^t \cdot c_1^X} = \frac{c_1^J}{1 - \dfrac{1}{\varepsilon_{mRNA_1}^d - \varepsilon_{mRNA_1}^t} \cdot \varepsilon_X^t \cdot \dfrac{1}{\varepsilon_x^2 - \varepsilon_x^1}} \tag{21}$$

The term:

$$\rho = c_t^{mRNA_1} \cdot \varepsilon_X^t \cdot c_1^X = \frac{1}{\varepsilon_{mRNA_1}^d - \varepsilon_{mRNA_1}^t} \cdot \varepsilon_X^t \cdot \frac{1}{\varepsilon_x^2 - \varepsilon_x^1} \tag{22}$$

is the circular causation term. It quantifies the regulation of enzyme 1 (through $mRNA_1$) by transcription, multiplied by the regulation of transcription by X, multiplied by the regulation of X by enzyme 1. If the circular causation term is −1) this halves the control the $V_{max}$ of enzyme 1 exerts on the flux, corresponding to a case of homeostatic regulation. If the circular causation term is plus 1, or higher, then the circular causation causes instability and perhaps self-organization, this is a bifurcation point (*cf.* chapter by Novak *et al.*).

This example shows how Hierarchical Control Analysis can deal with circular causation. An important aspect is that circular causation can have one out of various possibly strengths. Only for some such strengths, it may become difficult to analyze the system. But for most others, circular causation can be analyzed quantitatively and is seen to adjust the robustness, *i.e.,* homeostasis of the system.

## 5 Concluding remarks

What is systems biology then? And if there was systems biology *avant la lettre,* what new is there under the systems biology sun?

Systems biology studies, in a fully scientific manner, the functional properties that arise in the dynamic nonlinear interactions between the components of living systems. This implies that it is based both on experimentation and on strict criteria of scientific testing of theories. There should be no open ends, and the system under study should be characterized completely, if not immediately then at least ultimately. Therefore, although in the initial building stages, systems biology may limit itself to fairly autonomous parts of living cells (*cf.* www.siliconcell.net): this is only whilst *en route* to the analysis of the complete living cell. Systems biology is tied in strongly with genomics, proteomics, and metabolomics, in the sense of the ability of measuring all concentrations cell-wide.

Systems biology is a science in and of itself. Hence, it does not boil down to modelling of part of a living cell, or to measuring all metabolite concentrations in that cell, however important each may be. Systems biology tries to discover new principles behind the functioning of living organisms. Genome-wide experiments and models of parts of living cells are tools in that discovery, not aims in themselves.

The systems biology *avant la lettre* mentioned here has shown that important principles can indeed be discovered, but that appreciable parts of the living cell remain to be explored. In addition, the already discovered principles have not yet been examined in terms of their validity or usefulness for the genome size systems. And, inspection of the larger systems, in terms of their functionality may indeed lead to principles that do not reign in the smaller, theoretical systems studied thus far (as suggested for instance by the study of Reed *et al.* (2003).

## References

Carlson JM, Doyle J (2002) Complexity and robustness. Proc Natl Acad Sci USA 99 Suppl 1:2538-2545

Cortassa S, Aon MA, Westerhoff HV (1991) Linear non-equilibrium thermodynamics describes the dynamics of an autocatalytic system. Biophys J 60:794-803

Hill TL (1977) Free Energy Transduction in Biology. Academic Press New York

Hofmeyr J-HS, Westerhoff HV (2001) Building the cellular puzzle: Control in Multi-level reaction networks. J Theor Biol 20:261-285

Hornberg JJ, Bruggeman FJ, Binder B, Geest CR, Bij de Vaate AJM, Lankelma J, Heinrich R, Westerhoff HV (2005) Principles behind the multifarious control of signal transduction ERK phosphorylation and kinase/phosphatase control. FEBS J 1:244-258

Kacser H, Burns JA (1973) The Control of Flux. Symp Soc Exp Biol 27:65-104

Kahn D, Westerhoff HV (1991) Control theory of regulatory cascades. J Theor Biol 153:255-285

Katchalsky A, Curran P F (1967) Non-equilibrium thermodynamics in biophysics. Harvard University Press Cambridge MA, USA

Keizer J (1987) Statistical thermodynamics of non-equilibrium processes. Springer-Verlag Berlin

Maturana HR, Varela FJ (1980) Autopoiesis and cognition: The realisation of the living. D. Reidel Publishing Company Dordrecht

Mitchell P (1961) Coupling of phosphorylation to electron and hydrogen transfer by a chemiosmotic type of mechanism. Nature 191:144-148

Nicolis G and Prigogine I (1977) Self-organization in nonequilibrium systems. Wiley and Sons, New York

Reed JL Vo TD Schilling CH, Palsson BO (2003) An expanded genome-scale model of *Escherichia coli* K-12. Genome Biol 13:2423-2434

Rosen R (1991) Life itself. Columbia University Press New York

Schuster S, Fell DA, Dandekar T (2000) A general definition of metabolic pathways useful for systematic organization and analysis of complex metabolic networks. Nat Biotechnol 18:326-232

Savageau MA (1976) Biochemical systems analysis. Addison-Wesley Reading MA

Teusink B, Walsh MC, Van Dam K, Westerhoff HV (1998) The danger of metabolic pathways with turbo design. Trends Biochem Sci 23:162-169

Westerhoff HV, Palsson BOP (2004) The evolution of molecular biology into systems biology. Nature Biotechnol 42:1249-1252

Westerhoff HV, Van Dam K (1987) Thermodynamics and control of biological free-energy transduction. Elsevier Amsterdam

Wu L, Wang W, van Winden WA, van Gulik WM, Heijnen JJ (2004) A new framework for the estimation of control parameters in metabolic pathways using lin-log kinetics. Eur J Biochem 271:3348-3359

Hofmeyr, Jan-Hendrik S.
Dept. of Biochemistry, University of Stellenbosch, Private Bag X1, Matieland 7602, Stellenbosch, South Africa

Westerhoff, Hans V.
Institute for Molecular Cell Biology and Swammerdam Institute for Life Sciences, BioCentrum Amsterdam, De Boelelaan 1087, NL-1081 HV Amsterdam, EU
hans.westerhoff@falw.vu.nl

# Mechanistic and modular approaches to modeling and inference of cellular regulatory networks

Boris N. Kholodenko, Frank J. Bruggeman, and Herbert M. Sauro

## Abstract

In this chapter we wish to present a vision of systems biology as a discipline that generates new insight and knowledge with quantitative and predictive explanatory power at the system level. We will review work that combines quantitative measurements and mathematical and computational modeling of signaling networks to link molecular mechanisms to physiological responses. An emerging synergistic modeling approach is presented that uses novel approaches to network inference, molecular dynamic simulations, chemical kinetics and reaction-diffusion equations, and functional modularization in biochemical networks.

## 1 From molecular to systems biology

The past decades have witnessed an accelerating pace of acquiring biological information at the molecular level. Advances in genetics and molecular biology have permitted the sequencing of the genomes of a large number of species and the determination of a plethora of protein and lipid components of intracellular regulatory networks. In the 21st century, cell biology faces a new challenge of integrating all available molecular knowledge to quantitatively understand the behavior of a cell or organism at the systems level. This is where systems biology, and, more specifically, iterative cycles of modeling and quantitative experimentation, can be applied.

A goal of systems biology is to understand living organisms at the systems level, combining quantitative information on individual components in order to understand the emergent behaviors that result. Moreover, a quantitative approach gives us much greater predictive power at the systems level that in turn offers new approaches to drug target discovery and bioengineering. Thus, one of the hallmarks of systems biology is the application of quantitative methods and models to understanding biological systems. There are now a wide variety of approaches to quantitative modeling of biochemical reaction networks, including continuum methods such as differential equation models, or discrete probabilistic models based on stochastic dynamics. In addition, there are also sophisticated analytical methods that can be applied to continuum models such as Control Analysis (MCA

Topics in Current Genetics, Vol. 13
L. Alberghina, H.V. Westerhoff (Eds.): Systems Biology
DOI 10.1007/b136809 / Published online: 26 April 2005

– including the closely related Biochemical Systems Theory) (Kacser and Burns 1973; Savageau 1976).

Both the continuum and stochastic methods depends on stoichiometric and mechanistic kinetic models (Tyson and Novak 2001; Goldstein et al. 2004). Other modeling strategies such as Bayesian networks or semi-qualitative modeling strategies are much more abstract in their approach and do not necessarily depend on mechanistic and stoichiometric approaches. Such methods, which have yet to prove themselves, will not be discussed in this chapter.

All models depend on the availability of experimental data, to both construct the initial models, to refute existing models and to enhance and modify models. Modeling is a highly iterative process involving continual refinement as a result of new data. By it's nature, a model is never complete but instead represents the best understanding of the given state of experimental knowledge. Within the modeling process, there are at least two approaches to model building, termed mechanistic ("bottom-up") and modular ("top-down"). Both approaches can be used to describe interactions within and across metabolic, signaling and gene networks. In each cycle, models generate new hypotheses, which ideally should be used to refute the model, a model can never be proved to represent a real situation, merely made consistent with the known experimental data.

This chapter will cover three stages in a systems biology approach, (i) Network inference, (ii) Model building using a "bottom-up" approach with the particular application of quantification of receptor tyrosine kinase signaling in space and time, and (iii) Model rationalization in terms of modular decomposition.


# 2 Network inference

## 2.1 Modular or top-down approach and the quantification of the network architecture by "connection" coefficients

If all molecular interactions were identified by a mechanistic, bottom-up approach, we would be able to reconstruct the network behavior in terms of chemical kinetic rate equations and reaction-diffusion equations provided all forward and backward rate constants and diffusivities were known (Tyson et al. 1996; Kholodenko et al. 2000a; Schoeberl et al. 2002). This technique will be described in the next section, where exceedingly complex models of the EGFR signaling pathways that explicitly incorporate numerous reactions, binding and dissociation events, allosteric interactions and cross-talk will be explored. For other pathways, particularly genetic and many signaling pathways, available mechanistic information and even basic network topology is far from complete. When such information is absent, we require methods to uncover the network structure (network inference) and generic rate laws that can approximate the kinetic laws. This requires the development of modular top-down techniques (Brand 1996). As a necessary prerequisite to this approach, we developed a framework for a quantitative description of the interaction architecture of cellular networks (Brown et al. 1997; Kholodenko et al. 1997,

2002; Sontag et al. 2004). A basic concept is to quantify *direct* interactions between any two elements (nodes) by keeping all other variables "frozen" and analyzing the extent to which a small change in one node affects the level or activity of another. Subsequently, the steady-state input-output relations, which are referred to as global responses of an entire network, can be expressed in terms of "connection" coefficients characterizing the strengths of direct interactions between components (Bruggeman et al. 2002; Kholodenko et al. 2002).

Let us assume that the network architecture and dynamic behavior are determined by a set of differential equations:

$$d\mathbf{x}/dt = \mathbf{f}(\mathbf{x},\mathbf{p}), \quad \mathbf{x} = (x_1,..,x_n), \quad \mathbf{p} = (p_1,..,p_m) \tag{2.1}$$

where a single state variable $x_i$ is assigned to each network node, representing its concentration or activity level, and the corresponding function $f_i$ describes how the rate of change of $x_i$ depends on all other elements of the network and the parameters $\mathbf{p}$. A straightforward means of quantifying the strength of a direct influence of node $j$ on node $i$ is by introducing the dimensionless coefficient $r_{ij}$ as the $(i,j)$ element of the Jacobian matrix ( $\mathbf{F} = \partial\mathbf{f}(\mathbf{x},\mathbf{p})/\partial\mathbf{x}$ ) "normalized" by the $i$th diagonal element, that is to say, $r_{ij} = -(\partial f_i/\partial x_j)/(\partial f_i/\partial x_i)$ (Kholodenko et al. 2002). In biological terms, the connection coefficient $r_{ij}$ tells us how much $x_i$ will change in response to a causative change in $x_j$, when all other network nodes are kept constant, while node $i$ is relaxing to its new steady state (meaning a conceptual "isolation" of node $i$ from other network interactions while characterizing the direct influence of $x_j$). The ratio ( $\delta x_i/\delta x_j$ ) of these changes is determined from the steady-state condition for node $i$:

$$f_i(x_1,..,x_i,..,x_j,..,x_n,\mathbf{p}) = 0 \tag{2.2}$$

At this state, $x_i$ can be regarded as an implicit function of $x_j$. Taking the partial derivative with respect to $x_j$ at constant $x_k$ for $k \neq i,j$, we obtain that the ratio $\delta x_i/\delta x_j$, which quantifies the connection strength, is equal to $r_{ij}$ in the limit of infinitesimal changes. The connection coefficients $r_{ij}$ compose the matrix $\mathbf{r}$ referred to as the network *interaction map* (Bruggeman et al. 2002; Kholodenko et al. 2002; Sontag et al. 2004):

$$\mathbf{r} = -(dg\mathbf{F})^{-1}\mathbf{F} \tag{2.3}$$

The matrix $dg(\mathbf{F})$ is the diagonal matrix with diagonal elements $(\mathbf{F})_{ii}$ and all off-diagonal elements equal to zero. Unfortunately, network interaction maps cannot be experimentally captured using intact cells. In fact, any perturbation to a particular node will subsequently propagate through multiple interactions changing nodes, which are not directly influenced by the initially perturbed node. Consequently, to measure the kinetics of local interactions between two components directly, they should be isolated from the network. Sometimes the interaction of interest can be reconstituted "in vitro", but often only an entire system is accessible experimentally, and only so-called "global" responses of system variables $x_i$ to perturbations can be measured. These global responses correspond to the observations (experiments), where the entire system relaxes to a new steady state after a parameter perturbation.

We, thus, need a new method that can extract the connection coefficients from intact *in vivo* systems.


## 2.2 Modularization of cellular networks

In this section, we demonstrate how a mathematically rigorous analysis of the steady-state behavior of cellular systems can be carried out using a modular framework. We conceptually divide a cellular network, for example, a signaling pathway or gene network into distinct reaction groups, referred to as functional units or modules (Kholodenko et al. 1995b; Hartwell et al. 1999; Lauffenburger 2000; Krauss et al. 2001; Ravasz et al. 2002; Stark et al. 2003; Sauro and Kholodenko 2004). Each module can consist of several signaling, metabolic, and gene interactions and performs one or more identifiable tasks. Note that most modularization approaches are based on structural properties of cellular networks, for example, when a network is analyzed as a graph (Milo et al. 2002; Ravasz et al. 2002). A typical definition of a module takes as the criterion that intramodular molecules should have more interactions with each other than with extramodular molecules; that is, they should form a "clique". However, this criterion does not consider the functional properties of a module nor does it stipulate the conditions under which the mechanistic analysis can be reduced to the examination of interactions only between modules. Both these issues are addressed by the definition of network modules employed here, whose key features are the absence of moiety conservations that embrace distinct modules and the presence of explicit intermediates (variables) considered as the module outputs (Schuster et al. 1993; Kholodenko et al. 1995a; Kholodenko and Westerhoff 1995; Heinrich 1996; Bruggeman et al. 2002; Kholodenko and Sontag 2002). For instance, each of the three tiers of a MAPK cascade can be considered as a functional module, that involves unphosphorylated, mono- and bisphosphorylated forms of a protein kinase and the reactions converting these forms. Importantly, this modular description assumes that at each cascade level, the sequestration of the kinase at the preceding level is negligible. Modules need not be rigid, and entire MAPK cascades, for example, p38, JNK and ERK pathways, can serve as functional modules in a signaling network that involves growth factor and stress activated pathways. For gene networks, modules can involve mRNAs of a particular gene or gene cluster with regulatory interaction loops running through metabolic and signaling pathways (de la Fuente et al. 2002).

We will first analyze a system of $n$ simultaneous algebraic equations (2) that explicitly consider all $n$ network components. At the steady state, the global response coefficients ($R_{ij}$) of this network to perturbations are determined as the parameter sensitivities, $R_{ij} = \partial x_i / \partial p_j \,|_{\text{steady state}}$, which form the $n$ by $n$ global response matrix $\mathbf{R}$ (for perturbations of $n$ independent parameters, $p$). Using (2.2) and (2.3), we can express the matrix $\mathbf{R}$ in terms of the network interaction map (matrix $\mathbf{r}$) and the partial derivatives ($\partial \mathbf{f} / \partial \mathbf{p}$) of the rates $\mathbf{f}$ with respect to $p$ (local sensitivities) as follows (Bruggeman et al. 2002):

$$\mathbf{R} = \partial \mathbf{x} / \partial \mathbf{p} = -(\partial \mathbf{f} / \partial \mathbf{x})^{-1} (\partial \mathbf{f} / \partial \mathbf{p}) = -(\mathbf{F})^{-1} (dg\mathbf{F})(dg\mathbf{F})^{-1}(\partial \mathbf{f} / \partial \mathbf{p}) = -\mathbf{r}^{-1} \cdot \mathbf{w}$$

$$(2.4)$$

We will show now that equations, such as equation 4, describing steady-state responses, continue to apply within a modular framework. Suppose a network is divided into $m$ reaction groups (modules), which are not linked by moiety conservations. Assuming, for simplicity, that only a single intermediate, $x_i$, referred to as a "communicating" intermediate, serves as the output of module $i$, all intermediates of that module represent the variables $(x_i, y_{i1}, y_{i2}, ..., y_{il_i})$, where $y_{ij}$ are the chemical species, which transformations occur within module $i$. The numbers $l_i$ of these species are different for different modules. Equations that determine steady states of an entire network we separate into two sets of equations, one set for communicating intermediates $x_i$, and another set for "intermodular" species described by the vector variables $\mathbf{y}_i$:

$$dx_i / dt = 0 = \varphi_i (x_1, \mathbf{y}_1, ..., x_i, \mathbf{y}_i, ..., x_n, \mathbf{y}_n, \mathbf{p}), \quad \mathbf{y}_i = (y_{i1}, y_{i2}, ..., y_{il_i})$$

$$dy_{ij} / dt = 0 = g_{ij} (x_1, \mathbf{y}_1, ..., x_i, \mathbf{y}_i, ..., x_n, \mathbf{y}_n, \mathbf{p}), \quad i = 1, ..., m, \; j = 1, ..., l_i$$

Assuming that the Jacobian matrix of the set of equations for $m$ vector variables $y_i$ is nonsingular at the steady state corresponding to $p$, the variables $\mathbf{y}_i$ can be presented locally as implicit functions of communicating intermediates $\mathbf{x}$ and $p$. Substitution of $\mathbf{y}_i = \mathbf{h}_i(\mathbf{x}, \mathbf{p})$ in the function $\varphi_i$ gives the steady-state equation for the communicating intermediate $x_i$ that formally coincides with equation (2.2). We conclude that it is possible to apply our techniques in a modular framework, in which only the derivatives ($\partial f_i / \partial x_j$) with respect to communicating intermediates are calculated (Kholodenko and Sontag 2002). This top-down analysis "black-boxes" the molecular organization of network modules, explicitly considering communicating intermediates only. Importantly, the proof above implies that communicating intermediates can be selected based on the experimental practicality; they should not necessarily be the molecules that physically mediate interactions. For instance, mRNAs that can be conveniently monitored using gene arrays may serve as communicating intermediates, even though interactions between different mRNAs (modules) are mediated by protein products, such as transcription factors.

## 2.3 Inference of connections between network modules

With the advent of high-throughput technologies, the data on the expression levels of large gene sets and the activity states of multiple proteins become available, giving us snapshots of transcriptional and signaling behavior of living cells. However, the web of regulatory interactions in cellular networks remains largely unknown at the present time. For instance, to relate genes to one another, current techniques use clustering algorithms, which group together genes that appear to be coherently activated or inactivated (Niehrs and Pollet 1999). Genes with similar expression patterns can be placed together into a cluster, yet the functional interactions between these and other genes are unknown. Another example comes from biology of signaling networks. Although the basic architecture of MAPK cascades has been worked out, both a complete pattern of feedback regulatory loops and cross-talk between MAPK pathways and other signaling systems remain elusive.

The daunting challenge of inferring multiple interconnections in complex signaling and gene networks containing thousands of components can be facilitated by presenting networks as interacting functional modules and determining interactions between modules. Modules can be interconnected in multiple ways, many of which may be unknown, even when the network components are identified in genetic and biochemical studies. Once connections between modules are determined, connection maps within modules can be investigated. Thus, by drilling down the hierarchy of modules we can ultimately uncover the network at the molecular level. Here, we describe a method that determines the connection coefficients $r_{ij}$ between modules from steady-state responses $R_{ij}$ to successive perturbations applied to each module of a network (Kholodenko et al. 2002). For simplicity, we will allow for specific perturbations that affect single modules only. The generalization of the method for perturbations that may simultaneously affect multiple modules is published elsewhere (Kholodenko and Sontag 2002; Sontag et al. 2004).

Mathematically, we will consider a set of parameters, $p_i$, each of which affects only a particular module ($i$) and does not influence any other module ($j$) directly, that is $\partial f_i / \partial p_j = 0$, if $j \neq i$. When these specific perturbations are applied, the matrix of the partial derivatives ($\partial \mathbf{f} / \partial \mathbf{p}$) becomes diagonal and, therefore, the matrix $\mathbf{w} = (dg\mathbf{F})^{-1}(\partial \mathbf{f} / \partial \mathbf{p})$ in equation (2.4) is also diagonal. From equation (2.4), the matrix $\mathbf{r}$, which is the network interaction map, is expressed as follows, $\mathbf{r} = -\mathbf{w} \cdot \mathbf{R}^{-1}$. Because all the diagonal elements of the matrix, $\mathbf{r}$, are equal to -1 (see the definition (2.3)), one can write, $\mathbf{I} = \mathbf{w} \cdot dg(\mathbf{R}^{-1})$, where $\mathbf{I}$ is the identity matrix, and $dg(\mathbf{R}^{-1})$ is the diagonal matrix with diagonal elements $(\mathbf{R}^{-1})_{ii}$ and all off-diagonal elements equal to zero. Therefore, $\mathbf{w} = (dg(\mathbf{R}^{-1}))^{-1}$, and we finally express the network interaction map $\mathbf{r}$ through the global response matrix $\mathbf{R}$,

$$r = -\left(dg\left(\mathbf{R}^{-1}\right)\right)^{-1} \cdot \mathbf{R}^{-1} \tag{2.5}$$

This final expression gives us the answer we were looking for: if the global responses of a cellular network to specific perturbations to all modules have been measured, the network interaction map composed of connection coefficients can be retrieved simply by the inversion of the measured response matrix. Using models of MAPK pathways and gene networks *in silico*, the proposed method was tested computationally and demonstrated robust properties with respect to the perturbation magnitude (Kholodenko et al. 2002). Experimental application of this "unraveling" method is an attractive task for systems biology.

# 3 Bottom-up approach

The bottom-up approach is the next stage in a systems biology approach and it assumes that most of interactions in the network are characterized. The bottom-up approach involves building a kinetic model, attaching appropriate rate laws to each of the connections in the system network. In some systems, the kinetic rate laws are known with a fair degree of certainty. This is particularly true for metabolic networks where decades of classical enzyme kinetics carried out between the 1950s and 1970s, uncovered many kinetic properties of enzymes. Many kinetic models built today, depend heavily on this knowledge, however, as many have indicated, such kinetic studies were carried out on isolated enzymes under conditions which do not necessarily match those *in vivo*. There has, therefore, been some discussion on the validity of using such legacy data. However, it is remarkable that there are now a significant number of kinetic models of networks that are very reliably given these constraints.

## 3.1 Spatio-temporal patterns of growth factor signaling and cell fate decisions

Cells receive external information in the form of growth factors, hormones, neurotransmitters, and other environmental signals. Processing this information occurs through a multilevel signaling network that leads to important cellular decisions ranging from cell survival, growth and proliferation to growth arrest, differentiation or apoptosis. A first filter in cellular signaling networks includes various plasma membrane receptors, such as G-protein coupled receptors (GPCRs) and receptor tyrosine kinases (RTKs). Activated receptors induce signal processing through a second network layer. This involves a battery of protein-protein interactions and phosphorylation steps, the generation of intracellular messenger molecules and the translocation of signaling proteins to specific cellular locations. In a multitude of cell types, receptor stimulation is followed by activation of the mitogen activated protein kinase (MAPK) cascades, which appear to function as a central integration module in the processing of these signals (van Biesen et al. 1996;

Della Rocca et al. 1997; Gutkind 1998; Schlessinger 2000; Bogdan and Klambt 2001). The question arises as to how a cell maintains specificity in response to an individual signal, while multiple signaling pathways overlap at the level of adaptor and target proteins and the MAP kinase cascade. For instance, epidermal growth factor receptor (EGFR), platelet-derived growth factor receptor (PGFR), insulin-like growth factor-1 (IGF-1) receptor, insulin receptor (IR) and others, all can activate the small GTP-hydrolysing protein Ras, which leads to the activation of the MAPK, known as the extracellular signal regulated kinase (ERK).

Specificity of signals generated by the stimulation of the GPCRs and RTKs depends on the temporal and spatial organization of the activation of cellular signaling cascades (Marshall 1995; Shymko et al. 1997; McCawley et al. 1999). Differential activation patterns may eventually result in diversified gene expression by signal-regulated transcription factors. For instance, the activation of Raf-1 (a direct downstream effector of Ras and the kinase for MEK, which phosphorylates ERK) has been linked to such opposing responses as the induction of DNA synthesis and growth inhibition (Lloyd et al. 1997; Sewing et al. 1997). Experiments conducted with mouse NIH 3T3 cells demonstrated that the biological outcome depends on the level of Raf-1 activation (a low Raf-1 activity led to the cell cycle progression, whereas a higher Raf-1 activity inhibited proliferation (Woods et al. 1997)). Recent data on PC12 cells have shown that cellular decisions depend on the duration of ERK activation, a transient activation resulted in proliferation, whereas sustained activation led to cell differentiation (Marshall 1995; York et al. 1998). These and other data suggest that specificity of signaling is generated by the timing, duration and amplitude of activation of the different components of GPCR and RTK pathways.


## 3.2 Differential temporal patterns of signaling responses can be explained using kinetic modeling

To elucidate the control of complex spatio-temporal responses, we conducted combined experimental and computational analyses of the EGFR signaling network. The EGF receptor belongs to a large RTK family, which regulates cell growth, survival, proliferation and differentiation. Binding of a ligand, such as EGF, causes receptor dimerization and rapid stimulation of intrinsic tyrosine kinase activity followed by autophosphorylation of multiple tyrosine residues in the cytoplasmic domain of the receptor. Receptor phosphorylation creates docking sites for cytoplasmic target proteins that possess the Src homology 2 (SH2) or phosphotyrosine-binding (PTB) domains, such as growth factor receptor binding protein 2 (Grb2), src homology and collagen domain protein (Shc), phospholipase C-$\gamma$ (PLC$\gamma$), and phosphatidylinositol 3-kinase (PI3K) (e.g. Pawson 1995; Schlessinger 2000). Activation and tyrosine phosphorylation of these proteins leads to further signal propagation through multiple interacting pathways including the SOS/Ras/MAPK, PLC$\gamma$ and PI3K/Akt pathways.

Studies of EGFR-mediated signaling in liver cells demonstrated that despite continuous EGF stimulation, tyrosine phosphorylation of the receptor and selected targets, such as PLCγ, PI3K or EGFR-Grb2-SOS complexes was transient (a marked peak is reached before the response descends to a low sustained level), whereas phosphorylation of other targets, such as Shc or the concentration of Shc-Grb2-SOS complex, increased almost monotonically (Di Guglielmo et al. 1994; Saso et al. 1997; Kholodenko et al. 1999; Kong et al. 2000; Moehren et al. 2002). In order to understand this complexity, we developed a kinetic model of EGFR-mediated signaling (Kholodenko et al. 1999; Moehren et al. 2002; Markevich et al. 2004a). The model accounted for the experimentally observed transient and sustained activation patterns of multiple signaling proteins targeted by EGFR. The robust nature of the results with respect to variations in the elementary rate constants within the range of experimentally measured values allowed us to identify the most important molecular mechanisms that determine experimentally observed signaling kinetics.

A surprising prediction of our model was that EGFR-mediated phosphorylation of Shc on Tyr 317 reduces its affinity for the receptor and facilitates the dissociation of Shc from EGFR (Kholodenko et al. 1999; Moehren et al. 2002). The EGF receptor binds Shc through its N-terminal PTB domain and C-terminal SH2 domain. Tyr317 is located within the central collagen-homology (CH) linker region of Shc at a distance of 110 residues away from the PTB domain and 53 residues away from the SH2 region. At first sight, the prediction of our computational model is at odds with current paradigms that emphasize the domain structure of large adapter proteins, such as Shc. A typical protein domain ranges in size from 50 to 120 amino acids and is responsible for specific interactions with other proteins or phospholipids (Pawson and Nash 2003). The modular structure underlying signaling interactions may imply that the phosphorylation of residues located outside specialized domains should not influence interactions with partner proteins mediated by these domains. However, molecular dynamics simulations revealed that Tyr317 phosphorylation significantly affects Shc domain motions and decreases the flexibility of the PTB and SH2 domains, reducing the capacity of these domains to interact productively with the EGF receptor (Suenaga et al. 2004). These findings corroborated the prediction of the kinetic model.

In summary, the model of the EGFR pathway demonstrated that a differential time-course of activation of receptor targets is controlled by: (i) phosphorylation-induced changes in the binding affinities of signaling proteins to the receptor (Suenaga et al. 2004), (ii) the differences in the $V_{max}/K_m$ ratios for the phosphatases of the receptor and its target proteins, (iii) the relative fractions of signaling proteins that are complexed with the phosphatases before and after receptor stimulation (such as complexes of the phosphatases SHP-1/SHP-2 with PLCγ and PI3K), and (iv) the relative abundance of signaling proteins (Kholodenko et al. 1999; Moehren et al. 2002).

## 3.3 Membrane translocation of SOS and RasGAP shapes Ras activation patterns

The different temporal response patterns to receptor activation are further complicated by translocation of signaling proteins to diverse cellular locations and their colocalization with various scaffolding and cytoskeletal proteins (reviewed in Carraway and Carraway 1995; McLaughlin and Aderem 1995; Mochly-Rosen 1995; Haugh and Lauffenburger 1997; Kholodenko et al. 2000b). In mitogenic signalling, receptor stimulation is linked to the activation of the MAPK cascades through the cytoplasmic protein SOS and the membrane-anchored Ras GTPase. SOS is the Ras-specific guanine nucleotide exchange factor catalyzing the conversion of inactive Ras-GDP to active Ras-GTP. The adapter protein Grb2 mediates the binding of SOS to activated RTKs, such as EGFR. Due to high affinity of the binding of SH3 domains of Grb2 to prolin-rich regions of SOS, Grb2-SOS complexes exist in the cytoplasm of unstimulated cells. Upon stimulation, Grb2-SOS complex binds to the activated EGFR either directly or through tyrosine phosphorylated Shc. Importantly, binding to the receptor does not increase SOS catalytic activity towards Ras, but only localizes SOS to the membrane where Ras resides. Therefore, we are left with the question of how EGFR facilitates SOS catalysis. The kinetic model helps us understand why EGFR-Grb2-SOS and EGFR-Shc-Grb2-SOS complexes activate Ras, whereas the catalysis by cytoplasmic Grb2-SOS complexes appears to be ineffective. A piggy-back ride of SOS on the activated receptor concentrates the catalyst in a narrow layer below the plasma membrane, about 5-10 nm thick corresponding to the dimension of membrane-anchored proteins. The volume of that layer ($Vm$) is much smaller than the cytoplasmic volume ($Vc$). When SOS binds to activated EGFR, both the catalyst and Ras protein are confined to this volume, whereas in unstimulated cells, SOS and Grb2-SOS complexes delocalize throughout the cytosol. For a spherical cell with a radius of 10 μm and a proximal membrane layer of 10 nm, the ratio of cytosolic volume to proximal membrane volume ($Vc/Vm$) equals about 300. Therefore, the decrease in the reaction volume is estimated to be $10^2$- to $10^3$-fold, resulting in a $10^2$- to $10^3$-fold increase in the apparent affinity of SOS to Ras (Haugh and Lauffenburger 1997; Kholodenko et al. 2000b). Our computer simulations demonstrate that if the spatial organization of SOS signaling would not be taken into account, no appreciable stimulation of Ras would occur (Markevich et al. 2004b).

Signaling of activated Ras is turned off by the activation of GTPase-activating proteins (GAPs) (Denhardt 1996). Alternatively, inhibitory phosphorylation of SOS by MAPK may provide a mechanism for switching off Ras signaling (Cherniack et al. 1995; Langlois et al. 1995; Hu and Bowtell 1996; Brightman and Fell 2000; Asthagiri and Lauffenburger 2001). However, SOS phosphorylation is likely to be an additional mechanism to switch Ras off, whereas the predominant regulation is the GAP-mediated termination of Ras signaling by RasGAP. In fact, calculations by our kinetic model demonstrate, that even after SOS activity has already decreased to almost zero, it would take over 60 minutes to return to a low stationary level of Ras-GTP, if only intrinsic GTPase activity of Ras would be involved (Markevich et al. 2004b). We conclude that a rapid deactivation of Ras signaling observed experimentally (within the first 5–15 minutes of EGF stimulation) suggested that the RasGAP activity is a crucial controlling factor. Both SOS and RasGAP are cytoplasmic proteins. As in the case of SOS signals, receptor-mediated membrane relocation of RasGAP is necessary for an appreciable increase in the rate of GTP hydrolysis on Ras. If RasGAP would interact with Ras from the cytosol, a $10^2$- to $10^3$-fold increase in the cytosolic concentration of RasGAP would be required. Computer simulations illustrate that the termination of Ras signaling does not occur, unless RasGAP is effectively recruited to the plasma membrane, where Ras protein is located (Markevich et al. 2004b).

## 3.4 Heterogeneous spatial distribution is an additional factor controlling signaling cascades

A hallmark of many signal relay processes is the spatial separation of activation and deactivation of signaling proteins, for example, a protein is phosphorylated at the cell surface by a membrane-bound kinase and dephosphorylated in the cytosol by a cytosolic phosphatase. The results of our analysis of reaction-diffusion systems demonstrate that this spatial separation potentially results in precipitous gradients of phospho-proteins given measured values of protein diffusion coefficients and phosphatase and kinase activities (Brown and Kholodenko 1999; Kholodenko et al. 2000a). Since active Raf-1, which is the kinase of MEK, is localized to the cell membrane and the MEK and ERK phosphatases are delocalized throughout the cytosol, the concentrations of activated MEK and ERK can be high at the plasma membrane and low at a short distance from the membrane in the cytosol, resulting in a strong decrease in the phosphorylation signal towards the cell interior (Kholodenko 2002). When the phosphatase is far from saturation, our calculations estimate the distance from the plasma membrane, at which the phosphorylation signal is attenuated by a factor of ten, to be less than 10 µm (a typical value for a hepatocyte radius). When effective signal transduction is hampered by slow protein diffusion and rapid dephosphorylation, phospho-protein trafficking within endocytic vesicles may be an efficient way to propagate the signals (Kholodenko 2002). Additional mechanisms facilitating the information transfer can involve the assembly of MAP kinases on a scaffolding protein, active transport of signaling complexes by molecular motors, and traveling waves of phosphorylated kinases (Kholodenko 2002, 2003).

## 4 Rationalization of network function

One of the key issues that confronts signaling research today is what is all the complexity for? What are the networks attempting to achieve, what data integration is occurring and is there any advantage to drawing parallels between natural signaling networks and man-made networks? There is a growing appreciation that many of the designs and strategies that man has developed to manipulate information, particularly within the electronics world, are present in biological networks. The excellent reviews by Tyson and colleagues (2003) and Wolf and Arkin (2003) offer an exciting glimpse into the parallels between natural and man-made signaling and control systems. It is very likely that the engineering sciences, particularly electronic and control engineering will play an ever-increasing role on interpreting the complex networks we find.

It has been shown in previous work that the response of a simple cascade cycle is very similar to the response of an electronic transistor (Sauro and Kholodenko 2004), and one may speculate that the cascade cycle is a fundamental unit of signaling networks. From such an observation it is not difficult to discern a variety of functional units present in biological signaling networks, including amplifiers, switches, oscillators and most likely many more that are await discovery. It is by this approach that we can begin to rationalize the complexity we find in biological signaling networks. There are attempts to bring some of the key concepts in control engineering to biology. In particular, the work by Ingalls' (2004) efforts to merge control engineering to Metabolic Control Analysis. Adam Arkin (Wolf and Arkin 2003) has been one of the most prominent and successful advocates of employing control engineering in systems biology.

As a simple example, feedback control, one of the central concepts in control engineering appears also to be central to the control of biological networks. The functional significance of feedback in a great variety of its manifestation in biological networks is not fully appreciated. A simple analysis of a feedback circuit can yield interesting insights in the potential role of feedback (Sauro and Kholodenko 2004). For instance, feedback will endow a network with the following properties:

1. Robustness to internal variation (for example thermal noise or genetic variability)
2. Robustness of operation under demand varying conditions
3. Linear response between input and output signals
4. Increased responsiveness to input signal variation

These properties are used extensively in the analog control engineering. In biological systems, the ability to resist the effects of thermal noise may well in many instances crucial to the normal functioning of the network. More importantly, the ability of a network to operate correctly under varying demand conditions permits networks to modularize their functionality so that the response of a downstream module will not affect the operation of the upstream module which supplies the signal. This later property may well be important in the case of the MAPK pathway where the final signal, ERKPP must migrate to the nucleus in order to actuate gene expression changes. Such migration places a demand on the MAPK circuit which if not protected, could badly disturb the operation of the MAPK pathway. It is known that in a number of cases, negative feedback operates in the MAPK pathway. The presence of the negative feedback in this pathway is not easily rationalized biologically; however, with knowledge of control engineering it becomes clear that such a control motif has very clear advantages.

Molecular biology has traditionally been heavily biased to descriptive science with little in the way of quantification and systems analysis. With the availability of new techniques capable of making detailed measurements and the influx of new researchers from other disciplines into biology, it is likely that the traditional approach will be augmented with a systems approach. Details of how engineering is making our understanding of complex networks more manageable can be found in Wolf and Arkin (2003) and Sauro and Kholodenko (2004).

# 5 Outlook: The future of systems biology

The future presents both the opportunity and the challenge for systems biology. The opportunity is based on an ever-increasing pace of the development of high-throughput tools and methods and an explosive growth of data. The challenge lies in an increasing gap between the amount of biochemical and molecular genetic information and our understanding of the implication of these data for the cell and organism functioning. Employing both a molecular-level description and a modular framework, systems biology combines quantitative measurement with mathematical and computational models of complex cellular networks to achieve quantitative understanding of the cell life at the systems level. A combination of molecular and cellular biology tools with rigorous quantitative approaches can help filling a gap between the discoveries of molecular mechanisms and integrative understanding of complex cellular and multicellular systems.

# References


Asthagiri AR, Lauffenburger DA (2001) A computational study of feedback effects on signal dynamics in a mitogen-activated protein kinase (mapk) pathway model. Biotechnol Prog 17:227-239

Bogdan S, Klambt C (2001) Epidermal growth factor receptor signaling. Curr Biol 11:R292-R295

Brightman FA, Fell DA (2000) Differential feedback regulation of the MAPK cascade underlies the quantitative differences in EGF and NGF signalling in PC12 cells [In Process Citation]. FEBS Lett 482:169-174

Brown GC, Hoek JB, Kholodenko BN (1997) Why do protein kinase cascades have more than one level? Trends Biochem Sci 22:288

Brown GC, Kholodenko BN (1999) Spatial gradients of cellular phospho-proteins. FEBS Lett 457:452-454

Bruggeman FJ, Westerhoff HV, Hoek JB, Kholodenko BN (2002) Modular response analysis of cellular regulatory networks. J Theor Biol 218:507-520

Carraway KL, Carraway CA (1995) Signaling, mitogenesis, and the cytoskeleton: where the action is. Bioessays 17:171-175

Cherniack AD, Klarlund JK, Conway BR, Czech MP (1995) Disassembly of son-of-sevenless proteins from Grb2 during p21ras desensitization by insulin. J Biol Chem 270:1485-1488 de la Fuente A, Brazhnik P, Mendes P (2002) Linking the genes: inferring quantitative gene networks from microarray data. Trends Genet 18:395-398

Della Rocca GJ, van Biesen T, Daaka Y, Luttrell DK, Luttrell LM, Lefkowitz RJ (1997) Ras-dependent mitogen-activated protein kinase activation by G protein-coupled receptors. Convergence of Gi- and Gq-mediated pathways on calcium/calmodulin, Pyk2, and Src kinase. J Biol Chem 272:19125-19132

Denhardt DT (1996) Signal-transducing protein phosphorylation cascades mediated by Ras/Rho proteins in the mammalian cell: the potential for multiplex signalling. Biochem J 318:729-747

Di Guglielmo GM, Baass PC, Ou WJ, Posner BI, Bergeron JJ (1994) Compartmentalization of SHC, GRB2 and mSOS, and hyperphosphorylation of Raf-1 by EGF but not insulin in liver parenchyma. EMBO J 13:4269-4277

Goldstein B, Faeder JR, Hlavacek WS (2004) Mathematical and computational models of immune-receptor signalling. Nat Rev Immunol 4:445-456

Gutkind JS (1998) The pathways connecting G protein-coupled receptors to the nucleus through divergent mitogen-activated protein kinase cascades. J Biol Chem 273:1839-1842

Hartwell LH, Hopfield JJ, Leibler S, Murray AW (1999) From molecular to modular cell biology. Nature 402:C47-C52

Haugh JM, Lauffenburger DA (1997) Physical modulation of intracellular signaling processes by locational regulation. Biophys J 72:2014-2031

Heinrich R, Schuster S (1996) The regulation of cellular systems, First edn. Chapman & Hall, New York

Hu Y, Bowtell DD (1996) Sos1 rapidly associates with Grb2 and is hypophosphorylated when complexed with the EGF receptor after EGF stimulation. Oncogene 12:1865-1872

Ingalls BP (2004) Frequency domain approach to sensitivity analysis of biochemical systems. J Phys Chem 108:1143-1152

Kacser H, Burns JA (1973) The control of flux. Symp Soc Exp Biol 27:65-104

Kholodenko BN (2002) MAP kinase cascade signaling and endocytic trafficking: A marriage of convenience? Trends Cell Biol 12:173-177

Kholodenko BN (2003) Four-dimensional organization of protein kinase signaling cascades: the roles of diffusion, endocytosis and molecular motors. J Exp Biol 206:2073-2082

Kholodenko BN, Brown GC, Hoek JB (2000a) Diffusion control of protein phosphorylation in signal transduction pathways. Biochem J 350 Pt 3:901-907

Kholodenko BN, Demin OV, Moehren G, Hoek JB (1999) Quantification of short term signaling by the epidermal growth factor receptor. J Biol Chem 274:30169-30181

Kholodenko BN, Hoek JB, Westerhoff HV (2000b) Why cytoplasmic signalling proteins should be recruited to cell membranes. Trends Cell Biol 10:173-178

Kholodenko BN, Hoek JB, Westerhoff HV, Brown GC (1997) Quantification of information transfer via cellular signal transduction pathways [published erratum appears in FEBS Lett 1997 Dec 8;419(1):150]. FEBS Lett 414:430-434

Kholodenko BN, Kiyatkin A, Bruggeman FJ, Sontag E, Westerhoff HV, Hoek JB (2002) Untangling the wires: a strategy to trace functional interactions in signaling and gene networks. Proc Natl Acad Sci USA 99:12841-12846

Kholodenko BN, Molenaar D, Schuster S, Heinrich R, Westerhoff HV (1995a) Defining control coefficients in "non-ideal" metabolic pathways. Biophys Chem 56:215-226

Kholodenko BN, Schuster S, Rohwer JM, Cascante M, Westerhoff HV (1995b) Composite control of cell function: Metabolic pathways behaving as single control units. FEBS Lett 368:1-4

Kholodenko BN, Sontag ED (2002) Determination of Functional Network Structure from Local Parameter Dependence Data. arXiv: physics/0205003

Kholodenko BN, Westerhoff HV (1995) The macroworld versus the microworld of biochemical regulation and control. Trends Biochem Sci 20:52-54

Kong M, Mounier C, Wu J, Posner BI (2000) Epidermal Growth Factor-induced Phosphatidylinositol 3-Kinase Activation and DNA Synthesis. Idenification of Grb2-associated binder 2 as the major mediator in rat hepatocytes. J Biol Chem 275:36035-36042

Krauss S, Brand MD, Buttgereit F (2001) Signaling takes a breath - new quantitative perspectives on bioenergetics and signal transduction. Immunity 15:497-502

Langlois WJ, Sasaoka T, Saltiel AR, Olefsky JM (1995) Negative feedback regulation and desensitization of insulin- and epidermal growth factor-stimulated p21ras activation. J Biol Chem 270:25320-25323

Lauffenburger DA (2000) Cell signaling pathways as control modules: complexity for simplicity? Proc Natl Acad Sci USA 97:5031-5033

Lloyd AC, Obermuller F, Staddon S, Barth CF, McMahon M, Land H (1997) Cooperating oncogenes converge to regulate cyclin/cdk complexes. Genes Dev 11:663-677

Markevich NI, Hoek JB, Kholodenko BN (2004a) Signaling switches and bistability arising from multisite phosphorylation in protein kinase cascades. J Cell Biol 164:353-359

Markevich NI, Moehren G, Demin O, Kiyatkin A, Hoek JB, Kholodenko BN (2004b) Signal processing at the Ras circuit: What shapes Ras activation patterns? IEE Systems Biology 1:104-113

Marshall CJ (1995) Specificity of receptor tyrosine kinase signaling: transient versus sustained extracellular signal-regulated kinase activation. Cell 80:179-185

McCawley LJ, Li S, Wattenberg EV, Hudson LG (1999) Sustained activation of the mitogen-activated protein kinase pathway. A mechanism underlying receptor tyrosine kinase specificity for matrix metalloproteinase-9 induction and cell migration. J Biol Chem 274:4347-4353

McLaughlin S, Aderem A (1995) The myristoyl-electrostatic switch: a modulator of reversible protein- membrane interactions. Trends Biochem Sci 20:272-276

Milo R, Shen-Orr S, Itzkovitz S, Kashtan N, Chklovskii D, Alon U (2002) Network motifs: simple building blocks of complex networks. Science 298:824-827

Mochly-Rosen D (1995) Localization of protein kinases by anchoring proteins: a theme in signal transduction. Science 268:247-251

Moehren G, Markevich N, Demin O, Kiyatkin A, Goryanin I, Hoek JB, Kholodenko BN (2002) Temperature dependence of the epidermal growth factor receptor signaling network can be accounted for by a kinetic model. Biochemistry 41:306-320

Niehrs C, Pollet N (1999) Synexpression groups in eukaryotes. Nature 402:483-487

Pawson T (1995) Protein modules and signalling networks. Nature 373:573-580

Pawson T, Nash P (2003) Assembly of cell regulatory systems through protein interaction domains. Science 300:445-452

Ravasz E, Somera AL, Mongru DA, Oltvai ZN, Barabasi AL (2002) Hierarchical organization of modularity in metabolic networks. Science 297:1551-1555

Saso K, Moehren G, Higashi K, Hoek JB (1997) Differential inhibition of epidermal growth factor signaling pathways in rat hepatocytes by long-term ethanol treatment. Gastroenterology 112:2073-2088

Sauro HM, Kholodenko BN (2004) Quantitative analysis of signaling networks. Prog Biophys Mol Biol 86:5-43

Savageau MA (1976) Biochemical systems analysis: a study of function and design in molecular biology. Addison-Wesley Publ Co, London

Schlessinger J (2000) Cell signaling by receptor tyrosine kinases. Cell 103:211-225

Schoeberl B, Eichler-Jonsson C, Gilles ED, Muller G (2002) Computational modeling of the dynamics of the MAP kinase cascade activated by surface and internalized EGF receptors. Nat Biotechnol 20:370-375

Schuster S, Kahn D, Westerhoff HV (1993) Modular analysis of the control of complex metabolic pathways. Biophys Chem 48:1-17

Sewing A, Wiseman B, Lloyd AC, Land H (1997) High-intensity Raf signal causes cell cycle arrest mediated by p21Cip1. Mol Cell Biol 17:5588-5597

Shymko RM, De Meyts P, Thomas R (1997) Logical analysis of timing-dependent receptor signalling specificity: Application to the insulin receptor metabolic and mitogenic signalling pathways. Biochem J 326:463-469

Sontag E, Kiyatkin A, Kholodenko BN (2004) Inferring dynamic architecture of cellular networks using time series of gene expression, protein and metabolite Data. Bioinformatics 20:1877-1886

Stark J, Callard R, Hubank M (2003) From the top down: towards a predictive biology of signalling networks. Trends Biotechnol 21:290-293

Suenaga A, Kiyatkin AB, Hatakeyama M, Futatsugi N, Okimoto N, Hirano Y, Narumi T, Kawai A, Susukita R, Koishi T, Furusawa H, Yasuoka K, Takada N, Ohno Y, Taiji M, Ebisuzaki T, Hoek JB, Konagaya A, Kholodenko BN (2004) Tyr-317 phosphorylation increases Shc structural rigidity and reduces coupling of domain motions remote from the phosphorylation site as revealed by molecular dynamics simulations. J Biol Chem 279:4657-4662

Tyson JJ, Chen KC, Novak B (2003) Sniffers, buzzers, toggles and blinkers: dynamics of regulatory and signaling pathways in the cell. Curr Opin Cell Biol 5:221-231

Tyson JJ, Novak B (2001) Regulation of the eukaryotic cell cycle: molecular antagonism, hysteresis, and irreversible transitions. J Theor Biol 210:249-263

Tyson JJ, Novak B, Odell GM, Chen K, Thron CD (1996) Chemical kinetic theory: understanding cell-cycle regulation. Trends Biochem Sci 21:89-96

van Biesen T, Luttrell LM, Hawes BE, Lefkowitz RJ (1996) Mitogenic signaling via G protein-coupled receptors. Endocr Rev 17:698-714

Wolf DM, Arkin AP (2003) Motifs, modules and games in bacteria. Curr Opin Microbiol 6:125-134

Woods D, Parry D, Cherwinski H, Bosch E, Lees E, McMahon M (1997) Raf-induced proliferation or cell cycle arrest is determined by the level of Raf activity with arrest mediated by p21Cip1. Mol Cell Biol 17:5598-5611

York RD, Yao H, Dillon T, Ellig CL, Eckert SP, McCleskey EW, Stork PJ (1998) Rap1 mediates sustained MAP kinase activation induced by nerve growth factor. Nature 392:622-626

Bruggeman, Frank J.
  Molecular Cell Physiology & Integrative Bioinformatics, Biocentrum
  & Faculty of Earth and Life Sciences, De Boelelaan 1085, 1081 HV, Vrije
  Universiteit, Amsterdam, The Netherlands

Kholodenko, Boris N.
  Department of Pathology, Anatomy and Cell Biology, Thomas Jefferson University, 1020 Locust St., Philadelphia, PA 19107, USA
  Boris.Kholodenko@jefferson.edu

Sauro, Herbert M.
  Keck Graduate Institute, 535 Watson Drive, Claremont, CA 91106, USA, and
  Control and Dynamical Systems, California Institute of Technology, Pasadena,
  CA 91125, USA

# METABOLISM

# Modeling the *E. coli* cell: The need for computing, cooperation, and consortia


Barry L. Wanner, Andrew Finney, Michael Hucka


## Abstract


*Escherichia coli* K-12 is an ideal test bed for pushing forward the limits of our ability to understand cellular systems through computational modeling. A complete understanding will require arrays of mathematical models, a wealth of data from measurements of various life processes, and readily accessible databases that can be interrogated for testing our understanding. Accomplishing this will require improved approaches for mathematical modeling, unprecedented standardization for experimentation and data collection, completeness of data sets, and improved methods of accessing and linking information. Solving the whole cell problem, even for a simple *E. coli* model cell, will require the concerted efforts of many scientists with different expertise. In this chapter, we review advances in (i) computing for modeling cells, (ii) creating a common language for representing computational models (the Systems Biology Markup Language), and (iii) developing the International *E. coli* Alliance, which has been created to tackle the whole cell problem.


## 1 Introduction

Biology has come a long way since Robert Hooke first used the term "cell" to describe the basic structural unit of cork in 1665. The tiny, room-like structures he saw under his microscope had solid walls, but they were empty because the cork was dead. Today, we can describe in exquisite detail many of the molecular parts and processes that furnish biological rooms. The complete genetic blueprints are available for common and deadly microbes, for economically important animals and plants, and even for human beings. And yet, we still lack a comprehensive understanding of any living cell.

One of the most striking successes of twentieth-century biology has been the identification and characterization of the molecules of life. This has been brought about through the development of disciplines such as biochemistry, biophysics, cell biology, molecular biology, molecular genetics, structural biology, and others. A major challenge of the twenty-first century is to describe the dynamic interactions of these molecules of life in the complex processes that are the essence of a living cell. Meeting this challenge requires that we enhance the highly successful, but limiting, reductionist approaches of the last several decades by revisiting


Topics in Current Genetics, Vol. 13
L. Alberghina, H.V. Westerhoff (Eds.): Systems Biology
DOI 10.1007/b138743 / Published online: 25 May 2005
© Springer-Verlag Berlin Heidelberg 2005

themes first articulated early in the twentieth century by systems-oriented thinkers such as Bogdanov and Bertalanffy (Capra 1996), and somewhat later by Wiener, Kacser, Mesarovic, and others (Wiener 1961; Kacser 1957; Mesarovic 1968). Fundamentally, systems biology strives to augment reductionist molecule-by-molecule accounts of cells by embedding the components within accounts of the broader context of cells and cell systems. This approach is simply an acknowledgment that dynamical interactions between components give rise to new functional properties at all levels from the genome, to molecular modules and networks, up through entire cell systems and beyond, and that these interactions are measurable and quantifiable.

These themes are the core of the definition of systems biology by Alberghia and Westerhoff (this volume) and others (Hood 1998; Ideker et al. 2001; Kitano 2001, 2002). The contemporary resurgence of interest in systems biology can be attributed to at least three major factors. First, there is the explosion of data brought about by modern molecular techniques and the commensurate realization by many researchers that future progress in understanding biological function rests inescapably in the development and application of computational methods (Alm and Arkin 2003; Arkin 2001; Fraser and Harland 2000; Hartwell et al. 1999; Noble 2002; Tyson et al. 2001; Zerhouni 2003). Second, there is the vastly greater power afforded by modern information technology (Butler 1999), beckoning us to reattempt solutions to problems that were beyond reach in the mid-twentieth century. And third, only in recent decades has the mathematical theory of nonlinear systems and stochastic systems advanced sufficiently to allow us to handle the classes of systems that emerge naturally when describing complex biological processes in a detailed, mathematical fashion (Burns 1971; Gillespie 1977; Gillespie and Petzold 2003; Kacser and Burns 1967; Savageau 1969, 1970).

The contrast between our tremendously increased computational and biological powers and technologies on the one hand, and our continuing lack of understanding of any whole cell on the other, has been the major inspiration for the formation of the IECA—the International *E. coli* Alliance (Holden 2002). The mission of this alliance has been to coordinate global efforts to understand a living bacterial cell, the K-12 strain of *Escherichia coli* that has laid so many of the golden eggs of basic biochemistry, genetics, and molecular biology in the last half of the twentieth century (Kornberg 2003). Scientists around the world are working together to create a computer model of *E. coli*, integrating all of the dynamic molecular interactions required for the life of a simple, self-replicating cell. A whole cell model of *E. coli* would not only significantly advance the field of biology; it would also have immediate practical benefits as well, for everything from drug discovery to bioengineering.

The IECA effort is emblematic of systems biology as a whole. The ambitious goal of the IECA is beyond the means of any single investigator or laboratory (Crick 1973). It requires an integrative research program and collaborations between scientists with expertise in biology, chemistry, computer sciences, engineering, mathematics, and physics. Success will depend crucially on bringing to bear both social and technological tools: namely, consortia that help forge collaborations and common understanding, computational tools that permit analysis of vast and complex data, and agreed-upon standards and tools that enable researchers to communicate, integrate, and use their results in practical and unambiguous ways.

In this chapter, we discuss these topics in the context of the IECA effort. We begin in Section 2 by describing the kind of models ultimately sought by Systems Biologists and provide an overview of computational modeling. In Section 3, we survey some of the software tools available today to help with computing in systems biology and follow this in Section 4 with a discussion of the Systems Biology Markup Language (SBML) and its role as an enabling technology for modelers to share their models. Section 5 briefly describes the kinds of experimental standards envisioned by IECA that will be required for successful whole cell modeling. One way of carrying out such standardized experiments as a community is also given in Section 5. It is unrealistic and probably unwise to develop a single database encompassing all information, even for a single cell. Section 6 describes an alternative approach for creation of an accessible and interoperable database that would not only store massive amounts of data in different formats but would also have the capability of interrogating other meaningful databases. Consortia such as the International *E. coli* Alliance have been created as one way to meet this challenge (see Section 7). We close by bringing the discussion back to the IECA effort itself. Several contributors to this book are also participating in the IECA.

## 2 Quantitative, formal models are essential instruments in systems biology

Models, as abstractions representing observed or hypothesized phenomena, are nothing new to the life sciences, having long been used by life scientists as tools for organizing and communicating conceptual and factual information. However, the majority of models in biology traditionally have been expressed in natural language narratives, sometimes augmented with block-and-arrow diagrams (Bower and Bolouri 2001a). These certainly can be useful and important for describing hypotheses about a system's components and their interactions, but these types of models also have crucial limitations that make them inadequate as vehicles for describing and understanding large and complex systems (Bialek and Botstein 2004). A block-and-arrow diagram combined with verbal explanations and statements about observed effects, quantities of substances involved, and so forth, may appear detailed and precise, but in practice it leaves too much room for ambiguity, misinterpretation, and hidden complexity. More importantly, as Phair and Misteli wrote:

"... it often remains difficult to make quantitative predictions for a given experimental protocol with the use of diagrams alone. Scientific intuition has been successful when systems are limited to a few molecules and processes, but today's summary diagrams generally have many more molecules and arrows than this, and even simple systems often behave in surprising ways. What is needed is a way to know what the diagram predicts in a given experiment" (Phair and Misteli 2001).

This is not to say that narrative descriptions and diagrams should be abandoned; rather, they should be used as stepping-stones, not stopping points. Scientists must go further and express their models in such a way that each molecular entity is knowable and quantifiable in terms of empirical evidence and each process is expressed step-by-step in a formal, mathematical language. It is by systematizing how entities and processes are defined, represented, manipulated and interpreted, that *formal, quantitative models* can enable "meaningful comparison between the consequences of basic assumptions and the empirical facts" (May 2004).

## 2.1 Computational modeling is an extension of the scientific method

Computational models are simply formal models expressed in a form that can be manipulated by a computer. The resulting descriptions are more likely to be coherent and internally consistent, because computable representations must be precise and detailed—vague or incomplete elements will not do, else it will not be possible to simulate or analyze the model. While frustrating at times, this is exactly the reason why computational models are an invaluable tool in helping us understand phenomena. Only if one can express every step of a process in such detail that it can be simulated in a computer program can one justifiably claim to understand it very well. This is the fundamental premise for doing computational modeling in biology, and other fields.

Computational models also allow quantitative calculations to be done on a model, allowing researchers not only to test their understanding, but also to explore "what-if" scenarios and make testable predictions about the behavior of the system being studied. This is an essential requirement for being able to understand complicated systems that are replete with feedback mechanisms (the hallmark of biological systems), where the resulting behaviors are rarely predictable through intuitive reasoning alone. Even for the simplest components and systems, it can be impossible to predict such characteristics as sensitivity to exact parameter values without constructing and analyzing a model. Such analyses have shown that some systems are insensitive (e.g. Yi et al. 2000) whereas others are exquisitely sensitive (e.g. McAdams and Arkin 1999). Computational modeling is thus an extension of the scientific method (Phair and Misteli 2001; Fall et al. 2002; Slepchenko et al. 2002), providing the means to create precise, unambiguous, quantitative descriptions of biological phenomena that can be used to evaluate hypotheses systematically and to explore non-obvious dynamical behavior of a biological system (Hartwell et al. 1999; Csete and Doyle 2002; Endy and Brent 2001). For all of these reasons, the emphasis on developing and using models for quantitative predictions is one of the foundations of systems biology.

Life scientists sometimes object to the idea of modeling by arguing, "If you understand something so well that you can simulate it, why bother? You already understand it!" But, this argument misses the point of modeling. One does not begin creating models with understanding; indeed, the opposite is often the case—one begins with ignorance. It is the exercise of developing a model(s) that leads to understanding. Developing a computer model requires a greater degree of intellectual honesty than writing down an informal verbal model or drawing a block-and-arrow diagram. It is all too easy to imagine that one understands something, but it is quite another to make a computer model work.

Many sometimes feel that they cannot create a model because they do not have enough data. Here again we reiterate a basic premise of modeling that developing the model can be an extremely useful exercise for discovering what data are missing. "[The] complexity of biological systems makes it increasingly difficult to identify the next best experiment without such a tool" (Bower and Bolouri 2001b).

Finally, biologists often express the concern that computational modeling is a lot of work and that it requires an entirely different training than, e.g., "wet-bench" experimentation. Unfortunately, this is to a large extent still true. However, modern software tools can provide considerable assistance in developing, verifying, analyzing and sharing computational models (see Section 3).

## 2.2 Mechanistic models can serve as frameworks for organizing data and hypotheses

A spectrum of types of formal models exists (Gershenfeld 1998; Phair and Misteli 2001). On one end of the spectrum lie observational models: ones that characterize and quantify patterns in data, but in such a way that the elements and processes in the model are not directly related to the components of the underlying system. Curve-fitting models fall into this category. On the other end of the spectrum lie *mechanistic* models: ones in which the entities and processes correspond directly to hypothesized structures and processes in the biological system being modeled. While mechanistic models are much more difficult to develop, they are also more valuable for making predictions that can be related to empirical data.

Detailed, mechanistic models are designed to capture essential structural, biochemical, and genetic aspects of a biological system, staying faithful to chemical and physical laws. Many scientists have been developing such models for decades, and have long recognized the utility of computers for assisting with model simulation and analysis—in fact, the first simulations of biochemical reaction models were made before the advent of digital computers (Chance et al. 1940, 1952; Chance 1943, 1960; Garfinkel 1965). However, the power afforded by modern computers has made possible new levels of model detail and analysis.

What is sometimes lost in the excitement over the power of simulation and analysis is the value of computational models to serve as focal points for research in ways that databases of experimental data cannot. Mechanistic, computational models are specifically constructed to illuminate the functional implications of the data upon which they are built. A realistic computational model represents a modeler's understanding of the structure and function of part of a biological system. Models thus can serve not only as the point of entry for data; they can also serve as dynamic tools that can be used to understand its significance (Bailey 1998). As the number of researchers constructing realistic models continues to grow, and as the models become ever more sophisticated, they collectively represent a significant accumulation of knowledge about the structural and functional organization of the system. Moreover, the assimilation of new hypotheses and data into existing models can be done in a more systematic fashion because the additions must be fitted into the existing constructs using the same rules as for the models themselves. Computational models can thus be far more useful than just encapsulating one modeler's abstraction of a particular system: once properly constructed, the models become *a dynamic representation of our current state of understanding of a system* in a form that can facilitate communication between research groups and help to direct further experimental investigations.

## 3 A variety of software resources are available today for computational modeling

One of the great advantages of modern software packages for biological modeling is that they allow users to avoid having to work with formal mathematics directly. Although one can certainly use a general-purpose mathematical package for developing and working with computational models, specialized tools can offer dedicated user interfaces and functionality for model development, simulation, and analysis, as well as other capabilities designed to simplify the work of biological modeling. In this section, we survey some of the capabilities provided by different tools as a way of informing prospective modelers of the choices available.

The user-interface paradigm used by a software system is one major dimension along which biological modeling tools can differ. The four most popular types of interfaces offered by modeling tools today are the following:

*Diagrammatic*: the tool enables users to express models visually by placing or drawing elements, structures, and relationships on a digital canvas. Often this takes the form of a graph resembling the block-and-arrow diagrams commonly presented by biologists as depictions of metabolic or signaling pathways. Additional quantitative information about the model is usually obtained from the user using a small number of fill-in-the-blank forms. Examples of tools implementing this kind of interface include JDesigner (Sauro et al. 2003; Sauro 2001),

**Fig. 1 (overleaf).** (Top) Screen image of JDesigner (Sauro 2001), a program that provides a graph-based interface allowing users to "draw" models. Nodes represent chemical species and arcs represent chemical reactions. Assignments of chemical rate laws to the arcs and chemical values for concentrations and other parameters are made using pop-up dialogue boxes. (Bottom) Screen image of the JigCell Model Builder (Vass et al. 2004), a program that provides a spreadsheet-style interface. Users input chemical equations, but different parts go into separate columns; moreover, the program performs consistency checking on the user's input, helping to eliminate some common errors.

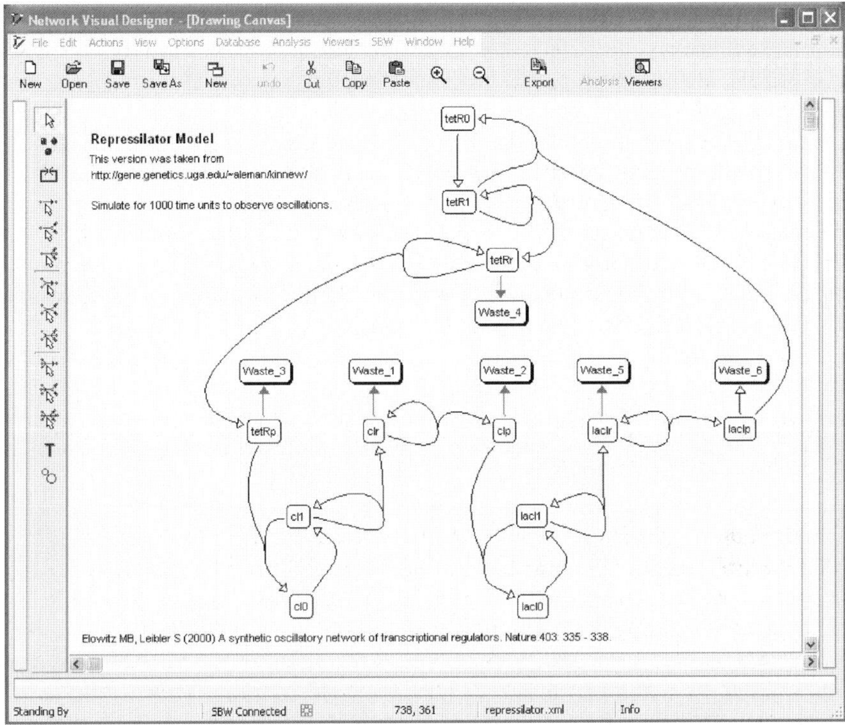


/home/sshealy/JigCell/models/frogegg.sbml

File  Options  Units  Help

| # | Reaction | Name | Type | Equation | Parameters |
|---|---|---|---|---|---|
| 1 |  | Michaelis-Menten | New | $k1*S1*M1/(J1+S1)$ |  |
| 2 |  | Michaelis_Menten | New | $k1*S1*M1/(J1+S1)$ |  |
| 3 |  | Mass_Action_1 | New | $kf*S1$ |  |
| 4 |  | Mass_Action_0 | New | $kf$ |  |
| 5 | Ma -> Mi | MPF inactivation | Mass_Action_1 | $kw*Ma$ | kf=kw |
| 6 | Mi -> Ma | MPF activation | Mass_Action_1 | $kc*Mi$ | kf=kc |
| 7 | Ca -> Ci | Cdc25 inactivation | Michaelis_Menten | $vcppp_*vc_*Ca*1.0/(kmcr_+Ca)$ | M1=1.0; J1=kmcr_; k1=vcppp_*vc_ |
| 8 | Ci -> Ca | Cdc25 activation | Michaelis_Menten | $vc_*Ci*Ma/(kmc_+Ci)$ | M1=Ma; J1=kmc_; k1=vc_ |
| 9 | Wa -> Wi | Wee1 inactivation | Michaelis_Menten | $ww_*Wa*Ma/(kmw_+Wa)$ | M1=Ma; J1=kmw_; k1=ww_ |
| 10 | Wi -> Wa | Wee1 activation | Michaelis_Menten | $ww_*wwppp_*Wi*1.0/(kmwr_+Wi)$ | M1=1.0; J1=kmwr_; k1=ww_*wwppp_ |
| 11 | L -> | Labelled inactive MPF affec... | Mass_Action_1 | $kc*L$ | kf=kc |
| 12 |  -> L2 | Labelled inactive MPF affec... | Local | $kw*(1.0-L2)$ |  |
| 13 | Cdc25Total_ |  | Species | Cdc25Total | Cdc25Total=Cdc25Total*Dilution |
| 14 | Wee1Total_ |  | Species | Wee1Total | Wee1Total=Wee1Total*Dilution |
| 15 | Cdc2Total_ |  | Species | Cdc2Total | Cdc2Total=Dilution |
| 16 | vcp_ |  | Species | vcp*Cdc25Total | Cdc25Total=Cdc25Total_; vcp=vcp |
| 17 | vcpp_ |  | Species | vcpp*Cdc25Total | Cdc25Total=Cdc25Total_; vcpp=vcpp |
| 18 | vcppp_ |  | Species | vcppp/Cdc25Total | Cdc25Total=Cdc25Total_; vcppp=vcppp |
| 19 | wwp_ |  | Species | wwp*Wee1Total | wwp=wwp; Wee1Total=Wee1Total_ |
| 20 | wwpp_ |  | Species | wwpp*Wee1Total | wwpp=wwpp; Wee1Total=Wee1Total_ |
| 21 | wwppp_ |  | Species | wwppp/Wee1Total | wwppp=wwppp; Wee1Total=Wee1Total_ |
| 22 | kmc_ |  | Species | kmc/Cdc25Total | Cdc25Total=Cdc25Total_; kmc=kmc |
| 23 | kmcr_ |  | Species | kmcr/Cdc25Total | Cdc25Total=Cdc25Total_; kmcr=kmcr |
| 24 | kmw_ |  | Species | kmw/Wee1Total | kmw=kmw; Wee1Total=Wee1Total_ |
| 25 | kmwr_ |  | Species | kmwr/Wee1Total | kmwr=kmwr; Wee1Total=Wee1Total_ |
| 26 | vc_ |  | Species | vc*Cdc2Total/Cdc25Total | vc=vc; Cdc2Total=Cdc2Total_; Cdc25Total=Cdc2... |
| 27 | ww_ |  | Species | ww*Cdc2Total/Wee1Total | Cdc2Total=Cdc2Total_; ww=ww; Wee1Total=Wee... |
| 28 | kc |  | Species | vcp*Ci+vcpp*Ca | vcp=vcp_; vcpp=vcpp_ |
| 29 | kw |  | Species | wwp*Wi+wwpp*Wa | wwp=wwp_; wwpp=wwpp_ |
| 30 |  |  |  |  |  |
| 31 |  |  |  |  |  |
| 32 |  |  |  |  |  |
| 33 |  |  |  |  |  |
| 34 |  |  |  |  |  |

CellDesigner (Funahashi et al. 2003; Kirkwood et al. 2003b), TERANODE Design Suite (Duncan et al. 2004; Teranode Inc. 2004), and the Virtual Cell (Schaff et al. 1997, 2001). The top half of Figure 1 shows an example screenshot from JDesigner.

*Spreadsheet*: the tool provides a multicolumn grid interface reminiscent of spreadsheet programs commonly offered in contemporary office productivity software suites. Information about reactions, species, and compartments typically are entered in separate spreadsheet areas, each having separate columns for different characteristics of the elements being entered. An example of a package providing this kind of interface is the JigCell Model Builder (Allen et al. 2003; Vass et al. 2004), a screenshot of which is shown in the bottom portion of Figure 1.

*Forms-based*: the tool prompts the user for information about a model using fill-in-the-blank forms or dialog boxes. An example of a tool implementing this kind of interface is COPASI (Mendes 2003) and its predecessor, Gepasi (Mendes 1993, 2001). Note that some tools take the information so gathered and display the resulting model using a diagram or a spreadsheet view but do not allow the user to edit the model directly using the diagram or spreadsheet, blurring the distinction somewhat.

*Text-based*: the tool enables users to define models using a formalized textual language and notation meant to be read and written by a human. Some of these languages mix constructs for defining models with directives for controlling simulations or other actions on the model. Some of the software packages provide a notation based on traditional chemical reaction style notation (e.g. $A + B \leftrightarrow C$), while others explore different notations. Examples of tools using this general user-interface paradigm include Cellerator (Shapiro et al. 2004b, 2003), Dizzy (Ramsey and Bolouri 2004), Jarnac (Sauro 2000b, 2000a), MathSBML (Shapiro 2004; Shapiro et al. 2004a), and WinSCAMP (Sauro and Fell 1991; Sauro et al. 2003).

Some packages provide more than one of these interface paradigms simultaneously, allowing users to switch between interface styles. An example in this category is TERANODE Design Suite.

Of course, the primary purpose of modeling tools is to allow users to perform analysis on the models created by users. Some software tools are dedicated model editors lacking built-in simulation and analysis capabilities; in these, users are expected to transfer the model to a separate analysis package. For this purpose, the Systems Biology Markup Language (SBML; see Section 4) is a popular model export format. Other tools provide built-in analysis capabilities.

In the context of simulation and analysis, software tools differ in the type of model representation *framework* they employ. The following are among the most popular types of frameworks in use today:

*Logical*: the tool converts the model description into a Boolean or extended logical representation (de Jong 2002). Certain classes of models, such as abstract models of regulatory networks, are more conveniently cast into this form than into, for example, differential-algebraic equations. An example of a tool in this category is NetBuilder (Brown et al. 2002; Schilstra and Bolouri 2002).

*Ordinary differential equations* (ODE): the tool converts the model description into a system of ordinary differential equations. This commonly involves one differential equation for each chemical species in the model. The ODE framework is the most popular one in use today for biochemical systems simulation. Representative examples include COPASI (Mendes 2003), Gepasi (Mendes 1993, 2001) and Jarnac (Sauro 2000b, 2000a).

*Differential-algebraic equations* (DAE): the tool converts the model into a system of ordinary differential equations with algebraic constraints. ODE representations are a popular framework, but complex models often include algebraic constraints and require the use of DAE representations. An example is a model that imposes constraints on species concentrations. The DAE framework subsumes the ODE framework. Because the DAE framework supports more of the constructs that modelers often want to express, it is a better match for modelers' needs. However, DAE solvers are more difficult to implement than ODE solvers, and fewer software packages provide full DAE support. An example of a tool that provides limited DAE support is Jarnac (Sauro 2000b, 2000a); an example of a tool providing a full DAE solver is MathSBML (Shapiro 2004; Shapiro et al. 2004a).

*Partial differential equations* (PDE): the tool converts the model into a system of partial differential equations. These arise when there is more than one independent variable in the system. For example, modeling spatial diffusion requires both time and space as independent variables. PDE solvers are much more difficult than ODE or DAE solvers to implement and use properly, which is why so few software tools use a PDE framework. One that does is the Virtual Cell (Schaff et al. 1997, 2001). (We note in passing that SBML does not currently have support to represent PDE-level models or chemical diffusion.)

*Hybrid*: the tool converts the model to a (continuous) differential equation framework that also supports time-dependent discontinuous events. Discontinuities can cause abrupt changes in the system of equations and the behavior of the system, and require specialised support in the model interpretation system. Hybrid modeling frameworks are necessary for properly handling such things as cell cycle models. Some packages providing hybrid simulators include E-Cell (Tomita et al. 1999; Tomita 2001), MathSBML (Shapiro 2004; Shapiro et al. 2004a), and TERANODE Design Suite (Duncan et al. 2004; Teranode Inc. 2004).

*Stochastic*: the tool casts the model as a set of discrete quantities (molecules or chemical species) and associated probabilities for interactions (reactions). Most such software uses the stochastic simulation algorithm by Gillespie (1977) or the Gibson-Bruck variant of Gillespie's algorithm (Gibson and Bruck 2000). Unlike differential-equation frameworks, stochastic frameworks do not approximate the model as a continuous, deterministic system. Instead, a stochastic framework treats the underlying biochemical reactions as random discrete processes in accordance with the chemical and physical properties of the component parts. In essence, stochastic frameworks more accurately represent true molecular interactions. However, the greater accuracy of stochastic frameworks comes at a high cost. Because the behavior of each chemical entity is individually modeled as a stochastic process, simulations are extremely demanding of computational resources. Some examples of systems implementing stochastic simulation capability include BASIS (Kirkwood et al. 2003a, 2003b), Dizzy (Ramsey and Bolouri

2004), E-Cell (Tomita et al. 1999; Tomita 2001), SigTran (DiValentin 2004) and StochSim (Morton-Firth and Bray 1998; Le Novere and Shimizu 2001).

Most of the packages discussed above are standalone applications (i.e. they can be installed and run locally on a computer), while a few are web-based, offering a service located on the Internet which users access remotely using a web browser. BASIS (Kirkwood et al. 2003a, 2003b) and the Virtual Cell (Schaff et al. 1997, 2001) are examples in the latter category.

A few software systems also provide database functionality. Some have an integrated database used to store models and model components in a form more organized than simply a collection of files. These systems sometimes also offer a means to share the database among different users. Examples in this category include Monod, TERANODE Design Suite (Duncan et al. 2004; Teranode Inc. 2004), and the Virtual Cell (Schaff et al. 1997, 2001). A few other systems provide a means to access third-party external repositories data, models or other information. An example in this category is E-Cell (Tomita et al. 1999; Tomita 2001).

Finally, most of the tools mentioned in this section are free for personal and/or educational use, although there may be costs for other users. Other packages, such as TERANODE Design Suite (Duncan et al. 2004; Teranode Inc. 2004), are commercial products.

## 4 Exchanging models between software tools: The Systems Biology Markup Language

To be useful as formal embodiments for understanding biological systems, computational models must be put into a format that can be communicated effectively between different software tools that work with them. The Systems Biology Markup Language (SBML) project is an effort to create a machine-readable format for representing computational models at the biochemical reaction level (Finney and Hucka 2003; Hucka et al. 2003). By supporting SBML as input and output formats, different software tools can operate on the identical representation of a model, removing chance for errors in translation and assuring a common starting point for analyses and simulations.

The SBML project is not an attempt to define a standard universal language for representing quantitative models; the fluid and rapidly evolving views of biological function, and the vigorous rate at which new computational techniques and individual tools are being developed today are incompatible with a one-size-fits-all concept of a universal language. Instead of trying to define how software tools should represent their models *internally*, the goal of the SBML project is to reach agreement on a format on how the tools communicate models *externally*. The SBML language allows software developers the freedom to explore different representations within their tools while still allowing some degree of interoperability between the tools. Such a format can serve as a *lingua franca* enabling communication of the most essential aspects of models between software systems in much the same way as "contact languages" first enabled human societies to communicate in the Mediterranean during the Middle Ages.

## 4.1 The general form of SBML

Although SBML models are intended to be read and written by software tools and not by humans, it is useful to overview the general characteristics of the representation in order to better understand how it organizes information about biological systems.

SBML is a machine-readable model definition language based upon XML, the eXtensible Markup Language (Bray et al. 2000; Bosak and Bray 1999), which is a simple and portable text-based substrate that has gained widespread acceptance in computational biology (Augen 2001; Achard et al. 2001). SBML can encode models consisting of biochemical entities (species) linked by reactions to form biochemical networks. An important principle in SBML is that models are decomposed into explicitly labeled constituent elements, the set of which resembles a verbose rendition of chemical reaction equations; the representation deliberately does not cast the model directly into a set of differential equations or other specific interpretations of the model. This decomposition makes it easier for a software tool to interpret the model and translate the SBML format into whatever internal form the tool actually uses.

SBML is being developed in levels, where each higher level adds richness to the model definitions that can be represented by the language. Level 2 is currently the highest level defined; it represents an incremental evolution of the language (Finney et al. 2003) resulting from the practical experiences of many users and developers, who have been working with Level 1 (Hucka et al. 2001, 2003). In SBML Level 2, the definition of a model consists of lists of one or more of the following components:

*Compartment*, a container of finite volume for homogeneously-mixed substances where reactions take place;

*Species*, a pool of a chemical substance located in a specific compartment, where this represents the concentration or amount of a substance and not a single molecule (example substances that form species are ions such as calcium and molecules such as ATP or DNA);

*Reaction*, a statement describing some transformation, transport or binding process that can change one or more species (each reaction is characterized by the stoichiometry of its products and reactants and optionally by a rate equation);

*Parameter*, a quantity that has a symbolic name, such as a frequently-used constant;

*Unit definition*, a name for a unit used in the expression of quantities in a model;

*Rule*, a mathematical expression that is added to the model equations constructed from the set of reactions (rules can be used to set parameter values, establish constraints between quantities, etc.);

*Function*, a named mathematical function that can be used in place of repeated expressions in rate equations and other formulas; and

*Event*, a mathematical formula evaluated at a specified moment in the time evolution of the system.

This simple formalism allows modeling of a wide range of biological phenomena, including cell signaling, metabolism, gene regulation, and others. Flexibility and power come from the ability to define arbitrary formulae for the rates of change of variables as well as the ability to express other constraints mathematically.

Many kinds of analyses can be applied to models in the elementary SBML format. The tools discussed in Section 3 are representative of the range of applications for which SBML is suitable.


## 4.2 The continued evolution of SBML

From its inception, SBML has been largely driven by practical needs of researchers interested in exchanging quantitative computational models between different software tools, databases, and other resources. The language reflects this, and in some respects exhibits the results of pragmatic choices more than elegant, top-down design. The development of SBML Level 2 benefited from two years of experience with SBML Level 1 by many modellers and software developers, and distils more effectively the fundamental needs of the biological network simulation community. It represents, in a concrete way, the consensus of a large segment of the modelling community about the intersection of features that should be possessed by a *lingua franca* for communicating models between today's software tools.

SBML's popularity has led to the formation of an active community of researchers and software developers who are now working together to push SBML in new directions. As a language that is an intersection rather than a union of features needed by all tools, SBML currently cannot support all the representational capabilities that all software systems offer to users. Some packages offer features that have no explicit equivalent in SBML Level 2, and those tools currently can only store those features as annotations in an SBML model. Yet, in many cases, those features could potentially be used by more than one tool, and thus it would be appropriate to have some agreed-upon representation for them in SBML. Using Level 2 as a starting point, the SBML community has been developing proposals and prototype implementations of many new capabilities that will become part of SBML Level 3.

Because of the demand-driven, consensus-oriented approach to SBML evolution, the features currently in SBML and in development for SBML Level 3 are a reflection of the state of computational modeling today. The list of planned features thus serves to foreshadow what is to come in terms of modeling capabilities in the near future:

*Composition*: The biochemical network models being constructed by modelers are becoming increasingly large and complex. Structuring the models in a modular fashion is an essential approach to managing their complexity. Composition, as its name suggests, involves composing a model out of a set of instances of submodels. The resulting model structure is hierarchical; for example, a model of a cell might be composed from a model of a nucleus, multiple model mitochondria, and various other model structures. The E-CELL (Tomita et al. 1999, 2001) and Pro-MoT/DIVA (Stelling et al. 2001) systems are examples of simulation tools that support composing models out of submodels. The addition of a modular composition facility into SBML will bring several benefits. First, it will allow a component submodel to be reused multiple times within a single (larger) model. Second, it will allow the creation of libraries of model components. In time, the systems biology field will be able to develop standard, vetted submodels for commonly-needed components, and eventually, modelers will be able to compose models using high-level components taken from libraries rather than have to re-create every piece from scratch themselves. And third, it will enable modelers to incorporate several alternative submodels for a given model instance, in which each alternative could contain a representation at a different level of detail and/or use a representation that is appropriate for a particular type of simulation algorithm.

*Multi-component species*: SBML Levels 1 and 2 can represent models in which the chemical species are treated as simple, indivisible biochemical entities having only one possible state. However, this approach becomes untenable when modeling systems in which the species have many possible internal states or the species are composed from subcomponents (Goldstein et al. 2002). An example of this situation involves a protein that can be phosphorylated at multiple locations: the possible phosphorylation combinations lead to a combinatorial explosion of states of the protein. Although currently this can be represented in SBML Levels 1 and 2 by treating each state or combination of subcomponents as a separately named chemical species, this approach is an awkward and limited solution. To address this problem, another current area of SBML development is a representation scheme in which the subcomponents of chemical species are the smallest logical entities, rather than whole species being the entities. The research task is to define a representation scheme that is flexible enough to represent all the relevant biochemical phenomena while remaining computationally feasible for simulation and analysis.

*Diagram Layout*: Biochemical models are often visualized and edited using software in diagrammatic form. Examples of software that enables this include: JDesigner (Sauro 2003, 2001) and CellDesigner (Funahashi et al. 2003, 2004). The diagram layout that the user creates with these programs is especially useful for interpreting models created with this software. Another active area of SBML development is extending SBML so that diagram information can be added to models in a standard form.

*Spatial geometry*: The spatial distribution and diffusion of chemical species in space can be highly significant (Fink et al. 2000) and often needs to be represented in models. Not all software tools today support the use of spatial information, but it is likely that more will in the near future.

*Alternative Mathematical Representations for Reactions*: The current definition of SBML is somewhat biased towards on ODE-based representation of biochemical models. While it is possible to transform a subset of models encoded in this representation into a form acceptable to stochastic simulators, this, unfortunately, does not allow expression of the complete range of facilities that are available in stochastic simulators. Similarly, while it is possible to describe deterministic discrete events explicitly in SBML Level 2, it is not possible to define a reaction that operates in this way. Addressing these and other issues are included in development for SBML Level 3.

## 5 Development of an *E. coli* systems biology project

A wealth of information has been gained from reductionist biology over the past fifty years. Reductionism has been especially rewarding when directed towards understanding highly amenable systems. Studies of *E. coli* and its phages have given birth to early concepts of the fine structure of the gene, co-linearity of gene structure and protein sequence, molecular mechanisms of suppression, gene regulation, transposition, and many other phenomena. *E. coli* is now the source for much of our information on biochemistry, molecular biology, metabolic pathways, and regulation, and it continues to be a source for new insights into how cells work. *E. coli* has served as a model for understanding innumerable fundamental processes like the mechanisms of DNA replication (Kornberg and Baker 1992) and DNA repair (Chen et al. 2001), DNA transcription, gene repression and activation, protein synthesis, protein folding, protein targeting, macromolecular assembly, signal transduction, the catalytic nature of disulfide bond formation, cell division, the function of catalytic and small regulatory RNAs, and other processes.

The decision to focus early studies of cell physiology on *E. coli* has often been credited to a well-known phrase by Jacques Monod dating from 1954 "Anything found to be true of *E. coli* must also be true of elephants." Early successes from *E. coli* research have also led many, most notably Sydney Brenner, to develop *E. coli*-like models for other processes (behavior, development, the immune response, multigene families, the nervous system, and many more). Many model organisms now exist for eukaryotic molecular biology (like yeast *Saccharomyces cerevisiae* and *S. pombe* and *Dictyostelium discoidium*), development and human disease (e.g. *Drosophila melanogaster*, *Caenorhabditis elegans*, *Fugu rubripes* (pufferfish), *Brachydanio rerio* (zebrafish), and human biology (*Mus musculus* (the laboratory mouse) and primates. *E. coli*-like models also exist for important processes in other bacteria (e.g. sporulation in *Bacillus subtilis*, cell division in *Caulobacter crescentus*, and development in *Myxococcus xanthus*) and for the Archae (e.g. *Haloferax volcani* and *Sulfolobus solfataricus*). Huge successes in pathogenic bacteriology have been more rapid for those bacteria most closely related to *E. coli*. The decision to create a *Shewanella* consortium for studying environmental bioremediation was based largely on its similarity with *E. coli*. Yet, no similar project exists for *E. coli* systems biology.

To meet this challenge, a small group of mostly *E. coli* biologists and modelers convened an informal workshop at the Intelligent Systems of Molecular Biology

Conference in Edmonton, Canada, in August 2002. Their meeting gave birth to the International *E. coli* Alliance (IECA), which was announced a few weeks later (Holden 2002). IECA was organized to help with the development of highly integrated and interdisciplinary research in bioinformatics, experimental, and modeling sciences that will be required to gain deeper understanding of cellular subsystems (gene regulatory, metabolic, and signaling networks), work that will contribute towards the development of a rudimentary whole cell model.

Subsequent meetings included discussions on how to organize a worldwide *E. coli* systems biology project. These were held in November 2002 at North Mymms, nearby London, UK, in February 2003 in San Diego, USA, and in March 2003 in Magdeburg, Germany. There was consensus that much work was needed. A standard strain would have to be selected, preferably based on data from rigorously controlled experiments. What kind? How many? Who would do them? New technologies would have to be developed. Metadata generated would be enormous. These data would need to be stored, disseminated, and modeled. We would need to reach agreements on data sharing and many other issues. Modelers were in a quandary about data formats and modeling languages, because modeling uses different kinds of data depending upon the approach, as described above. Committees were formed on strain and experimental standards, metabolic measurement and nomenclature, and modeling.

If our objective were modeling of *E. coli*, then experimentalists and modelers would need to work together from the start. This would require cooperation and collaborations among scientists with diverse interests and expertise. Experimentalists and modelers would need to be equally represented. To further promote *E. coli* systems biology research, the First IECA Conference was held in June 2003 at the Institute for Advanced Biosciences, Tsuruoka, Japan. The Second IECA Conference was held in June 2004 in Banff, Canada. More than one hundred international scientists have attended. Plans are now underway to hold the Third IECA Conference in September 2006 in Korea.

A major modeling problem is biological variability, even for experiments with the "same" strain by various investigators in the same or different laboratories. One way to overcome this hurdle would be to grow cells for modeling at a central location and to provide samples from standardized cultures to other researchers for an assortment of measurements. Predictive quantitative modeling is also often beyond the comprehension and belief of many biologists. Indeed, it is difficult to find examples in which modeling has given predictive outcomes where the results had not been known beforehand or could not have been inferred solely on the basis of prior experimental knowledge. It will be necessary to coordinate new experimentation with mathematical modeling as a means to validate or refute the predictive value of different modeling approaches for understanding new features of *E. coli* biology.

Foremost, a standard strain must be chosen that conforms as close to wild type as possible. This will probably be the *E. coli* K-12 sequenced strain MG1655 (Blattner et al. 1997). The finding of discrepancies for the "same" standard strain, e.g. Corbin et al. (2003), in different labs gives impetus to the concept for

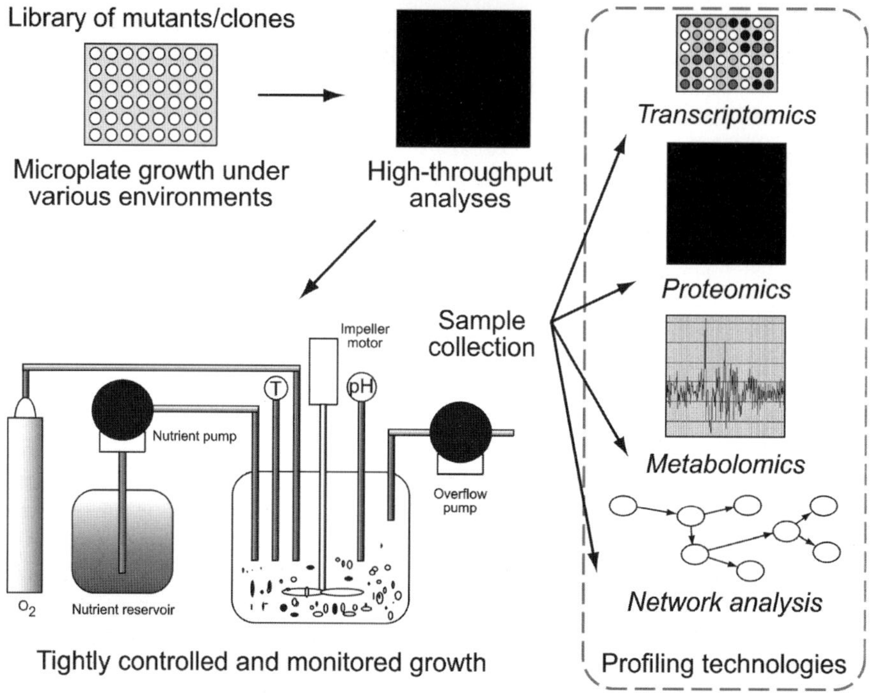



**Fig. 2.** Schematic of a centralized microbial growth facility. Normal *E. coli*, specific mutant *E. coli*, or *E. coli* cells identified from screening mutant libraries in microplates and characterized by high throughput techniques would be examined. Strains possessing an interesting phenotype would be selected for growth under standardized, rigorously controlled conditions. The fermentation would be continuous mode and samples would be collected and immediately frozen for further analysis by collaborators.

development of a standardized growth facility. To be sure, others had found discrepancies between east and west coast variants of *E. coli* K-12 AB1157 (Verma and Egan 1985). Comparisons of RNA polymerase sigma factor subunits of *E. coli* K-12 W3110 samples revealed multiple variants existed between labs in Japan (Jishage and Ishihama 1997).

Accordingly, a consortium may grow standard cells at a community microbial growth facility (MGF), collect samples, and distribute them to researchers with special expertise in conducting measurements. This should permit doing "community experiments" that capture the interest and expertise of many talented investigators in different fields, regardless of their affiliation with consortia. This should also foster an open data-sharing policy between members and rapid release to the entire scientific community. Numerous kinds of measurements (e.g. transcriptome, proteome, metabolome, and interactome analyses) require diverse expertise that seldom can be found at one location (Fig. 2). These new technologies are also rapidly evolving. Ideally, all measurements to develop, verify or refute quantitative models should be made on the same culture. Thus, a central source for generation of samples under rigorously optimized and standardized growth and harvesting procedures may be a key to success of a whole cell *E. coli* systems biology project.

## 6 An integrated *E. coli* database for community research and systems biology

One of the requirements of an *E. coli* systems biology project is the establishment of an information center where all data on *E. coli* and related cells are integrated. Several gene, protein, or function-specific *E. coli* databases now contain vast information on gene structure, metabolic pathways, gene regulation, protein function, and other processes, e.g., ASAP (Glasner et al. 2003), ColiBase (Chaudhuri et al. 2004), Colibri (Medigue et al. 1993), EcoCyc (Karp et al. 2002), EcoGene (Rudd 2000), Ecoli Genome (www.genome.wisc.edu), Genobase (http://ecoli.aist-nara.ac.jp/GB5/), GenProtEC (Serres et al. 2004), RegulonDB (Salgado et al. 2004), and others. Links to these and other databases can be found at www.EcoliCommunity.org. Yet, none of these is comprehensive and substantial gaps exist. Also, many contain redundant information that has often been acquired from other databases, sometimes without proper attribution. Considerable biological resources (e.g. mutants, clones, fusions, etc.) now exist for systematic, genome-wide studies of *E. coli* (Mori et al. 2000; Baba et al. 2005; Kang et al. 2004), however, access to information about them is often unavailable or hard to find.

Whole cell modeling will require the application of new systems approaches as well as continual reductionist experimentation of the *E. coli* cell, especially for processes that are still poorly understood. New computational and experimental resources are needed. These resources should support both the development of an *E. coli* systems biology project and enhancement of the highly successful biochemistry, biophysics, molecular biology, molecular genetics, and physiology research now being done by the *E. coli* community. One way to strengthen both community and consortia research would be to develop a federated *E. coli* database, which for the purpose of discussion we will call EcoliBase.

A model organism database such as the envisioned EcoliBase should contain all available information on *E. coli*, a repository of computational and modeling tools, database(s) of all experimental resources for studying *E. coli* and their availability, and a data warehouse for storage, manipulation, and analysis of diverse kinds of high-throughput data.

The development of a new experimental resources database would be of value to *E. coli* systems biology, as well as the *E. coli* community, including both experimentalists and computational scientists. This database should be an integral component of EcoliBase (Fig. 3). EcoliBase should be accessible via a web browser, so that researchers can easily view, retrieve, and exchange data. It should be designed so that it can be queried by typing or clicking on a scrollable genome map, as well as being accessible by modeling software. Various kinds of data

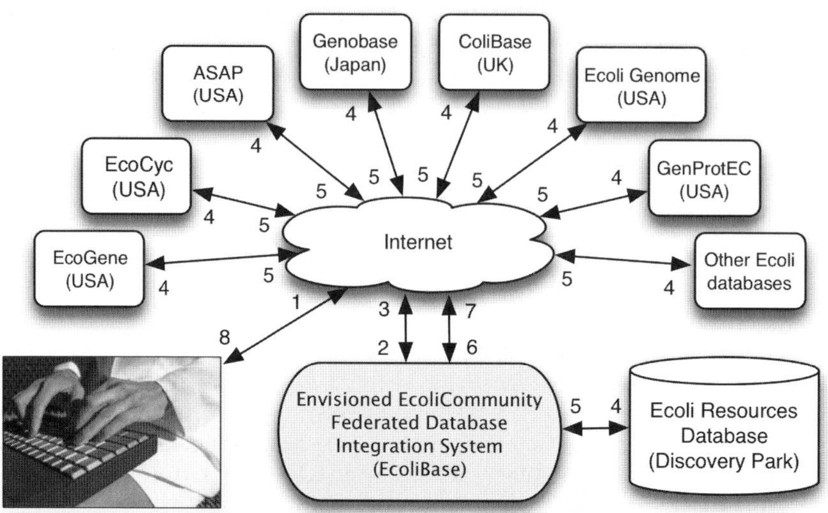



**Fig. 3.** Steps envisioned when a user submits a request to EcoliBase. A user would submit a request over the web to EcoliBase (Steps 1 and 2). EcoliBase would compile the request, decompose it into multiple sub-queries, and then submit the sub-queries to interoperating, participant databases including the *E. coli* Resources Database (Steps 3 and 4). Each participating database would evaluate the query and submit the answer back to EcoliBase (Steps 5 and 6). EcoliBase would compile and integrate the results, possibly addressing conflicts, and submit the compiled answer to the initiating user (Steps 7 and 8).

should be visualized for integration. An important aim of centralized databases is to set standards for the format of various data. Many kinds of data are essential to bring *E. coli* research to the goals of the next level, the foundation of systems biology and cell simulation of this organism. A few categories that would be stored in the database are discussed below.

The core and most important basis of a database is the list of parts determined by the genomic sequence, accurately annotated. A serious problem with current sequence annotation databases, including the new UniProt database (a combination of the ExPASy (Swiss-Prot), TrEMBL, and PIR databases, which were most commonly used) is that the source of annotation information is sometimes unclear. Since a new genome is usually annotated by homology searching against existing sequence databases, once wrong information is contaminated in the sequence database, the error can be propagated to another gene. Unfortunately, this error propagation is frequently observed in the current databases (Galperin and Koonin 1998; Gilks et al. 2002). In the database that we envision, the annotation of genes will clearly indicate the source of the information (history tracking), by indicating whether it is from experimental evidence or prediction by sequence similarity. In the former case, minimally a link pointing to the relevant literature should be added. In the latter case, a gene(s) with high sequence similarity to the gene of interest should be shown together with the score and homologous regions between the two to clarify where the predicted function originates so that one can trace the annotation history of genes to a set of genes with experimental evidence. To implement this annotation chain management system, we would first need to select only proteins annotated by experimental evidence from the UniProt database then repeat the sequence annotation procedure again. Once a new experiment provides new function information of a gene, the updated information would be passed to "downstream" genes that are annotated from the gene by tracking the annotation chain. These clean annotation data would be valuable not only for the *E. coli* research community but also for all bioinformatics research dealing with gene function.

Any cis-acting regulatory sites associated with a sequence, the boundaries of protein coding and structural RNA genes would also be included in the annotation with information as to how they were determined. Also included should be all other sequence features within the genome: replication origins, repeat elements, non-coding and structural RNAs, prophages (as intact elements as well as component parts). Transcriptional and non-transcriptional regulatory information, at the level of the gene, operon, regulon, and other regulatory circuits would also be described together with experimental information. Non-transcriptional regulation would include allostery and feedback inhibition, translational regulation, modifications, and protein degradation.

Annotations should aim to describe what is known about the gene and encoded protein, as well as any known interactions with other functions, defined genetically, biochemically, and by regulatory patterns. Some of this information would appear in other forms in other parts of the set of interoperating databases, but cross-annotation to the particular gene would be important as well.

As high throughput experiments continue to accumulate, an accessible and searchable repository for these data would be critical for allowing researchers to make correlations and do preliminary tests of hypotheses. These data would be deposited from collaborators around the world. In all cases, clear indications of the strains and growth conditions used and how the data were collected would need to be available, to allow the user to have sense of the reliability of the data. In many cases, the information would be linked to publications describing it. It is expected that this category of information would grow at the greatest rate and thus would require attention to simplify the access to new data as it becomes available, including the discussion of possible templates for experimental protocols and data analysis to allow comparisons.

Many groups have undertaken computational methods for predicting not only genes and the families of the predicted proteins, but sites, non-coding RNAs and secondary structure elements such as terminators. Such studies, with information about the nature of the predictions, would provide investigators with the ability to incorporate these predictions into their work. Combined with some large-scale experimental data, these analyses will give a system-wide view of the organism. Several groups have already begun collaborations to identify all probable regulatory motifs in the *E. coli* genome by using a variety of approaches. These would be shown with a confidence score in the database. Predicted protein tertiary structure (Kihara and Skolnick 2004) and protein localization would be included.

**Table 1.** Features of an envisioned *E. coli* federated database integrated system.

| EcoliBase Integrated Tools | EcoliBase Federated Database Engine |
| --- | --- |
| Data-mining | Data importing/exporting |
| Bioinformatics & statistics | Metadata/version management |
| Microarray analysis | Schema mapping, evolution & integration |
| Data visualization | Multi-DB query translation & integration |
| | Access control & backup |
| | Annotation management |
| | Federated DBMS engine |
| | Web manager and user interfaces |

Comparative information allows one to leverage the whole genome information available for *E. coli* to the understanding of other organisms. Orthologous and paralogous genes in other organisms would be listed from an *E. coli* gene. BLAST/FASTA methods, inference of phylogenetic trees and studies of within-species variability are powerful methods of DNA and protein sequence analysis that allow predicting functions of genes and proteins based upon experimentally determined functions in *E. coli* and tracing the evolutionary transformations of functions (gene duplications, genome organization, pseudogenes, etc.). Bioinformatics analyses would be made to the other organisms to allow a comparative study.

Standard sequence analysis tools, such as homology search, motif search, protein secondary structure prediction, should be available by simple manipulation from each gene. Experimental data, such as microarray data, would be linked to analysis tools so that it can be analyzed instantly and in a standard way. Some pathway simulators (Mendes and Kell 2001; Shapiro et al. 2003; Takahashi et al. 2003) would be made available on the web, or if not, at least downloadable.

Not only public domain databases, such as PDB (protein structure), UniProt (proteins), PROSITE (motifs), EcoCyc and KEGG (pathways), but also other existing *E. coli* databases would be integrated as much as possible by collaboration. We would need a unified *E. coli* database (EcoliBase) that would be designed for interoperability so that it can be linked transparently to a larger database structure in the future. We envision EcoliBase to be a web-based interoperable federated database integration system.

All interoperating participant *E. coli* databases including the *E. coli* Resources Database would be registered with EcoliBase. Users would have a web interface to access EcoliBase (Fig. 3). A user may issue a request to EcoliBase via the web. EcoliBase translates and decomposes the submitted requests into sub-queries, then submits the sub-queries to the corresponding and interoperating participant databases. The results of each sub-query are integrated inside EcoliBase and are returned to the user. It is expected that EcoliBase will have the functional components shown in Table 1.

# 7 Putting models to work: The International *E. coli* Alliance

From the dawn of modern biology, the intestinal bacterium *E. coli* has been the most intensively studied organism. Many basic molecular events, best understood in *E. coli*, are universal throughout the natural world. *E. coli* has laid so many of the golden eggs of basic biochemistry, genetics, and molecular biology that no doubt it will lay even more. Our present day level of basic understanding of natural phenomenon far exceeds the imagination of even the most creative scientists a few decades ago. New tools for gaining even more biological information ensure future revelations will continue to be uncovered at an ever-increasing pace. Although creating a truly virtual cell may be far in the future, the place to start is with a well understood system for which there are tools for deepening our knowledge. Systems biology approaches are needed for conceptualizing and testing our interpretations of these data.

It was with these concepts in mind that IECA was formed as a worldwide alliance for the purpose of constructing a large-scale model of a simple, self-replicating cell. Bringing such a dream to fruition requires not only computational and experimental tools, but also changes in how we do science – the human factor. Many impediments must be overcome. Large-scale experimentation is new to biologists. Other fields of science, most notably areas of physics requiring huge and expensive resources, have dealt with issues now facing systems biology. Much more time is spent planning and designing major experiments in physics than seems to be the norm in systems biology. As in many present day physics projects, systems biology projects of the future will depend more and more on large numbers of researchers working together in distantly located teams. How to achieve this through collaboration and building consortia will be challenging. Funding agencies must also find creative ways of encouraging scientists with diverse expertise to work together in teams to reach a common goal.

Like the physicists' goal for a complete understanding of the world from the inner workings of an atom to the motion and expansion of the universe, the goal of IECA is the complete modeling of a whole cell. Perhaps, modeling a cell is itself a bit too ambitious. However, the time to start is now. A practical way to do this would be to begin by studying modules, like regulatory systems or metabolic or signaling pathways, then to build these into networks that can then be joined together at an ever higher level. Surely, a computerized *E. coli* virtual cell will add powerful new tools to our existing arsenal of discovery, including virtual experimentation and mathematical simulation. These biological and computational tools promise to be useful for everything from drug discovery to bioengineering.

# Acknowledgements

Authors are supported by the National Institutes of Health GM62662 (B.L.W.) and GM70923 (A.F. and M.H.).

## References


Achard F, Vaysseix G, Barillot E (2001) XML, bioinformatics and data integration. Bioinformatics 17:115-125

Allen NN, Calzone L, Chen KC, Ciliberto A, Ramakrishnan N, Shaffer CA, Sible JC, Tyson JJ, Vass MT, Watson LT, Zwolak JW (2003) Modeling regulatory networks at Virginia Tech. OMICS 7:285-299

Alm E, Arkin AP (2003) Biological networks. Curr Opin Struct Biol 13:193-202

Arkin AP (2001) Simulac and deduce. http://gobi.lbl.gov/~aparkin/Stuff/Software.html.

Augen J (2001) Information technology to the rescue! Nat Biotechnol 19:BE39-BE40

Baba T, Ara T, Okumura Y, Hasegawa M, Takai Y, Baba M, Oshima T, Datsenko KA, Tomita M, Wanner BL, Mori H (2005) Systematic construction of single gene deletions mutants in *Escherichia coli* K-12, submitted

Bailey JE (1998) Mathematical modeling and analysis in biochemical engineering: Past accomplishments and future opportunities. Biotechnol Prog 14:8-20

Bialek W, Botstein D (2004) Introductory science and mathematics education for 21st-century biologists. Science 303:788-790

Blattner FR, Plunkett G III, Bloch CA, Perna NT, Burland V, Riley M, Collado-Vides J, Glasner JD, Rode CK, Mayhew GF, Gregor J, Davis NW, Kirkpatrick HA, Goeden MA, Rose DJ, Mau B, Shao Y (1997) The complete genome sequence of *Escherichia coli* K-12. Science 277:1453-1462

Bosak J, Bray T (1999) XML and the second-generation web. Sci Am May

Bower JM, Bolouri H (2001a) Computational modeling of genetic and biochemical networks. MIT Press, Cambridge, Mass

Bower JM, Bolouri H (2001b) Introduction: understanding living systems. In: Bower, James M and Bolouri H (eds) Computational modeling of genetic and biochemical networks. MIT Press, Cambridge, Mass., p xiii-xx

Bray T, Paoli J, Sperberg-McQueen CM, Maler E (2000) Extensible markup language (XML) 1.0 Second Edition: http://www.w3.org/TR/1998/REC-xml-19980210

Brown CT, Rust AG, Clarke PJC, Pan Z, Schilstra MJ, De Buysscher T, Griffin G, Wold BJ, Cameron RA, Davidson EH, Bolouri H (2002) New computational approaches for analysis of cis-regulatory networks. Dev Biol 246:86-102

Burns JA (1971) Studies on complex enzyme systems. University of Edinburgh

Butler D (1999) Computing 2010: from black holes to biology. Nature 402:C67-C70

Capra F (1996) The Web of Life: A new scientific understanding of living systems. Anchor Books, New York

Chance B (1960) Analogue and digital representations of enzyme kinetics. J Biol Chem 235:2440-2443

Chance B (1943) The kinetics of the enzyme-substrate compound of peroxidase. J Biol Chem 151:553-577

Chance B, Brainerd JG, Cajori FA, Millikan GA (1940) The kinetics of the enzyme-substrate compound of peroxidase and their relation to the Michaelis theory. Science 92:455

Chance B, Greenstein DS, Higgins J, Yang CC (1952) The mechanism of catalase action. II. Electric analog computer studies. Arch Biochem Biophys 37:322-339

Chaudhuri RR, Khan AM, Pallen MJ (2004) coliBASE: an online database for *Escherichia coli*, *Shigella* and *Salmonella* comparative genomics;

http://colibase.bham.ac.uk/about/index.cgi?help=about&frame=genomechoose.  Nucleic Acids Res 32:D296-D299

Chen S, Bigner SH, Modrich P (2001) High rate of CAD gene amplification in human cells deficient in MLH1 or MSH6. Proc Natl Acad Sci USA 98:13802-13807

Corbin RW, Paliy O, Yang F, Shabanowitz J, Platt M, Lyons CE Jr, Root K, McAuliffe J, Jordan MI, Kustu S, Soupene E, Hunt DF (2003) Toward a protein profile of *Escherichia coli*: comparison to its transcription profile. Proc Natl Acad Sci USA 100:9232-9237

Crick FHC (1973) Project K: "The complete solution of *E. coli*". Perspect Biol Med 67-70

Csete ME, Doyle JC (2002) Reverse engineering of biological complexity. Science 295:1664-1669

de Jong H (2002) Modeling and simulation of genetic regulatory systems: a literature review. J Computat Biol 9:67-103

DiValentin, P SigTran (2004) http://csi.washington.edu/teams/modeling/projects/sigtran/

Duncan J, Arnstein L, Li Z (2004) Teranode corporation launches first industrial-strength research design tools for the life sciences at DEMO: http://www.teranode.com/about/pr_2004021601.php

Endy D, Brent R (2001) Modelling cellular behaviour. Nature Suppl 409:391-395

Fall C, Marland ES, Wagner JM, Tyson JJ (2002) Computational cell biology. Springer-Verlag, New York

Fink CC, Slepchenko B, Moraru II, Watras J, Schaff JC, Loew LM (2000) An image-based model of calcium waves in differentiated neuroblastoma cells. Biophys J 79:163-83

Finney A, Hucka M, Sauro H, Bolouri H, Funahashi A, Bornstein B, Kovitz B, Matthews J, Shapiro BE, Keating S, Doyle J, Kitano H (2003) The systems biology workbench (SBW) Version 1.0: Framework and modules. Hawaii, USA. Pacific symposium on biocomputing 2003

Finney AM, Hucka M (2003) Systems Biology Markup Language: Level 2 and beyond. Biochem Soc Trans 31:1472-1473

Fraser SE, Harland RM (2000) The molecular metamorphosis of experimental embryology. Cell 100:41-55

Funahashi A, Tanimura N, Morohashi M, Kitano H (2003) CellDesigner: a process diagram editor for gene-regulatory and biochemical networks. BioSilico 1:159-162

Funahashi A, Tanimura N, Morohashi M, Kitano H (2004) CellDesigner; http://www.systems-biology.org/002/

Galperin MY, Koonin EV (1998) Sources of systematic error in functional annotation of genomes: domain rearrangement, non-orthologous gene displacement and operon disruption. In Silico Biol 1:55-67

Garfinkel D (1965) Simulation of biochemical systems. In: Stacy, Ralph W and Waxman, BD (eds) Computers in biomedical research. Academic Press, New York, pp 111-134

Gershenfeld NA (1998) The nature of mathematical modeling. Cambridge University Press, Cambridge

Gilks WR, Audit B, De Angelis D, Tsoka S, Ouzounis CA (2002) Modeling the percolation of annotation errors in a database of protein sequences. Bioinformatics 18:1641-9

Gillespie DT (1977) Exact stochastic simulation of coupled chemical-reactions. J Phys Chem 81:2340-2361

Gillespie DT, Petzold LR (2003) Improved leap-size selection for accelerated stochastic simulation. J Chem Phys 119:8229-8234

Glasner JD, Liss P, Plunkett G III, Darling A, Prasad T, Rusch M, Byrnes A, Gilson M, Biehl B, Blattner FR, Perna NT (2003) ASAP, a systematic annotation package for community analysis of genomes; https://asap.ahabs.wisc.edu/annotation/php/home.php?formSubmitReturn=1. Nucleic Acids Res 31:147-151

Goldstein B, Faeber JR, Hlavacek WS, Blinov ML, Redondo A, Wolfsy C (2002) Modeling the early signaling events mediated by FceRI. Mol Immunol137:1-7

Hartwell LH, Hopfield JJ, Leibler S, Murray AW (1999) From molecular to modular cell biology. Nature 402:C47-C52

Holden C (2002) Cell biology: Alliance launched to model *E. coli*. Science 297:1459-1460

Hood L (1998) Systems biology: New opportunities arising from genomics, proteomics, and beyond. Exp Hematol 26:681

Hucka M, Finney A, Sauro HM, Bolouri H (2001) Systems Biology Markup Language (SBML) Level 1: Structures and facilities for basic model definitions; http://www.sbml.org/

Hucka M, Finney A, Sauro HM, Bolouri H, Doyle JC, Kitano H, Arkin AP, Bornstein BJ, Bray D, Cornish-Bowden A, Cuellar AA, Dronov S, Gilles ED, Ginkel M, Gor V, Goryanin II, Hedley WJ, Hodgman TC, Hofmeyr JH, Hunter PJ, Juty NS, Kasberger JL, Kremling A, Kummer U, Le Novere N, Loew LM, Lucio D, Mendes P, Minch E, Mjolsness ED, Nakayama Y, Nelson MR, Nielsen PF, Sakurada T, Schaff JC, Shapiro BE, Shimizu TS, Spence HD, Stelling J, Takahashi K, Tomita M, Wagner J, Wang J (2003) The Systems Biology Markup Language (SBML): a medium for representation and exchange of biochemical network models. Bioinformatics 19:524-531

Ideker T, Galitski T, Hood L (2001) A new approach to decoding life: systems biology. Annu Rev Genomics Hum Genet 2:343-372

Jishage M, Ishihama A (1997) Variation in RNA polymerase sigma subunit composition within different stocks of *Escherichia coli* W3110. J Bacteriol 179:959-963

Kacser H (1957) Appendix: Some physico-chemical aspects of biological organisation. In: Waddington CH (ed) The strategy of the genes: A discussion of some aspects of theoretical biology. George Allen and Unwin Ltd, London, pp 191-249

Kacser H, Burns JA (1967) Causality, complexity and computers. In: Locker A (ed) Quantitative biology of metabolism. Springer-Verlag, New York, NY, pp 11-23

Kang Y, Durfee T, Glasner JD, Qiu Y, Frisch D, Winterberg KM, Blattner FR (2004) Systematic mutagenesis of the *Escherichia coli* genome. J Bacteriol 186:4921-4930

Karp PD, Riley M, Saier M, Paulsen IT, Collado-Vides J, Paley SM, Pellegrini-Toole A, Bonavides C, Gama-Castro S (2002) The EcoCyc database; http://ecocyc.org/. Nucleic Acids Res 30:56-58

Kihara D, Skolnick J (2004) Microbial genomes have over 72% structure assignment by the threading algorithm PROSPECTOR_Q. Proteins 55:464-473

Kirkwood TBL, Boys R, Wilkinson D, Gillespie C, Proctor C, Hanley D (2003a) BASIS; http://www.basis.ncl.ac.uk/. 3-19-2004a

Kirkwood TBL, Boys RJ, Gillespie CS, Proctor CJ, Shanley DP, Wilkinson DJ (2003b) Towards an e-biology of ageing: integrating theory and data. Nature Reviews Molecular Cell Biology 4:243-249

Kitano H (2002) Computational systems biology. Nature 420:206-210

Kitano H (2001) Foundations of systems biology. MIT Press, Cambridge, MA

Kornberg A (2003) Ten commandments of enzymology, amended. Trends Biochem Sci 28:515-517

Kornberg A, Baker TA (1992) DNA replication. WH Freeman and Company, San Francisco, California

Le Novere N, Shimizu TS (2001) STOCHSIM: modelling of stochastic biomolecular processes. Bioinformatics 17:575-576

May RM (2004) Uses and abuses of mathematics in biology. Science 303:790-793

McAdams HH, Arkin A (1999) It's a noisy business! Genetic regulation at the nanomolar scale. Trends Genet 15:65-69

Medigue C, Viari A, Henaut A, Danchin A (1993) Colibri: a functional data base for the *Escherichia coli* genome. Microbiol Rev 57:623-654

Mendes P (2001) Gepasi 3.21; http://www.gepasi.org

Mendes P (2003) COPASI: Complex pathway simulator; http://mendes.vbi.vt.edu/tiki-index.php?page=COPASI

Mendes P (1993) Gepasi - a software package for modeling the dynamics, steady-states and control of biochemical and other systems. Comput Appl Biosci 9:563-571

Mendes P, Kell DB (2001) MEG (Model Extender for Gepasi): a program for the modelling of complex, heterogeneous, cellular systems. Bioinformatics 17:288-289

Mesarovic MD (1968) Systems theory and biology. Proceedings of the 3rd Systems Symposium at Case Institute of Technology. Springer-Verlag, Berlin, New York

Mori H, Isono K, Horiuchi T, Miki T (2000) Functional genomics of *Escherichia coli* in Japan. Res Microbiol 151:121-128

Morton-Firth CJ, Bray D (1998) Predicting temporal fluctuations in an intracellular signalling pathway. J Theor Biol 192:117-128

Noble D (2002) The rise of computational biology. Nat Rev Mol Cell Biol 3:460-463

Phair RD, Misteli T (2001) Kinetic modelling approaches to *in vivo* imaging. Nat Rev Mol Cell Biol 2:898-907

Ramsey S, Bolouri H (2004) Dizzy; http://labs.systemsbiology.net/oolouri/software/Dizzy/

Rudd KE (2000) EcoGene: a genome sequence database for *Escherichia coli* K-12. Nucleic Acids Res 28:60-64

Salgado H, Gama-Castro S, Martinez-Antonio A, Diaz-Peredo E, Sanchez-Solano F, Peralta-Gil M, Garcia-Alonso D, Jimenez-Jacinto V, Santos-Zavaleta A, Bonavides-Martinez C, Collado-Vides J (2004) RegulonDB (version 4.0): transcriptional regulation, operon organization and growth conditions in *Escherichia coli* K-12; http://www.cifn.unam.mx/Computational_Genomics/regulondb/. Nucleic Acids Res 32:D303-D306

Sauro HM (2003) WinScamp; http://www.cds.caltech.edu/~hsauro/Scamp/scamp.htm

Sauro HM (2000b) Jarnac; http://www.cds.caltech.edu/~hsauro

Sauro HM (2000a) Jarnac: A system for interactive metabolic analysis. Snoep JL, Hofmeyr JH, and Roywer JM; Animating the Cellular Map: Proceedings of the 9th International Meeting on BioThermoKinetics. Stellenbosch University Press

Sauro HM, Fell DA (1991) SCAMP: A metabolic simulator and control analysis program. Mathl Comput Modelling 15:15-28

Sauro HM, Hucka M, Finney A, Wellock C, Bolouri H, Doyle J, Kitano H (2003) Next generation simulation tools: The systems biology workbench and BioSPICE integration. OMICS 7:355-372

Sauro HS (2001) JDesigner: A simple biochemical network designer; http://members.tripod.co.uk/sauro/biotech.htm

Savageau MA (1969) Biochemical systems analysis.1. Some mathematical properties of rate law for component enzymatic reactions. J Theor Biol 25:365-366

Savageau MA (1970) Biochemical systems analysis .3. Dynamic solutions using a power-law approximation. J Theor Biol 26:215

Schaff J, Fink CC, Slepchenko B, Carson JH, Loew LM (1997) A general computational framework for modeling cellular structure and function. Biophys J 73:1135-1146

Schaff J, Slepchenko B, Morgan F, Wagner J, Resasco D, Shin D, Choi YS, Loew L, Carson J, Cowan A, Moraru I, Watras J, Teraski M, Fink C (2001) Virtual Cell; http://www.nrcam.uchc.edu

Schilstra M, Bolouri H (2002) NetBuilder; http://strc.herts.ac.uk/bio/maria/NetBuilder/index.html

Serres MH, Goswami S, Riley M (2004) GenProtEC: an updated and improved analysis of functions of *Escherichia coli* K-12 proteins; http://www.genprotec.mbl.edu/. Nucleic Acids Res 32:D300-D302

Shapiro BE (2004) MathSBML; http://sbml.org/mathsbml.html

Shapiro BE, Hucka M, Finney A, Doyle JC (2004a) MathSBML: A package for manipulating SBML-based biological models. Bioinformatics 20:2829-2831

Shapiro BE, Levchenko A, Meyerowitz EM, Wold BJ, Mjolsness ED (2003) Cellerator: extending a computer algebra system to include biochemical arrows for signal transduction simulations. Bioinformatics 19:677-678

Shapiro BE, Mjolsness E, Levchenko A (2004b) Cellerator; http://www-aig.jpl.nasa.gov/public/mls/cellerator/

Slepchenko BM, Schaff JC, Carson JH, Loew LM (2002) Computational cell biology: Spatiotemporal simulation of cellular events. Annu Rev Biophys Biomol Struct 31:423-441

Stelling J, Kremling A, Ginkel M, Bettenbrock K, Gilles E (2001) Towards a virtual biological laboratory. In: Kitano H (ed) Foundations of systems biology. MIT Press, Cambridge, MA, pp 189-212

Takahashi K, Ishikawa N, Sadamoto Y, Sasamoto H, Ohta S, Shiozawa A, Miyoshi F, Naito Y, Nakayama Y, Tomita M (2003) E-Cell 2: Multi-platform E-Cell simulation system. Bioinformatics 19:1727-1729

Teranode Inc. (2004)VLX Design Suite

Tomita M (2001) Towards computer aided design (CAD) of useful microorganisms. Bioinformatics 17:1091-1092

Tomita M, Hashimoto K, Takahashi K, Shimizu TS, Matsuzaki Y, Miyoshi F, Saito K, Tanida S, Yugi K, Venter JC, Hutchison CA, III (1999) E-CELL: software environment for whole-cell simulation. Bioinformatics 15:72-84

Tomita M, Nakayama Y, Naito Y, Shimizu T, Hashimoto K, Takahashi K, Matsuzaki Y, Yugi K, Miyoshi F, Saito Y, Kuroki A, Ishida T, Iwata T, Yoneda M, Kita M, Yamada Y, Wang E, Seno S, Okayama M, Kinoshita A, Fujita Y, Matsuo R, Yanagihara T, Watari D, Ishinabe S, Miyamoto S (2001) E-CELL; http://www.e-cell.org/

Tyson JJ, Chen K, Novak B (2001) Network dynamics and cell physiology. Nat Rev Mol Cell Biol 2:908-916

Vass M, Shaffer CA, Tyson JJ, Ramakrishnan N, Watson LT (2004) The JigCell model builder: a tool for modeling intra-cellular regulatory networks. Bioinformatics 20:3680-3681

Verma M, Egan JB (1985) Phenotypic variations in strain AB1157 cultivars of *Escherichia coli* from different sources. J Bacteriol 164:1381-1382

Wiener N (1961) Cybernetics; or, control and communication in the animal and the machine, 2nd edn. MIT Press, New York

Yi TM, Huang Y, Simon MI, Doyle J (2000) Robust perfect adaptation in bacterial chemotaxis through integral feedback control. Proc Natl Acad Sci USA 97:4649-4653

Zerhouni E (2003) The NIH roadmap. Science 302:63-64



Wanner, Barry L.

Department of Biological Sciences, Purdue University, West Lafayette, IN 47907-2054 USA

blwanner@purdue.edu

Andrew Finney

Science and Technology Research Institute, University of Hertfordshire, Hatfield, AL10 9AB, UK

Michael Hucka

Control and Dynamical Systems, California Institute of Technology, Pasadena, CA 91125-8100 USA

# Metabolic flux analysis: A key methodology for systems biology of metabolism

Uwe Sauer

## Abstract

Genome-wide analyses of mRNA, protein, or metabolite complements of biological systems produce unprecedented data sets. In contrast to such cellular composition data, *in vivo* quantification of molecular fluxes through metabolic networks links genes and proteins to higher-level functions that result from biochemical and regulatory interactions between network components. By unraveling novel or unexpected pathways in microbes, metabolic flux analyses begin to question the ability of well-known 'textbook' pathways to portray flux through complex networks. Accumulating data on flux responses to genetic or environmental changes reveal general design principles and system properties of metabolic network operation. Beyond such discoveries, flux data assume increasingly important roles in completing network models and verifying or refuting their predictions. With recent advances in analytical accuracy, mathematical frameworks, available software, and experimental throughput, steady state flux analysis became a key methodology for metabolic systems biology.

## 1 Complex systems – Systems biology

The essence of biology is to understand evolved living systems that range from molecular assemblies to cells, tissues, multi-cellular organisms, and finally entire ecosystems. To unravel causal relationships within such inherently complex systems, the successful approach has been – out of necessity – decomposition into molecular or cellular constituents. There is ample experimental and theoretical evidence, however, that this reductionist approach does not necessarily assess the behavior of the whole - and almost never in a quantitatively rigorous sense. The dissonance between our understanding of the isolated parts and the whole is readily apparent from experience in such diverse fields as metabolic engineering (Bailey 1999; Koffas et al. 1999; Stephanopoulos et al. 2004), ecology (Purugganan and Gibson 2003), neurobiology (Stewart 2004), molecular and cell biology (Arkin 2001; Marcotte 2001), functional genomics (Bailey 2001a; Covert et al. 2001), and pharmaceutical research (Hellerstein 2003).

What is the underlying reason for the apparent difficulty to assemble the whole from the parts? It would be feasible if the system was merely *complicated* like a sophisticated mechanical watch - although agreeably difficult if the main analyti-

Topics in Current Genetics, Vol. 13
L. Alberghina, H.V. Westerhoff (Eds.): Systems Biology
DOI 10.1007/b136810 / Published online: 26 April 2005

cal tool was to 'knockout' single components by shooting at the system and recording malfunction 'phenotypes' (Lazebnik 2002). Reconstituting the whole from the parts fails, however, for extensively interconnected *complex* systems. Topology, strength, and dynamics of interactions between the components give rise to an adaptive and self-organizing entity with emergent new properties; a hallmark of many technical systems such as the internet and essentially all biological systems. In a typical cell, gene, protein, and metabolite components are connected by highly interactive information and biochemical networks that involve numerous control circuits with different quantitative strengths and often acting in different directions. More than a century of reductionist study has identified and characterized many key components and delineated some of their direct interactions. Biological systems, however, cannot be fully understood from this qualitative knowledge because much of their properties and behavior is the integration of often non-linear interactions that are not represented in the components themselves, and the questions migrate to realms that are more congenial to mathematicians and system theoreticians (Kitano 2002b).

The nascent field of systems biology (Ideker et al. 2001a; Kitano 2002a, 2002b; Aggarwal and Lee 2003; Selinger et al. 2003; Westerhoff and Palsson 2004) attempts to identify and quantify how new properties arise from the interactions that are important for the functioning of living systems. We stand today at the crossroads that provide unprecedented opportunity to make these connections because much about the molecular components is known, global explorations become tractable thanks to 'omics' technologies, quantitative analytical methods with high temporal and spatial resolution are being developed, and computational tools and mathematical frameworks that can integrate the data are becoming available. In this (post)genomic age with its inherently new perceptions of genetic and metabolic networks (Shen-Orr et al. 2002; Stelling et al. 2002; Bar-Joseph et al. 2003; Wolf and Arkin 2003; Barabasi and Oltavi 2004; Covert et al. 2004; Papin et al. 2004), few would argue that forthcoming applications in many areas require more quantitative analysis and comprehensive description of biological systems (Gallagher and Appenzeller 1999; Bray 2001; Nurse 2003; Covert et al. 2004; Stewart 2004). There can also be little doubt that biology is about to face a disciplinary 'blurring', with particularly strong input and insights from mathematics and computer science. Many envisage systems biology as a unifying theme of approaches – rather than as a codified scientific discipline. Whether or not it matures or remains a theme will, to a large extent, depend on making otherwise not possible scientific discoveries and to provide the promised understanding of operation and essential features in biological systems – in a way that it becomes useful to experimentalists and theoreticians alike.

Sidestepping the terminological minefield of an all-encompassing definition, the key features of any systems biology approach (Kitano 2002a, 2002b) are quantitative data *and* their integration in computer models that allow to study emergent new properties from the interaction between components. The models must be capable (or at least be constructed with the potential) to quantitatively *predict* the systems behavior upon genetic or environmental perturbation, thereby, specifying testable hypotheses and pivotal experiments for their verification in an iterative process. A frequent misconception is that collecting extensive 'omics data and their bioinformatic mining equates to systems biology. Obviously, this is important in its own right and a rich source *for* systems biology, but the data-driven approach captures only statistically relevant features. Since this approach does not attempt to represent causal relationships within a model, it does not lead to any higher level understanding all by itself (Bailey 2001b; Kitano 2002a). The iterative process of (model)prediction-driven experimentation and model development aims to accrue knowledge and understanding from data (Bray 2001), thus, in some sense, reintroduces scientific hypothesis testing into (post)genomic data gathering.

## 2 Accessing metabolic network operation through steady state flux analysis

Much of the present emphasis in systems biology is on the development of theoretical frameworks and new computational/mathematical approaches that can integrate massive and heterogeneous data in a quantitative fashion (Varner 2000; Arkin 2001; Covert and Palsson 2002; Kholodenko et al. 2002; Kitano 2002a; Wiechert 2002; Ideker and Lauffenburger 2003; Zak et al. 2003). Since data are the precursor to any model, the other focus area is development of novel methodologies that provide (i) quantitative data on the components with high spatial and temporal resolution or (ii) quantitative and functional insight into *in vivo* operation of assembled systems. Mature examples of the latter type are rare, but one that fulfills the criteria is metabolic flux analysis, an off-spring of metabolic engineering (Bailey 1991; Stephanopoulos 1999; Rohlin et al. 2001). The essential concept is that biochemical reactions are considered together as a metabolic network rather than individually. The concept is hardly new and has occupied the attention of systems-oriented biochemists for decades (Kacser and Burns 1973; Heinrich and Rapoport 1974). What is new though, is the experimental capability to quantify fluxes through complex networks. The goal of this article is to review methods for quantification of *in vivo* molecular fluxes through microbial metabolism during steady state conditions and to illustrate how such results contribute to systems biology.

Metabolic adaptation to ever-changing environmental conditions is brought about by delicate and dynamic regulation at several levels that include transcription and translation but also enzyme kinetics and allosteric control. The integration of all regulation processes ultimately determines the material fluxes through networks of up to 1,000 reactions in a typical microbe. Systems biology assumes the view that metabolic systems are fundamentally composed of two types of information: genes, encoding the molecular machines that execute the functions of life, and networks of regulatory interactions that control the hierarchical and interactive flow of information from DNA->mRNA->protein->metabolites->fluxes->phenotype (Fig. 1). While advances in global mRNA and protein analyses shed

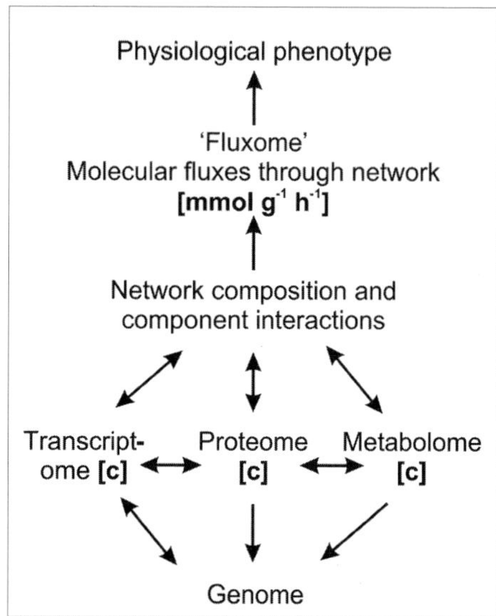



**Fig. 1.** Interactions between compositional and functional units in metabolic networks.

light on certain aspects of this complexity, genome-wide data on network composition provide no direct link to the connectivity, interactions, and dynamic properties of metabolic systems (Bailey 1999; Stephanopoulos 1999; Hellerstein 2003). The functional determinants of cellular physiology are *in vivo* molecular fluxes through fully assembled metabolic networks, because they reflect the integration of genetic and metabolic regulation and interactions (Fig. 1). In sharp contrast to measurements of mRNA and protein abundance that assess transcriptional and translational regulation, metabolism-wide quantification of fluxes, the fluxome (Sauer et al. 1999), does not distinguish between different regulatory mechanisms *per se* but assesses the culmination of all regulatory processes that occur in a metabolic network (Koffas et al. 1999; Sanford et al. 2002; Sauer 2004). In linking genes and proteins to higher-level biological functions, molecular fluxes through the network may also be regarded as operational units of function that determine the systemic phenotype – a consequence of genome, transcriptome, proteome, and metabolome interaction (Hellerstein 2003) (Fig. 1).

Intracellular fluxes, or *in vivo* reaction rates, are *per se* non-measurable and must be estimated from measurable quantities by methods of metabolic flux analysis (Fig. 2). In its simplest incarnation, flux analysis relies on flux (i.e. metabolite) balancing in a given stoichiometric reaction model that is largely based on assumed 'textbook' biochemistry (Varma and Palsson 1994). Applying mass balances around intracellular metabolites, fluxes are, thus, estimated from experimentally determined biosynthetic requirements, nutrient uptake and product secretion rates. Since the biosynthetic reactions to cellular macromolecules such as RNA or protein are biochemically characterized linear reaction sequences, precursor and cofactor fluxes for biomass formation can be estimated reliably with detailed models of cellular composition (Neidhardt et al. 1990). Intermediary metabolism, in contrast, is redundant with multiple combinations of pathways and reactions that potentially lead to the formation of identical products or intermediates. Limited data and stoichiometric constraints then lead to underdetermined systems of linear equations that do not allow to resolve the flux distribution uniquely; i.e., an infinite number of flux distribution scenarios fulfill the available constraints (Bonarius et al. 1997; Price et al. 2003). Thus, either objective functions on network operations are defined (e.g. maximize growth or energy production) or additional constraints (e.g. on the stoichiometry of energy metabolism) are derived from various assumptions on biological functions. The flux estimates then critically depend on the validity of rather arbitrary assumptions, in particular when genetic perturbations are introduced. While the flux balancing approach is valid for anaerobically growing microbes with little metabolic freedom (Daran-Lapujade et al. 2004; Sonderegger et al. 2004), it is typically not very informative for higher cells or aerobically growing microbes (Wiechert 2001). In particular, it fails to resolve many important fluxes through alternative pathways and its capability to unravel novel causal relationships is rather limited.

To increase the reliability and resolution of such flux balancing approaches, additional constraints on intracellular flux partitioning can be derived from appropriately designed stable isotopic tracer experiments (Fig. 2). Typically, $^{13}$C-labeled substrates are administered and products of metabolism are analyzed by nuclear magnetic resonance (NMR) (Szyperski 1998; Grivet et al. 2003; Sherry et al. 2004) or mass spectrometry (MS) (Christensen and Nielsen 1999; Hellerstein and Neese 1999; Wittmann 2002; Hellerstein 2003). Thus, discriminated isotope isomer (isotopomer) patterns reflect carbon backbone modifications that result from the activity of alternative pathways or reactions.

Ideally, isotopomer patterns are detected in a large number of intermediate metabolites, but their low concentration and high turnover numbers pose significant technical challenges for measurement and rapid quenching of metabolism. While this approach was used successfully to resolve specific flux responses in mammalian cells (Shulman and Rothman 2001; Hellerstein 2003) and yeast (van Winden et al. 2005), microbial cells with metabolite turnover rates in the seconds range are typically investigated in physiological steady state during exponential growth in batch or continuous cultures. For steady state analyses, $^{13}$C-labeling patterns are investigated in accumulated extracellular or intracellular products of metabolism (Park et al. 1997; Wittmann and Heinzle 2002). The most frequently employed procedure exploits $^{13}$C-pattern in amino acids that provide access to several key branch points in metabolism (Szyperski 1995; Marx et al. 1996; Christensen et al. 2002; Fischer and Sauer 2003a; Hua et al. 2003). Safely 'stored' in cellular protein, these proteinogenic amino acids reflect the carbon isotopomer pattern of the reaction intermediates that they are derived from. In contrast to the intermediates themselves, the concentration of amino acids in protein hydrolyzates is high and their isotopomer pattern is not affected by the harvesting procedures (Sauer et al. 1999; Fischer and Sauer 2003a).

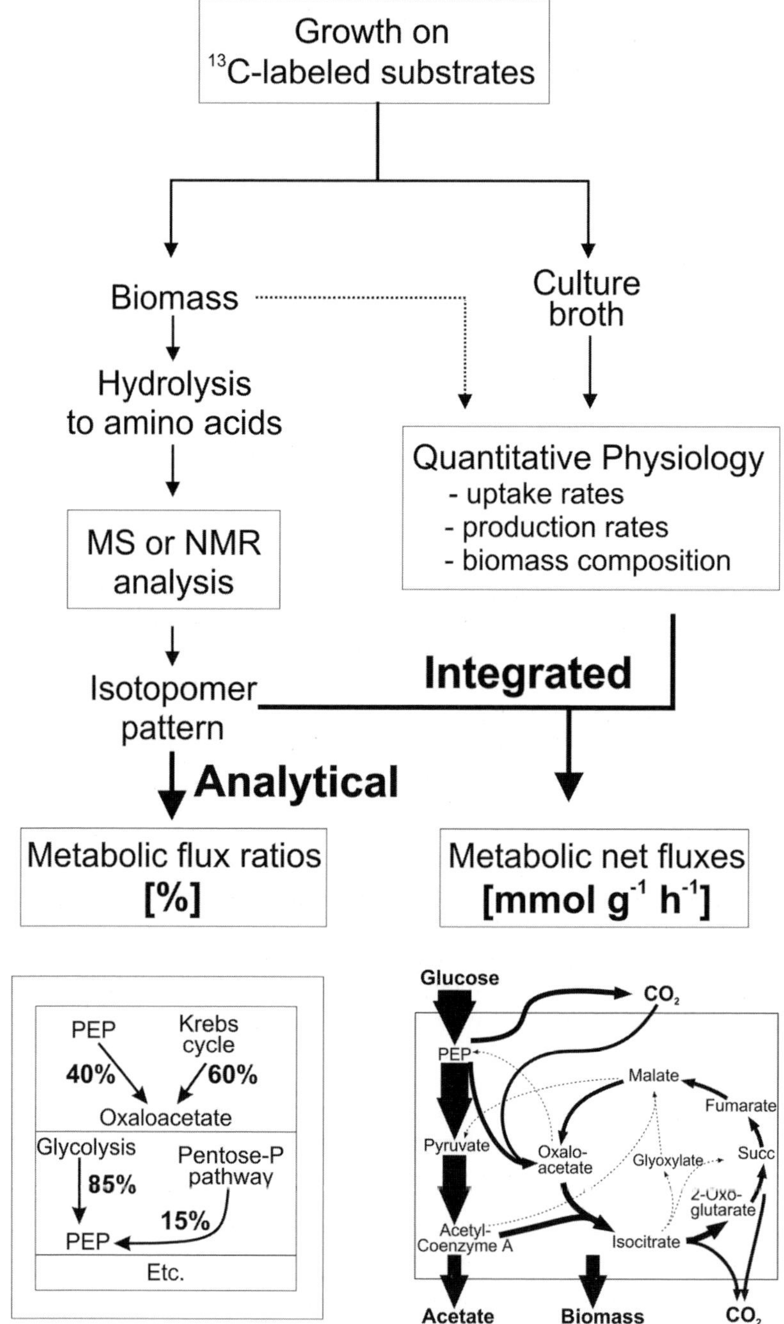



**Fig. 2.** Schematic flow chart of metabolic flux analysis.

Once NMR or MS spectra are recorded, two general approaches for isotopomer pattern interpretation from tracer experiments can be distinguished: analytical and integrated (Fig. 2). Direct analytical interpretation of selected $^{13}$C-labeling pattern in pathway intermediates or products, thereof, has been used extensively for pathway elucidation or flux quantification of individual pathways (Szyperski 1998; Christensen et al. 2001; Portais and Delort 2002; Grivet et al. 2003). Generally, algebraic or probabilistic equations relate $^{13}$C-patterns to flux partitioning ratios. Such analytically deduced flux ratios were also used successfully as constraints for metabolic flux analysis (Walsh and Koshland 1984; Sauer et al. 1997; Fischer et al. 2004). A particularly informative methodology is metabolic flux ratio (METAFoR) analysis that identifies simultaneously the network topology and quantifies the relative contribution of multiple converging pathways and reactions to the formation of target metabolic intermediates from $^{13}$C-pattern in proteinogenic amino acids (Szyperski et al. 1999). Originally developed for NMR-based analysis (Szyperski 1995; Sauer et al. 1997; Maaheimo et al. 2001; Yang et al. 2003), it was recently modified to quantify more than 10 independent flux ratios from GC-MS analysis of [1-$^{13}$C] and [U-$^{13}$C]glucose experiments with bacteria or yeast (Fischer and Sauer 2003a; Blank and Sauer 2004).

In the integrated approach, all available isotopomer data, extracellular material fluxes, and biosynthetic requirements are simultaneously interpreted with metabolic models of varying complexity. For this purpose, the labeling state of metabolic intermediates is balanced within a computational model to map metabolic fluxes in an iterative fitting procedure on the entire isotopomer pattern of network metabolites (Schmidt et al. 1997; Zupke et al. 1997; Dauner et al. 2001a; Wiechert et al. 2001). The total set of $^{12}$C and $^{13}$C carbon isotopomers of a metabolite of $n$ carbon atoms comprises $2^n$ species that are all possible position-dependent combinations of the two different isotopes. Since they are rarely determined completely from the available NMR or MS data, the computations are often simplified by using models that represent only subsets of the isotopomer distribution; i.e. cumomers (sets of molecules that are $^{13}$C-labeled at a certain position, while the remaining positions may be $^{12}$C or $^{13}$C) (Wiechert et al. 1999), bondomers (molecules that vary only in numbers and positions of intact C-C bonds) (van Winden et al. 2002), or summed fractional labels (Christensen et al. 2002). Such rigorous and comprehensive accounting of all available physiological and isotopomer data from a single experiment retrieves the maximum information through data integration. The *in vivo* carbon flux distribution in different microorganisms has been quantified by this approach with $^{13}$C data from NMR analysis (Petersen et al. 2000; Emmerling et al. 2002; van Winden et al. 2003; Yang et al. 2003), MS analysis (Wittmann and Heinzle 2002; Fischer and Sauer 2003b; Klapa et al. 2003), or both (Zhao et al. 2003).

Although proper statistics are mandatory (Dauner et al. 2001a; Wiechert et al. 2001), the evidence for a particular flux remains indirect in the integrative approach and not all fluxes may be observable from a given set of data (van Winden et al. 2001b; Wittmann and Heinzle 2001). Since all data are globally integrated into a single flux solution, measurement errors or incomplete network models affect the entire flux solution. Hence, alternative methods must be invoked to obtain independent and direct confirmation of key fluxes. Qualitative confirmation may be derived from *in vitro* enzyme activities or mRNA abundance data if a flux is estimated to be on/off, but such data do not account for kinetic regulation and may be misleading. More reliable are analytically determined flux ratios that provide direct and quantitative evidence for the *in vivo* operation of a particular flux. Since such METAFoR analysis is a local approach, the determined flux ratios are independent of each other and a given ratio is not adversely affected by an incomplete model structure or measurement errors elsewhere (Szyperski 1998; Fischer and Sauer 2003a). Moreover, the flux ratios are completely independent of the physiological data. Consequently, METAFoR analysis has been used successfully to confirm flux estimates from isotopomer balancing approaches in a number of cases (Emmerling et al. 2002; Fischer and Sauer 2003b; Hua et al. 2003; Yang et al. 2003; Fischer et al. 2004). The apparent favorable agreement between global and local flux estimates also demonstrates the reliability of either approach.


# 3 Metabolic networks in motion: Flux analysis in systems biology

Biochemical reactions do not occur as isolated processes but function as components in highly organized and regulated networks. While static measures of mRNA, protein, and metabolite concentrations assess certain aspects of biochemical component interactions and regulation, flux analysis is presently the exclusive methodology that provides a global perspective on network operation and the integrated regulation at all levels (Hellerstein 2003) (Fig. 1). It cannot, however, decipher between the different levels of regulation, e.g. genetic or biochemical. Notably, flux analysis quantifies *in vivo* reaction velocities (enzyme activities), thus, is not influenced by uncertainties in effector/free metabolite concentrations that typically qualify conclusions derived from *in vitro* enzyme activity measurements.


## 3.1 Identification of unexpected or novel pathways and reactions

Quantitative data on carbon traffic are pivotal to understand governing principles of network operation and can provide a basis for metabolic model building and predictions. Much of the present focus is on flux quantification upon selected genetic or environmental modifications. While the results are mostly descriptive, some are rather surprising and identify new features of network operation that cannot be obtained by alternative methods. One class of examples reveals how effective bypass reactions contribute to genetic network robustness (Gu et al. 2003). Examples include (i) a bypass of pyruvate kinase via PEP carboxylase and malic enzyme in *Escherichia coli* (Sauer et al. 1999; Emmerling et al. 2002) or via the reverse PEP carboxykinase reaction in *Bacillus subtilis* (Zamboni et al. 2004b), and (ii) a bypass of phosphoglucoisomerase via the pentose phosphate pathway in several bacteria (Fischer and Sauer 2003a; Marx et al. 2003) but not in the yeast

*Saccharomyces cerevisiae* (Fiaux et al. 2003). At genome-scale, however, only about $^1/_4^{th}$ of the non-lethal metabolic knockouts in *S. cerevisiae* are bypassed through alternative pathways, while network redundancy through duplicate genes was the major cause of genetic robustness (Blank et al. 2005a).

Another case is unexpected operation of pathways that, based on qualitative genetic or *in vitro* evidence, were considered inactive; e.g. the Entner-Doudoroff pathway in an actinomycete (Gunnarsson et al. 2004) or the TCA cycle in glucose-grown *S. cerevisiae* batch cultures with low growth rates that were induced by suboptimal the chemical conditions (including pH, osmolarity, and decouplers) (Blank and Sauer 2004). Although gluconeogenic gene expression is repressed during growth on glucose, $^{13}$C-flux analysis frequently reveals gluconeogenic fluxes in microbial cultures on glucose-containing media (Marx et al. 1996; Christensen et al. 2002; Christiansen et al. 2002; Moreira dos Santos et al. 2003a; Zamboni and Sauer 2003; Fuhrer et al. 2005). Simultaneous operation of glycolytic and gluconeogenic reactions effectively constitutes futile cycles between the involved metabolites, thereby, dissipating ATP of up to one molecule of ATP per consumed molecule of glucose (or about 10% of the energy requirements for biomass formation) (Sauer et al. 1997; Emmerling et al. 2002; Yang et al. 2003). A particularly extreme example is the extensive futile cycling between $C_3$ carboxylation and $C_4$ decarboxylation in *Corynebacterium glutamicum* during growth on glucose and lactate (Petersen et al. 2000); possibly a consequence of lactate-triggered gluconeogenesis (Klapa et al. 2003). Futile cycle-based ATP dissipation does not, however, appear to be a relevant energy-dissipating mechanism under conditions of carbon and energy excess (Dauner et al. 2001b).

Since employed network models for flux analysis are mathematical abstractions that do not necessarily reflect biochemical reality, the identification of novel reactions or pathways is inherently difficult with global flux and isotopomer balancing approaches when data are fitted to the assumed model. Deleted genes are typically removed from the flux model because the estimated flux through the inactivated reaction is rarely zero (Emmerling et al. 2002), but newly activated reactions may be overlooked by this approach. One example is a potential bypass of the conversion from glucose-6-P to 6-P-gluconate that is catalyzed by the gene product of *zwf* in *E. coli* (Fig. 3). When the reaction is removed from the model, global flux estimates for *zwf* mutants show, of course, complete absence of the *in vivo* reaction (Hua et al. 2003; Sauer et al. 2004). While local METAFoR analysis confirms the radical flux rerouting, it shows also that not a full 100% of glucose catabolism proceeds through glycolysis (Fischer and Sauer 2003a; Sauer et al. 2004). This would suggest a small catabolic contribution of the gluconate bypass via glucose dehydrogenase in *zwf* mutants, and this was indeed demonstrated in *B. subtilis* (Zamboni et al. 2004a). In some microbes, an extracellular glucose dehydrogenase is the primary route of glucose degradation (Fuhrer et al. 2005).

Such results also question the ability of existing 'textbook' model pathways to portray flux through complex reaction networks: the pathway is an abstraction, whereas the combination of all pathways establish the real network of fluxes (Marcotte 2001). Complementary to theoretical approaches (Schuster et al. 2000; Stelling et al. 2002; Papin et al. 2003), metabolic flux analysis can make important

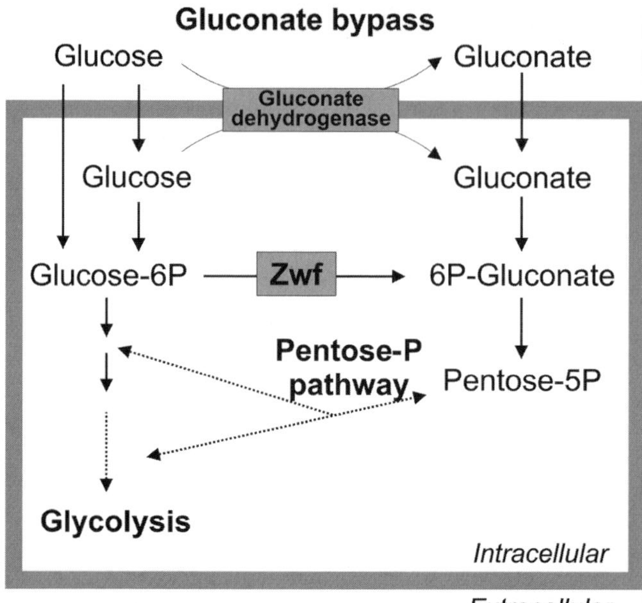


**Fig. 3.** Metabolic reactions involved in the conversion of glucose to pentose-5P.

inroads to a systems-level understanding by unraveling entirely new reactions or new modes of operation of known reactions. An example to the latter is the recent discovery of the novel PEP-glyoxylate cycle that catalyzes glucose oxidation in *E. coli* under strict glucose limitation (Fischer and Sauer 2003b); a property that was considered to be exclusive to the ubiquitous TCA cycle. Key reactions are the constituents of the well-known glyoxylate shunt and PEP carboxykinase, whose conjoint operation in this bi-functional catabolic and anabolic cycle is in sharp contrast to their generally recognized functions (Fig. 4). The operation and function of this novel cycle cannot be understood from the analysis of the isolated enzymes. While the PEP-glyoxylate cycle was computed before as one of several elementary flux modes (Schuster et al. 1999), such new modes of network operation are exclusively accessible through experimental flux analytical methods and this may explain why the PEP-glyoxylate pathway was only identified very recently (Fischer and Sauer 2003b; Hua et al. 2003; Sauer et al. 2004). As it may have become clear, experimental analysis of metabolic fluxes is yet in its infancy with much of the present focus on checking consistency between 'textbook' knowledge on reactions and pathways and the behavior of networks as a whole. The current emphasis is on correct assembly of the metabolic systems and on providing baseline data that allow constructing comprehensive models.

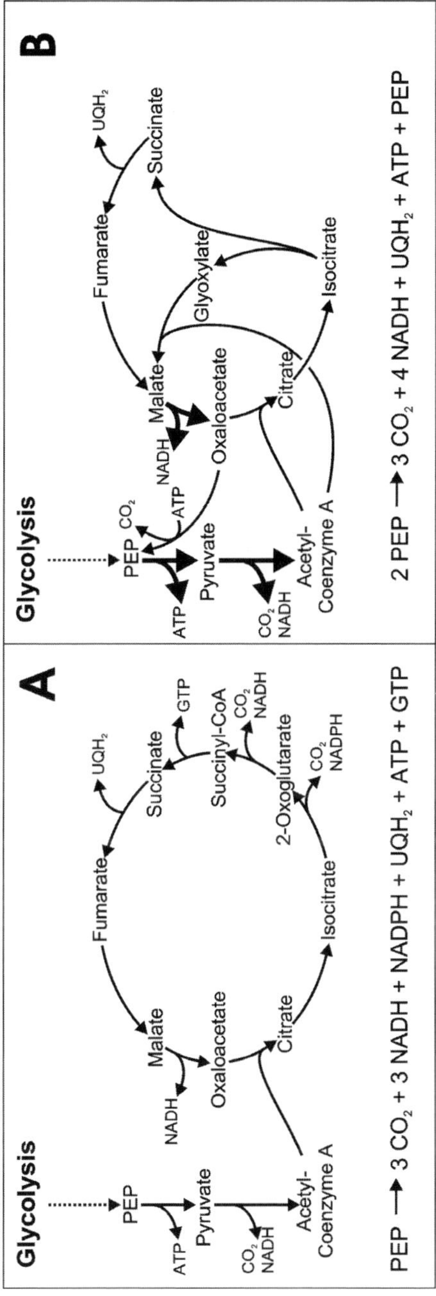


**Fig. 4.** Stoichiometry of the TCA cycle (A) and the PEP-glyoxylate cycle (B) in *E. coli* (Fischer and Sauer 2003b). Solid arrows indicate reactions that are required twice per cycle.

## 3.2 Identification of metabolic systems properties

Beyond discovery of novel or unexpected modes of metabolic network operation, flux analysis has great potential for hypothesis-driven research by unraveling systems properties that emerge from biochemical component interactions. One example of such interactions in the realm of metabolic networks is production and consumption of enzymatic cofactors that participate in hundreds of reactions. The redox cofactors NADH and NADPH, for example, serve distinct biochemical functions and must be supplied at appropriate rates and stoichiometries for balanced growth. Whereas NADH is the main respiratory cofactor with a primary role in energy metabolism, NADPH exclusively drives anabolic reductions, thus, linking the fundamental processes of catabolism and anabolism (Fig. 5). Quantitative understanding of the key players in NADPH metabolism cannot be attained through detailed knowledge on individual enzymes but only through comprehensive investigation of their biochemical interaction in the network, and flux responses to perturbations in NADPH metabolism have been quantified in a number of microbes (Marx et al. 1999; Canonaco et al. 2001; Moreira dos Santos et al. 2003b). While some organisms such as *S. cerevisiae* cannot tolerate imbalances between catabolic NADPH production and anabolic NADPH consumption (Nissen et al. 2001; Fiaux et al. 2003), others have apparently significant flexibility (Dauner et al. 2001a; Marx et al. 2003; Blank et al. 2005b).

In *E. coli*, $^{13}$C-flux-based *in vivo* quantification of NADPH production and consumption rates in all major reactions identified metabolic situations where either more (Emmerling et al. 2002; Yang et al. 2003) or less NADPH (Fischer et al. 2004; Sauer et al. 2004) was produced than required for biosynthesis. Hypothesis-driven knockouts of key enzymes and their flux responses then identified divergent functions of the two transhydrogenases PntAB and UdhA (Sauer et al. 2004). While the energy-dependent transhydrogenase PntAB exclusively catalyzes electron transfer from NADH to NADP$^+$, the soluble and energy-independent isoform UdhA catalyzes the reverse reaction. The former supplies about one third of the required NADPH during standard batch growth on glucose, and the latter becomes relevant under conditions with excess NADPH formation from substrate catabolism. Such excess NADPH conditions are growth on acetate or in glucose-limited, aerobic continuous cultures with their extensive TCA cycle fluxes that produce much NADPH in the isocitrate dehydrogenase reaction (Emmerling et al. 2002; Yang et al. 2003). Additionally, the aforementioned PEP-glyoxylate cycle provides a possibility to generate less NADPH because it achieves the same stoichiometry of glucose oxidation to $CO_2$ as the TCA cycle without the concomitant NADPH formation (Fig. 4). Consistently, operation of this cycle was observed in NADPH-overproducing phosphoglucose isomerase mutants of *E. coli* (Fischer and Sauer 2003b; Hua et al. 2003; Sauer et al. 2004). While the two transhydrogenases endow *E. coli* metabolism with an extraordinary flexibility to cope with varying catabolic and anabolic demands, only few organisms contain both isoforms and some none at all (Dauner et al. 2001a; Marx et al. 2003). Novel, transhydrogenase-independent NADPH balancing mechanisms such as enzymes

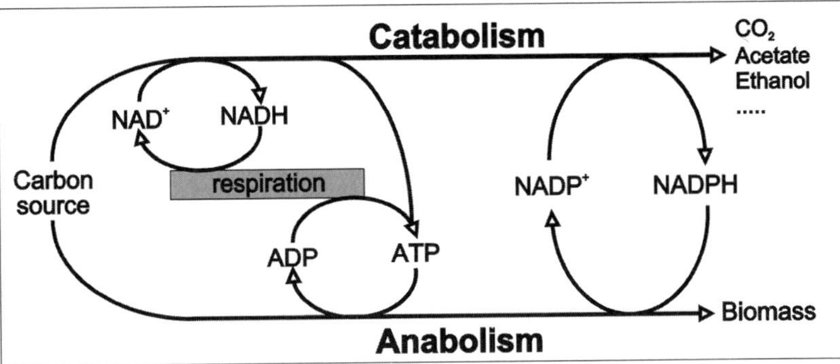


**Fig. 5.** Metabolic cofactor balancing.

with high affinity for both redox cofactores (Verho et al. 2002) have profound impact on network operation. Metabolic flux analysis is then a key methodology to i) identify apparently 'unbalanced' cases where such mechanisms must operate (Blank et al. 2005b; Fuhrer et al. 2005), and ii) to unravel the underlying mechanism in combination with other methods.

# 4 Recent developments and future needs in metabolic flux analysis

Methods for the quantification of *in vivo* molecular fluxes have attained a high level of sophistication and new experimental discoveries on metabolic systems operation and design principles of flux traffic are beginning to precipitate (Fischer and Sauer 2005). However humble such discoveries may appear in view of the grand goal of understanding and *predicting* cellular behavior, they are the necessary fundament for correct assembly of realistic networks that reflect the potential capacity to manage carbon traffic beyond the mere network stoichiometry. The increasing confidence in experimental flux results is fueled by confirmatory results obtained with the complementary methods for local and global analyses (Christensen et al. 2001; Emmerling et al. 2002), MS and NMR spectroscopy (Zhao et al. 2003), and the availability of a universal framework for $^{13}$C-flux analysis (Wiechert et al. 2001). Moreover, major breakthroughs were achieved in the throughput of flux methods that pave the road to large-scale functional studies (Sauer 2004; Blank et al. 2005a; Fischer and Sauer 2005), and hundreds or thousands of mutant/condition analyses will be realized in the next years. At present, flux data are used primarily to complete metabolic models or to verify predictions from computational analyses of network properties (Segre et al. 2002; Almaas et al. 2004; Barabasi and Oltavi 2004; Burgard et al. 2004; Wiback et al. 2004), but they will likely become a more integrated component in the iterative systems biology cycle.

The obvious next step in flux method development is direct detection of $^{13}$C-pattern in pathway intermediates rather than proteinogenic amino acids or accumulated extracellular metabolites. This is done for few selected metabolites to quantify selected fluxes in higher cells (Shulman and Rothman 2001; Hellerstein 2003; see also the entire first issue of *Metabolic Engineering* 6 in 2004). Building on recent advances in metabolome analyses (Buchholz et al. 2001; Weckwerth and Fiehn 2002), it may become feasible to develop a more general methodology that allows to resolve many fluxes, and a first application to the upper part of metabolism has been published (van Winden et al. 2005). Generalization of the approach will greatly extend the applicability of flux analysis beyond microbial growth on single substrates and allow resolution of dynamic responses of metabolism. Current methods discriminate the impact of perturbations on steady state fluxes, but more rapid responses to environmental perturbations, such as pulses of limiting substrates or depletion of carbon or energy sources during the transition from exponential growth to stationary phase, are beyond the present capabilities.

Potential problems that have not yet been addressed much are (i) changes in enzyme cofactor preference when metabolite concentrations change in perturbed cells and (ii) so-called underground metabolism; i.e. enzymes catalyzing reactions with alternative substrates (D'Ari and Casadesus 1998; van Winden et al. 2001a). Such metabolic 'inaccuracy' could lead to actual *in vivo* network operations that cannot be inferred from genome data without detailed biochemical knowledge – which is not available for most enzymes. Some computed flux results would also be affected by this problem because the assumed reaction stoichiometry does not hold true, at least not completely. When combined with metabolite measurements and biochemical data, however, flux analysis is also a route to identify imprecise reactions or altered cofactor usage in a hypothesis-driven approach akin to that described for unexpected pathways (see Section 3.1).

Quantitative monitoring of whole network operation by metabolic flux analysis provides a global perspective on the integrated regulation at all levels, including gene expression, enzyme kinetics, and allosteric regulation. To distinguish between the different levels of regulation and to identify the global metabolic control structure, flux data must be combined with transcriptome, proteome, and metabolome data. Ideally, such data are generated from the same experiment and first attempts have already been made to collect flux data in combination with global mRNA abundance (Oh and Liao 2000; Daran-Lapujade et al. 2004; Sonderegger et al. 2004) or intracellular metabolite concentrations (Yang et al. 2003). Unfortunately, these descriptive data have only been interpreted verbally but not quantitatively within computer models. Generations of such data from a number of environmental or genetic conditions and their quantitative interpretation are of immediate relevance for a system-based understanding of regulatory networks in a broader sense. Applied to steady state cultures, it could reveal, for example, precisely how control in complex metabolic networks is shared between different regulatory mechanisms (ter Kuile and Westerhoff 2001).

# 5 Quo vadis metabolic systems biology?

As a discovery tool, flux(ome) analysis can probe network behavior, identify the role of individual proteins within the network, and identify network topology by quantifying the key variable of metabolic network *operation*. The two key variables of network *composition* - the concentrations of enzymes and metabolites - may be assessed globally through proteome and metabolome analyses. Comprehensive network analysis requires quantitative data on the dynamic response of the operational (flux) and compositional (concentrations) variables to perturbations, together with knowledge on parameters like enzyme kinetics and network topology. Integration of metabolite concentrations and enzyme kinetics within kinetic models allows to predict short-term flux responses (Chassagnole et al. 2002; Moritz et al. 2002). To understand the global regulatory control structure or to predict metabolic network operation on longer time scales when genetic *and* metabolic regulation is relevant, compositional data on mRNA and protein abundance must be incorporated. The goal of such comprehensive models is to predict metabolic phenotypes and, thus, also molecular fluxes from compositional data (Fig. 1). While phenotypic predictions are easily verified or refuted through standard physiological experiments, experimental flux data can quantify the deviation of predictions from reality at the molecular level. Hence, flux data may either be used to identify model parameters or as a molecular verification of molecular predictions, as was described in several recent cases (Segre et al. 2002; Almaas et al. 2004; Barabasi and Oltavi 2004; Burgard et al. 2004; Wiback et al. 2004).

As a typical (post)genomic science (Kell and Oliver 2003), systems biology follows either the inductive, top-down or the hypothesis-driven, bottom-up approach. The former is founded on massive functional genomic data (Auffray et al. 2003) and the models may be genome-scale representations of metabolic network topology (Price et al. 2003). While such models represent only the stoichiometric aspect of metabolism, they allow to assess certain static network properties and to qualitatively predict metabolic phenotypes; currently with an accuracy of about 80% (Edwards and Palsson 2000; Famili et al. 2003). To increase the predictive power, the present challenge is to extend such models by adding regulation so that large-scale 'omics data may be incorporated (Covert et al. 2004) or to develop novel mathematical frameworks that predict regulation (Stelling et al. 2002). The complementary bottom-up approach affords more detailed mathematical descriptions of particular subsystems that may be individual genetic circuits (Isaacs et al. 2003), metabolic (Chassagnole et al. 2002) or signal transduction pathways (Bhalla et al. 2002; Kholodenko et al. 2002). While the represented kinetic detail requires extensive experimental data and may be computationally demanding, kinetics-based models enable dynamic and quantitative simulation of systems behavior. Out of necessity, the bottom-up approach focuses on computationally and experimentally manageable sub-systems, with the inherent understanding that more comprehensive models are assembled in a modularized fashion (Csete and Doyle 2002; Kholodenko et al. 2002; Ginkel et al. 2003; Wolf and Arkin 2003; Castellanos et al. 2004).

Data-driven approaches will continuously provide large data sets. Nevertheless, it is impossible to measure everything and luckily this is not necessary. Irrespective of the systems biology approach taken, one key contribution of modeling is to identify key variables and to specify testable hypotheses that go beyond bioinformatic data correlation. Such *quantitative* model-based hypotheses are in sharp contrast to *qualitative* intuitive or logical conclusions that drive the reductionist approach. For this purpose, it is not always necessary that all model parameters can be determined from the measured variables. It would suffice, for example, if simulations suggest critical experiments that could discriminate between different biological mechanisms, interactions, or behavior that are consistent with the presently available data but are mutually exclusive. Thus, genome-wide system response monitoring must be complemented with (model hypothesis-driven) specific experiments of high temporal and spatial resolution. In particular the exact cellular location of proteins, their interaction in supramolecular structures, or reliable protein-protein interaction data will be of utmost importance. Most likely, functional genomic data generation and modeling will coexist somewhat independently for some time, but model-suggested experiments that iteratively feedback into the model will mark the breakthrough of systems biology. Given the complexity of biological systems, the iterative systems biology cycle appears more attractive, in the long run, than exclusive reductionist or inductive, data-driven approaches to finally arrive at a quantitative understanding.

Eventually, developed computer models will serve as experimental objects *in lieu* of the real system (thing or process) they represent. Developing whole cell models for quantitative simulation is an ambitious goal (Holden 2002; Palsson 2002), and the jury will remain out for some time whether this is feasible at all (Bray 2001; Nurse 2003). Nevertheless, there is a high probability that the endeavor as such will aid tremendously in understanding governing principles of biological systems operation that cannot be obtained by studying sub-systems or their components in isolation. Despite an emphasis on either experimental or computational analyses, promising first success of systems-approaches was demonstrated already in a number of cases that include signal transduction pathways (Bhalla and Iyengar 1999; Bhalla et al. 2002; Kholodenko et al. 2002; Schoeberl et al. 2002; Kikuchi et al. 2003; Lee et al. 2003), developmental networks (von Dassow et al. 2000), genetic modules (Isaacs et al. 2003), chemotaxis (Barkai and Leibler 1997), and metabolic networks (Ideker et al. 2001b; Ibarra et al. 2002; Stelling et al. 2002; Francke et al. 2003; Sauer et al. 2004).

The undebatable necessity for iterations makes much of the current practice appear suboptimal: experimentalists generate data that modelers download from databases, extract from papers, or in rare cases, obtain from a distant collaborator. To a large extent, experimentalists and theorists have remained unrelated species! It comes hardly as a surprise though that experimentalists do not unanimously embrace computational approaches, as has been described so entertainingly by Bray (2001) and Lazebnik (2002). Nevertheless, much of the present debate on systems biology is semantic and a normal process of defining a new field and its essential concepts. To enable the conceptual transition from reductionism to systems biology, experimentalists, and theorists must engage in deep and continued collaborations, develop a common, quantitative language (Lazebnik 2002), and breed new scientists that are equally comfortable at the bench and the computer (Bialek and Botstein 2004). While much of this is commonplace, it implies a slow shift in academic paradigms. Systems biologists must exploit, value, and further develop the quantitative rigor of engineering methods and theoreticians must solve actual biological problems at the necessary scientific depth, rather than 'postdict' available data.

## Acknowledgements

I am grateful to Jörg Stelling for fruitful comments on the manuscript.

## References

Aggarwal K, Lee KH (2003) Functional genomics and proteomics as a foundation for systems biology. Br Funct Genomics Prot 2:175-184

Almaas E, Kovacs B, Vicsek T, Oltavi ZN, Barabasi AL (2004) Global organization of metabolic fluxes in the bacterium *Escherichia coli*. Nature 427:839-843

Arkin AP (2001) Synthetic cell biology. Curr Opin Biotechnol 12:638-644

Auffray C, Imbeaud S, Roux-Rouquie M, Hood L (2003) From functional genomics to systems biology: concepts and practices. C R Biol 326:879-892

Bailey JE (1991) Toward a science of metabolic engineering. Science 252:1668-1675

Bailey JE (1999) Lessons from metabolic engineering for functional genomics and drug discovery. Nat Biotechnol 17:616-618

Bailey JE (2001a) Complex biology with no parameters. Nature Biotechnol 19:503-504

Bailey JE (2001b) Reflections on the scope and the future of metabolic engineering and its connections to functional genomics and drug discovery. Metab Eng 3:111-114

Bar-Joseph Z, Gerber GK, Lee TI, Rinaldi NJ, Yoo JY, Robert F, Gordon DB, Fraenkel E, Jaakkola TS, Young RA, Gifford DK (2003) Computational discovery of gene modules and regulatory networks. Nat Biotechnol 21:1337-1342

Barabasi AL, Oltavi ZN (2004) Network biology: Understanding the cell's functional organization. Nat Rev Genet 5:101-105

Barkai N, Leibler S (1997) Robustness in simple biochemical networks. Nature 387:913-917

Bhalla US, Iyengar R (1999) Emergent properties of networks of biological signaling pathways. Science 283:381-387

Bhalla US, Ram PT, Iyengar R (2002) MAP kinase phosphatase as a locus of flexibility in a mitogen-activated protein kinase signaling network. Science 297:1018-1023

Bialek W, Botstein D (2004) Introductory science and mathematics education for 21st-century biologists. Science 303:788-790

Blank LM, Kuepfer L, Sauer U (2005a) Large-scale $^{13}$C-flux analysis reveals mechanistic principles of metabolic network robustness to null mutations in yeast. Genome Biol (in press)

Blank LM, Lehmbeck F, Sauer U (2005b) Metabolic flux and network analysis in 14 hemi-ascomycetous yeasts. FEMS Microbiol Lett (ePub ahead of print)

Blank LM, Sauer U (2004) TCA cycle activity in *Saccharomyces cerevisiae* is a function of the environmentally determined specific growth and glucose uptake rate. Microbiology 150:1085-1093

Bonarius HPJ, Schmid G, Tramper J (1997) Flux analysis of underdetermined metabolic networks: the quest for the missing constraint. Trends Biotechnol 15:308-314

Bray D (2001) Reasoning for results. Nature 412:863

Buchholz A, Takors R, Wandrey C (2001) Quantification of intracellular metabolites in *Escherichia coli* K12 using liquid chromatographic-electrospray ionization tandem mass spectrometric techniques. Anal Biochem 295:129-137

Burgard AP, Nikolaev EV, Schilling CH, Maranas CD (2004) Flux coupling analysis of genome-scale metabolic network reconstructions. Genome Res 14:301-314

Canonaco F, Hess TA, Heri S, Wang T, Szyperski T, Sauer U (2001) Metabolic flux response to phosphoglucose isomerase knock-out in *Escherichia coli* and impact of overexpression of the soluble transhydrogenase UdhA. FEMS Microbiol Lett 204:247-252

Castellanos M, Wilson DB, Shuler ML (2004) A modular minimal cell model: purine and pyrimidine transport and metabolism. Proc Natl Acad Sci USA 101:6681-6686

Chassagnole C, Noisommit-Rizzi N, Schmid JW, Mauch K, Reuss M (2002) Dynamic modeling of the central carbon metabolism of *Escherichia coli*. Biotechnol Bioeng 79:53-73

Christensen B, Christiansen T, Gombert AK, Nielsen J (2001) Simple and robust method for estimation of the split ratio between the oxidative pentose phosphate pathways and the Embden-Meyerhof-Parnas pathway in microorganisms. Biotechnol Bioeng 74:517-523

Christensen B, Gombert AK, Nielsen J (2002) Analysis of flux estimates based on $^{13}$C-labeling experiments. Eur J Biochem 269:2795-2800

Christensen B, Nielsen J (1999) Metabolic network analysis. Adv Biochem Eng Biotechnol 66:209-231

Christiansen T, Christensen B, Nielsen J (2002) Metabolic network analysis of *Bacillus clausii* on minimal and semirich medium using $^{13}$C-labeled glucose. Metab Eng 4:159-169

Covert MW, Knight EM, Reed JL, Herrgard MJ, Palsson BO (2004) Integrating high-throughput and computational data elucidates bacterial networks. Nature 429:92-96

Covert MW, Palsson BO (2002) Transcriptional regulation in constraints-based metabolic models of *Escherichia coli*. J Biol Chem 277:28058-28064

Covert MW, Schilling CH, Famili I, Edwards JS, Goryanin II, Selkov E, Palsson BO (2001) Metabolic modeling of microbial strains *in silico*. Trends Biochem Sci 26:179-186

Csete ME, Doyle JC (2002) Reverse engineering of biological complexity. Science 295:1664-1669

D'Ari R, Casadesus J (1998) Underground metabolism. Bioessays 20:181-186

Daran-Lapujade P, Jansen ML, Daran JM, van Gulik W, De Winde JH, Pronk JT (2004) Role of transcriptional regulation in controlling fluxes in central carbon metabolism of *Saccharomyces cerevisiae*, a chemostat culture study. J Biol Chem 279:9125-9138

Dauner M, Bailey JE, Sauer U (2001a) Metabolic flux analysis with a comprehensive isotopomer model in *Bacillus subtilis*. Biotechnol Bioeng 76:144-156

Dauner M, Storni T, Sauer U (2001b) *Bacillus subtilis* metabolism and energetics in carbon-limited and carbon-excess chemostat culture. J Bacteriol 183:7308-7317

Edwards JS, Palsson BO (2000) The *Escherichia coli* MG1655 *in silico* metabolic phenotype: its definition, characteristics, and capabilities. Proc Natl Acad Sci USA 97:5528-5533

Emmerling M, Dauner M, Ponti A, Fiaux J, Hochuli M, Szyperski T, Wuthrich K, Bailey JE, Sauer U (2002) Metabolic flux responses to pyruvate kinase knockout in *Escherichia coli*. J Bacteriol 184:152-164

Famili I, Förster J, Nielsen J, Palsson BO (2003) *Saccharomyces cerevisiae* phenotypes can be predicted by using constraint-based analysis of a genome-scale reconstructed metabolic network. Proc Natl Acad Sci USA 100:13134-13139

Fiaux J, Çakar ZP, Sonderegger M, Wüthrich K, Szyperski T, Sauer U (2003) Metabolic flux profiling of the yeasts *Saccharomyces cerevisiae* and *Pichia stipitis*. Euk Cell 2:170-180

Fischer E, Sauer U (2003a) Metabolic flux profiling of *Escherichia coli* mutants in central carbon metabolism by GC-MS. Eur J Biochem 270:880-891

Fischer E, Sauer U (2003b) A novel metabolic cycle catalyzes glucose oxidation and anaplerosis in hungry *Escherichia coli*. J Biol Chem 278:46446-46451

Fischer E, Sauer U (2005) Large-scale *in vivo* flux analysis reveals rigidity and sub-optimal performance of *Bacillus subtilis* metabolism. Nat Genet (in press)

Fischer E, Zamboni N, Sauer U (2004) High-throughput metabolic flux analysis based on gas chromatography-mass spectrometry derived $^{13}$C constraints. Anal Biochem 325:308-316

Francke C, Postma PW, Westerhoff HV, Blom JG, Peletier MA (2003) Why the phosphotransferase system of *Escherichia coli* escapes diffusion limitation. Biophys J 85:612-622

Fuhrer T, Fischer E, Sauer U (2005) Experimental identification and quantification of glucose metabolism in seven bacterial species. J Bacteriol 187:1581-1590

Gallagher R, Appenzeller T (1999) Beyond reductionism. Science 284:79

Ginkel M, Kremling A, Nutsch T, Rehner R, Gilles ED (2003) Modular modeling of cellular systems with ProMoT/Diva. Bioinformatics 19:1169-1176

Grivet J-P, Delort A-M, Portais J-C (2003) NMR and microbiology: from physiology to metabolomics. Biochimie 85:823-840

Gu Z, Steinmetz LM, Gu X, Scharfe C, Davis RW, Li W-H (2003) Role of duplicate genes in genetic robustness against null mutations. Nature 421:63-66

Gunnarsson N, Mortensen UH, Sosio M, Nielsen J (2004) Identification of the Entner-Doudoroff pathway in an antibiotic-producing actinomycete species. Mol Microbiol 52:895-902

Heinrich R, Rapoport TA (1974) A linear steady-state treatment of enzymatic chains. General properties, control and effector strength. Eur J Biochem 42:39-95

Hellerstein MK (2003) *In vivo* measurement of fluxes through metabolic pathways: The missing link in functional genomics and pharmaceutical research. Annu Rev Nutr 23:379-402

Hellerstein MK, Neese RA (1999) Mass isotopomer distribution analysis at eight years: theoretical, analytical, and experimental considerations. Am J Physiol 276:E1146-E1170

Holden C (2002) Cell biology: alliance launched to model *E. coli*. Science 297:1459-1460

Hua Q, Yang C, Baba T, Mori H, Shimizu K (2003) Responses of the central carbon metabolism in *Escherichia coli* to phosphoglucose isomerase and glucose-6-phosphate dehydrogenase knockouts. J Bacteriol 185:7053-7067

Ibarra RU, Edwards JS, Palsson BO (2002) *Escherichia coli* K-12 undergoes adaptive evolution to achieve *in silico* predicted optimal growth. Nature 420:186-189

Ideker T, Galitski T, Hood L (2001a) A new approach to decoding life: systems biology. Annu Rev Genomics Hum Genet 2:343-372

Ideker T, Lauffenburger D (2003) Building with a scaffold: emerging strategies for high- to low-level cellular modeling. Trends Biotechnol 21:255-262

Ideker T, Thorsson V, Ranish JA, Christmas R, Buhler J, Eng JK, Bumgarner R, Goodlett DR, Aebersold R, Hood L (2001b) Integrated genomic and proteomic analyses of a systematically perturbed metabolic network. Science 292:929-934

Isaacs FJ, Hasty J, Cantor CR, Collins JJ (2003) Prediction and measurement of an autoregulatory genetic module. Proc Natl Acad Sci USA 100:7714-7719

Kacser H, Burns JA (1973) The control of flux. Symp Soc Exp Biol 27:65-104

Kell DB, Oliver SG (2003) Here is the evidence, now what is the hypothesis? The complementary roles of inductive and hypothesis-driven science in the post-genomic era. Bioessays 26:99-105

Kholodenko BN, Kiyatkin A, Bruggeman FJ, Sontag E, Westerhoff HV, Hoek JB (2002) Untangling the wires: a strategy to trace functional interactions in signaling and gene networks. Proc Natl Acad Sci USA 99:12841-12846

Kikuchi S, Fujimoto K, Kitagawa N, Fuchikawa T, Abe M, Oka K, Takei K, Tomita M (2003) Kinetic simulation of signal transduction system in hippocampal long-term potentiation with dynamic modeling of protein phosphatase 2A. Neural Netw 16:1389-1398

Kitano H (2002a) Computational systems biology. Nature 420:206-210

Kitano H (2002b) Systems biology: a brief overview. Science 295:1662-1664

Klapa MI, Aon JC, Stephanopoulos G (2003) Systematic quantification of complex metabolic flux networks using stable isotopes and mass spectrometry. Eur J Biochem 270:3525-3542

Koffas M, Roberge C, Lee K, Stephanopoulos G (1999) Metabolic engineering. Annu Rev Biomed Eng 1:535-557

Lazebnik Y (2002) Can a biologist fix a radio? - Or, what I learned while studying apoptosis. Cancer Cell 2:179-182

Lee E, Salic A, Kruger R, Heinrich R, Kirschner MW (2003) The roles of APC and axin derived from experimental and theoretical analysis of the Wnt pathway. PLoS Biol 1:E10-E15

Maaheimo H, Fiaux J, Çakar ZP, Bailey JE, Sauer U, Szyperski T (2001) Central carbon metabolism of *Saccharomyces cerevisiae* explored by biosynthetic fractional $^{13}$C labeling of common amino acids. Eur J Biochem 268:2464-2479

Marcotte EM (2001) The path not taken. Nature Biotechnol 19:626-627

Marx A, de Graaf AA, Wiechert W, Eggeling L, Sahm H (1996) Determination of the fluxes in the central metabolism of *Corynebacterium glutamicum* by nuclear magnetic resonance spectroscopy combined with metabolite balancing. Biotechnol Bioeng 49:111-129

Marx A, Eikmanns BJ, Sahm H, de Graaf AA, Eggeling L (1999) Response of the central metabolism in *Corynebacterium glutamicum* to the use of an NADH-dependent glutamate dehydrogenase. Metabolic Eng 1:35-48

Marx A, Hans S, Möckel B, Bathe B, de Graaf AA (2003) Metabolic phenotype of phosphoglucose isomerase mutants of *Corynebacerium glutamicum*. J Biotechnol 104:185-197

Moreira dos Santos M, Gombert AK, Christensen B, Olsson L, Nielsen J (2003a) Identification of *in vivo* enzyme activities in the cometabolism of glucose and acetate by *Saccharomyces cerevisiae* by using $^{13}$C-labeled substrates. Eukar Cell 2:599-608

Moreira dos Santos M, Thygesen G, Kötter P, Olsson L, Nielsen J (2003b) Aerobic physiology of redox-engineered *Saccharomyces cerevisiae* strains modified in the ammonium assimilation for increased NADPH availability. FEMS Yeast Res 4:59-68

Moritz B, Striegel K, de Graaf AA, Sahm H (2002) Changes of pentose phosphate pathway flux *in vivo* in *Corynebacterium glutamicum* during leucine-limited batch cultivation as determined from intracellular metabolite concentration measurements. Metab Eng 4:295-305

Neidhardt FC, Ingraham JL, Schaechter M (1990) Physiology of the bacterial cell: a molecular approach. Sinauer Associates, Inc., Sunderland, Mass

Nissen TL, Anderlund M, Nielsen J, Villadsen J, Kielland-Brandt MC (2001) Expression of a cytoplasmic transhydrogenase in *Saccharomyces cerevisiae* results in formation of 2-oxoglutarate due to depletion of the NADPH pool. Yeast 18:19-32

Nurse P (2003) Understanding cells. Nature 424:883

Oh M-K, Liao JC (2000) Gene expression profiling by DNA microarrays and metabolic fluxes in *Escherichia coli*. Biotechnol Prog 16:278-286

Palsson BO (2002) *In silico* biology through "omis". Nature Biotechnol 20:649-650

Papin JA, Price ND, Wiback SJ, Fell DA, Palsson BO (2003) Metabolic pathways in the post-genome era. Trends Biochem Sci 28:250-258

Papin JA, Stelling J, Price ND, Klamt S, Schuster S, Palsson BO (2004) Comparison of network-based pathway analysis methods. Trends Biotechnol 22:400-405

Park SM, Shaw-Reid C, Sinskey AJ, Stephanopoulos G (1997) Elucidation of anaplerotic pathways in *Corynebacterium glutamicum* via $^{13}$C-NMR spectroscopy and GC-MS. Appl Microbiol Biotechnol 47:430-440

Petersen S, de Graaf AA, Eggeling L, Möllney M, Wiechert W, Sahm H (2000) *In vivo* quantification of parallel and bidirectional fluxes in the anaplerosis of *Corynebacterium glutamicum*. J Biol Chem 275:35932-35941

Portais J-C, Delort A-M (2002) Carbohydrate cycling in micro-organisms: what can $^{13}$C-NMR tell us? FEMS Microbiol Rev 26:375-402

Price ND, Papin JA, Schilling CH, Palsson BO (2003) Genome-scale microbial *in silico* models: the constraints-based approach. Trends Biotechnol 21:162-169

Purugganan M, Gibson G (2003) Merging ecology, molecular evolution, and functional genetics. Mol Ecol 12:1109-1112

Rohlin L, Oh M-K, Liao JC (2001) Microbial pathway engineering for industrial processes: evolution, combinatorial biosynthesis and rational design. Curr Opin Microbiol 4:330-335

Sanford K, Soucaille P, Whited G, Chotani G (2002) Genomics to fluxomics and physiomics - pathway engineering. Curr Opin Microbiol 5:318-322

Sauer U (2004) High-throughput phenomics: experimental methods for mapping fluxomes. Curr Opin Biotechnol 15:58-63

Sauer U, Canonaco F, Heri S, Perrenoud A, Fischer E (2004) The soluble and membrane-bound transhydrogenases UdhA and PntAB have divergent functions in NADPH metabolism of *Escherichia coli*. J Biol Chem 279:6613-6619

Sauer U, Hatzimanikatis V, Bailey JE, Hochuli M, Szyperski T, Wüthrich K (1997) Metabolic fluxes in riboflavin-producing *Bacillus subtilis*. Nat Biotechnol 15:448-452

Sauer U, Lasko DR, Fiaux J, Hochuli M, Glaser R, Szyperski T, Wuthrich K, Bailey JE (1999) Metabolic flux ratio analysis of genetic and environmental modulations of *Escherichia coli* central carbon metabolism. J Bacteriol 181:6679-6688

Schmidt K, Carlsen M, Nielsen J, Villadsen J (1997) Modeling isotopomer distributions in biochemical networks using isotopomer mapping matrices. Biotechnol Bioeng 55:831-840

Schoeberl B, Eichler-Jonsson C, Gilles ED, Muller G (2002) Computational modeling of the dynamics of the MAP kinase cascade activated by surface and internalized EGF receptors. Nature Biotechnol 20:370-375

Schuster S, Dandekar T, Fell DA (1999) Detection of elementary flux modes in biochemical networks: a promising tool for pathway analysis and metabolic engineering. Trends Biotechnol 17:53-60

Schuster S, Fell DA, Dandekar T (2000) A general definition of metabolic pathways useful for systematic organization and analysis of complex metabolic networks. Nature Biotechnol 18:326-332

Segre D, Vitkup D, Church GM (2002) Analysis of optimality in natural perturbed metabolic networks. Proc Natl Acad Sci USA 99:15112-15117

Selinger DW, Wright MA, Church GM (2003) On the complete determination of biological systems. Trends Biotechnol 21:252-254

Shen-Orr SS, Milo R, Mangan S, Alon U (2002) Network motifs in the transcriptional regulation network of *Escherichia coli*. Nature Genet 31:64-68

Sherry AD, Jeffrey FM, Malloy CR (2004) Analytical solutions for $^{13}$C isotopomer analysis of complex metabolic conditions: substrate oxidation, multiple pyruvate cycles, and gluconeogenesis. Metab Eng 6:12-24

Shulman RG, Rothman DL (2001) $^{13}$C NMR of intermediary metabolism: implications for systemic physiology. Annu Rev Physiol 63:15-48

Sonderegger M, Jeppsson M, Hahn-Hägerdal B, Sauer U (2004) The molecular basis for anaerobic growth of *Saccharomyces cerevisiae* on xylose, investigated by global gene expression and metabolic flux analysis. Appl Environ Microbiol 70:2307-2317

Stelling J, Klamt S, Bettenbrock K, Schuster S, Gilles ED (2002) Metabolic network structure determines key aspects of functionality and regulation. Nature 420:190-193

Stephanopoulos G (1999) Metabolic fluxes and metabolic engineering. Metabolic Eng 1:1-11

Stephanopoulos G, Alper H, Moxley J (2004) Exploiting biological complexity for strain improvement through systems biology. Nat Biotechnol 22:1261-1267

Stewart I (2004) Networking opportunity. Nature 427:601-604

Szyperski T (1995) Biosynthetically directed fractional $^{13}$C-labeling of proteinogenic amino acids: an efficient analytical tool to investigate intermediary metabolism. Eur J Biochem 232:433-448

Szyperski T (1998) $^{13}$C-NMR, MS and metabolic flux balancing in biotechnological research. Q Rev Biophys 31:41-106

Szyperski T, Glaser RW, Hochuli M, Fiaux J, Sauer U, Bailey JE, Wuthrich K (1999) Bioreaction network topology and metabolic flux ratio analysis by biosynthetic fractional $^{13}$C-labeling and two-dimensional NMR spectroscopy. Metabolic Eng 1:189-197

ter Kuile BH, Westerhoff HV (2001) Transcriptome meets metabolome: hierarchical and metabolic regulation of the glycolytic pathway. FEBS Lett 200:169-171 van Winden W, Verheijen P, Heijnen JJ (2001a) Possible pitfalls of flux calculations based on $^{13}$C-labeling. Metabolic Eng 3:151-162

van Winden WA, Heijnen JJ, Verheijen PJ (2002) Cummulative bondomers: A new concept in flux analysis from 2D [$^{13}$C, $^1$H] COSY NMR data. Biotechnol Bioeng 80:731-745

van Winden WA, Heijnen JJ, Verheijen PJ, Grievink J (2001b) A priori analysis of metabolic flux identifiability from $^{13}$C-labeling data. Biotechnol Bioeng 74:505-516

van Winden WA, van Dam JC, Ras C, Kleijn RJ, Vinke JL, van Gulik WM, Heijnen JJ (2005) Metabolic-flux analysis of *Saccharomyces cerevisiae* CEN.PK113-7D based on mass isotopomer measurements of $^{13}$C-labeled primary metabolites. FEMS Yeast Res 5: 559-568

van Winden WA, van Gulik WM, Schipper D, Verheijen PJ, Krabben P, Vinke JL, Heijnen JJ (2003) Metabolic flux and metabolic network analysis of *Penicillium chrysogenum* using 2D [$^{13}$C, $^1$H] COSY NMR measurements and cumulative bondomer simulation. Biotechnol Bioeng 83:75-92

Varma A, Palsson BO (1994) Metabolic flux balancing: Basic concepts, scientific, and practical use. Bio/Technol 12:994-998

Varner J (2000) Large-scale prediction of phenotype: concept. Biotechnol Bioeng 69:664-678

Verho R, Richard P, Jonson PH, Sundqvist L, Londesborough J, Penttilä M (2002) Identification of the fist fungal NADP-GAPDH from *Kluveromyces lactis*. Biochemistry 41:13833-13838

von Dassow G, Meir E, Munro EM, Odell GM (2000) The segment polarity network is a robust developmental module. Nature 406:188-192

Walsh K, Koshland DE Jr (1984) Determination of flux through the branch point of two metabolic cycles. J Biol Chem 259:9646-9654

Weckwerth W, Fiehn O (2002) Can we discover novel pathways using metabolomic analysis? Curr Opin Biotechnol 13:156-160

Westerhoff HV, Palsson BO (2004) The evolution of molecular biology into systems biology. Nat Biotechnol 22:1249-1252

Wiback SJ, Mahadevan R, Palsson BO (2004) Using metabolic flux data to further constrain the metabolic solution space and predict internal flux pattern: the *Escherichia coli* spectrum. Biotechnol Bioeng 86:317-331

Wiechert W (2001) $^{13}$C metabolic flux analysis. Metabolic Eng 3:195-206

Wiechert W (2002) Modeling and simulation: tools for metabolic engineering. J Biotechnol 94:37-63

Wiechert W, Möllney M, Isermann N, Wurzel M, de Graaf AA (1999) Bidirectional reaction steps in metabolic networks: III. Explicit solution and analysis of isotopomer labeling systems. Biotechnol Bioeng 66:69-85

Wiechert W, Möllney M, Petersen S, de Graaf AA (2001) A universal framework for $^{13}$C metabolic flux analysis. Metab Eng 3:265-283

Wittmann C (2002) Metabolic flux analysis using mass spectroscopy. Adv Biochem Eng Biotechnol 74:39-64

Wittmann C, Heinzle E (2001) Modeling and experimental design for metabolic flux analysis of lysine-producing *Corynebacteria* by mass spectrometry. Metabolic Eng 3:173-191

Wittmann C, Heinzle E (2002) Genealogy profiling through strain improvement by using metabolic network analysis: metabolic flux genealogy of several generations of lysine-producing Corynebacteria. Appl Environ Microbiol 68:5843-5859

Wolf DM, Arkin AP (2003) Motifs, modules and games in bacteria. Curr Opin Microbiol 6:125-134

Yang C, Hua Q, Baba T, Mori H, Shimizu K (2003) Analysis of *Escherichia coli* anaplerotic metabolism and its regulation mechanisms from the metabolic responses to altered dilution rates and phosphoenolpyruvate carboxykinase knockout. Biotechnol Bioeng 84:129-144

Zak DE, Gonye GE, Schwaber JS, Doyle III FJ (2003) Importance of input perturbations and stochastic gene expression in the reverse engineering of genetic regulatory networks: insights from an identifiability analysis of an *in silico* network. Genome Res 13:2394-2405

Zamboni N, Fischer E, Laudert D, Aymerich S, Hohmann H-P, Sauer U (2004a) The *Bacillus subtilis yqjI* gene encodes the $NADP^+$-dependent 6-P-gluconate dehydrogenase in the pentose phosphate pathway. J Bacteriol 186:4528-4534

Zamboni N, Maaheimo H, Szyperski T, Hohmann H-P, Sauer U (2004b) The phosphoenolpyruvate carboxykinase also catalyzes $C_3$ carboxylation at the interface of glycolysis and the TCA cycle of *Bacillus subtilis*. Metab Eng 6:277-284

Zamboni N, Sauer U (2003) Knockout of the high-coupling cytochrome $aa_3$ oxidase reduces TCA cycle fluxes in *Bacillus subtilis*. FEMS Microbiol Lett 226:121-126

Zhao J, Baba T, Mori H, Shimizu K (2003) Analysis of metabolic and physiological responses to *gnd* knockout in *Escherichia coli* by using C-13 tracer experiment and enzyme activity measurements. FEMS Microbiol Lett 220:295-301

Zupke C, Tompkins R, Yarmush D, Yarmush M (1997) Numerical isotopomer analysis: estimation of metabolic activity. Anal Biochem 247:287-293

## Abbreviations

GC: gas chromatography
NMR: nuclear magnetic resonance
METAFoR: metabolic flux ratio
mRNA: messenger RNA
MS: mass spectrometry
NAD(P)H: Nicotinamide adenine dinucleotide (phosphate)
PEP: phosphoenolpyruvate
TCA: tricarboxylic acid

Sauer, Uwe
    Institute of Molecular Systems Biology, ETH Zürich, CH-8093 Zürich, Switzerland
    sauer@biotech.biol.ethz.ch

# Metabolic networks: biology meets engineering sciences

A. Kremling, J. Stelling, K. Bettenbrock, S. Fischer, and E.D. Gilles

## Abstract

A hallmark of systems biology is the interdisciplinary approach to the complexity of biological systems, in which mathematical modeling constitutes an important part. Here, we use the example of sugar metabolism in the simple bacterium *Escherichia coli* and its associated control to illustrate the process of model development. Even for this well-characterized biological system, a close interaction between experimentation and theoretical analysis revealed novel, unexpected features. Additionally, the example shows how concepts from engineering sciences can facilitate the formal investigation of biological networks. More generally, we argue that analogies between complex biological and technical systems such as modular structures and common design principles provide crystallization points for fruitful research in both domains.

## 1 Systems biology: an interdisciplinary approach

For the past 30 years, it has been characteristic for biology to be qualitative and descriptive, directed towards the understanding of the molecular detail. However, for the understanding of complex system properties like optimal control, adaptation and memory, both the systems' components and their interactions have to be considered. Primarily the new 'omic technologies now make the complete determination of biological systems a realistic goal (Selinger et al. 2003). As a result, biology moves from the focus on few components to the study of networks of molecular interactions that give rise to complex physiological functions (Alm and Arkin 2003). Systems biology adopts this holistic view on biological function. However, several characteristics distinguish it from, and extend bioinformatics approaches to network analysis. A hallmark of many cellular networks, such as the intricate networks in cellular regulation, is that they respond dynamically towards extra- and intracellular conditions and signals. Only by means of a quantitative description of the systems' constituents and interactions, the resulting behavior can be understood in terms of the quantitative dynamics.

Furthermore, achieving this goal requires a theory-based approach to the complexity not understandable by intuition alone. Mathematical modeling of complex biological systems plays a central role in systems biology because it allows for a formalized treatment of biological networks in the computer, using tools from mathematics and systems sciences (Kitano 2002b; Tyson et al. 2001). Ideally, mathematical modeling requires and entails a precise representation of the knowledge on the

Topics in Current Genetics, Vol. 13
L. Alberghina, H.V. Westerhoff (Eds.): Systems Biology
DOI 10.1007/4735_88 / Published online: 6 July 2005
© Springer-Verlag Berlin Heidelberg 2005

system, and of hypotheses for unknown mechanisms. It allows one to apply formal methods of analysis. Mainly these two characteristics are expected to lead to a deepened understanding of the biological systems under consideration (Endy and Brent 2001; Gilman and Arkin 2002). Consequently, the efforts directed towards a quantitative, system-level understanding in biology rely on an interdisciplinary approach combining concepts from biology, information sciences and systems engineering. A central objective of systems biology is finally to develop virtual representations of cells and organisms. These representations should allow for computer experiments similar to experiments with real biological systems. Thereby, the way for a predictive biology can be paved, which will enhance, for instance, the understanding and the treatment of human diseases (Kitano 2002a; Stelling et al. 2001). There are already some examples of systems biological approaches that successfully couple experimental and theoretical approaches. They cover a broad spectrum of organisms and systems (www.siliconcell.net). The analysis of bacterial chemotaxis can be regarded as a paradigm of such an approach. The extensive experimental and theoretical analysis has helped substantially in the understanding of the system (Barkai and Leibler 1997). Currently, however, the knowledge on virtually any biological system does not permit to detail a complete list of parts, interactions and mechanisms, on which 'true' mathematical representations could be built. Instead, despite considerable progress in high-throughput experimentation, the resulting networks are still incomplete and bear inaccuracies (von Mering et al. 2002). Under these circumstance, an often encountered argument is that theoretical analysis should await an – in some sense – complete biological knowledge before becoming meaningful. We and others, however, argue that only an iterative cycle of experimentation and theory will be able to fulfill the promises of systems biology. Experiments generate data and hypotheses, subsequent mathematical modeling allows to assess the compatibility of both, and to derive novel or alternative explanations that can be evaluated in new experiments (Kitano 2002b; Stelling et al. 2001).

'Traditional' biology integrates new findings into cartoons of pathways or regulatory networks, or uses new knowledge to revise these representations. Similarly, mathematical models are 'work in progress' (Lee et al. 2003). In this process, however, unbiased predictions from formal representations can reveal unexpected properties of, or critical components in biological systems as in a recent experimental and theoretical analysis of the Wnt signaling pathway (Lee et al. 2003). In another case, mathematical modeling suggested a bistable trigger as a core element of cell cycle regulation a long time before an experimental confirmation of the mechanism was obtained (Novak and Tyson 1993; Pomerening et al. 2003; Sha et al. 2003). Here, we use the control of sugar uptake in the simple bacterium *Escherichia coli* to show that an iterative cycle of experimentation and model development can yield deeper insight into apparently well-understood systems. In particular, our background in engineering sciences provides concepts and methods for this study. We will focus the discussion of recent developments and future challenges in systems biology on potential (further) contributions that engineering could make to understand complex biological systems.

# 2 Model set-up

Starting point of every model developing procedure is the biological knowledge available from literature or text books. For *Escherichia coli*, knowledge on metabolism as well as for genetic regulation is rich and especially the lactose operon and its control has been investigated for a long time. Starting with the pioneering work of Jacques Monod who proposed the concept of defining operons as a sequence of genes that were expressed in a coordinated manner, current research in molecular biology has revealed a number of further strategies of cellular systems to adapt very efficiently to alterations of environmental conditions. Here, we used the lactose metabolism, i.e., the uptake of lactose and its break down to precursors, as an origin for the model set up presented. In successive steps we extend the model to cope with further environmental situations like different carbon sources to show how the individual pathways are organized to fulfill their physiological task and how the cells arrange the interaction of different pathways on a higher level of control. This approach differs from previous studies and modeling efforts on the PTS mainly in that it aims to an understanding of the interactions of genetic regulation and metabolism. Previous approaches mainly dealt with small subsystems covering either only metabolic reactions or only genetic regulation. A very seminal example is the work of Rohwer et al. (2000) who set up a detailed kinetic model of the PTS phosphorylation chain. This model gives interesting insights into the effects of complex formation, molecular crowding and flux response coefficients of these reactions but as the system is uncoupled from metabolism and genetic regulation it is not suitable for the understanding of the coupling of both levels.

## 2.1 Environment – the liquid phase

Considering a bio-reactor, the environment of the cells is described with the concentration of the carbon source $S$ in the liquid phase. Since below, the focus is on the cellular interior, the overall biomass $X$ is taken as the macroscopic variable. Growth of the biomass is coupled to the uptake of the carbon sources via yield coefficient $Y$ and uptake rate $r$. The uptake rates are functions of concentration of substrate in the bio-reactor and the concentration of the respective transport system which is located in the cytoplasmic membrane. For one substrate the respective equations read for a batch process:

$$\dot{X} = \mu \cdot X$$
$$\dot{S} = -r \cdot mg \cdot X \tag{1}$$

with the specific growth rate $\mu$ is given by $\mu = Y * r$ and $mg$ is the molecular weight of the carbon source (fluxes are given in [$\mu$mol/g DW h] and concentrations in the liquid phase are given in [g/l]). The equations in (1) are very general and are widely used in bioprocess engineering, since they describe the overall behavior of the biomass and the substrate in a simple manner. To describe the uptake reactions in a more detailed way, biological knowledge on the individual pathways has to be incorporated. As an example, the lactose pathway is considered in the following.

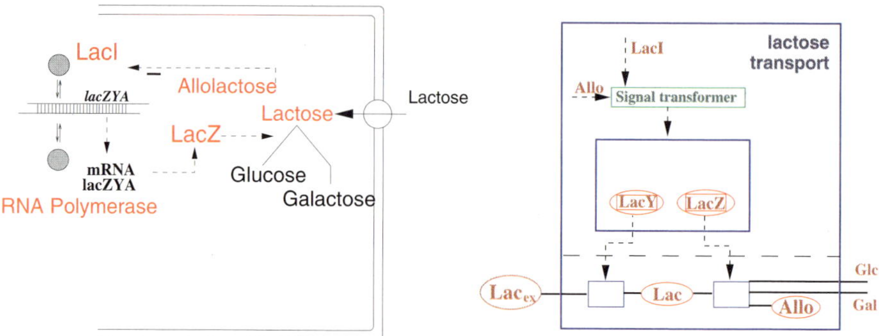


**Fig. 1.** Lactose uptake and metabolism. Left: Schematic diagram of lactose induction. Right: Representation of the sub-model with modeling objects.

## 2.2 Lactose pathway

Lactose is taken up via the lactose permease LacY (gene *lacY*). The permease works as a symporter, i.e., for every molecule lactose that is taken up, some molecules of $H^+$ is also taken up from the medium. Intracellular lactose is split into glucose and galactose by the $\beta$-galactosidase enzyme LacZ (gene *lacZ*). One by-product of this reaction is allolactose, the natural inducer of the lactose operon. If allolactose is present inside the cell, it deactivates the lactose repressor LacI, which blocks the binding of the RNA polymerase and therefore prevents the synthesis of the mRNA. A further transcription factor, Crp, which is activated by cAMP, activates the transcription of the operon. As can be seen in Figure 1, the lactose pathway represents a loop with positive return. The more allolactose is present, the more protein can be synthesized. With increasing amounts of the respective enzymes, allolactose is also degraded faster and a steady-state can be reached. From the scheme, it becomes also clear that the initial conditions for all components could not be zero, if the system should be inducible. If lactose is not present in the medium, a few molecules are necessarily available in each cell and will allow induction by lactose. In Figure 1, the modeling objects for the lactose pathway are shown. For the enzymatic reactions, simple Michaelis-Menten type kinetics are used. To describe the transcription efficiency, a reasonable approach is the choice of the fraction of free promoter binding sites with respect to all available promoter binding sites for the lactose operon. In comparison to the model equation system (1) the uptake of the carbohydrate is described more realistic since the synthesis of the transport system is included, which leads to a short delay of uptake.

## 2.3 Glucose uptake

To extend the scheme for a further carbohydrate, here glucose, knowledge on the transport system on the metabolic and genetic level is incorporated. A very important experimental observation is diauxic growth, if glucose and lactose are provided at the

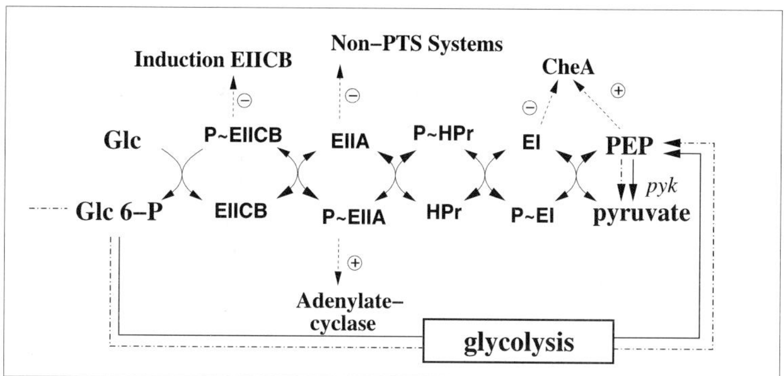


**Fig. 2.** Schematic representation of the glucose PTS. Inputs are the entire concentrations of EI, HPr, EIIA, EIICB, PEP, pyruvate, and extracellular glucose. Important outputs are the phosphorylated and unphosphorylated forms of EIIA. These two conformations are measured in several experiments. Solid lines represent metabolic reactions and dashed lines signal outputs of the PTS. Dash-dot lines represents metabolic flux in case of no PTS transport system. Since PEP is also converted by the pyruvate kinase reaction (gene *pyk*) to pyruvate, the degree of phosphorylation is strongly influenced by PEP and pyruvate, even if the PTS is not active (after Kremling et al. 2004).

same time in the bio-reactor. Therefore, the model must be set up in such a way, that this behavior is reproduced. Starting with metabolism, besides the uptake reactions (Fig. 2), which comprise four proteins, glycolytic reactions are also included, since the energy for the transport comes from phosphoenolpyruvate (PEP). Glucose is taken up by the phosphoenolpyruvate dependent glucose phosphotransferase system (PTS). In a sequence of five steps, the high energy bond is translocated to the incoming substrate that appears in its phosphorylated form inside the cell. Connecting both pathways only on the metabolic level, does not lead to the required behavior in a simulation study. Therefore, knowledge on the genetic level of control has to be included also.

Initially, knowledge on genetic regulation was restricted to a cAMP·Crp dependent induction of the gene *ptsG* which codes for the actual transport system EIICB$^{Glc}$. Transcription factor Crp is called a global transcription factor since it is involved in the expression of nearly 200 genes. Since the lactose operon is also under control of cAMP·Crp the question arose, in which way the local control by LacI and the global control via Crp have to be modeled. Years ago, Lee and Bailey (1984ab) proposed a method where the transcription efficiency $\eta$ is proportional to the fraction $\psi_P$ of occupied promoters. The influence of an inhibitor, e.g., a repressor, blocking the promoter is taken into account with $1 - \psi_R$ which represents the free sites. Activators are taken into account by parameter $\alpha$ in the term $(1 + \alpha\,\psi_A)$. For transcription efficiency, the following equation holds according to the method of Lee and Bailey

$$\eta = \psi_P\,(1 - \psi_R)\,(1 + \alpha\,\psi_A)\,. \tag{2}$$

The proposed method is limited to the consideration of single operons. To be more

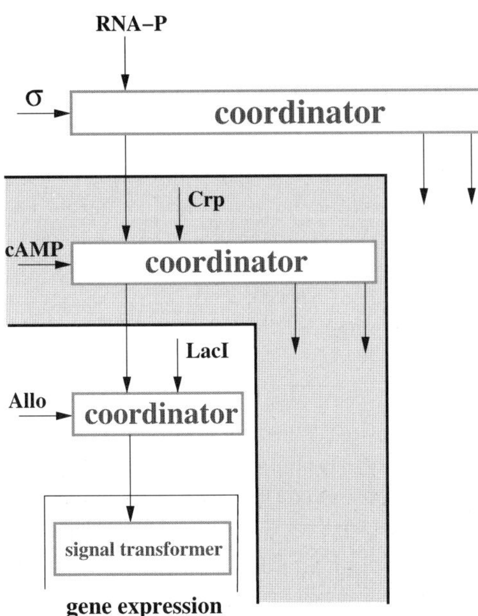


**Fig. 3.** Hierarchical set up of the genetic regulation network. Signals are transduced from the top level to the lower level but not vice versa. The lowest level represents individual pathways, the second level represents global transcription factors while the highest level is reserved for the RNA polymerase.

flexible and to allow model extensions in a very simple way, we proposed a new method with focus on the hierarchical set up of the genetic regulation (Kremling and Gilles 2001). For this method, the transcription factors are assigned to different levels in the hierarchy. The lowest level is represented by individual pathways, e.g., the lactose repressor LacI, which is involved only in lactose metabolism. The second level is represented by global transcription factors, e.g., Crp, which control a number of pathways. The highest level is reserved for the RNA polymerase, which is involved in nearly all transcription processes (Fig. 3).

As far as we have described the details of metabolism and genetic control, the picture is not yet completed and simulation results does not show a diauxic growth behavior. The missing link between both pathways is the interaction between the PTS, here output protein EIIA$^{Glc}$ (gene *crr*) and (i) the lactose permease and (ii) the activation of the cAMP generating enzyme adenylate cyclase CyaA. In the follo-wing, both effects are analyzed in a detailed way (Fig. 4). Protein EIIA$^{Glc}$ is expected to be in either of two states: phosphorylated or unphosphorylated. It is known that unphosphorylated EIIA$^{Glc}$ is able to inhibit the lactose permease (as well as some more enzymes in different carbohydrate uptake pathways). This is referred to as "in-ducer exclusion" since it prevents the entry of the substrates. On the other hand, the phosphorylated form of EIIA$^{Glc}$ is able to activate CyaA and therefore activates the synthesis of cAMP. However, the degree of phosphorylation of EIIA$^{Glc}$ depends on

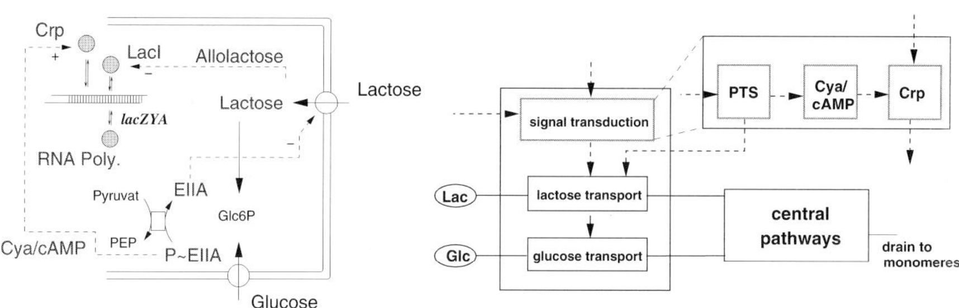


**Fig. 4.** Interaction between the PTS and the lactose pathway. Left: Schematic diagram. Right: Modeling objects.

a number of input variables as can be seen in Figure 2. In the case that the PTS is not active, the degree of phosphorylation depends only on the concentration of PEP and pyruvate (Kremling et al. 2004). Different PEP and pyruvate concentrations resulting in different $EIIA^{Glc}$ phosphorylation states have already been demonstrated for varying growth substrates indicating that this imput is the most important one (Hogema et al. 1998). Another major input is the dephosphorylation of the PTS proteins by incoming substrates. Model analysis by dynamical simulation studies with the proposed model structure gives interesting insights in the dynamics of the intracellular components. We started with a batch experiment where glucose and lactose are provided from the beginning. Figure 5 shows the time course of selected state variables (all model equations and parameters are summarized in (Kremling et al. 2001)). As expected, glucose is taken up while lactose is not. After the run out of glucose, the PTS protein $EIIA^{Glc}$ shows a very quick switch from the unphosphorylated to the phosphorylated form. This abolishes the inhibition of the lactose permease and furthermore leads to an activation of gene expression by the cAMP·Crp complex. cAMP is very low during the glucose uptake and rises in the lactose phase as a consequence of the degree of phosphorylation of $EIIA^{Glc}$. For the simulation, it was assumed that some molecules of $EIICB^{Glc}$ were available from the beginning. Since the promoter of *ptsG* has a high basal activity, the concentration during the glucose phase remains nearly constant. However, in the lactose phase, the concentration of $EIICB^{Glc}$ rises due to the higher cAMP levels. Since glucose is no longer available for uptake and growth, the further synthesis of $EIICB^{Glc}$ seemed not to be meaningful. In fact, during the time period when the model was developed, genetic research revealed that, a so far unknown transcription factor, Mlc (also called DgsA) is involved in the specific control of $EIICB^{Glc}$ (Plumbridge 1998). The repressor is active if no glucose is present in the medium and leads to a shut off of gene expression during growth on lactose. Since the detailed mechanism was unclear, a simple model for repression of the *ptsG* gene (Kremling et al. 2001) shows a satisfactory behavior. This can also be seen in Figure 5.

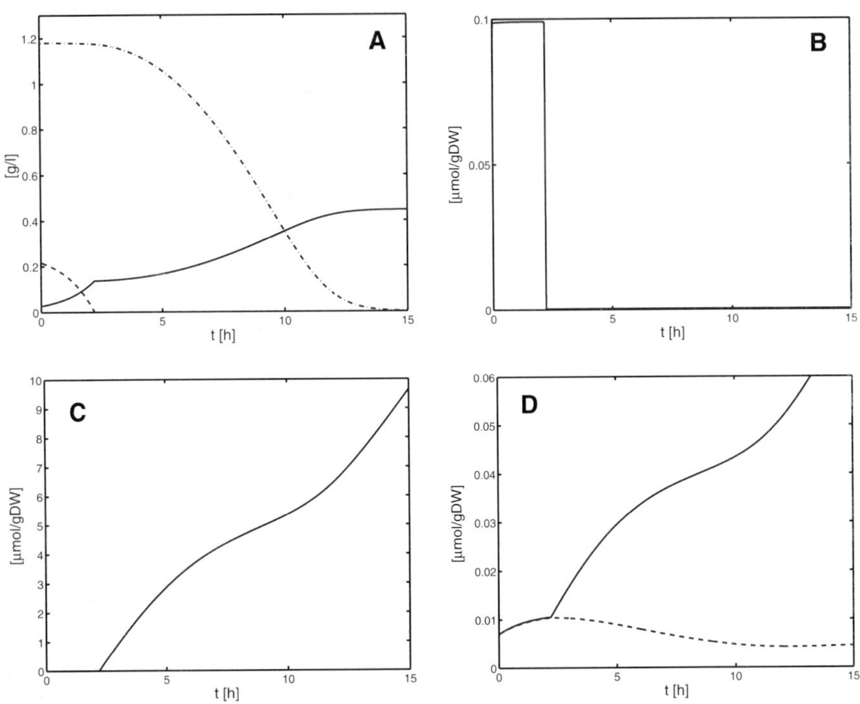

**Fig. 5.** Simulation results of selected state variables. **A** Biomass (solid), extracellular glucose (dashed) and lactose (dash-dot). **B** EIIA. **C** cAMP. **D** EIICB$^{Glc}$. Comparison of two model variants. Repression of EIICB$^{Glc}$ is not included (solid) and included with a simple model (dashed).

## 2.4  More detailed description of regulatory phenomena

Comparing the simulation results of the proposed model with experimental data, we noticed that the model was not able to reproduce the intracellular dynamics: (i) In Inada et al. (1996) the time course of intracellular cAMP was measured. In an experiment using glucose and lactose as carbon sources, cAMP shows an adaptive behavior, i.e., after a steep rise at the end of the glucose uptake phase, the concentration of cAMP goes back to the values observed in the glucose phase. To reproduce this behavior, we included knowledge on the genetic control of the proteins involved in the signal transduction pathway, Cya and Crp, respectively. While Cya is negatively controlled by the cAMP·Crp complex, Crp is auto-controlled. The proposed mechanism is rather complex: for low cAMP·Crp concentrations transcription is inhibited while for larger concentrations an activation is proposed (Hanamura and Aiba 1992). In Figure 6, the impact of the model extension is shown. Now, the qualitative behavior is reproduced correctly. (ii) The second model extension focuses on the kinetics of the inducer exclusion. A common approach for modeling enzymatic kinetics is to use Michaelis-Menten type kinetics. A mechanism to describe inhibi-

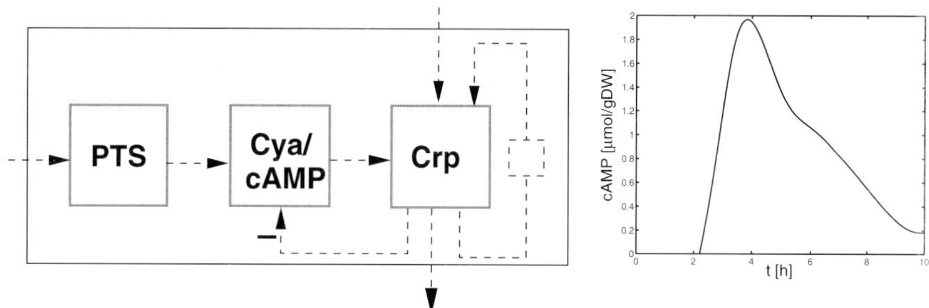

**Fig. 6.** Model extension and simulation results of cAMP. The feedback loop to the adenylate cyclase Cya leads to an adaptive behavior. Crp is auto-controlled. This is indicated by the dashed box.

tion extends the simple Michaelis-Menten equation by additional factors. A widely accepted assumption, hereby, is, that the amount of inhibitor (normally a metabolite) does not change significantly during binding at the enzyme since the concentration of the enzyme is much lower than the concentration of the metabolite. In Rohwer et al. (1998) an interesting experiment is described where it is shown that the proportion of the concentration of enzyme and inhibitor is near one, depending on the experimental design used. In our model, protein $EIIA^{Glc}$ interacts with the lactose pathway. Interestingly, inhibition occurs only if lactose is present in the medium. To include these facts into the model, the inhibitor $EIIA^{Glc}$ (unphosphorylated) is assumed to be in two conformations that are in equilibrium:

$$EIIA^{f} \quad \overset{K}{\rightleftharpoons} \quad EIIA \cdot LacY \cdot Lac_{ex} \tag{3}$$

with $EIIA^{f}$ being the free form, $EIIA \cdot LacY \cdot Lac_{ex}$ being a ternary complex of EIIA, lactose permease LacY and external lactose, and $K$ being the overall binding affinity. In the model equations for the PTS, only the free form is used as the driving potential.

(iii) Own measurements during the glucose/lactose diauxie experiment revealed some interesting dynamics of the degree of phosphorylation of protein $EIIA^{Glc}$ during the second growth phase. As shown in Figure 5, $EIIA^{Glc}$ is in its phosphorylated form during growth on lactose. Our experimental observation, however, indicates a slow rise of the unphosphorylated form for two hours and afterwards a slow decrease. It was speculated that the splitting of intracellular lactose into galactose and glucose and subsequent phosphorylation of intracellular glucose in glucose 6-phosphate is involved in the dephosphorylation of $EIIA^{Glc}$. As sketched in Figure 7, glucose has two possibilities to get phosphorylated: on the one hand, a glucokinase phosphorylates intracellular glucose with ATP or, on the other hand, intracellular glucose gets phosphorylated by the PTS. In the former case, protein $EIIA^{Glc}$ remains in the phosphorylated state while in the second case, $EIIA^{Glc}$ gets more and more dephosphorylated depending on the accumulation of intracellular glucose.

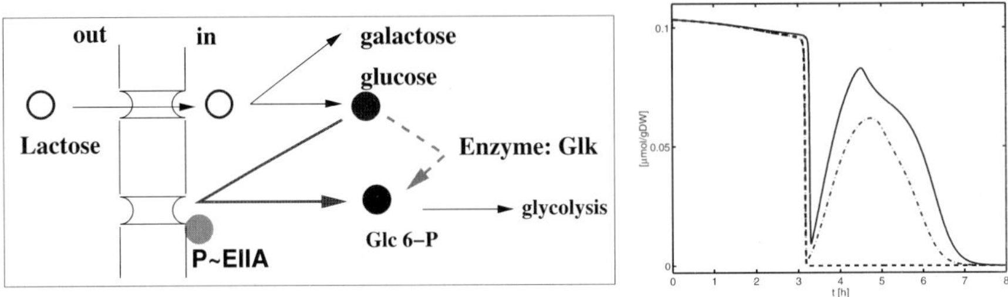


**Fig. 7.** Model extension and simulation results of EIIA$^{Glc}$. During the second growth phase, phosphorylation by either the glucokinase or the PTS is possible. The simulation on the right side show the expected results: Flux only via the PTS (solid line), only via the glucokinase (dashed line) or a mixture form both possibilities (dash-dot line).

## 2.5 Regulation by Mlc

The simulation shown above indicates that the PTS phosphorylates the intracellular glucose. To further verify this hypothesis, experiments with mutant strains were designed that differ only in one gene of interest (isogenic mutants). Strain Glk⁻ misses the glucokinase enzyme while strain Mlc⁻ misses the specific repressor for EIICB$^{Glc}$ and the other PTS proteins HPr and EI. Simulation results show that the Mlc⁻ strain should show lower values of EIIA$^{Glc}$ in the lactose phase since higher levels of the PTS proteins are expected to phosphorylate EIIA$^{Glc}$ in a more efficient way. In Figure 8, the dynamics of protein EIIA$^{Glc}$ (unphosphorylated) is shown. A good agreement between the simulation results and the experimental data is observed. After fitting the parameters of the model, all experiments could be described with a single set of parameters (publication in preparation). Note, that for parameter identification experimental data for other state variables like biomass, extracellular substrates, extracellular cAMP and LacZ was used. Figure 9 shows the time course of these state variables for the wild type strain during the batch experiment.

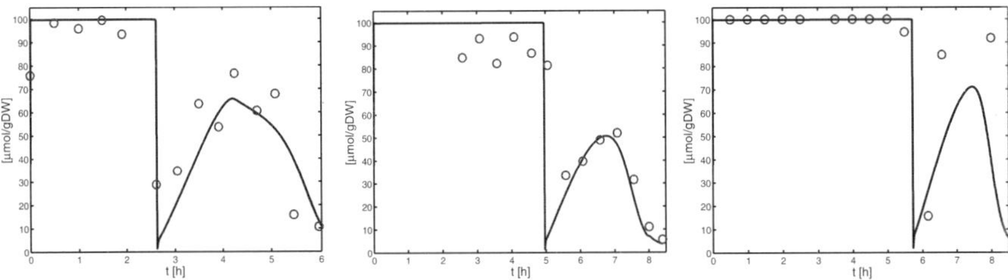


**Fig. 8.** Simulation (solid line) and experimental data (circles) for three batch experiments using wild type strain (left), Mlc⁻ strain (middle) and Glk⁻ (right). The model was fitted to the data; the experiments could be described with one set of parameters (publication in preparation).

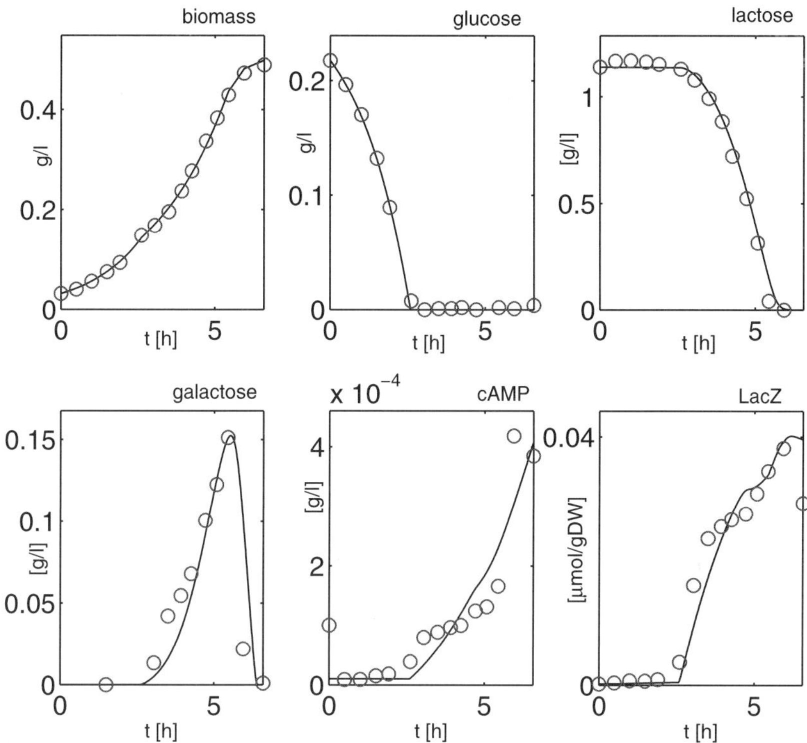

**Fig. 9.** Simulation (solid line) and experimental data (circles) for the wild type strain during the batch experiment. Glucose is taken up right from the beginning while lactose is taken up in the second growth phase. Interestingly, galactose, a product from the LacZ reaction is excreted in the medium at the beginning of the second growth phase. When lactose runs out, *E. coli* uses galactose as additional carbohydrate source. cAMP is also excreted in large amount during the second growth phase.

## 2.6 Model analysis – implications for diauxic growth

The model described so far was extended step by step by incorporating pathways for additional carbohydrates. The current version is able to describe the uptake of six sugars, glucose, lactose, galactose, glycerol, glucose 6-phosphate, and sucrose (*E. coli* wild type strain is not able to grow on sucrose; therefore, a mutant strain with a sucrose PTS was constructed and analyzed (Kremling et al. 2004)). To fit the parameters, experiments under different environmental conditions, experiments with mutant strains, and experiments with different pre-culture conditions were performed (publication in preparation). The model has 50 state variables and needs 300 parameters. For nearly all parameters values were found in literature. Based on a sensitivity and parameter analysis, 60 parameters could be estimated from the experimental data.

The key elements of the model are summarized for a PTS carbohydrate and a non PTS carbohydrate in Figure 10. The transport system are normally under dual

**Table 1.** Summary of functional units, number of parameters and number of estimated parameters. About 20 different experiments are used for parameter fitting. [a] Parameters estimated with Metabolic Flux Analysis.

| module name | param. | param. estimated | number of states | type |
|---|---|---|---|---|
| PTS (general) | 21 | 9 | 9 | ODE |
| PTS Glc | 12 | 4 | 1 | ODE |
| Cya | 9 | 2 | 2 | ODE |
| Crp | 17 | 3 | 1 | ODE |
| 2nd Glc transporter | 18 | 3 | 3 | ODE |
| Lac transporter | 16 | 7 | 4/2 | ODE/ algebraic |
| Scr transporter | 26 | 9 | 6 | ODE |
| Gly transporter | 24 | 5 | 5 | ODE |
| Gal transporter | 43 | 4 | 11/2 | ODE/ algebraic |
| Catabolic reactions | 51 | 11 | 8 | ODE |
| Monomer synthesis | 7 | $4^a+3$ | 1 | ODE |
| Liquid phase | 7 | 5 | 8 | ODE |

control. Besides a carbohydrate specific control by repressors like LacI, GalR, or GlyR, most systems are under control of the global regulator Crp, thereby, depending on the degree of phosphorylation of the PTS protein EIIA$^{Glc}$. In this case, the advantages of a systems biological approach become obvious. Because of the wealth of important and interacting regulations, metabolite concentrations and protein states, only a quantitative systems oriented approach will help in the understanding and will be able to identify the abilities of the system. It can also help to identify some general properties of the system.

Diauxic growth is observed for a number of couples of carbohydrates. With the model at hand and the simplified scheme in Figure 10, some general conclusions could be drawn; there is no unique control circuit that leads to diauxic behavior. Rather, diauxic growth is the result of a number of different control schemes and kinetic parameter constellation. So, a PTS sugar does not repress the uptake of a non PTS sugar in general. Own measurements with glucose and glucose 6-phosphate (a non PTS sugar with an uptake system that is also under control of the cAMP·Crp complex) show that the uptake of glucose is repressed while glucose 6-phosphate is taken up immediately. Measurement of the synthesis of the glucose transporter EIICB$^{Glc}$ by LacZ fusion revealed that EIICB$^{Glc}$ is no more synthesized although glucose is present in the medium. In Morita et al. (2003), it is speculated that high concentrations of glycolysis intermediates like glucose 6-phosphate or fructose 6-phosphate may be involved in the down regulation of the EIICB$^{Glc}$ messenger RNA. With the model, simulation studies can be done to verify the hypothesis. Figure 11 shows simulation results for glucose 6-phosphate uptake. If it is assumed that high levels of intracellular glucose 6-phosphate is able to inhibit the synthesis of protein EIICB$^{Glc}$ the uptake of glucose is inhibited in the first growth phase. Since glucose 6-

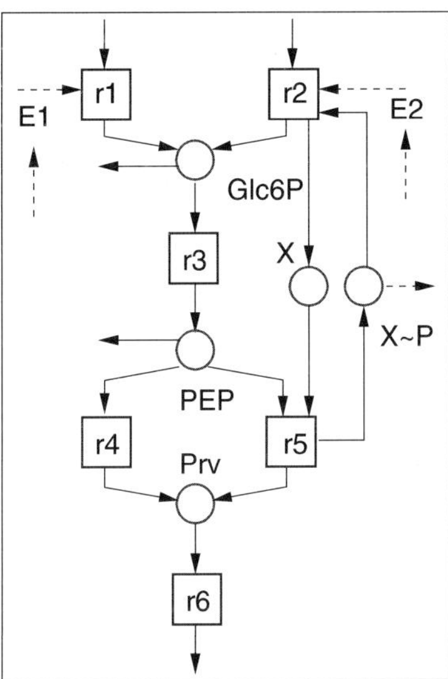


**Fig. 10.** Key elements of the carbohydrate uptake systems. r1 represents the uptake system for a non PTS sugar, r2, together with r5 the uptake system for a PTS sugar, $X$ and $X{\sim}P$ represent a PTS protein, r3 glycolysis, r4 the pyruvate kinase reaction, and r6 the drain of pyruvate (Prv). $E1$ and $E2$ are the respective proteins for the transporter. Both are subject to control. Most of the carbohydrate transport systems are controlled by Crp that is activated by cAMP. In the scheme this is represented by the signal arrow coming from a PTS representative $(X{\sim}P)$.

phosphate does not accumulate any longer, transporter $EIICB^{Glc}$ can be synthesized again for uptake of glucose.

# 3 Recent developments and future challenges

The example of modeling carbohydrate uptake in *E coli* showed that close interactions between experimentation and theoretical analysis may yield novel insight into an 'old' biological system. Apparently, for less well-characterized cellular systems, the question of how to best organize these interactions is of even more relevance. Besides discussing recent developments and challenges in this aspect of systems biology, we will broaden our view to more general principles of organization and function. In all cases, engineering sciences offer concepts and methods that can help in understanding biology. We will draw on analogies between complex biological and technical systems to illustrate this point.

## 3.1 Experimentation and Theory

The characterization of network components and interactions in qualitative and quantitative terms is a prerequisite for an integrated understanding as well as for realistic

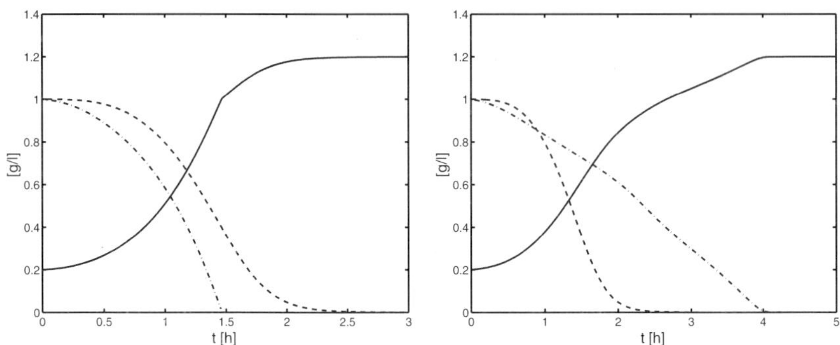

**Fig. 11.** Simulation study for glucose 6-phosphate uptake. Left: In the uncontrolled system both sugars are taken up in parallel. Right: Assuming a feedback loop from intracellular glucose 6-phosphate to the synthesis of the glucose uptake system, the uptake of glucose is inhibited. Biomass (solid line), glucose (dash-dot), glucose 6-phosphate (dashed).

mathematical models of biological systems (Kitano 2002b). For instance, determining all interactions between the components in an organism of low complexity such as *E. coli* has been estimated to require between 50 and 40,000 microarray experiments (Selinger et al. 2003). Hence, optimizing the way in which these experiments are conducted holds great promises for the efficiency of systems approaches. In experimental biology, educated guesses in 'traditional' hypothesis-driven research and, more recently, comprehensive studies using, for instance, systematic gene knockouts prevail. Systems engineering offers a large body of theory for the identification problem (Ljung 1999) that can be employed to assess the information content of experimental data (as for the estimation of parameters in our *E. coli* example), and to suggest efficient strategies for generating quantitative data. For instance, a recent study applied tools from systems sciences to an artificial gene network in order to analyze the effect of (inherent) stochastic fluctuations and (purpose-driven) input perturbations on the identification of model parameters (Zak et al. 2003). We believe that, besides specific predictions leading to new experiments once a mathematical model is available, systematic investigations of this type can result in more general guidelines for experimental strategies to quantitatively characterize biological networks.

Many biological systems of interest, however, are not yet amenable to this approach relying on detailed mathematical models, for instance, owing to an incomplete and/or inaccurate knowledge on components and interactions. There, the challenge is to derive the system's working principles from the observable behavior. This reverse-engineering usually entails the discrimination between a large number of hypothetical mechanisms to infer the causal relationships. For the analysis of gene networks, for instance, several theoretical approaches to the problem have been suggested. They range from boolean networks that consider only the 'on' and 'off' states of genes to detailed dynamical models (D'haeseleer et al. 2000). A great challenge for the future obviously is to assess the relative power of the methods, and their data requirements.

More generally, however, it will be crucial to find a common basis for theoretical approaches at different resolution that are currently incompatible with each other. Only such a unification will allow for the desired gradual transition from coarse to very detailed representations of complex biological networks, depending on the knowledge on the system and the specific interest of the investigator (Ideker and Lauffenburger 2003). Here, for instance, general systems theory provides a theoretical framework (Willems 1991) that could be built upon. In brief, it regards systems (and models as their representation) as functional entities that simply map a set of inputs to a set of outputs. As such, it enables a general treatment of models similar to the ideas outlined in Selinger et al. (2003).

## 3.2 Modules and hierarchies

One parallel between biological and technical systems is particularly striking and can greatly facilitate the systems biology approach: it is increasingly accepted that both types of systems are composed of semi-autonomous modules that perform a specific function. Biological modules acting as switches, triggers, amplifiers, and other functional units are paralleled by similar devices in, for instance, electrical and control engineering (Hartwell et al. 1999; Nurse 2003). Modularity in general, and these analogies in particular, have at least three important implications for our ability to understand integrated biological systems: (i) they allow for the decomposition of complex networks into manageable units, which can later be reassembled to obtain the whole picture, (ii) corresponding modular concepts for mathematical modeling and formal analysis facilitate theoretical investigations in systems biology as illustrated by our *E. coli* example, and (iii) it will be possible to draw on the large repertoire of methods and insights from engineering sciences by elaborating common operating principles of prototypical technical and biological (sub)systems (Csete and Doyle 2002; Stelling et al. 2001).

A major current challenge for elucidating and exploiting modularity in biology, however, is to find objective criteria for the demarcation of modules. Several approaches have been suggested in the literature, for instance, regarding the dissection of complex metabolic networks into simpler modules (Schuster et al. 1993). Most intuitively, functional units can be characterized as performing a common physiological task and belonging to the same genetic unit and/or signal processing entity (Kremling et al. 2000). Yet, similar to the delineation of pathways from complex interaction maps in traditional biology, in many cases these 'soft' criteria prevent an unambiguous assignment of modules. Methods from graph theory that analyze the components (nodes) and their interactions (links) in networks yielded statistically overrepresented 'motifs' in transcriptional networks (Shen-Orr et al. 2002). A particular functionality could be assigned to some of these recurring small networks of interactions through more detailed dynamic analysis. For instance, a three-gene circuit termed the 'feed-forward motif' specifically either accelerates or delays transcriptional responses (Mangan and Alon 2003). These analyses are confined to small patterns of interactions, and their role in the larger system is unclear at present. At a larger scale, graph-theoretical approaches revealed a hierarchical ordering of modules for the genome-wide metabolic network of *E. coli* (Ravasz et al. 2002). However, graph models may be too coarse to reflect biological functionality (Arita 2004).

Furthermore, concepts exist that explicitly take function into account from the beginning. For instance, metabolic pathway analysis identifies the smallest functional units in metabolism, but these units are usually overlapping (Rohwer et al. 1996; Schuster et al. 2000). The search for coregulated genes in libraries on gene expression data obtained by microarrays showed common patterns of hierarchical modularity in different organisms, yet the resolution of individual modules is influenced by adjustable parameters of the analysis method (Bergmann et al. 2004). Finally, a recent proposal concerns the demarcation of modules based on a criterion from systems theory, namely the absence of retro-activity (Saez-Rodriguez et al. 2004). In summary, thus, albeit a multitude of methods to analyze modularity in biological systems exists, their caveats do not allow to conclusively specify modules – or to prove their existence. Apparent next steps could consist in, for instance, a systematic comparison of the analysis results for a model system. It will be tempting to develop hybrid approaches taking into account multiple criteria for delineating modules. In addition, a hierarchical structure of biological networks raises important, largely unaddressed questions on the role of hierarchies in the coordination of cellular functions. Modularity and hierarchies open new directions for the multi-level analysis of biological systems, for which, for instance, electric circuit engineering provides suitable paradigms (Nurse 2003). Not only systems biology, but also engineering theory will benefit from analogies between biological and technical systems.

## 3.3 Functions and design principles

The notion of function is a common denominator of biological and engineered systems. In contrast, physical systems may show equally complex networks, resulting in complex behavior. However, they arise without purpose, and are not driven by evolution or voluntary engineering as for the first two classes of systems (Hartwell et al. 1999). The crucial point here is that, to perform similar functions, biological and synthetic systems use similar design principles. Negative feedback, for instance, serves to maintain homeostasis in both domains. Consequently, translation of engineering principles into the realm of biology will have a major impact on understanding the structure and function of complex biological systems (Csete and Doyle 2002). At a detailed level, two directions of future research appear obvious. As for perfect adaptation in bacterial chemotaxis, mapping a complicated biological network to a well known engineering principle   integral feedback control in this case   may explain the observed behavior (Yi et al. 2000). Conversely, necessary conditions for achieving a particular function in engineered systems can guide detailed investigations in biology. For example, methods from control theory were recently employed to provide an analytical method for deciding whether positive feedback in biology leads to bistable switching (Angeli et al. 2004); when such a behavior is observed *in vivo*, 'missing links' in the assumed circuit diagram could, hence, be identified. This kind of study, however, is only at the beginning. Important avenues of future research will be to examine control-theoretical concepts such as (structural) identifiability and controllability with respect to their applicability to biological systems.

Systems biology and engineering alike are presumably most challenged by the need to understand and/or to optimize highly integrated systems with a large number of interacting components. In both domains, robustness, that is, resistance to perturbations and failures constitutes a prominent design goal. Some ingredients for achieving this property such as feedback control, modularity, and hierarchies are known in engineering, and engineered systems were highly optimized in this regard (Csete and Doyle 2002). However, it seems reasonable to assume that evolution in biology came up with more efficient and/or alternative solution to the problem. Hence, in our opinion, analyzing the design principles of biology in this respect will prove beneficial both for systems biology and for engineering. Model-based analyses of metabolic networks in bacteria already revealed parts of the control logic: whereas control at the level of fluxes ensures optimal growth for each particular situation the organism encounters (Ibarra et al. 2002), the control of metabolic gene expression seems to trade-off the efficiency in this situation, and the organism's flexibility to respond to environmental changes (Stelling et al. 2002). Although seemingly being at completely different levels of abstraction, the search for design principles profoundly feeds back on the interactions between experimentation and theory. For instance, the insight into metabolic control was obtained by using the structure of metabolic networks alone. Hence, theoretical investigations may help to decipher information from well-known properties, and to indicate less rewarding, in addition to new directions of experimental research.

# 4 Conclusions

Biological complexity is the substrate for the emerging field of systems biology, with the aim of an integrated understanding of complex biological systems as its driving force. Beyond this, however, we believe that a main characteristic of the systems biology approach is its interdisciplinary nature that combines methods and concepts from biology, information sciences and engineering. In particular, mathematical modeling of biological systems will serve to achieve the goals of systems biology, and to help establishing a more quantitative biology. As our example of sugar uptake in *E. coli* and its control showed, a close interaction between experimental biology and computational analysis is able to establish quantitative and predictive mathematical models. Such models can, for instance, be employed to reveal inconsistencies in the current knowledge on a system, assess the explanatory power of alternative hypotheses, and ultimately suggest new experiments that verify or falsify the model predictions. We believe that this iterative cycle, combining experimentation and theory will be essential for the success of systems biology.

A major current challenge, thus, is to increase the efficiency of the interactions. In this case, as for other fields that warrant more intense research, engineering can provide well-established theoretical concepts. Analogies between complex biological and technical systems are obvious, for instance, their modular and hierarchical structure, the notion of functions that have been optimized, and the underlying general design principles. Future research in these fields can be anticipated to yield operating principles that will increase our comprehension of how complex systems in generals are designed and perform. Moreover, for biology, such design principles will guide detailed investigations of specific biological systems. For engineering, they can provide new paradigms (or revive old ones), for instance, regarding the efficient control of integrated technical systems. A major obstacle on the way to gain these potential benefits from systems biology, however, is the still existing 'clash of civilizations' (Huntington 1993) between the sciences in biology and engineering. Finding a common language and educating a new breed of scientists that are familiar with both fields (Lazebnik 2002), thus, should be a central objective of all initiatives in systems biology.

# References


Alm E, Arkin AP (2003) Biological networks. Curr Opin Struct Biol 13:193–202

Angeli D, Ferrell Jr. JE, Sontag ED (2004) Detection of multistability, bifurcations, and hysteresis in a large class of biological positive-feedback systems. Proc Natl Acad Sci USA 101:1822–1827

Arita M (2004) The metabolic world of *Escherichia coli* is not small. Proc Natl Acad Sci USA 101:1543–1547

Barkai N, Leibler S (1997) Robustness in simple biochemical networks. Nature 387

Bergmann S, Ihmels J, Barkai N (2004) Similarities and differences in genome-wide expression data of six organisms. PLoS Biol 2:E9

Csete ME, Doyle JC (2002) Reverse engineering of biological complexity. Science 295:1664–1669

Endy D, Brent R (2001) Modelling cellular behaviour. Nature 409:391–395

Gilman A, Arkin AP (2002) Genetic "code": representations and dynamical models of genetic components and networks. Annu Rev Genomics Hum Genet 3:341–369

Hanamura A, Aiba H (1992) A new aspect of transcriptional control of the *Escherichia coli* crp gen: positive autoregulation. Mol Microbiol 6:2489–2497

Hartwell LH, Hopfield JJ, Leibler S, Murray AW (1999) From molecular to modular cell biology. Nature 402 (Supp.):C47–C52

Hogema BM, Arents JC, Bader R, Eijkemanns K, Yoshida H, Takahashi H, Aiba H, Postma PW (1998) Inducer exclusion in *Escherichia coli* by non-PTS substrates: the role of the PEP to pyruvate ratio in determining the phosphorylation state of enzyme IIA$^{Glc}$. Mol Microbiol 30:487–498

Huntington SP (1993) The clash of civilizations. Foreign Affairs 72:22–28

Ibarra RU, Edwards JS, Palsson BO (2002) *Escherichia coli* K-12 undergoes adaptive evolution to achieve in silico predicted optimal growth. Nature 420:186–89

Ideker T, Lauffenburger D (2003) Building with a scaffold: emerging strategies for high- and low-level cellular modeling. Trends Biotechnol 21:255–262

Inada T, Kimata K, Aiba H (1996) Mechanism responsible for glucose-lactose diauxie in *Escherichia coli*: challenge to the camp model. Genes Cells 1:293–301

Kitano H (2002a) Computational systems biology. Nature 420:206–210

Kitano H (2002b) Systems biology: a brief overview. Science 295:1662–1664

Kremling A, Bettenbrock K, Laube B, Jahreis K, Lengeler JW, Gilles ED (2001) The organization of metabolic reaction networks: III. Application for diauxic growth on glucose and lactose. Metab Eng 3(4):362–379

Kremling A, Fischer S, Sauter T, Bettenbrock K, Gilles ED (2004) Time hierarchies in the *Escherichia coli* carbohydrate uptake and metabolism. BioSystems 73(1):57–71

Kremling A, Gilles ED (2001) The organization of metabolic reaction networks: II. Signal processing in hierarchical structured functional units. Metab Eng 3(2):138–150

Kremling S, Jahreis K, Lengeler JW, Gilles ED (2000) The organization of metabolic reaction networks: A signal-oriented approach to cellular models. Metab Eng 2(3):190–200

Lazebnik Y. (2002) Can a biologist fix a radio? – Or what I learned while studying apoptosis. Cancer Cell 2:179–182

Lee E, Salic A, Kruger R, Heinrich R, Kirschner MW (2003). The roles of APC and axin derived from experimental and theoretical analysis of the Wnt pathway. PLoS Biol 1:E10

Lee SB, Bailey JE (1984a) Genetically structured models for *lac* promotor-operator function in the *Escherichia coli* chromosome and in multicopy plasmids: *lac* operator function. Biotechnology and Bioengineering 26:1372–1382

Lee SB, Bailey JE (1984b) Genetically structured models for *lac* promotor-operator function in the *Escherichia coli* chromosome and in multicopy plasmids: *lac* promotor function. Biotechnology and Bioengineering 26:1383–1389

Ljung L (1999) System identification: theory for the user. 2nd edn. Prentice Hall PTR, Upper Saddle River, NJ

Mangan S, Alon U (2003) Structure and function of the feed-forward loop network motif. Proc Natl Acad Sci USA 100:11980–11985

D'haeseleer P, Liang S, Somogy R (2000) Genetic network inference: from co-expression clustering to reverse engineering. Bioinformatics 16:707–726

von Mering C, Krause R, Snel B, Cornell M, Oliver SG, Fields S, Bork P (2002) Comparative assessment of large-scale data sets of protein-protein interactions. Nature 417:399–403

Morita T, El-Kazzar W, Tanaka Y, Inada T, Aiba H (2003) Accumulation of glucose 6-phosphate or fructose 6-phosphate is responsible for destabilization of glucose transporter mRNA in *Escherichia coli*. J Biol Chem 278(18):15608–15614

Novak B, Tyson JJ (1993) Numerical analysis of a comprehensive model of M-phase control in Xenopus oocyte extracts and intact embryos. J Cell Sci 106:1153–1168

Nurse P (2003) Understanding cells. Nature 424:883

Plumbridge J (1998) Expression of *ptsG*, the gene for the major glucose pts transporter in *Escherichia coli*, is repressed by Mlc and induced by growth on glucose. Mol Microbiol 29(4):1053–1063

Pomerening JR, Sontag ED, Ferrell Jr. JE (2003) Building a cell cycle oscillator: hysteresis and bistability in the activation of Cdc2. Nat Cell Biol 5: 346–351

Ravasz E, Somera AL, Mongru DA, Oltvai ZN, Barabási A-L (2002) Hierarchical organization of modularity in metabolic networks. Science 297:1551–1555

Rohwer JM, Bader R, Westerhoff HV, Postma PW (1998) Limits to inducer exclusion: Inhibition of the bacterial phosphotransferase system by glycerol kinase. Mol Microbiol 29:641–652

Rohwer JM, Meadow ND, Roseman S, Westerhoff HV, Postma PW (2000) Understanding glucose tranport by the bacterial phosphoenolpyruvate:glycose phosphotransferase system on the basis of kinetic measurements in vitro. J Biol Chem 275:34909–34921

Rohwer JM, Schuster S, Westerhoff HV. How to recognize monofunctional units in a metabolic system. Journal of Theoretical Biology 179:214–228

Saez-Rodriguez J, Kremling A, Gilles ED (2004) Dissecting the puzzle of life: Modularization of signal transduction networks. Computers & Chemical Engineering, accepted

Schuster S, Fell DA, Dandekar T (2000) A general definition of metabolic pathways useful for systematic organization and analysis of complex metabolic networks. Nat Biotechnol 18:326–332

Schusterm S, Kahn D, Westerhoff HV (1993) Modular analysis of the control of complex metabolic pathways. Biophys Chem 48:1–17.

Selinger DW, Wright MA, Church GM (2003) On the complete determination of biological systems. Trends Biotechnol 21:251–254

Sha W, Moore J, Chen K, Lassaletta AD, Yi CS, Tyson JJ, Sible JC (2003) Hysteresis drives cell-cycle transitions in *Xenopus laevis* egg extracts. Proc Natl Acad Sci USA 100

Shen-Orr SS, Milo R, Mangan S, Alon U (2002) Network motifs in the transcriptional regulation network of *Escherichia coli*. Nat Genet 31(1):64–68

Stelling J, Klamt S, Bettenbrock K, Schuster S, Gilles ED (2002) Metabolic network structure determines key aspects of functionality and regulation. Nature 420:190–193

Stelling J, Kremling A, Ginkel M, Bettenbrock K, Gilles ED (2001) Towards a Virtual Biological Laboratory. In: Kitano H (ed), Foundations of Systems Biology, pp 189–212. MIT Press, Cambridge, MA

Tyson JJ, Chen K, Novak B (2001) Network dynamics and cell physiology. Nat Rev Mol Cell Biol 2:908–916

Willems JC (1991) Paradigms and puzzles in the theory of dynamical systems. IEEE Transac Automat Control 36(3):259–294

Yi T-M, Huang Y, Simon MI, Doyle J (2000) Robust perfect adaptation in bacterial chemotaxis through integral feedback control. Proc Natl Acad Sci USA 97(9):4649–4653

Zak DE, Gonye GE, Schwaber JS, Doyle III FJ (2003) Importance of input perturbations and stochastic gene expression in the reverse engineering of genetic regulatory networks: Insights from an identifiability analysis of an in silico network. Genome Res 13:2396–2405

Bettenbrock, Katja
   Systems biology group, Max-Planck-Institut für Dynamik, komplexer technischer Systeme, Sandtorstr. 1, 39106 Magdeburg, Germany

Fischer, Sophia
   Systems biology group, Max-Planck-Institut für Dynamik, komplexer technischer Systeme, Sandtorstr. 1, 39106 Magdeburg, Germany

Gilles, Ernst D.
   Systems biology group, Max-Planck-Institut für Dynamik, komplexer technischer Systeme, Sandtorstr. 1, 39106 Magdeburg, Germany
   gilles@mpi-magdeburg.mpg.de

Kremling, Andreas
   Systems biology group, Max-Planck-Institut für Dynamik, komplexer technischer Systeme, Sandtorstr. 1, 39106 Magdeburg, Germany

Stelling, Jörg
   Institut für Computational Science, ETH Zentrum HRS H 28, Hirschengraben 84, 8092 Zürich, Switzerland

# Integration of metabolic and signaling networks


Dirk Müller, Luciano Aguilera-Vázquez, Matthias Reuss, Klaus Mauch


## Abstract


This contribution addresses the construction of mathematical models that provide a combined description of metabolic and regulatory processes within cells. In the first part of the article, strategies for reconstruction of metabolic and signaling networks are outlined followed by a discussion of their characteristic properties. The second part focuses on the development of integrated models of metabolism and signal transduction. The case of yeast cyclic AMP (cAMP) signaling and its interaction with energy metabolism and elements of the cell cycle machinery is used to exemplify this approach.


## 1 Rationale

Today, systems biology is driven by a series of developments (see Table 1). High-throughput experimental technologies are generating biological data at unprecedented rates and increasing accuracy. Standard operation protocols ensure data comparability whereas the bioinformatics infrastructure makes data available. In addition, we observe an ever increasing computer power, new modeling tools and algorithms. Standardized representations of network models broaden their usability as well as extensibility within various computational platforms. Both at the wet lab and the dry lab, costs for obtaining and processing data decrease significantly. Finally, the field of systems biology is pulled by the need of biotechnological and biopharmaceutical research for describing, discovering, and predicting cellular properties based on global experimental data.

Metabolic and regulatory network models offer descriptive as well as predictive characteristics like, for example, to what extent a gene knockout effects cellular growth. If predictions are to be based upon genome-wide experimental data, genome-wide network models capable of integrating the data are required. Genome-wide models in the literal sense, that is to say, models which capture the whole genomic information of the combined metabolic and regulatory networks, are not yet available, whereas in a few cases genome-wide models of either metabolism or genetic regulation have been proposed. This reflects the fact that concepts for coupling regulatory and metabolic networks are still at their early stages.


Topics in Current Genetics, Vol. 13
L. Alberghina, H.V. Westerhoff (Eds.): Systems Biology
DOI 10.1007/b136529 / Published online: 7 April 2005
© Springer-Verlag Berlin Heidelberg 2005

**Table 1.** Driving forces of systems biology.

| Wet Lab | Dry Lab |
| --- | --- |
| Increasing data availability | Increasing computing power |
| Increasing data accuracy | New tools and algorithms |
| Standard operation protocols | Standardized and open interfaces |
| Falling costs (per data point) | Falling costs (per bit) |

In the following, strategies for reconstructing cellular networks are discussed and exemplified by a genome-wide network of *Helicobacter pylori* and TGF-β signaling, respectively. The second part of the article focuses on the construction of a modular model which describes the combined dynamics of energy metabolism, cell cycle progression, and cAMP signal transduction in *Saccharomyces cerevisiae*.

# 2 *In-silico* reconstruction of cellular networks

## 2.1 Top-down versus bottom-up

It is widely accepted that the set-up of networks should be guided by a clearly stated purpose. But in many cases it is *a priori* far from clear which of the thousands of cellular components and their interactions must be part of a network model for accurately describing processes like cell cycle control or apoptosis. Network reconstruction *in silico* may, therefore, follow two strategies: the top-down and the bottom-up approach, both of which have their advantages and disadvantages, and which, in many cases, show complementary features. Today, virtually no model is fully bottom-up or top-down. Figure 1 summarizes both strategies.

At the core of the bottom-up approach is the idea of aggregating detailed biological knowledge on individual components and their molecular interactions into appropriate modules and then to interconnect these into architectures suitable for holistic analysis. The top-down approach starts with an entire system with individual components described less detail. Through global optimization ('reverse engineering'), structural and kinetic properties of network components are identified. Biological insight is created by processing information obtained from the behavior of the entire system right from the beginning.

Without an idea of the complete picture, the initial design of experiments takes place blindly. Thus, the risk for overlooking important parts is substantial. Moreover, when setting up a network using a bottom-up approach, the appropriate degree of detail for the network components is initially unknown. Subject to these limitations, a top-down approach seems to be more promising compared to a bottom-up approach. On the other hand, using a holistic description right at the beginning requires a considerable computational effort. Furthermore, it might turn out to be difficult, if not impossible, to set-up a large-scale globally stable dynamic network right from the start. In general, the larger the size of cellular

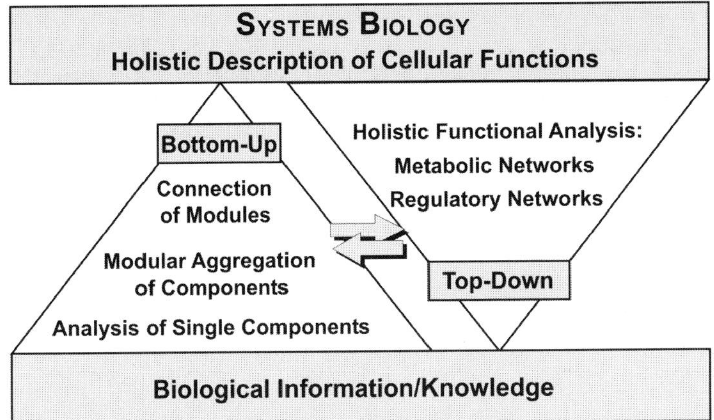


**Fig. 1.** Bottom-up and top-down approach to systems biology.

networks, the less important is the behavior of an individual component for the overall network behavior. This experience is backed by the finding that in metabolic networks control is usually distributed among several transformation steps. From this it follows that in large-scale networks kinetic behavior can often be described in a relatively simple way without losing much of the systems diversity. This outcome might facilitate the transition from a bottom-up to a top-down strategy.

## 2.2 Reconstruction of large-scale cellular networks

For a number of organisms like, for instance, *Escherichia coli* and *Saccharomyces cerevisiae* the genome is functionally well-characterized and stored in publicly accessible databases (e.g. EcoCyc, SGD). The reconstruction of metabolic networks is further simplified through links to protein databases (e.g. Swiss-Prot). In contrast to metabolic networks, relatively little is known on signaling networks. Moreover, reconstruction of signaling networks *in silico* is hampered by the fact that the functionality of signaling components has to be extracted directly from primary literature sources in most cases. Databases for the field of signaling research are only emerging now frequently led by international initiatives like the Alliance for Cellular Signaling (AfCS) (cf. http://www.signaling-gateway.org).

Consistent elementary and charge balances of the various transformation steps are a precondition for almost every method in the systems biology tool box that is based on balancing chemical species. This is the reason why cellular networks have first to be tested against a compound database providing a unique elemental composition. Modern computational tools (see Fig. 2) facilitate reconstruction and management of networks through interactive editors. Topological analyses like, for example, detection of unused transformation steps, dead-ends, parallel routes,

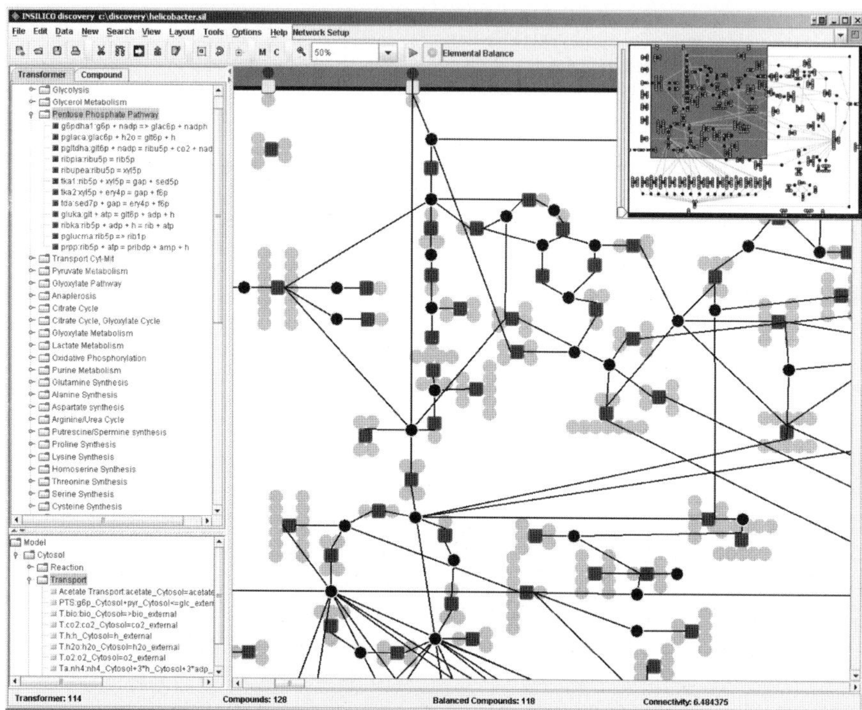

**Fig. 2.** Network reconstruction *in silico*. (1) Network components are dragged from databases to the design board. (2) Computer aided layout (spring, hierarchical, circular) facilitates the reconstruction with balancing and checking of model consistency being performed in the background (INSILICO discovery; http://www.insilico-biotechnology.com). Grey rectangles indicate transformations steps (reactions, transport steps, polymerization), circles indicate compounds.

and elementary flux modes (Mauch et al. 2002) are valuable methods for reconstructing genome-wide network models. Furthermore, automated layout proves to be helpful especially for the reconstruction of large-scale networks.

Figure 3 shows a genome-wide metabolic network of *Helicobacter pylori* reconstructed from the MetaCyc database (http://biocyc.org/meta). Historically defined 'pathways' like, for example 'glycolysis', 'TCA-cycle', and the 'pentose-phosphate-shunt' are aggregated in groups. Note that these groups do not necessarily coincide with pathways derived through flux balancing.

A signaling network with receptors for transforming growth factor beta (TGF-β) is shown in Figure 4. The network components were extracted from primary literature. An introduction is given by ten Dijke (2000), whereas Massague (2002) reviews recent findings in detail. The number of network compounds within the TGF-β signaling network (561) is of the same order of magnitude as the number of compounds in the metabolic network (461). However, the topologies of the

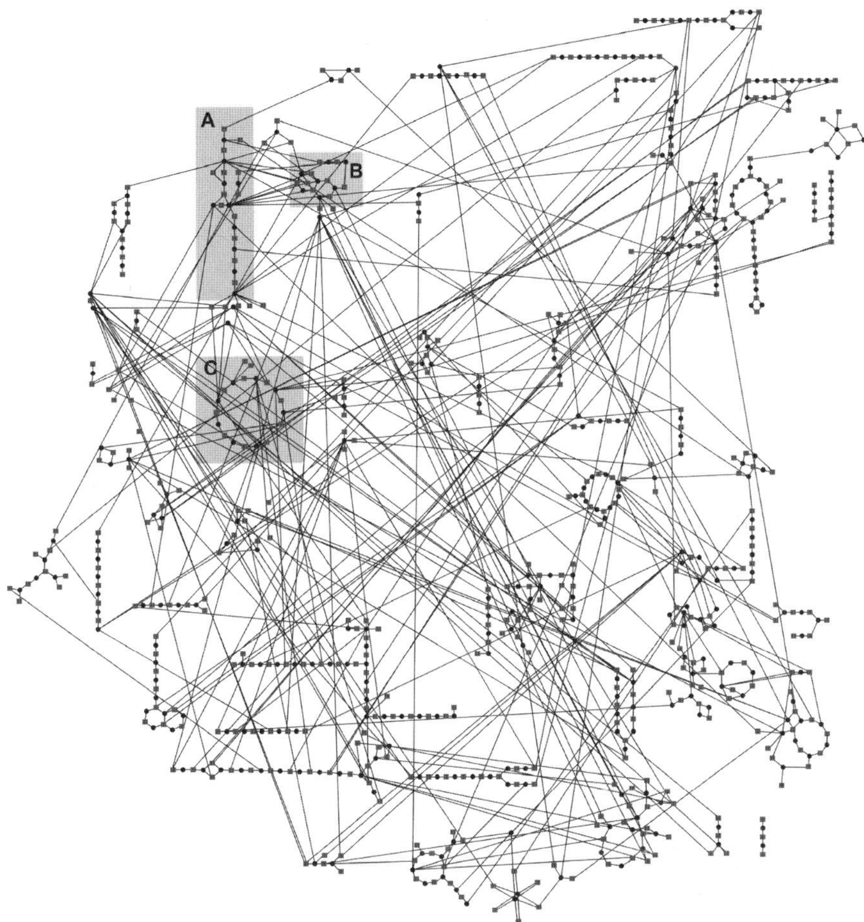

**Fig. 3.** Genome-wide metabolic network of *Helicobacter pylori* in a pathway-oriented set-up. The network consists of 450 reactions and 461 metabolites. (A) Glycolysis (B) Pentose-Phosphate-Pathway (C) TCA Cycle. Database: MetaCyc (http://biocyc.org/meta). The model of the *H. pylori* network can be downloaded as SBML file from http://sbml.org/models/.

metabolic and the signaling network are apparently different from each other. Aspects of the topological differences are briefly discussed below.

## 2.3 Topological properties of metabolic and signaling networks

Within both, the signaling and the metabolic network, modular organization, and clear boundaries between sub-networks are not immediately apparent. In fact, recent studies (Eéka et al. 2000; Ravasz et al. 2002; Barabási and Oltvai 2004) have demonstrated that metabolic networks have a scale-free topology. As shown in

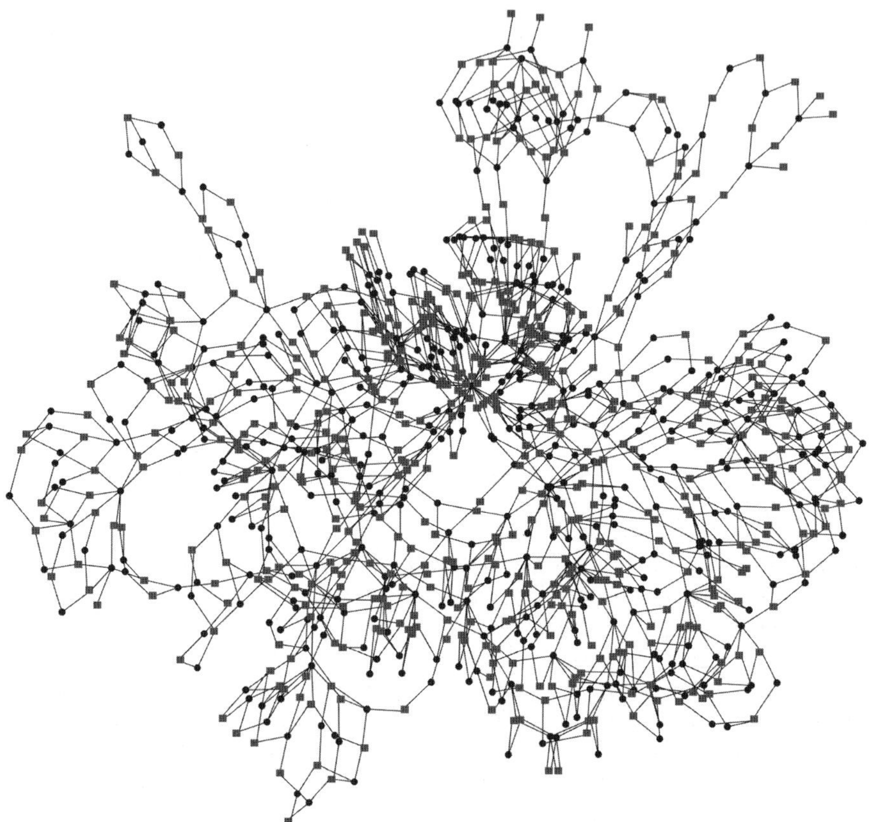

**Fig. 4.** Signal transduction network with receptors for transforming growth factor β (TGF-β). The network consists of 553 reactions and 561 compounds. Most of the compounds are proteins. The model of the TGF-β network can be downloaded as SBML file from http://sbml.org/models/.

Figure 5, the metabolic network of *Helicobacter pylori* possesses a scale-free topology with degree exponent $\gamma = 2.4$. A distinguishing feature of such scale-free networks is the existence of a few hubs, such as pyruvate, NADPH, or ATP, which participate in a very large number of links. These hubs integrate all substrates into a single integrated web in which the existence of fully separated modules is prohibited. In contrast to the metabolic network, a significantly smaller degree exponent $\gamma = 1.5$ is determined for the TGF-β signaling network. This indicates that the signaling network architecture is kept together through relatively fewer highly connected nodes (e.g. MKK1, PDK1, PI3K).

When decomposing signaling networks into elementary flux modes, usually no direct solution connecting the input (binding receptor-ligand) with the output (e.g. activation of a transcription factor) is detected. Instead, the network is fragmented into a number of small regions, each of which can be attributed to classes of signal

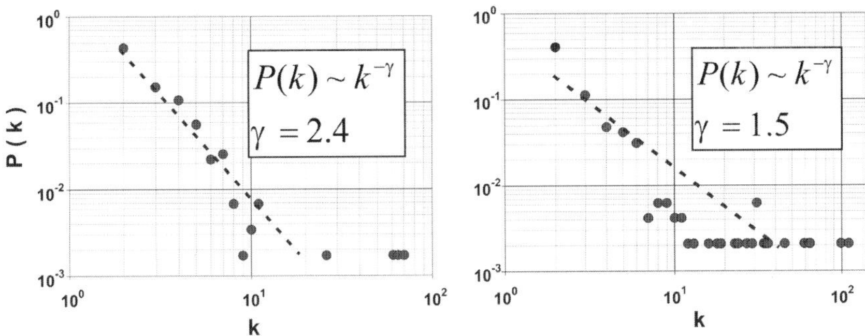


**Fig. 5.** The connectivity distribution P(k). Here, k signifies the number of connections per node and P(k) denotes the associated probability density function of nodes with k connections in the network. Left: Distribution of pool connectivities in a metabolic network (*H. pylori*). Right: Distribution of pool connectivities in a signaling network (TGF-β signal transduction)

transduction units. Members of those units are, among others, association and dissociation of protein complexes, activation, and inactivation of enzymes or transport processes across compartment borders. Many of the regions identified belong to cascades (Brightman and Fell 2000; Kholodenko et al. 2000; Heinrich et al. 2002) which, when connected with each other, allow for signal transduction across larger areas of the network. Whereas a mass flux within the cascades enables the regeneration of network compounds, no net flux from the input to the output takes place (see Fig. 6). This is less surprising since the main function of signaling networks is to transfer a signal rather than to maintain a steady-state mass flow. Interestingly, regeneration of network compounds within the cascades is essential for using the signaling network for a consecutive signal transduction. Another characteristic difference between signaling and metabolic networks arises from the fact that the role of compounds in signaling networks can be twofold. Whereas in metabolic network a compound (metabolite) only serves as substrate or product and never catalyzes another reaction, a compound in a signaling network (usually a protein) is able to act as (i) reactant and (ii) catalyst for yet another transformation (see also Table 2).

**Table 2.** Characteristics of metabolic and signaling networks

| **Metabolic Networks** | **Signaling Networks** |
|---|---|
| Flux of matter | Information transfer |
| The steady state behavior plays a major role in the functioning of metabolic networks | The transient behavior in signaling networks determines their function |
| The enzymes do not participate in the transformations steps as reactants | Compounds that are transformed frequently act as catalysts in other reactions |

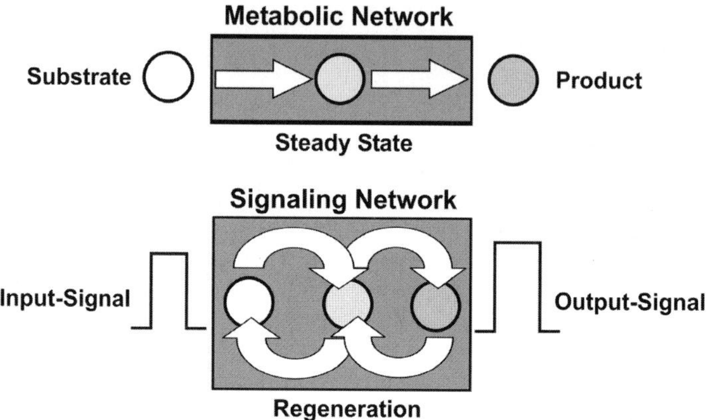


**Fig. 6.** Mass transfer (metabolic network) and information transfer (signaling network).

# 3 Marrying diverse partners – integrated models of signaling and metabolism

## 3.1 Motivation

Real cell populations are inherently heterogeneous (Levsky and Singer 2003). Thus, it is of major importance to assess the impact of this heterogeneity on the resulting dynamics observed at the population level, that is to say, to what extent does the behavior of the individual cell during its lifeline differ from the population average? This question is of particular relevance in the field of biomedicine and biotechnology: certain subpopulations of cells exhibiting different susceptibility to drugs, often due to minor variations in the genetic make-up or the cell cycle position of the cells, may greatly influence the success of drug application schemes and the development of resistance, for example, in the case of antibiotics or cancer therapeutics (Smith et al. 2000; Kitano 2003). A similar argument holds for biotechnological production processes, where the individual plasmid content or cellular state of the cells constituting the population can significantly affect overall productivity and product yield (Kromenaker and Srienc 1994; Altintas et al. 2001; Kacmar et al. 2004). In many cases, subtle differences in the interplay between the associated metabolic and signaling networks of the cells give rise to population heterogeneity. The central role played by the second messenger cyclic AMP (cAMP) in coordinating energy metabolism and cell division in the yeast *Saccharomyces cerevisiae* serves to illustrate the difference between individual cell dynamics and the behavior observed at the population level. At the same time, this example highlights the need for integrated models that describe system behavior as the outcome of the combined dynamics of its metabolic and signaling networks.

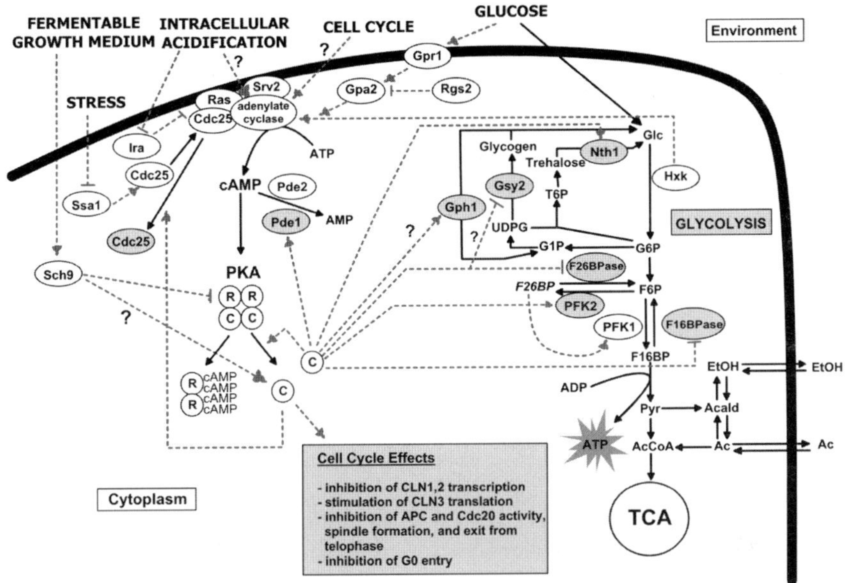


**Fig. 7.** Scheme of the cAMP-PKA signal transduction cascade showing selected interactions with energy metabolism and the cell cycle. Solid lines represent reactions, broken lines stimulatory (arrow heads) or inhibitory signals (blunt ends), respectively. Ellipses denote proteins, shaded ones signify those phosphorylated by PKA. Question marks indicate postulated phosphorylations. Abbreviations: Ac: acetate, Acald: acetaldehyde, AcCoA: acetyl-CoA, APC: anaphase promoting complex, ATP: adenosine triphosphate, C: catalytic subunit of PKA, EtOH: ethanol, F6P: fructose-6-phosphate, F16BP: fructose-1,6-bisphosphate, F26BP: fructose-2,6-bisphosphate, F26BPase: fructose-2,6-bisphosphatase, Glc: glucose, G1P: glucose-1-phosphate, G6P: glucose-6-phosphate, Hxk: hexokinase, PDC: pyruvate decarboxylase, PDH: pyruvate dehydrogenase, PFK1: phosphofructokinase 1, PFK2: phosphofructokinase 2, Pyr: pyruvate, R: regulatory subunit of PKA, TCA: tricarboxylic acid cycle, UDPG: UDP-glucose.

## 3.2 Coupling cell cycle progression and energy metabolism in *Saccharomyces cerevisiae*

Activation of protein kinase A (PKA) via a cAMP-dependent signaling cascade fosters the glycolytic rate and regulates the dynamics of storage carbohydrate metabolism in yeast (Fig. 7, for a review see Thevelein and de Winde 1999; Francois and Parrou 2001). Cyclic-AMP pathway hyperactivation also influences the dynamics of G1 cyclins (Baroni et al. 1994; Tokiwa et al. 1994; Hall et al. 1998), which determine the critical cell size required for budding. It also affects the timing of mitotic exit and, thus, the cell size at division by inhibiting the activity of the anaphase promoting complex (APC), at least in part through inhibition of Cdc20 (Anghileri et al. 1999; Bolte et al. 2003; Searle et al. 2004). Hence, a differential regulation of the intracellular cAMP concentration may be expected during the cell cycle. However, no such dynamics can be observed in exponentially growing asynchronous cultures where the fraction of cells in the different stages of the cell cycle remains nearly constant over time (Smith et al. 1990). The application of synchronous cultures where the cells of a population have been synchronized with respect to their cell cycle position proves favorable in this case. Synchronous cultures allow for analysis at the quasi-single cell level while still providing sufficient biomass for most biochemical assays. They are preferably obtained by selection techniques like centrifugal elutriation that minimizes perturbation of cellular state (Futcher 1999). In combination with rapid sampling techniques (Theobald et al. 1993; Schaefer et al. 1999; Buziol et al. 2002), it is then possible to follow the cell-cycle dynamics of both intra- and extracellular metabolites and second messengers like cAMP, respectively. This technique has been applied previously to validate models of central carbon metabolism based on measurements of the dynamic response of metabolites in asynchronous continuous cultures challenged with a glucose pulse (Rizzi et al. 1997; Vaseghi et al. 1999; Buchholz et al. 2002). In addition, quantitative microscopic analysis of the culture employing fluorescent stains can be used, for example, to determine the genealogical age distribution (Calcofluor) or the distribution of cell cycle phases of the population based on nuclear morphology (DAPI). Performing these experiments under controlled growth conditions in a bioreactor serves to minimize activation of the signaling pathway by unwanted secondary stimuli.

Investigations of this kind have demonstrated distinct dynamics of cAMP and downstream targets of PKA in energy metabolism during the cell cycle (Silljé et al. 1997; Müller et al. 2003), which are also in accord with the presumed role of PKA during mitosis. Figure 8 shows the cell-cycle behavior of the cAMP concentration in a glucose-limited synchronous culture: the cAMP concentration increases near the G1 to S transition, which is accompanied by a stimulation of oxygen uptake (low dissolved oxygen) as well as trehalose and glycogen breakdown (Silljé et al. 1997). During mitosis, the cAMP level transiently drops transiently, which is paralleled by a decrease in oxygen uptake (high dissolved oxygen) and supposedly a reduced the inhibition of the APC. Corresponding analyses in partially synchronized chemostat cultures have confirmed that the dynamics of cAMP and dissolved oxygen at the G1 to S transition are indeed accompanied by a periodic activation of the PKA target enzyme trehalase, the stimulation of glycolytic flux and the occurrence of overflow metabolism (Müller et al. 2003). However, the upstream signal responsible for the differential activation of the cAMP signaling pathway during the cell cycle has remained elusive so far. Preliminary evidence indicates that the cAMP cell cycle dynamics observed under glucose-limited conditions is accompanied by a corresponding change in adenine and guanine nucleotide concentrations (A. Niebel and D. Müller, unpublished results), which apparently feed back on the activation state of the signaling cascade (Thevelein and de Winde 1999; Rudoni et al. 2001). This mechanism may serve to link a varying energy demand during cell cycle progression to the available energy supply in the form of ATP.

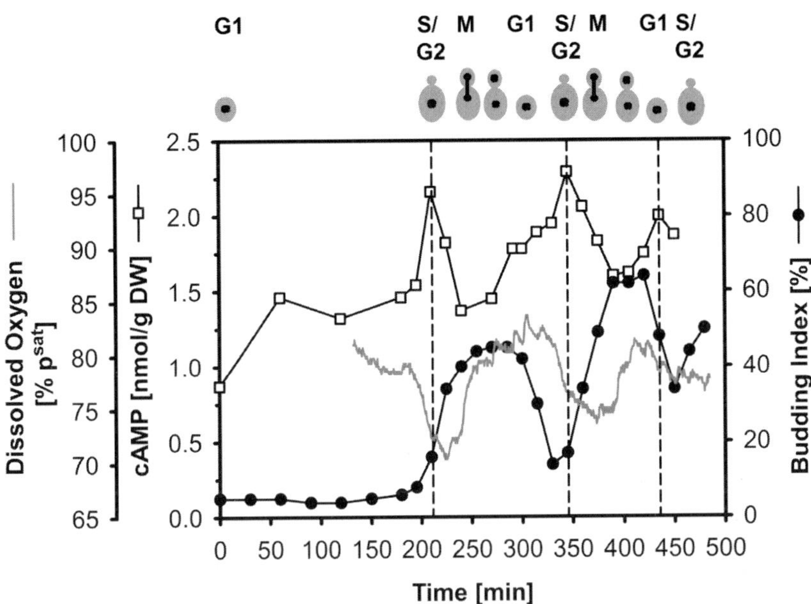



**Fig. 8.** Dynamics of the intracellular cAMP concentration, dissolved oxygen tension (DOT), and the budding index (BI) of the population in a glucose-limited synchronous culture. A synchronous culture obtained from centrifugal elutriation was aerobically grown in a fed batch (30°C, pH 5.0) in defined mineral medium at a specific growth rate of $\mu=0.15$ h$^{-1}$ (Müller et al. 2003). The yeast cells depicted as cartoons on top of the graph indicate the cell cycle position of the cells as determined from DAPI staining.

## 3.3 Establishing a modular model

If we want to describe the impact of cAMP on the coordination of cell growth and division at the population level we need to begin with a model of these processes as they occur in the single cell. In the following, the development of such a single-cell model following the bottom-up approach discussed above (cf. Section 2.1) is outlined.

A straightforward approach is to employ a modular structure for this purpose, where each functional module is represented by a different submodel. But what actually defines a functional module? Chemical isolation resulting either from spatial localization or from the specificity of chemical interactions has been suggested to be the main characteristic of a module. Additionally, functional modules should comprise only a small part of the total system and perform a largely autonomous function. Hartwell et al. (1999) have presented a number of criteria that may be used to delineate these modules: typically, modules can be recognized as such if either (i) they can be reconstituted *in vitro*, (ii) they still function when transplanted into other organisms, or when (iii) their behavior can be adequately

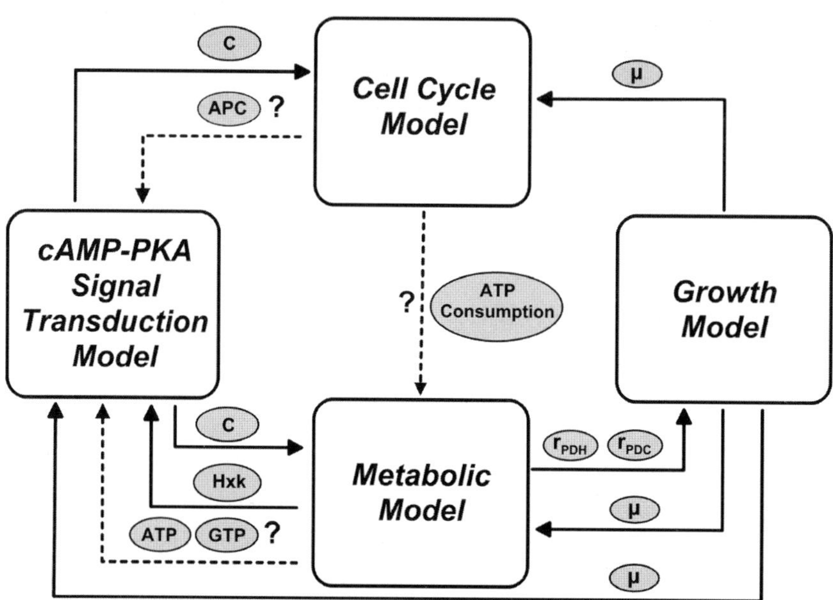


**Fig. 9.** Sketch of the integrated single cell model consisting of mutually interlinked modules for cAMP signal transduction, energy metabolism, cell growth, and the cell cycle. Solid lines indicate known interactions; broken lines and question marks denote hypothetical effects. Abbreviations: APC: anaphase promoting complex, ATP: adenosine triphosphate, C: catalytic subunit of PKA, GTP: guanosine triphosphate, Hxk: hexokinase, PDC: pyruvate decarboxylase, PDH: pyruvate dehydrogenase, r: enzymatic rate, μ: specific cellular growth rate.

described by a computational model including only module components when compared to experimental data - a process termed '*in silico* reconstitution.' The obvious advantage of such a modular structure is that the constituting modules of an integrated model may differ in mathematical nature and level of detail. Also, single modules can be easily replaced by other descriptions of the same process, for example, to test alternative model hypotheses or mathematical representations.

Developing the single cell model in this fashion makes it possible to verify whether or not the concept of functional modules can applied to the system at hand. Discrete submodules have been defined for metabolism and cell growth, cell cycle progression, and cAMP signal transduction. These modules are connected via mutual feedback effects to yield an integrated model of the overall process (Fig. 9). The signaling module describes the dynamics of cAMP synthesis and degradation as well as the resulting PKA activity (Müller et al. "A Dynamic Model of cAMP Signal Transduction in Yeast," manuscript in preparation). An adaptive response of this signaling pathway to persistent stimuli constitutes a key feature that needs to be captured by the model. This characteristic is achieved

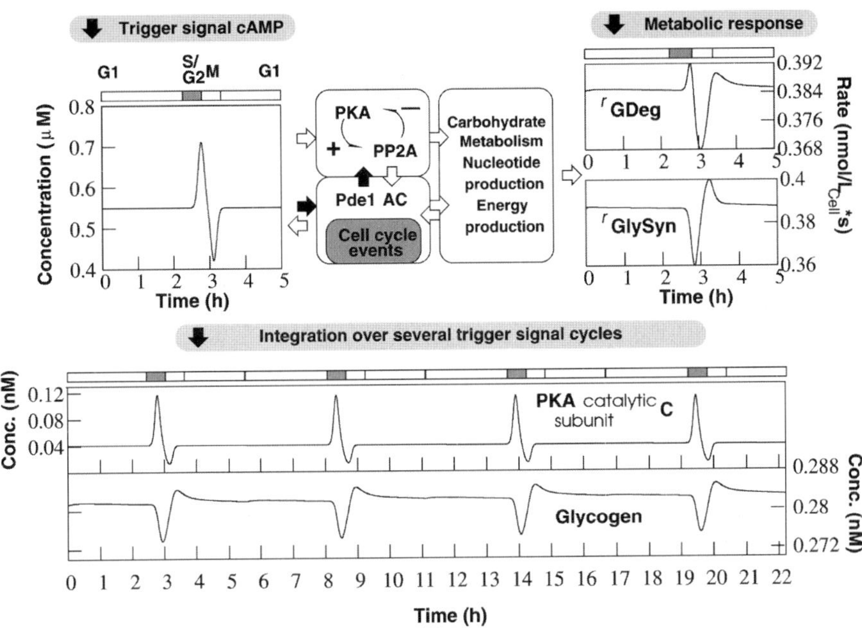



**Fig. 10.** Simulated response of PKA and the enzymes producing and degrading glycogen in the metabolic module to a cAMP time course as observed during the cell cycle. AC stands for adenylate cyclase while Pde1 denotes the low-affinity phoshodiesterase. White arrows represent stimulatory; black ones inhibitory effects. Intracellular concentrations (Conc.) are related to cytosolic volume. $r_{GDeg}$ signifies the rate of glycogen phosphorylase, $r_{GlySyn}$ that of glycogen synthase.

by accounting for two negative feedback loops involving the free PKA catalytic subunits. One route directly leads to cAMP degradation via PKA-stimulated activation of the phosphodiesterase Pde1 (Ma et al. 1999) while PKA also downregulates cAMP production via a strong negative feedback (Nikawa et al. 1987), most probably by decreasing the concentration of membrane-bound adenylate cyclase complexes (Engelberg et al. 1990; Gross et al. 1992; Pardo et al. 1993). Experimental evidence suggests that only these signaling complexes are able to transduce extracellular signals, such as an increase in the extracellular glucose concentration (Thevelein and de Winde 1999).

A previously developed model of glycolysis and the pentose phosphate pathway (Rizzi et al. 1997; Vaseghi et al. 1999) has been extended to include the dynamics of the storage carbohydrates trehalose and glycogen along with their associated regulation by PKA (Aguilera-Vázquez et al. "A Dynamic Model of Storage Carbohydrate Metabolism in *Saccharomyces cerevisiae*," manuscript in preparation). Cell growth is described by a stoichiometric expression based on the output from the metabolic module while the influence of PKA activity on cell cycle progression has been integrated into a literature model of the yeast cell cycle (Chen et al. 2000; Cross 2003). Parameter estimation has been performed on the basis of own and published experimental data, including information from global datasets and public databases. This guarantees the intimate connection between experiments and model development characteristic of the systems biology approach.

Figure 10 shows representative simulation results of the cAMP-induced dynamics of the metabolic module. An analytic function mimicking the cAMP time course observed in synchronous culture (cf. Fig. 8) has been used to trigger the combined system describing PKA activation and inactivation as well as the response of the metabolic module, here, considering the case of glycogen dynamics as example. As can be seen from the figure, the simulation predicts that the increase in cAMP at the G1/S transition is accompanied by a parallel rise in the concentration of free PKA catalytic subunits, that is to say, PKA activity. This is followed by an increased phosphorylation of PKA target proteins leading to activation of glycogen phosphorylase and a corresponding inhibition of glycogen synthase, respectively (Fig. 10), which results in a net decrease of glycogen levels. By contrast, when the cAMP concentration decreases during M phase, the simulated PKA activity declines while that of PP2A rises. Consequently, glycogen degradation is lowered and synthesis is increased, such that the glycogen concentration inclines again and returns to its initial level near the M/G1 transition. This behavior is qualitatively in accord with experimental findings (Silljé et al. 1997; Müller et al. 2003), although the quantitative effect predicted by the model is much lower. This may be ascribed to the fact that regulation of the glycogen pathway exerted by other signaling cascades, for example, Pho85 and Snf1 (Francois and Parrou 2001) has been neglected. Also, possible changes in protein levels of the modeled species during the cell cycle are not included in the current version of these two modules either, where the focus has been on rapid processes like signaling and metabolic regulation. However, incorporating quantitative information on protein levels would constitute an important extension en route to a more comprehensive model with improved predictive capabilities for long-term effects, which will be feasible once the necessary proteome data has become available.

The integrated single cell model yields a dynamic description of the cAMP-dependent regulation of metabolism and cell cycle progression during the different cell cycle phases. Its modular structure allows to simultaneously account for such diverse processes as signal transduction, metabolism, and cell cycle dynamics. In addition, the single cell model can also serve as a starting point for a segregated description of cell populations, for example, using a cell-ensemble approach (Domach and Shuler 1984; Henson et al. 2002). This provides the necessary modeling basis to capture population heterogeneity, which is a crucial factor in many applications in biotechnology and medicine as pointed out in the beginning. The preliminary model presented here certainly represents only a first step towards a truly quantitative description of this system. Despite neglecting regulatory actions not mediated by PKA and although suitable quantitative data are not yet available for all reactions, the model properly captures the qualitative system behavior under the given conditions. Further model refinements and incorporation of novel quantitative experimental data in the future can serve to progressively enhance the predictive capabilities of the model in an iterative fashion, which is characteristic of the systems biology approach.

Although the modular approach is promising, it is not without limitations as outlined above (cf. Section 2.1). It needs to be stressed that while the definition of functional modules greatly facilitates model development for cellular processes, for example, a metabolic pathway or a signal transduction cascade, one must not forget that the definition of these modules is not unique and mainly serves to delineate system boundaries when developing models for small subsystems. Also, some cellular compounds may belong to different functional modules depending on time and subcellular localization. Thus, these modules cannot be likened to electric circuit elements with single input – single output characteristics, that can be assembled into larger models simply by inserting them in the correct position in the - static - 'circuit' because they do not act in isolation within the living cell. Instead, molecular and regulatory interactions can exist at multiple levels within the module, providing links to cell constituents not included in the module itself. This yields a multi-input multi-output block with an intricate and dynamical wiring pattern to different parts of the metabolic and regulatory networks, respectively.

Assigning highly-connected compounds, such as ATP in the above example, to a specific module represents a further challenge, considering that this metabolite is produced and consumed in a vast number of intracellular reactions, in processes as diverse as respiration, biosynthetic reactions, solute transport, cytoskeletal dynamics, or cell cycle progression. When formulating a model for one of these processes as an isolated functional module it will, thus, frequently be difficult to obtain a satisfactory description of ATP dynamics solely on the basis of its role within the module when compared to *in vivo* data. Here, it may prove useful to combine the detailed representation of a functional module with a less detailed large-scale model to describe the dynamics of highly-connected module compounds. Such a large-scale model necessarily extends beyond the module boundaries and it can be constructed in a first step using only topological information about the underlying metabolic or signaling network.

# 4 Future directions – Or – How to catch a black cat in a dark room?

Systems biology still faces a number of challenges en route to a comprehensive description of systems behavior. Continued efforts are required to develop experimental techniques (wet lab) and computational tools (dry lab) that not only allow to perform quantitative experiments on a global scale, but also permit to thoroughly analyze their outcome. In addition, further efforts are needed to arrive at suitable mathematical frameworks to integrate the resulting diverse types of data for modeling purposes.

## 4.1 Wet lab

Regarding the experimental aspect, DNA chips have become a standard technology for analyzing transcriptome dynamics. However, the situation is different when looking at the levels of the proteome and metabolome because development of similar generic methods is hindered by the fact that metabolites and proteins are chemically much more diverse than nucleic acids. Nevertheless, the advent of protein microarrays and the development of powerful methods combining gel electrophoresis or LC with various MS setups now allow us to quantitatively determine the composition and dynamics of proteins and protein complexes in many cases, or even to specifically follow the fate of low-abundance signaling compounds carrying a certain post-translational modification (Aebersold and Mann 2003; Mann and Jensen 2003; Melton 2004). Recent studies in yeast have demonstrated the feasibility of determining both protein localization and concentration at a genome-wide scale under *in vivo* conditions by combining fluorescent fusion proteins with an affinity tag (Ghaemmaghami et al. 2003; Huh et al. 2003). Similarly, Schubert (2003) has employed a combination of fluorescent markers in an approach termed 'topological proteomics' to observe in 3D the spatiotemporal dynamics of many proteins simultaneously in a living cell. Since these imaging techniques are applied to single cells, they are not only capable of yielding spatial information that is crucial, for example, to correctly describe the functioning of signaling pathways, but they can also serve to assess cell population heterogeneity as well. Multiparametric flow cytometry, as routinely used in immunophenotyping, constitutes a complementary approach that can provide a wide variety of information about the population distribution of quantities like metabolic activity, specific protein levels, or the occurrence of signaling events based on a large number of single-cell measurements. These latter two features are still widely absent from most other large-scale approaches.

At the metabolite level, various methods based on mass spectrometry (e.g. GC/MS, LC/MS, or CE/MS), NMR, and other spectroscopic techniques are now widely employed for simultaneous identification and quantification of several tens of metabolites (Weckwerth 2003; Goodacre et al. 2004). Although many challenges still lie ahead in improving specificity, accuracy, and comprehensiveness of these global methods, they promise to aid significantly in obtaining quantitative global datasets that are crucial for the validation of large-scale network models.

## 4.2 Dry lab

Aside from the experimental challenges present today, also the development of suitable computational tools requires further progress. Continuing efforts are needed to efficiently manage the vast amount of data generated by global analyses. Moreover, curated, web-accessible databases should permit to integrate the various data types, preferably using a standardized representation that allows for easy exchange.

Development of mathematical models is crucial to obtain a unified description of the system under investigation while at the same time providing a check of (in)consistency of the available data. Since the lack of knowledge about many qualitative, let alone quantitative, details of biological processes will remain a problem for many years to come, it represents a major challenge to devise strategies that permit incorporation of this uncertainty into the model formulation while still yielding a robust model. Once developed, integrated models of regulatory and metabolic networks need to be implemented that describe cellular processes extending over many orders of magnitude in temporal and spatial dimensions. Even in the simplest case, when a continuous, deterministic representation is chosen, this results in large systems of stiff coupled differential-algebraic equations which are not easily solved numerically. Thus, advances in the fields of algorithm development and simulation software are important, especially to enable construction of hybrid models that combine, for example, a deterministic model of metabolism with a stochastic description of a gene regulatory network or a signaling cascade. Several computational tools addressing this problem have just recently been presented (You et al. 2003; Puchalka and Kierzek 2004; Takahashi et al. 2004; Vasudeva and Bhalla 2004). Moreover, generic mathematical methods suited for the analysis of hybrid models and associated tasks such as identifying sensitive parameters and estimating their values from data are still underdeveloped. The simulation of models including spatial information is numerically even more demanding, but these constitute a key step toward a more realistic description of the cell. Whole-organ simulations of the heart (Noble 2002) or dedicated simulation engines like, for example, VIRTUAL CELL, MCELL, or STOCHSIM (reviewed in Slepchenko et al. 2002) represent encouraging examples in this field, many of which build on a long-standing tradition of spatial modeling in neurobiology. Last but not least, methods based on inductive inference like network identification by 'reverse engineering' create formidable demands for computing power and will probably necessitate the extensive use of parallel computing architectures. In the future, web services complemented by application service providers should allow for the usage of centralized computational resources.


# 5 Concluding remarks

Given that progress in systems biology is heavily dependent on a combination of experimental and computational state-of-the-art techniques, many of which are still relatively unexplored, it is obvious that describing the behavior of whole cells or organs based on a mechanistic description of the spatiotemporal dynamics of constituents will not be achieved anytime soon. However, first steps in this direction have already been made and although the pace will probably be slower than expected by many, the multi-disciplinary approach taken and the combined treatment of cellular processes that have so far mostly been studied in isolation will add new facets to our understanding of biological systems. The promise that we will arrive at more comprehensive representations and a significant expansion of our biological knowledge that may also be exploited for practical applications, for example, to preselect new drug candidates, to optimize a drug dosage or to improve biotechnological production processes, will certainly make this effort worthwhile.

## Acknowledgment

The authors acknowledge support for this work by the DFG within the collaborative research center SFB 495, by the BMBF initiative "Systems of Life - Systems Biology," and by the DAAD to L. A.-V. (grant A/99/03632). We thank Andreas Politzer and Henning Schladebach for preparing the network graphs.

## References

Aebersold R, Mann M (2003) Mass spectrometry-based proteomics. Nature 422:198-207

Altintas MM, Kirdar B, Onsan ZI, Ulgen KO (2001) Plasmid stability in a recombinant *S. cerevisiae* strain secreting a bifunctional fusion protein. J Chem Technol Biotechnol 76:612-618

Anghileri P, Branduardi P, Sternieri F, Monti P, Visintin R, Bevilacqua A, Alberghina L, Martegani E, Baroni MD (1999) Chromosome separation and exit from mitosis in budding yeast: Dependence on growth revealed by cAMP-mediated inhibition. Exp Cell Res 250:510-523

Barabási A-L, Oltvai ZN (2004) Network biology: understanding the cell's functional organization. Nat Genet 5:101-113

Baroni MD, Monti P, Alberghina L (1994) Repression of growth-regulated G1 cyclin expression by cyclic AMP in budding yeast. Nature 371:339-342

Bolte M, Dieckhoff P, Krause C, Braus GH, Irniger S (2003) Synergistic inhibition of APC/C by glucose and activated Ras proteins can be mediated by each of the Tpk1-3 proteins in *Saccharomyces cerevisiae*. Microbiology (UK) 149:1205-1216

Brightman FA, Fell DA (2000) Differential feedback regulation of the MAPK cascade underlies the quantitative differences in EGF and NGF signaling in PC12 cells. FEBS Lett 482:169-174

Buchholz A, Hurlebaus J, Wandrey C, Takors R (2002) Metabolomics: quantification of intracellular metabolite dynamics. Biomol Eng 19:5-15

Buziol S, Bashir I, Baumeister A, Claassen W, Noisommit-Rizzi N, Mailinger W, Reuss M (2002) New bioreactor-coupled rapid stopped-flow sampling technique for measurements of metabolite dynamics on a subsecond time scale. Biotechnol Bioeng 80:632-636

Chen KC, Csikasz-Nagy A, Gyorffy B, Val J, Novak B, Tyson JJ (2000) Kinetic analysis of a molecular model of the budding yeast cell cycle. Mol Biol Cell 11:369-391

Cross FR (2003) Two redundant oscillatory mechanisms in the yeast cell cycle. Dev Cell 4:741-752

Domach MM, Shuler ML (1984) A finite representation model for an asynchronous culture of *Escherichia coli*. Biotechnol Bioeng 26:877-884

Eéka A, Hawoong J, Barabási A-L (2000) Error and attack tolerance of complex networks. Nature 406:378-382

Engelberg D, Simchen G, Levitzki A (1990) *In vitro* reconstitution of Cdc25-regulated *Saccharomyces cerevisiae* adenylyl cyclase and its kinetic properties. EMBO J 9:641-651

Francois J, Parrou JL (2001) Reserve carbohydrates metabolism in the yeast *Saccharomyces cerevisiae*. FEMS Microbiol Rev 25:125-145

Futcher B (1999) Cell cycle synchronization. Methods Cell Sci 21:79-86

Ghaemmaghami S, Huh W, Bower K, Howson RW, Belle A, Dephoure N, O'Shea EK, Weissman JS (2003) Global analysis of protein expression in yeast. Nature 425:737-741

Goodacre R, Vaidyanathan S, Dunn WB, Harrigan GG, Kell DB (2004) Metabolomics by numbers: acquiring and understanding global metabolite data. Trends Biotechnol 22:245-252

Gross E, Goldberg D, Levitzki A (1992) Phosphorylation of the *Saccharomyces cerevisiae* Cdc25 in response to glucose results in its dissociation from Ras. Nature 360:762-765

Hall DD, Markwardt DD, Parviz F, Heideman W (1998) Regulation of the Cln3-Cdc28 kinase by cAMP in *Saccharomyces cerevisiae*. EMBO J 17:4370-4378

Hartwell LH, Hopfield JJ, Leibler S, Murray AW (1999) From molecular to modular cell biology. Nature 402:C47-C52

Heinrich R, Neel BG, Rapoport T (2002) Mathematical models of protein kinase signal transduction. Mol Cell 9:957-970

Henson MA, Müller D, Reuss M (2002) Cell population modelling of yeast glycolytic oscillations. Biochem J 368:433-446

Huh WK, Falvo JV, Gerke LC, Carroll AS, Howson RW, Weissman JS, O'Shea EK (2003) Global analysis of protein localization in budding yeast. Nature 425:686-691

Kacmar J, Zamamiri A, Carlson R, Abu-Absi NR, Srienc F (2004) Single-cell variability in growing *Saccharomyces cerevisiae* cell populations measured with automated flow cytometry. J Biotechnol 109:253-268

Kholodenko BN, Demin OV, Moehren G, Hoek JB (2000) Quantification of short term signaling by the epidermal growth factor receptor. Biol Chem 274:30169-30181

Kitano H (2003) Cancer robustness: tumour tactics. Nature 426:125

Kromenaker SJ, Srienc F (1994) Cell-cycle kinetics of the accumulation of heavy and light-chain immunoglobulin proteins in a mouse hybridoma cell-line. Cytotechnology 14:205-218

Levsky JM, Singer RH (2003) Gene expression and the myth of the average cell. Trends Cell Biol 13:4-6

Ma P, Wera S, Van Dijck P, Thevelein JM (1999) The PDE1-encoded low-affinity phosphodiesterase in the yeast *Saccharomyces cerevisiae* has a specific function in controlling agonist-induced cAMP signaling. Mol Biol Cell 10:91-104

Mann M, Jensen ON (2003) Proteomic analysis of post-translational modifications. Nat Biotechnol 21:255-261

Massague J (2002) How cells read TGF-beta signals. Nat Rev Mol Cell Biol 3:169-178

Mauch K, Buziol S, Schmid JW, Reuss M (2002) Computer aided design of metabolic networks. AIChE Symposium Series 98:82-91

Melton L (2004) Proteomics in multiplex. Nature 429:101-107

Müller D, Exler S, Aguilera-Vázquez L, Guerrero-Martín E, Reuss M (2003) Cyclic AMP mediates the cell cycle dynamics of energy metabolism in *Saccharomyces cerevisiae*. Yeast 20:351-367

Nikawa J, Cameron S, Toda T, Ferguson KM, Wigler M (1987) Rigorous feedback control of cAMP levels in *Saccharomyces cerevisiae*. Genes Dev 1:931-937

Noble D (2002) Modeling the heart - from genes to cells to the whole organ. Science 295:1678-1682

Pardo LA, Lazo PS, Ramos S (1993) Activation of adenylate cyclase in Cdc25 mutants of *Saccharomyces cerevisiae*. FEBS Lett 319:237-243

Puchalka J, Kierzek AM (2004) Bridging the gap between stochastic and deterministic regimes in the kinetic simulations of the biochemical reaction networks. Biophys J 86:1357-1372

Ravasz E, Somera L, Mongru DA, Oltvai ZN, Barabási A-L (2002) Hierarchical organization of modularity in metabolic networks. Science 297:1551-1555

Rizzi M, Baltes M, Theobald U, Reuss M (1997) *In vivo* analysis of metabolic dynamics in *Saccharomyces cerevisiae*. 2. Mathematical model. Biotechnol Bioeng 55:592-608

Schaefer U, Boos W, Takors R, Weuster-Botz D (1999) Automated sampling device for monitoring intracellular metabolite dynamics. Anal Biochem 270:88-96

Schubert W (2003) Topological proteomics, toponomics, MELK-technology. Adv Biochem Eng Biotechnol 83:189-209

Silljé HH, ter Schure EG, Rommens AJ, Huls PG, Woldringh CL, Verkleij AJ, Boonstra J, Verrips CT (1997) Effects of different carbon fluxes on G1 phase duration, cyclin expression, and reserve carbohydrate metabolism in *Saccharomyces cerevisiae*. J Bacteriol 179:6560-6565

Slepchenko BM, Schaff JC, Carson JH, Loew LM (2002) Computational cell biology: Spatiotemporal simulation of cellular events. Annu Rev Biophys Biomolec Struct 31:423-441

Smith DM, Gao G, Zhang X, Wang G, Dou QP (2000) Regulation of tumor cell apoptotic sensitivity during the cell cycle. Int J Mol Med 6:503-507

Smith ME, Dickinson JR, Wheals AE (1990) Intracellular and extracellular levels of cyclic AMP during the cell cycle of *Saccharomyces cerevisiae*. Yeast 6:53-60

Takahashi K, Kaizu K, Hu B, Tomita M (2004) A multi-algorithm, multi-timescale method for cell simulation. Bioinformatics 20:538-546

ten Dijke P, Miyazono K, Heldin CH (2000) Signaling inputs converge on nuclear effectors in TGF-beta signaling. Trends Biochem Sci 25:64-70

Theobald U, Mailinger W, Reuss M, Rizzi M (1993) *In vivo* analysis of glucose-induced fast changes in yeast adenine nucleotide pool applying a rapid sampling technique. Anal Biochem 214:31-37

Thevelein JM, de Winde JH (1999) Novel sensing mechanisms and targets for the cAMP-protein kinase A pathway in the yeast *Saccharomyces cerevisiae*. Mol Microbiol 33:904-918

Tokiwa G, Tyers M, Volpe T, Futcher B (1994) Inhibition of G1 cyclin activity by the Ras/cAMP pathway in yeast. Nature 371:342-345

Vaseghi S, Baumeister A, Rizzi M, Reuss M (1999) *In vivo* dynamics of the pentose phosphate pathway in *Saccharomyces cerevisiae*. Metab Eng 1:128-140

Vasudeva K, Bhalla US (2004) Adaptive stochastic-deterministic chemical kinetic simulations. Bioinformatics 20:78-84

Weckwerth W (2003) Metabolomics in systems biology. Annu Rev Plant Biol 54:669-689

You LC, Hoonlor A, Yin J (2003) Modeling biological systems using Dynetica - a simulator of dynamic networks. Bioinformatics 19:435-436

## Abbreviations

Ac: acetate
AC: adenylate cyclase
Acald: acetaldehyde
AcCoA: acetyl CoA
APC: anaphase promoting complex
BI: budding index
C: catalytic subunit of protein kinase A
cAMP: adenosine 3',5'-cyclic monophosphate
CE/MS: capillary electrophoresis/mass spectrometry
Conc: intracellular concentration related to cytosolic volume
DAPI: 4',6-diamidino-2-phenylindole
F6P: fructose-6-phosphate
F16BP: fructose-1,6-bisphosphate
F26BP: fructose-2,6-bisphosphate
F26Bpase: fructose-2,6-bisphosphatase
G1: G1 phase of the mitotic cell cycle
G1P : glucose-1-phosphate
G2: G2 phase of the mitotic cell cycle
G6P: glucose-6-phosphate
GC/MS: gas chromatography/mass spectrometry
Glc: glucose
GlySyn: glycogen synthase
Gdeg: glycogen phosphorylase
Hxk: hexokinase
LC : liquid chromatography
LC/MS: liquid chromatography/mass spectrometry
M: mitotic phase of the cell cycle
NMR: nuclear magnetic resonance
PDC: pyruvate decarboxylase
Pde1: low-affinity 3',5'-cyclic-nucleotide phosphodiesterase
PDH: pyruvate dehydrogenase
PFK1: phosphofructokinase 1
PFK2: phosphofructokinase 2
PKA: protein kinase A
PP2A: protein phosphatase 2A
Pyr: pyruvate
r: enzymatic rate
R: regulatory subunit of protein kinase A
S: phase of the mitotic cell cycle during which DNA is synthesized

TCA: tricarboxylic acid cycle
UDPG: UDP-glucose
$\mu$: specific growth rate

Aguilera-Vázquez, Luciano
   Institute for Biochemical Engineering, Allmandring 31, 70569 Stuttgart, Germany

Mauch, Klaus
   Institute for Biochemical Engineering, Allmandring 31, 70569 Stuttgart, Germany

Müller, Dirk
   Institute for Biochemical Engineering, Allmandring 31, 70569 Stuttgart, Germany
   Mueller@ibvt.uni-stuttgart.de

Reuss, Matthias
   Institute for Biochemical Engineering, Allmandring 31, 70569 Stuttgart, Germany

# SIGNAL TRANSDUCTION

# Mathematical modelling of the Wnt-pathway

Reinhart Heinrich

## Abstract


A kinetic model of the Wnt-signal transduction pathway is presented which includes the main molecular components of this system. It is based on a set of differential equations reflecting the interactions of the proteins Dishevelled, GSK3, Axin, APC, PP2A, and ß-catenin. The predictions for an unstimulated and a stimulated state as well as for time dependent states concerning the stability of ß-catenin are compared with experimental data obtained for *Xenopus* egg extracts. By applying concepts of metabolic control analysis conclusions are drawn concerning the robustness of the pathway with respect to parameter changes. The role of the scaffolds Axin and APC are discussed in detail. It is shown that the low concentration of Axin is an important design feature, which may support the modularity of the pathway. It is proposed that the degradation of Axin may be considered to be an effective drug target for reducing the concentration of β-catenin.


## 1 Introduction

The Wnt-signal transduction pathway regulates cell fate in embryonal development and plays a crucial role in the formation of cancer by controlling the concentration of β-catenin (Wodarz and Nusse 1998; Polakis 2000; Salic et al. 2000). The main components of the Wnt-signalling pathway are the frizzled receptor (Frz), the scaffold proteins Axin and adenomatous polyposis coli (APC), the glycogen synthase 3 kinase (GSK3), the protein Dishevelled (Dsh), the phosphatase PP2A as well as the transcription factors ß-catenin and TCF. The pathway controls the concentration of ß-catenin via assembly and disassembly of a destruction complex consisting of Axin, GSK3, APC, and PP2A. In the absence of the Wnt-signal ß-catenin binds to this complex, and becomes, after phosphorylation, a substrate for ubiquitination leading to its degradation by proteasomes. In this way, the concentration of newly synthesized β-catenin is kept low. In the presence of the Wnt-signal the Dsh protein is activated, resulting in an inhibition of GSK3 and in turn in a reduction of ß-catenin phosphorylation and degradation. The build-up of ß-catenin in the presence of a Wnt-signal leads to transcription of specific genes (Nusse 1999). The present papers gives an overview on the assumptions and the results of a first mathematical model of the Wnt-signal transduction pathway, which was developed by the author in close collaboration with Ethan Lee, Adrian Salic, Roland Krüger, and Marc W. Kirschner (Lee et al. 2003).


Topics in Current Genetics, Vol. 13
L. Alberghina, H.V. Westerhoff (Eds.): Systems Biology
DOI 10.1007/b136811 / Published online: 26 April 2005
© Springer-Verlag Berlin Heidelberg 2005

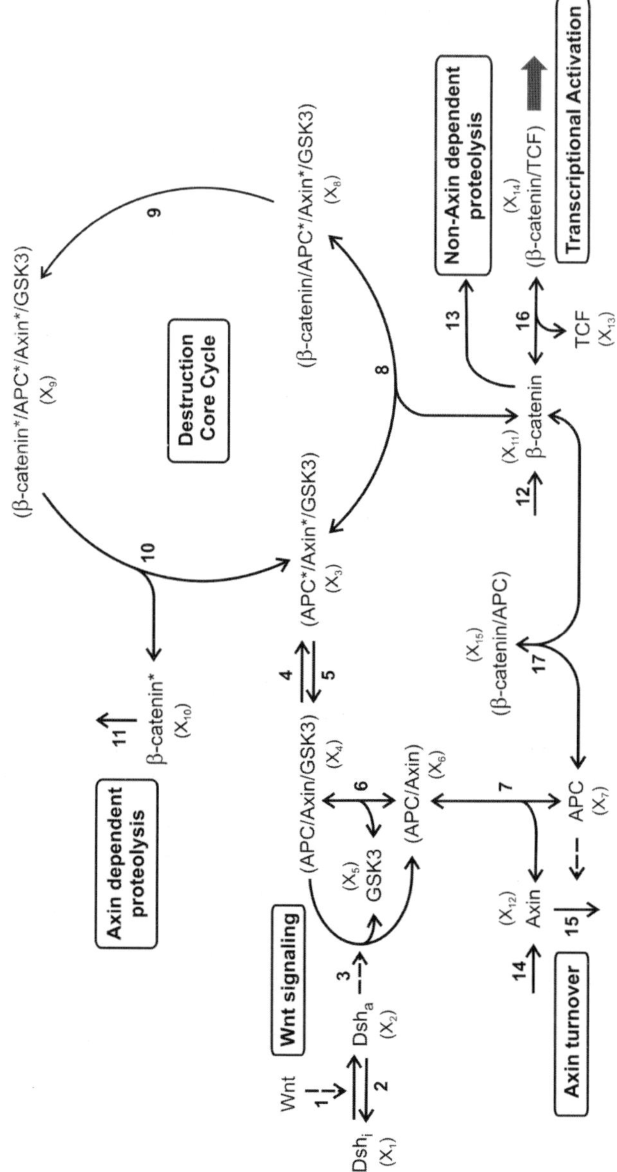


**Fig. 1.** Reaction scheme of the Wnt-pathway model. The model includes the activation of Dsh after Wnt-stimulation resulting in phosphorylation of Wnt, phosphorylation of Axin and APC (reaction 4), phosphorylation of β-catenin (reaction 9), dephosphorylations (reactions 2 and 5), protein-protein interactions (reactions 3, 6, 7, 8, 10, 16, and 17), protein syntheses (reactions 12 and 14) as well as protein degradations (reaction 11, 13, and 15). Phosphorylated compounds are marked by an asterisk.

## 2 Detailed reaction scheme

The model is based on the reaction scheme shown in Figure 1. The core of this signalling network is the destruction complex to which unphosphorylated ß-catenin binds (reaction 8). After phosphorylation (reaction 9) ß-catenin is released from this complex (reaction 10) and degraded by proteasomes (reaction 11). GSK3 also phosphorylates the two scaffold proteins Axin and APC (reaction 4). The latter process is counteracted by PP2A (reaction 5). The destruction complex is formed by binding of Axin to APC (reaction 7) and subsequent binding of GSK3 (reaction 6). Dsh is activated upon stimulation of the frizzled-receptor by Wnt (reaction 1). Reaction 2 denotes inactivation of Dsh. Inactivation of GSK3 is described by a release of GSK3 from the destruction complex (reaction 3). Reactions 14 and 15 denote the synthesis and degradation of Axin, respectively. Newly synthesized ß-catenin (reaction 12) is not only degraded after binding to the destruction complex, but also via a non-Axin dependent proteolysis (reaction 13). In addition, the model takes into account complex formation of ß-catenin with TCF (reaction 16) and with APC (reaction 17).

## 3 Systems equations

The model consists of a set of differential equations governing the time dependent changes of the concentrations of proteins, either in their free form or as protein complexes (Table 1). In order to allow for a compact representation these concentrations are denoted by $X_i$, instead by the chemical names of the compounds (see Fig. 1). The indices at the reaction rates $V_i$ correspond to the number of the reactions shown in Figure 1. In each case the reaction rates are described in a most simple way by using linear or bilinear functions for forward and backward reactions listed in Table 2.

Inspection of the system of differential equations reveals the existence of conservation quantities whose concentrations are independent of time. These quantities correspond to the total concentrations of those proteins whose synthesis and degradation is not taken into account. Specifically, the following conservation relations hold true:

$$X_1 + X_2 = Dsh^0 = const. \tag{1}$$

$$X_3 + X_4 + X_5 + X_8 + X_9 = GSK^0 = const. \tag{2}$$

$$X_3 + X_4 + X_6 + X_7 + X_8 + X_9 + X_{15} = APC^0 = const. \tag{3}$$

$$X_{13} + X_{14} = TCF^0 = const. \tag{4}$$

The existence of 4 conservations quantities implies that the number of differential equations can be reduced from 17 equations to 13 independent differential equations. The conservation relations for GSK and APC can be considerably simplified by taking into account that the concentration of Axin is extremely low

**Table 1.** Balance equations for the concentrations of proteins and protein complexes of the Wnt-pathway according to the reaction scheme depicted in Figure 1. The rate equations $V_i$ are listed in Table 2.

| Compound | Balance equation | Number |
|---|---|---|
| $Dsh_i$ | $\dfrac{dX_1}{dt} = -V_1 + V_2$ | (1) |
| $Dsh_a$ | $\dfrac{dX_2}{dt} = V_1 - V_2$ | (2) |
| $APC*/Axin*/GSK3$ | $\dfrac{dX_3}{dt} = V_4 - V_5 - V_8 + V_{10}$ | (3) |
| $APC/Axin/GSK3$ | $\dfrac{dX_4}{dt} = -V_3 - V_4 + V_5 + V_6$ | (4) |
| $GSK3$ | $\dfrac{dX_5}{dt} = V_3 - V_6$ | (5) |
| $APC/Axin$ | $\dfrac{dX_6}{dt} = V_3 - V_6 + V_7$ | (6) |
| $APC$ | $\dfrac{dX_7}{dt} = -V_7 - V_{17}$ | (7) |
| $\beta\text{-}catenin/APC*/Axin*/GSK3$ | $\dfrac{dX_8}{dt} = V_8 - V_9$ | (8) |
| $\beta\text{-}catenin*/APC*/Axin*/GSK3$ | $\dfrac{dX_9}{dt} = V_9 - V_{10}$ | (9) |
| $\beta\text{-}catenin*$ | $\dfrac{dX_{10}}{dt} = V_{10} - V_{11}$ | (10) |
| $\beta\text{-}catenin$ | $\dfrac{dX_{11}}{dt} = -V_8 + V_{12} - V_{13} - V_{16} - V_{17}$ | (11) |
| $Axin$ | $\dfrac{dX_{12}}{dt} = -V_7 + V_{14} - V_{15}$ | (12) |
| $TCF$ | $\dfrac{dX_{13}}{dt} = -V_{16}$ | (13) |
| $\beta\text{-}catenin/TCF$ | $\dfrac{dX_{14}}{dt} = V_{16}$ | (14) |
| $\beta\text{-}catenin/APC$ | $\dfrac{dX_{15}}{dt} = V_{17}$ | (15) |

compared to the concentration of all other compounds (see below for experimental data). Accordingly, we neglect in the conservation relations for GSK and APC the concentration of all complexes, which contain Axin. This leads to the simplified relations

$$X_3 = GSK^0 \quad , \quad X_7 + X_{15} = APC^0 \tag{5a,b}$$

**Table 2.** Rate equations for the processes involved in the Wnt-pathway. $k_i$ and $k_{\pm i}$ denote rate constants. Except of binding processes all reactions are considered to be irreversible. $W$ denotes the normalized concentration of the Wnt ligand.

| Step | Rate equation | Process |
|---|---|---|
| 1 | $V_1 = k_1 W X_1$ | activation of Dsh by Wnt |
| 2 | $V_2 = k_2 X_2$ | deactivation of Dhs |
| 3 | $V_3 = k_3 X_2 X_4$ | dissociation of GSK3 from the destruction complex |
| 4 | $V_4 = k_4 X_4$ | phosphorylation of Axin and APC |
| 5 | $V_5 = k_5 X_3$ | dephosphorylation of Axin and APC |
| 6 | $V_6 = k_{+6} X_5 X_6 - k_{-6} X_4$ | binding of GSK3 to the (APC/Axin) complex and |
| 7 | $V_7 = k_{+7} X_7 X_{12} - k_{-7} X_6$ | binding of Axin to APC |
| 8 | $V_8 = k_{+8} X_3 X_{11} - k_{-8} X_8$ | binding of β-catenin to the destruction complex |
| 9 | $V_9 = k_9 X_8$ | phosphorylation of β-catenin |
| 10 | $V_{10} = k_{10} X_9$ | dissociation of phosphorylated β-catenin |
| 11 | $V_{11} = k_{11} X_{10}$ | degradation of phosphorylated β-catenin |
| 12 | $V_{12} = const$ | synthesis of β-catenin |
| 13 | $V_{13} = k_{13} X_{11}$ | degradation of β-catenin |
| 14 | $V_{14} = const$ | synthesis of Axin |
| 15 | $V_{15} = k_{15} X_{12}$ | degradation of Axin |
| 16 | $V_{16} = k_{+16} X_{11} X_{13} - k_{-16} X_{14}$ | binding of β-catenin to TCF |
| 17 | $V_{17} = k_{+17} X_7 X_{11} - k_{-17} X_{15}$ | binding of β-catenin to APC |

The model is further simplified by performing a rapid equilibrium approximation (Heinrich and Schuster 1996) for most binding processes. Binding of GSK to (APC/Axin) was excluded from this approximation since otherwise Dishevelled-mediated dissociation of GSK from the (APC*/Axin*/GSK3), reaction 3, could not compete with the dissociation resulting from direct interaction of GSK3 with the (APC/Axin) complex. The rate constants $k_{\pm i}$ of fast binding processes ($i = 7$, 8, 16, and 17 enter the model equations only in the form of dissociation constants $K_i = k_{-i}/k_{+i}$ (for details, see Krüger and Heinrich 2004).

# 4 Model reference state

The reference state is defined as the unstimulated state corresponding to the absence of Wnt ($Wnt = 0$). It is a stationary state where Dsh is inactive and does not affect the degradation complex. β-catenin concentration is kept low by continuous

**Table 3.** Complete list of model parameters of the Wnt-signal transduction model. The rate constants marked with (#) play a role only in stimulated states where $W \neq 0$.

| Parameter | Value | |
|---|---:|---|
| **Conservation quantities** | | |
| $Dsh^0$ | 100 | |
| $APC^0$ | 100 | nM |
| $TCF^0$ | 15 | nM |
| $GSK3^0$ | 50 | nM |
| **Dissociation constants** | | |
| $K_7$ | 50 | nM |
| $K_8$ | 120 | nM |
| $K_{16}$ | 30 | nM |
| $K_{17}$ | 1200 | nM |
| **Rate constants** | | |
| $k_1$ (#) | 0.182 | $\text{min}^{-1}$ |
| $k_2$ (#) | $1.82 \cdot 10^{-2}$ | $\text{min}^{-1}$ |
| $k_3$ (#) | $5.00 \cdot 10^{-2}$ | $\text{nmol}^{-1}\text{min}^{-1}$ |
| $k_4$ | 0.267 | $\text{min}^{-1}$ |
| $k_5$ | 0.133 | $\text{min}^{-1}$ |
| $k_6$ | $9.09 \cdot 10^{-2}$ | $\text{nmol}^{-1}\text{min}^{-1}$ |
| $k_{-6}$ | 0.909 | $\text{min}^{-1}$ |
| $k_9$ | 206 | $\text{min}^{-1}$ |
| $k_{10}$ | 206 | $\text{min}^{-1}$ |
| $k_{11}$ | 0.417 | $\text{min}^{-1}$ |
| $k_{13}$ | $2.57 \cdot 10^{-4}$ | $\text{min}^{-1}$ |
| $k_{15}$ | 0.167 | $\text{min}^{-1}$ |
| **Synthesis fluxes** | | |
| $V_{12}$ | 0.423 | $\text{nmol min}^{-1}$ |
| $V_{14}$ | $8.22 \cdot 10^{-5}$ | $\text{nmol min}^{-1}$ |

phosphorylation and degradation. The reference state is characterized by the special values of the system parameters, which are the rates of protein syntheses, the rate constants of irreversible reactions, and the equilibrium constants for fast binding processes as well as by the conservation quantities. Solution of the systems equations under time independent conditions leads to values of the systems variables which are the concentrations of proteins and protein complexes. In combination with the rate equations (Table 2) also steady state fluxes, for example, rates of protein degradation, can be calculated. Table 3 presents a complete list of the system parameters. Some of these parameters have been measured directly whereas

**Table 4.** Concentrations of proteins and protein complexes in the unstimulated reference state $Wnt = 0$ and the stimulated state $Wnt = 1$.

| | Concentration (nM) | |
| --- | --- | --- |
| | $W = 0$ | $W = 1$ |
| $Dsh_i$ | 100 | 9.09 |
| $Dsh_a$ | 0 | 90.9 |
| Compound | $9.66 \cdot 10^{-3}$ | $1.46 \cdot 10^{-3}$ |
| | $4.83 \cdot 10^{-3}$ | $7.29 \cdot 10^{-4}$ |
| (APC/Axin) | $9.66 \cdot 10^{-4}$ | $9.75 \cdot 10^{-4}$ |
| APC | 98.0 | 88.7 |
| ($\beta$-catenin/APC) | 2.05 | 11.3 |
| ($\beta$-catenin/APC*/Axin*/GSK3) | $2.02 \cdot 10^{-3}$ | $1.86 \cdot 10^{-3}$ |
| ($\beta$-catenin*/APC*/Axin*/GSK3) | $2.02 \cdot 10^{-3}$ | $1.86 \cdot 10^{-3}$ |
| $\beta$-catenin* | 1.00# | 0.92 |
| $\beta$-catenin | 25.1 | 153 |
| Axin | $4.93 \cdot 10^{-4}$ | $4.93 \cdot 10^{-4}$ |
| TCF | 8.17 | 2.46 |
| ($\beta$-catenin.TCF) | 6.83 | 12.5 |
| Total concentrations: | | |
| $\beta - catenin^0$ | 35.0# | 178 |
| $Axin^0$ | $2.00 \cdot 10^{-2}\#$ | $7.28 \cdot 10^{-3}$ |

others have been obtained by fitting the model results to experimental data (see Lee et al. 2003). Parameters marked in Table 3 by (#) do not play a role in the reference state but are necessary describing the effect of Wnt-stimulation (see below). The values of all variables in the reference state are listed in the first column of Table 4. These values represent the steady state solutions of system equations using the data in Table 2 as input quantities with the value of Wnt set at $Wnt = 0$.


# 5 The stimulated state


Using the reference state as a starting point, other steady states can be calculated when the pathway is permanently stimulated. To characterize the strength of Wnt-stimulation, a dimensionless parameter $W = Wnt \big/ Wnt^0$ is introduced representing the ratio of the Wnt concentration with respect to its concentration $Wnt^0$ in a "standard" signalling state. Accordingly, $W = 0$ and $W = 1$ characterize the reference state and a standard stimulated state, respectively. The calculation of concentrations in the stimulated state requires additional input parameters, which are marked in Table 3 by an asterisk. Compared to the concentrations in the unstimulated state the stimulated state is described by the following properties: (1) The concentration of free unphosphorylated ß-catenin is increased by a factor of approximately 6, from 25 to 153 nM. (This increase reflects the decrease in the degradation rate of ß-catenin caused by the reduction in the ability of GSK3 to phosphorylate it); (2) The free phosphorylated ß-catenin concentration decreases by 8% from 1 nM to 0.92 nM; (3) The concentration of the (ß-catenin/TCF) complex increases by a factor of 1.8; (4) The large increase in the ß-catenin concentration shifts the binding equilibrium between APC and ß-catenin and the concentration of free APC falls slightly; (5) The total Axin concentration is decreased by a factor of 2.7 since the higher value of active Dishevelled decreases the concentrations of the various Axin containing complexes, making Axin susceptible to degradation.

When simulating the concentration of ß-catenin as a function of the strength of the Wnt-stimulus for ($0 \leq W \leq 1$) reveals a nearly hyperbolic saturation curve. Moreover, in the whole range of Wnt concentrations the steady state concentration of Axin is affected in an opposite direction to the concentration of ß-catenin (see Lee et al. 2003).

## 6 Comparison of theory and experiment

To test whether the mathematical model may represent correctly the dynamical properties of Wnt-pathway, it was used to simulate experimental data for the time courses for ß-catenin degradation under a variety of conditions, such as increased Axin concentration, increased Dsh concentration, inhibition of GSK3, and increased TCF concentration. The simulations and experimental results are each shown in plots of total β-catenin concentration versus time (Fig. 2A and 2B).

The straight line for $t < 0$ in Figure 2A represents the reference state. The simulated reference state curve (Fig. 2A, curve a) for β-catenin degradation is calculated for $t > 0$ at which there is an absence of protein synthesis for Axin ($V_{14} = 0$) and β-catenin ($V_{12} = 0$). This reference curve is in close agreement with the experimental data (Fig. 2B, curve a') with identical half-lives for β-catenin degradation (theoretical value of $t_{1/2} = 60.2 \, \mathrm{min}$ versus experimental value of $t_{1/2} = 60 \, \mathrm{min}$). Curve b in Figure 2A, describes the effect of an increased amount of endogenous Axin (increase from 0.02 nM to 0.2 nM). The additional Axin markedly accelerated β-catenin degradation ($t_{1/2} = 11.8 \, \mathrm{min}$) in agreement with the experimentally obtained values (Fig. 2B, curve b'; $t_{1/2} = 12 \, \mathrm{min}$). Theoretically the effect of Axin on β-catenin degradation is primarily due to the large concentration difference between the two scaffold proteins, APC and Axin. Due to the high concentration of APC, an increase in Axin concentration results in a sharp increase in the concentration of the (APC/Axin) complex thereby accelerating ß-catenin binding to the destruction complex.

Curves c and c' in Figures 2A and 2B, respectively, show the effect of addition of Dsh in inhibiting the degradation of β-catenin. Both curves show an initial rapid decrease in the β-catenin concentration in the first 30 minutes to 1 hour followed by a much slower decrease. Such a feature can be explained by the fact that Dsh acted only on the dephosphorylated complex (through step 3) to remove GSK3

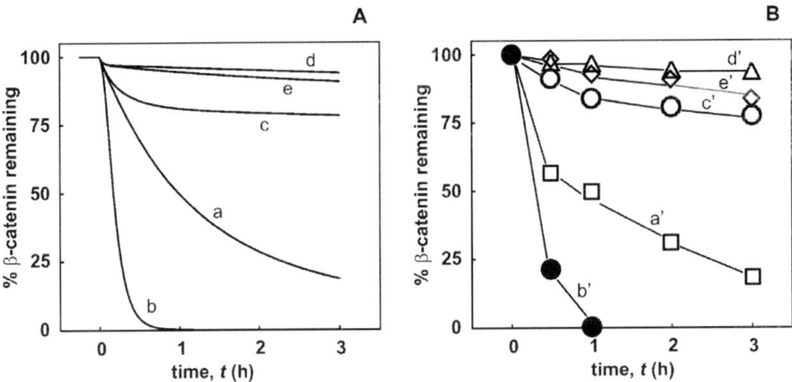

**Fig. 2.** Kinetics of β-catenin degradation. Simulations (A) refer to the following cases. Curve a: reference case (no addition of further components); Curve b: addition of 0.2 nM Axin; Curve c: addition of 1 μM activated Dishevelled; Curve d: inhibition of GSK3; Curve e: Addition of 1 μM TCF. Experimental data (B, curves a' - e') are obtained from time courses of β-catenin degradation in *Xenopus* egg extracts (Lee et al. 2003).

and thus block phosphorylation of the complex. In other words, Dsh inhibits the phosphorylation of the scaffold complex by GSK3 but does not inhibit the phosphorylation of ß-catenin. When Dsh binds, the complex can go around many times binding and phosphorylating ß-catenin before it dissociates and is inhibited by Dsh. One hour after the addition of Dsh, β-catenin degradation is significantly inhibited due to the removal of a significant pool of GSK3 from the degradation complex over time (through the action of Dsh). As a result, the scaffold protein Axin is dephosphorylated by the phosphatase (step 5). Dephosphorylated Axin is rapidly ubiquitinated and degraded when the ß-catenin degradation normally stops. The small decrease in β-catenin levels in Figure 2 (curve c) after a one hour incubation with Dsh is due to degradation of β-catenin via non-Wnt-pathway mechanisms (reaction 13, cf. Fig. 1 and Table 1), which was incorporated into the model.

Curve d in Figure 2 shows the effect of inhibiting GSK3 on β-catenin degradation. This effect is produced in the simulation by inhibiting GSK3 activity (steps 4 and 9). Only a small fraction of β-catenin (phosphorylated β-catenin) is available for degradation after complete inhibition of β-catenin phosphorylation (step 9), so inhibition is rapid. This is in complete agreement with the experimental data in which degradation is essentially blocked after inhibiting GSK3 activity by lithium (curve d' in Fig. 2B).

Curve e in Figure 2A predicts that β-catenin degradation is strongly inhibited after the addition of 1 μM TCF. It has been shown previously that β-catenin is sequestered by TCF, thereby resulting in a significant decrease in free β-catenin (Lee et al. 2001). The addition of TCF would be expected to decrease the rate of β-catenin phosphorylation (step 9) and subsequently β-catenin-degradation. This is also seen experimentally (Fig. 2B, Curve e').

## 7 Effect of APC on the β-catenin concentration

The scaffold APC is known to act as a potential tumour suppressor as it destabilizes the concentration of β-catenin. Loss of function mutations of APC result in an increase of the β-catenin concentration, which is thought to contribute to the development of colorectal cancer.

The model was used, therefore, to calculate for the unstimulated state the effect of changes in the concentrations of APC on the steady state concentrations of β-catenin. The results are shown in Figure 3 (solid line). It is seen that a decrease of the APC concentration has a strong effect on the β-catenin level (by a factor of more than three when APC is decreased by a factor of four), which supports the hypothesis that APC acts as a tumour suppressor.

It was recently shown that the degradation of Axin (reaction 15) is dependent on APC (Lee et al. 2003). In an extension of the basic model this effect was incorporated and the degradation rate of Axin is described by the following rate equation:

$$V_{15} = \frac{k'_{15} APC \cdot Axin}{K_M + APC} \tag{6}$$

where $K_M$ represents a half saturation constant for the activating effect of APC. It turns out that such an activation of Axin degradation by APC represents a "regulatory loop" exerting a homeostatic effect on the β-catenin concentration. When incorporating this regulatory loop a loss of APC affects the β-catenin concentration to a much less extent (broken line in Fig. 3). This effect is easily understood by the fact in the presence of the regulatory loop a decrease in the concentration of APC inhibits the degradation of Axin, thereby promoting the formation of the degradation complex.

## 8 Transient stimulation of the pathway

Wnt-stimulation *in vivo* is transient, likely due to receptor inactivation/internalization and/or other downregulatory processes. In the simplest way, transient Wnt-stimulation can be described by an exponential decay:

$$W = \begin{cases} 0 & , \quad \text{for } t < t_0 \\ e^{-\lambda(t-t_0)} & , \quad \text{for } t \geq t_0 \end{cases} \tag{7}$$

where the reciprocal of $\lambda$ represents the characteristic life time $\tau_W$ of receptor stimulation and $t_0$ denotes the onset of signalling. The concentration changes of all other pathway compounds resulting from Wnt-stimulation can be calculated by numerical solution of the system equations with initial values of the variables corresponding to the reference state. Figure 4 shows the time dependent behaviour of the total concentration of β-catenin and the total concentration of Axin upon transient Wnt-stimulation. The concentration of β-catenin increases temporarily and

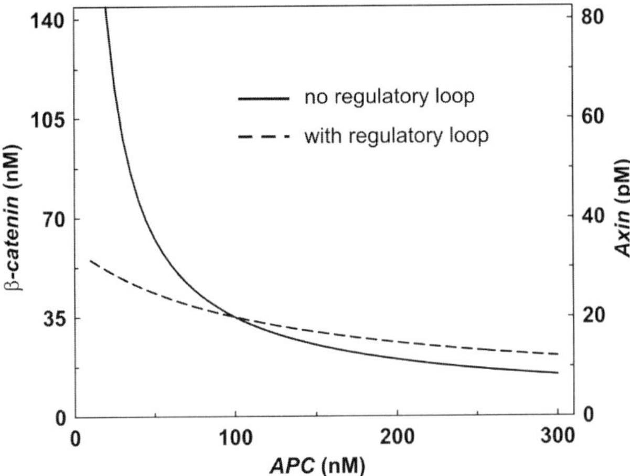

**Fig. 3.** Effect of APC on the concentrations of β-catenin. The solid line shows the results of the basic model, whereas as the broken line was calculated by including a regulatory loop where Axin degradation is dependent on the APC concentration (see Equation (6)). Parameter values: $K_M = 98.0\,\text{nM}$, $k'_{15} = 0.33\,\text{min}^{-1}$ ).

then returns to its initial value. In contrast, the concentration of Axin is temporarily downregulated. Curve a for β-catenin and curve a' for Axin were calculated by using the reference set of parameter values given in Table 3. The other curves were obtained by changed turnover rates of Axin. It is seen that the rate of Axin turnover (Yamamoto et al. 1999) sharply affects the dynamics of the response of Wnt-signalling. The curves b and b' and the curves c and c' are obtained for the case where both the synthesis rate $V_{14}$ and the degradation rate constant $k_{15}$ of Axin are increased by a factor of 5 and decreased by a factor of 5, respectively. Interestingly, an increase in the turnover rate of Axin leads to higher amplitudes and shorter durations of the β-catenin signal. This can be explained by the faster degradation of Axin after its Dsh mediated release from the destruction complex. Thus, β-catenin degradation is effectively inhibited for a certain time period due to a reduced availability of the scaffold Axin. Furthermore, a fast Axin turnover favours rapid replenishment of the Axin pool after decline of the Wnt-stimulus and, in this way, fast recovery of the destruction complex. This explains why the β-catenin signal is not only amplified but becomes more spike-like. Increasing the turnover rate of Axin affects the response of Axin to temporary Wnt-stimulation in a similar way as the response of β-catenin, i.e., the signal is amplified and sharpened (Fig. 4). Slowing down the Axin turnover results in opposite effects, i.e., the signals of β-catenin and Axin become less pronounced with lower amplitude and show a slower return to the initial state.

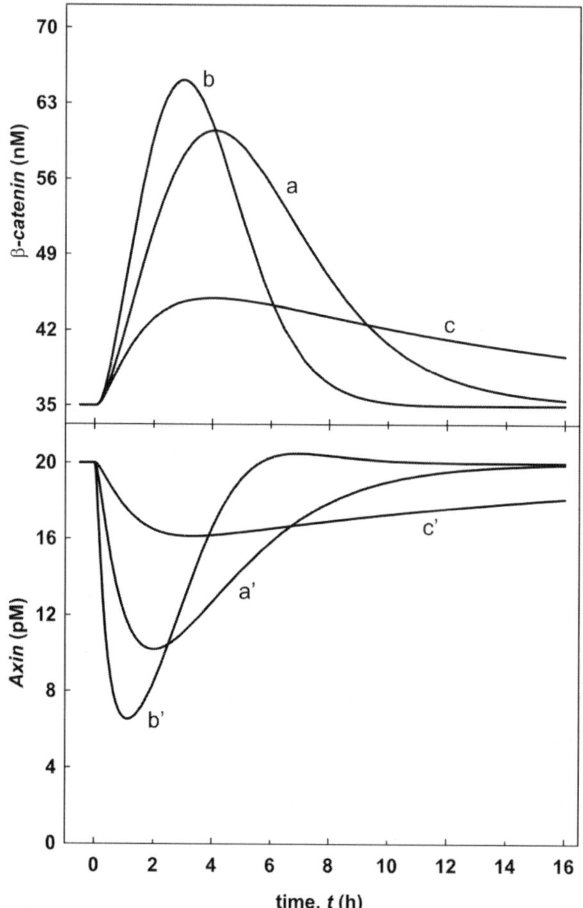

**Fig. 4.** Time courses of β-catenin and Axin concentrations resulting from a transient Wnt-stimulation. The curves a and a' were calculated by using the reference parameter values given in Table 3. For the characteristic life time of receptor stimulation a value $\tau_W = 20\,\text{min}$ was used. The various curves differ in the turnover rate of Axin. For curves b the parameters $V_{14}$ and $k_{15}$ were increased by a factor of five, and for curves c these parameters were reduced by a factor of 5.

## 9 Control and robustness of the Wnt-pathway

The model contains many parameters that affect the system behaviour in different ways and to various extents. These parameter effects can be systematically studied by using control coefficients as introduced in the framework of metabolic control analysis (Heinrich and Rapoport 1974; Heinrich and Schuster 1996; Fell 1997).

For example, the effects of the rate constants $k_{+i}$ and $k_{-i}$ on the total concentration of β-catenin is described by the coefficients

$$C_{\pm i}^{\beta\text{-}catenin} = \frac{k_{\pm i}}{\beta\text{-}catenin} \frac{\partial(\beta\text{-}catenin)}{\partial k_{\pm i}} \ . \tag{8}$$

Similar, the control coefficients for other compounds are defined. The control coefficients for β-catenin and Axin in the reference state are listed in Table 5.

There are 6 steps exerting strong negative control on the total β-catenin concentration. To this group belong the reactions participating in the assembly of the destruction complex (APC*/Axin*/GSK3). The corresponding parameters involve the rate constants $k_7$ for the binding of Axin to APC, $k_6$ for the association of GSK3 to the (APC/Axin) complex, and $k_4$ for the phosphosphorylation of Axin and APC in the destruction complex. Similar strong negative control is exerted by β-catenin binding to the phosphorylated destruction complex (rate constant: $k_8$ ), the phosphorylation of β-catenin in the destruction complex (rate constant: $k_9$ ), and the synthesis of Axin $(V_{14})$.

Six other reactions exert strong positive control in the reference state on the total concentration of β-catenin. To this group belong the reactions participating in the disassembly of the destruction complex (APC*/Axin*/GSK3) which are described by the rate constants $k_{-7}$ for the dissociation of the (APC/Axin) complex, $k_{-6}$ for the dissociation of GSK3 from the destruction complex, and $k_5$ for the dephosphorylation of the APC and Axin in the destruction complex. Other steps with a high positive control are the dissociation of β-catenin from the destruction complex (rate constant: $k_{-8}$ ), Axin degradation (rate constant: $k_{15}$ ), and β-catenin synthesis $(V_{12})$.

There are many reactions exerting in the reference state almost no control on β-catenin levels. This group includes binding of β-catenin to TCF and APC ( $k_{16}$ and $k_{17}$ ), and the corresponding dissociation processes ( $k_{-16}$ and $k_{-17}$ ). Interestingly, the effects of the two degradation processes of β-catenin (rate constants $k_{11}$, $k_{13}$ ) are also small.

The effects of parameter changes on Axin are generally opposite to those on β-catenin, that is processes with a positive control coefficient for β-catenin have negative control coefficients for Axin and vice versa. A significant exception is the synthesis of β-catenin, which exerts a positive control not only on β-catenin but also on Axin.

Closer inspection of Table 5 reveals that the values of the control coefficients for the rate constants sum up to zero. This fact is known as the summation theorem for concentration control (Heinrich and Rapoport 1974) and is valid for all reaction networks at steady state. This result finds its explanation in the invariance of the steady state concentrations against simultaneous change of all rate constants by the same factor. Interestingly, in the present case there are subgroups of processes whose control coefficients separately sum up to zero, indicating a modular

**Table 5.** Control coefficients for the total concentrations of β-catenin and Axin for the un-stimulated reference state

| | Parameter of step $j$ | $C_j^{\beta\text{-}catenin}$ | $C_j^{axin}$ |
|---|---|---|---|
| Kinase/ | $k_4$ | $-0.885$ | $0.502$ |
| phosphatase module | $k_5$ | $0.885$ | $-0.502$ |
| β-catenin module | $k_9$ | $-0.885$ | $-0.082$ |
| | $k_{10}$ | $-(10^{-5})$ | $-0.101$ |
| | $k_{11}$ | $-0.029$ | $0$ |
| | $V_{12}$ | $0.928$ | $0.186$ |
| | $k_{13}$ | $-0.014$ | $-0.003$ |
| Axin module | $V_{14}$ | $-0.885$ | $0.817$ |
| | $k_{15}$ | $0.885$ | $-0.817$ |
| Binding, dissociation | $k_6, k_{-6}$ | $\mp 0.885$ | $\pm 0.744$ |
| | $k_7, k_{-7}$ | $\mp 0.885$ | $\pm 0.792$ |
| | $k_8, k_{-8}$ | $\mp 0.885$ | $\pm 0.019$ |
| | $k_{16}, k_{-16}$ | $\pm 0.106$ | $0$ |
| | $k_{17}, k_{-17}$ | $\pm 0.076$ | $\mp 0.016$ |
| | $\sigma$ | $0.726$ | $0.413$ |

structure of the pathway. In Table 5, the control coefficients of the different modules are indicated by their names. The main three subgroups are the kinase/phosphatase module, the β-catenin module and the Axin module. Any combination of rate constants for binding and corresponding dissociation form subgroups for its own, which means that control coefficients of the binding reactions are opposite to those of the corresponding dissociation reactions.

Considering that positive control coefficients for β-catenin indicate an increase in the concentration of β-catenin upon activation of the corresponding process, reactions $i$ with $C_i^{\beta\text{-}catenin} > 0$ have a potential oncogenic effect whereas reactions with $C_i^{\beta\text{-}catenin} < 0$ have a potential tumour suppressor effect. Due to the summation theorem for control coefficients mentioned above we may conclude that

$$\sum \text{oncogenic effects} = \sum \text{tumor suppressor effects} \; . \tag{9}$$

Usually, oncogenic effects and tumour suppressor effects are discussed in terms of effects of the concentrations of proteins. However, since protein concentrations in signalling pathways are in general systems variables which can only be varied by changing reaction rates it may be more clear to speak about oncogenic reactions instead of oncogenic genes. Genetic defects can be considered in terms of changes in the activities of transcription, translation, or proteolysis, that is, by changes of reaction rates.

The control coefficients may be used to characterize the robustness of the pathway toward changes in kinetic parameters. Clearly, the robustness of a variable towards parameter changes is higher, the lower the corresponding concentration control coefficient. To arrive at an estimation of the overall effects of parameter perturbations on the system as a whole one may consider the standard deviation $\sigma$ of the control coefficients from their mean value. According to the summation theorem the mean value of all control coefficients for a given concentration is zero. Thus, one gets for the standard deviation for the control coefficients of β-catenin:

$$\sigma^{\beta\text{-}catenin} = \sqrt{\frac{1}{n}\sum_{j=1}^{n}\left(C_j^{\beta\text{-}catenin}\right)^2} \tag{10}$$

where the summation is performed over all $n$ reactions including forward and backward steps of fast equilibria. High values of $\sigma$ indicate that the given variable is on average very sensitive towards changes of rate constants. The standard deviations $\sigma$ of the control coefficients and the $\rho$ values for β-catenin and Axin are presented in the last two rows of Table 5. Because many control coefficients are close to zero and since the absolute values of the others hardly exceed unity the $\sigma$ values for β-catenin as well as for Axin are rather small. Since all values for $\sigma$ are lower than unity a one percent change in a rate constant leads, on average, to a response of less than one percent in the overall level of ß-catenin. The total concentration of Axin is more robust against parameter perturbations than the total concentration of β-catenin.

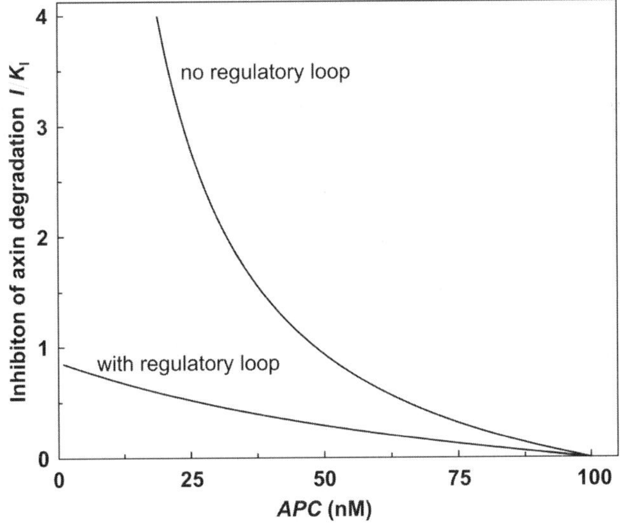


**Fig. 5.** Axin degradation as potential drug target for reducing the β-catenin concentration. Plotted is the concentration of a potential proteasome inhibitor I (scaled to the inhibition constant $K_I$), which is necessary for reducing β-catenin concentration to its original level depending on the APC concentration.

Due to the fact that Axin degradation has a strong effect on the concentration of $\beta$-catenin $C_{Axin-degr}^{\beta\text{-}catenin} = 0.89$ this process may be considered to be an effective drug target for reducing the concentration of $\beta$-catenin. Figure 5 shows that after loss of APC the concentration of $\beta$-catenin can be reduced to its original level by a potential proteasome inhibitor I (see also Tolwinski and Wieschaus 2004 for discussion of that effect).

## 10 Discussion

The model presented here is based on a simplified scheme of the Wnt/$\beta$-catenin signalling pathway. The scheme is still incomplete in the sense that there are other components of Wnt-signalling, such as caseine kinase, which have not been included yet. Caseine kinase is known to contribute to a multiple phosphorylation of $\beta$-catenin, which is not described in the present model. However, it could be easily extended in such a direction. Moreover, an extended model should include the nuclear transport of $\beta$-catenin, which is known to be mediated by APC.

The analysis revealed that the low concentration of Axin is a very important characteristic of Wnt-signalling pathway in *Xenopus* oocytes (0.02 nM of Axin compared to concentrations > 10 nM for most other compounds). This low concentration of a scaffold protein may indicate a very general and important design feature in the modularity of signalling pathways. Axin is a critical node point for the control of ß-catenin levels through the regulation of ß-catenin but it also interacts with components that are shared with many other essential systems. As the binding of these components such as GSK3$\beta$ fluctuate due to Wnt-signals (reflecting changes in binding as well as changes in axin concentration), other components important in other pathways would also have to fluctuate. Yet because the concentration of axin is so low there will be no appreciable changes in the overall levels of GSK3, Dsh, or APC. Hence, the very low axin concentration isolates the Wnt-pathway from perturbing other systems, a simple mechanism to achieve modularity. Other scaffold proteins may serve similar functions for other pathways. One may conclude that quantitative and kinetic data may be important in detecting modules. Inspection of circuit diagrams of signal transduction may, therefore, be not enough to identify modules in signalling networks.

The analysis demonstrates that the concept of metabolic control theory can be applied to understand the regulation of a specific signal transduction pathway. Whereas for metabolic pathways flux-control coefficients are mainly used to quantify rate limitation by enzymes, in signal transduction pathways concentration-control coefficients may play a more important role. For example, they can be used to express quantitatively the effects of a particular reaction on a given transcription factor. The sign and magnitude of these control coefficients give some indication what gene products could be oncogenes or tumour suppressors. As our understanding of pathways improve, the effect of mutation or pharmacologic inhibition could be estimated quantitatively using control coefficients.

# References


Bienz M, Clevers H (2000) Linking colorectal cancer to Wnt-signaling. Cell 103:311-320

Fell D (1997) Understanding the control of metabolism. Portland Press, London

Heinrich R, Rapoport TA (1974) A linear steady-state treatment of enzymatic chains. General properties, control and effector strengths. Eur J Biochem 42:89-95

Heinrich R, Schuster S (1996) The Regulation of Cellular Systems. Chapman and Hall, New York

Krüger R, Heinrich (2004) Model reduction and analysis of robustness for the Wnt/β-catenin signal transduction pathway. Genome Informatics 15:138-148

Lee E, Salic A, Kirschner MW (2001) Physiological regulation of β-catenin stability by Tcf3 and CK1ε. J Cell Biol 154:983-993

Lee E, Salic A, Krüger R, Heinrich R, Kirschner MW (2003) The roles of APC and Axin derived from experimental and theoretical analysis of the Wnt-pathway. PLoS Biol 1:116-132

Nusse R (1999) Wnt targets. Repression and activation. Trends Genet 15:1-3

Polakis P (2000) Wnt-signaling and cancer. Genes Dev 14:1837-1851

Tolwinski NS, Wieschaus E (2004) Rethinking Wnt-signaling. Trends Genet 20:177-181

Wodarz A, Nusse R (1998). Mechanisms of Wnt-signaling in development. Annu Rev Cell Dev Biol 14:59-88

Yamamoto H, Kishida S, Kishida, M, Ikeda S, Takada S, Kikuchi A (1999). Phosphorylation of Axin, a Wnt-signal negative regulator, by glycogen synthase kinase-3β regulates its stability. J Biol Chem 274:10681-10684



Heinrich, Reinhart
Humboldt University Berlin, Institute of Biology, Department of Theoretical Biophysics, Invalidenstraße 42, 10115 Berlin, Germany
reinhart.heinrich@biologie.hu-berlin.de

# Modelling signalling pathways - a yeast approach

Bodil Nordlander, Edda Klipp, Bente Kofahl, Stefan Hohmann

## Abstract

MAP kinase pathways are conserved signalling systems in eukaryotes that control stress responses, cell growth, and proliferation, as well as differentiation. Here, we discuss and compare the feedback control mechanisms of two very well studied yeast signalling systems: the pheromone response pathway and the osmosensing HOG pathway. Mathematical models have recently been generated, allowing *in silico* analysis of signalling properties of both pathways. To advance our understanding of pathway control and to make modelling less dependent on parameter estimation, quantitative time course data of high precision and resolution need to be generated in the future and implemented into mathematical models. We expect that a combination of quantitative analyses and modelling/simulation will provide novel insight into the rules with which signalling pathways control cellular processes.

## 1 Introduction

Signal transduction pathways are the cellular information routes by which cells monitor their surroundings, as well as their own state, and adjust to environmental changes or hormonal stimuli. Signalling encompasses the processes with which cells sense changes, generate intracellular signals, transduce the signals and ultimately mount responses. In doing so, signal transduction pathways orchestrate cellular metabolism, establish stress tolerance, control growth, proliferation and development, and determine morphogenesis.

Typically, signalling pathways consist of sensors or receptors, signal transducing molecules and proteins mediating responses. Sensors or receptors of environmental stimuli are often transmembrane proteins localised at the cell surface. Binding of hormone or physical stimulation generates a signal that alters activity of an intracellular enzyme involved in signal transduction. This enzyme either produces an internal second messenger or affects the activity of further signalling enzymes, such as protein kinases or phosphatases, thereby, activating the signalling system. Signalling cascades in eukaryotic cells may encompass a large number of steps, often more than five, from sensing to response. Such large numbers of steps are needed to allow transfer of the signal from one cellular compartment to another (typically from cell surface via the cytoplasm to the nucleus) as well as to generate specific system properties and targets for regulation. The response, the

Topics in Current Genetics, Vol. 13
L. Alberghina, H.V. Westerhoff (Eds.): Systems Biology
DOI 10.1007/b106656 / Published online: 20 January 2005

output of the signalling pathway, is typically mediated by altering the activity of the transcriptional machinery or of cellular enzymes.

Presently, our level of understanding is restricted, at best, to the wiring schemes of signalling pathways. Studies that aim at elucidating the wiring scheme usually monitor on/off states of a given signalling pathway. However, the parameters that are crucial for the operation of signalling pathways are the amplitude and the period of pathway activity, the spatial organisation and subcellular movements of the signalling components as well as their assembly into larger complexes that insulate signalling and control crosstalk between pathways. These quantitative, temporal, and spatial aspects of signal transduction have to be addressed by a combination of front-line experimental measurements as well as mathematical modelling and simulation.

While turning signalling pathways on is critical, it is of equally importance to turn signalling off again. Inappropriate, prolonged, or enhanced activation of signalling systems will cause fatal responses, such as cell death or continuous proliferation of cancer cells. Here, we will review the mechanisms of pathway activation and, in particular, deactivation. As an example, we will use the two best studied MAP kinase pathways: the yeast pheromone response pathway and the HOG (High Osmolarity Glycerol) pathway. Both pathways have developed highly effective mechanisms of feedback control but the final outcome is very different: the pheromone pathway becomes desensitised, rendering the cell unable to respond to stimulus for a certain period of time. In contrast, the HOG pathway can be reactivated by additional stimulation, but how this is achieved; why it is important will be discussed later in this review.


## 2 Yeast MAPK pathways

MAPK (Mitogen Activated Protein Kinase) pathways are conserved throughout eukaryotes (Gustin et al. 1998; Widmann et al. 1999). They consist of three tiers of protein kinases that sequentially activate each other: a MAPKKK, a MAPKK, and a MAPK. This conserved module is controlled by systems consisting of sensors/receptors, G-proteins and/or protein kinases, which can be of different types. MAPKs mediate responses by controlling further protein kinases and transcriptional regulators, again of different types. Protein phosphatases are negative regulators of MAPK pathways and both feed-forward and feedback loops can be involved in pathway regulation. Each cell type expresses numerous MAPK pathways, which interact and crosstalk in highly complex manners, and often share components. For this reason, the specific signalling routes are impossible to predict on the basis of protein sequence conservation, which otherwise allows prediction of MAPK pathway components. The yeast *Saccharomyces cerevisiae* has a relatively simple, though still complex, network of MAPK kinase pathways that mediate morphogenic and stress responses upon stimulation by pheromone, nutrient starvation, cell wall perturbation, or osmotic changes (Gustin et al. 1998). Based on physiological responses and genetic studies, a total of six distinct pathways, controlled by four MAPKKKs, four MAPKKs, and five MAPKs, are generally distinguished (Gustin et al. 1998; de Nadal et al. 2002; Hohmann 2002; O'Rourke et al. 2002). Thanks to the power of yeast genetics, the pheromone response pathway (Elion 2000) and the HOG pathway (Gustin et al. 1998; de Nadal et al. 2002; Hohmann 2002; O'Rourke et al. 2002) are the best studied MAPK pathways, and they have recently been subjected to modelling.

## 3 The yeast pheromone response pathway

*Saccharomyces cerevisiae*, although unicellular, is a sexual organism (Marsh et al. 1991; Sprague and Thorner 1992; Herskowitz 1995). Haploids are either of mating type MATa or MATα. Haploids of opposite sexes mate by fusion to form a diploid, which, in response to starvation, can undergo meiosis and sporulation to form four haploid spores, closing the life cycle. Communication between cells of different mating types occurs via pheromones, small peptides akin to peptide hormones. MATa cells produce a-factor pheromone and express the receptor for α-factor, which is produced by MATα cells, which in turn express a-factor receptors. The mating process is orchestrated by the pheromone response signalling pathway. The upstream part of this pathway has two roles: to re-orient cell polarity and cell growth towards the mating partner as well as to activate the MAPK cascade. The MAPK cascade will support an enhancement of the mating response and lead to production of proteins important for cell-cell adhesion and cell fusion. Most importantly, activation of the MAPK cascade mediates a G1 cell cycle arrest, thereby, synchronising the two haploid cells for successful mating (Marsh et al. 1991; Sprague and Thorner 1992; Herskowitz 1995).

In analogy to peptide hormone signalling, the pheromone receptors are G-protein-coupled receptors, namely Ste2 (for α-factor) and Ste3 (for a-factor). Upon pheromone binding, interaction between the receptor and the Gα subunit (Gpa1) leads to a series of conformational changes, allowing the release of GDP from Gα and the association of GTP (Dohlman and Thorner 2001; Yi et al. 2003; Bardwell 2004). This in turn leads to the release of Gβγ (Ste4 and Ste18, respectively). Free Gβγ interacts with the guanine nucleotide exchange factor (GEF) Cdc24, the p21-activated protein (PAK) Ste20, and the scaffold proteins Ste5 and Far1. The binding of Gβγ to Cdc24 is required for polarised growth and formation of mating projections and involves other proteins such as Ste20 and Bem1. Binding of Gβγ to the PAK Ste20 initiates signal transduction (Dohlman and Thorner 2001; Bardwell 2004). Activation of the MAPKKK Ste11 requires the Ste11-associated protein Ste50. Ste11 phosphorylates and activates the MAPKK Ste7, which in turn phosphorylates and activates the MAPK Fus3 (Gustin et al. 1998; Elion 2000; Dohlman and Thorner 2001). Fus3 and the MAPK Kss1 have some overlapping, although also clearly distinct functions. Kss1 is the MAPK of a different pathway, the filamentous growth pathway, which shares most components with the pheromone pathway (except for the receptor and associated G-protein)

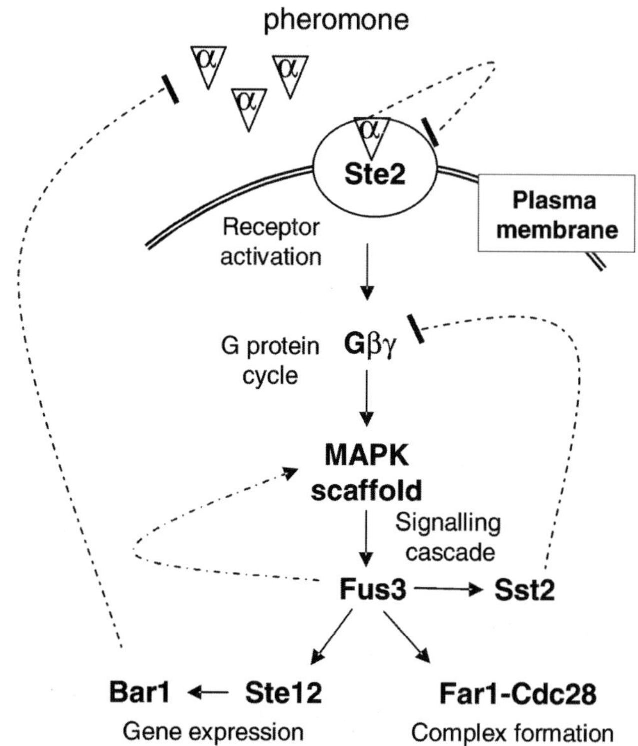



**Fig. 1.** Feedback loops in the pheromone pathway.

and controls the transition to filamentous growth upon nutrient starvation. To determine signal specificity, phosphorylated Kss1 is attenuated by activated Fus3 directly, as well as at the level of downstream transcription factors (Zeitlinger et al. 2003; Bardwell 2004; Kusari et al. 2004).

The transcriptional response to pheromone is mediated by the Ste12 transcriptional activator. In the resting state, the repressors Dig1/Rst1 and Dig2/Rst2 inhibit Ste12 by direct binding and the complex is stabilised by unphosphorylated Kss1. After release of phosphorylated Kss1 and following Fus3 dependent phosphorylation of the repressors and of Ste12, free phosphorylated Ste12 binds to pheromone response elements and activates transcription of target genes (Gustin et al. 1998; Elion 2000; Bardwell 2004). An important pathway target is phosphorylation of Far1, which is involved in chemotropic growth but also in cell cycle regulation. Far1 is a cyclin dependent kinase inhibitor (CKI) and causes cell cycle arrest at START in late G1 by inhibiting the activity of G1-cyclin Cdc28 complexes (Herskowitz 1997; Elion 2000).

A critical role in signalling is played by the scaffold protein Ste5, which binds the proteins of the pheromone response MAPK cascade (Elion 2000, 2001; van Drogen and Peter 2001, 2002; van Drogen et al. 2001; Sprague et al. 2004). Following pathway stimulation, Ste5 is tethered to the plasma membrane by binding to the Gβγ dimer. By localising to the plasma membrane, the scaffold protein brings its associated kinases into proximity of the membrane-associated Ste20, allowing Ste20 to phosphorylate Ste11. While activated Fus3 dissociates rapidly from Ste5, the scaffold protein remains bound at the plasma membrane, forming a platform for activation of many Fus3 molecules before it dissociates from Gβγ, which may lead to an amplification of the signal. The scaffold protein seems to be crucial for signal and substrate specificity, by preventing crosstalk to other pathways, accumulating the associated proteins in specific locations of the cell, bringing them in appropriate orientation to each other, preventing the influence of negative regulators on the bound kinases and suppressing the auto-inhibitory conformation of Ste11. The scaffold Ste5 is, hence, needed for signal transmission as well as pathway specificity, since Ste5 is specific for the pheromone pathway and not involved in other pathways that use the same MAPKKK and MAPKK (Elion 2000, 2001; van Drogen and Peter 2001, 2002; van Drogen et al. 2001; Sprague et al. 2004).

Activation of the pheromone pathway is counteracted by a series of feedback mechanisms that ensure effective pathway downregulation and recovery from pheromone-induced growth arrest. The relative importance of these mechanisms will be discussed in more detail below. Briefly, those mechanisms encompass (i) α-factor pheromone degradation by the secreted Bar1 endoprotease, whose production is stimulated by the pheromone response pathway; (ii) internalisation and degradation of pheromone-receptor complexes; (iii) stimulation of the GTPase activity of Gα via activated Fus3 and the RGS Sst2; (iv) dephosphorylation of Fus3 by the protein phosphatases Ptp2, Ptp3, and Msg5; (v) enhanced turnover of activated Ste11 due to ubiquitin-dependent degradation and enhanced transcription.


## 3.1 Simulating feedback control mechanisms of the pheromone response pathway

The dynamics of the pheromone pathway has recently been described in a mathematical model (Kofahl and Klipp 2004). It covers the following modules of the pathway (Fig. 1): (i) the activation of the receptor Ste2 by α-factor, and its subsequent degradation; (ii) the G-protein cycle triggered by active receptor; (iii) the stimulation-independent formation of the MAP kinase scaffold complex; (iv) the signal passage through the MAP kinase cascade, and (v) activation of downstream components including Ste12 and the formation of Far1-Cdc28 and Far1-Gβγ complexes as well as the stimulated expression and export of the protease Bar1. The corresponding set of 35 ordinary differential equations with 47 parameters has been implemented in Mathematica, Wolfram Research. Simulation results have been qualitatively compared with experimental results for 16 mutants described in the literature. In the following we present a more detailed look at the different feedback loops of the pheromone response pathway, employing simulations with this model in its original form and comparing the results with a second stimulation with the same amount of α-factor after 15, 30, and 45 minutes.

### 3.1.1 *α–factor degradation*

The pheromone response pathway mediates stimulated expression of the secreted endopeptidase Bar1, which inactivates α-factor (Manney 1983). This mechanism, thereby, leads to reduction of the signal itself and, hence, to signal cessation. The feedback loop involves a large number of steps from signalling through the pathway, gene and protein expression as well as vesicular transport to export Bar1 and for this reason takes a considerable period of time to close. Conceivably, the function of this feedback is to prolong refractory time rather than actual pathway downregulation. For instance, pheromone degradation could assist preventing re-stimulation of newly synthesized receptor.

In a simulation of the pheromone response removal of the Bar1 feedback loop has an effect on pathway performance. Fus3 is activated for a somewhat longer period of time while the formation of the complex Far1-Cdc28 is not affected at all (Fig. 2).

### 3.1.2 *Receptor internalisation*

Pheromone-binding to the receptor enhances the rate of receptor degradation (Elion 2000; Versele et al. 2001). In contrast to Bar1-mediated pheromone degradation this feedback mechanism is so short that it barely qualifies as a "loop" and constitutes an intrinsic mechanism for pathway downregulation, as described in the models of Yi et al. (2003) and of Kofahl and Klipp (2004). Krauss (2004) describes it as a longer loop where, upon activation, the βγ-complex associates with the receptor kinase and recruits it to the membrane. Furthermore, after phosphorylation of the activated receptor by the receptor kinase, the receptor becomes internalised, dephosphorylated, and at a later stage recycled or bound to the protein arrestin. The interaction with the G protein gets interrupted and the signal transduction is attenuated.

In simulations, elimination of receptor internalisation shows clear effect on the activation/deactivation profile. Pathway activation is stronger and prolonged (Fig. 2). The elimination of both pheromone degradation and receptor internalisation has even stronger effects on the activity profile of the pathway.

### 3.1.3 *GTP hydrolysis*

After receptor activation, the G protein is split into the βγ-complex and a GTP-bound Gα (Dohlman and Thorner 2001). The subunits can only re-associate after hydrolysis of the Gα-bound GTP to GDP. Spontaneous hydrolysis is very slow. The βγ-complex activates the MAPK cascade resulting in activation of the Fus3 MAPK, which in turn activates the transcription of Sst2 and stabilizes the protein Sst2 by phosphorylation. Active Sst2 accelerates the hydrolysis of Gα-bound GTP to GDP, resulting in downregulation of the MAPK cascade.

In order to avoid this feedback loop, Yi et al. (2003) made measurements with constantly active or constantly inactive Sst2. In our simulations, absence of the activation of GTP hydrolysis has a strong impact and impedes downregulation and

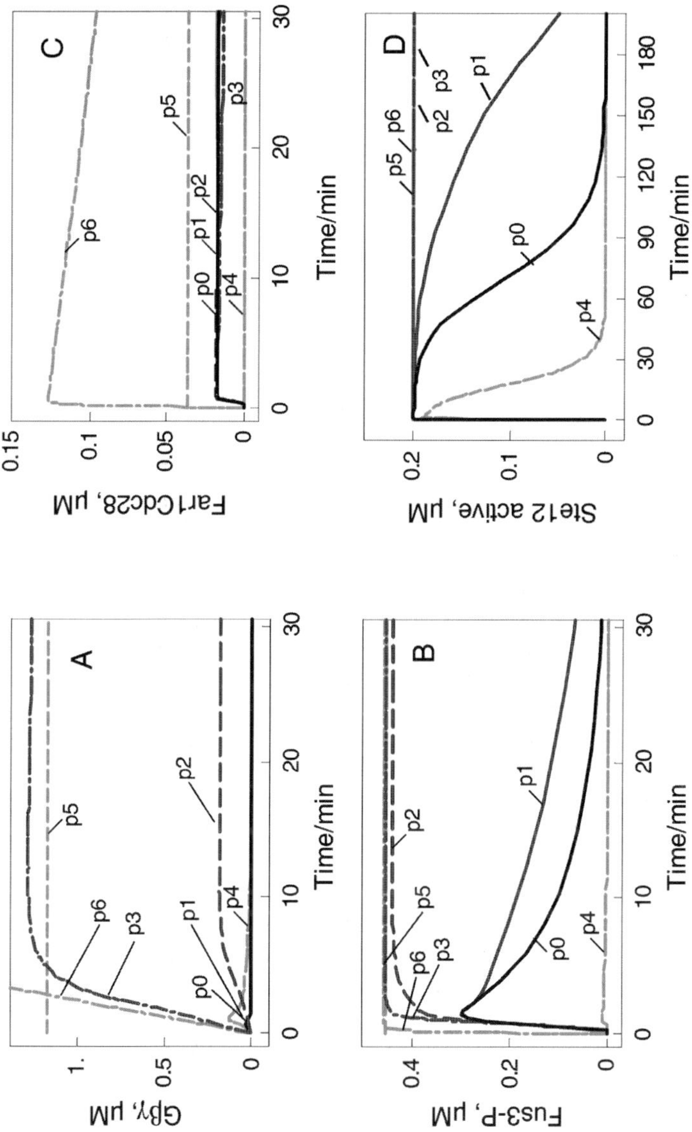

**Fig. 2.** Simulation of the temporal behaviour of key components of the pheromone pathway in wild type cells and in cells with artificially interrupted feedback loops. Panel A: Gβγ, Panel B: phosphorylated Fus3; Panel C: complex Far1Cdc28. Panel D: active Ste12. Curves in each panel: (p0): Wild type. (p1): Without action of endonuclease Bar1. (p2): Without degradation of activated receptor. (p3): Without Sst2 activated GαGTP hydrolysis. (p4): Without reuse of scaffold complex. (p5): Artificial stimulation by the addition of Gβγ. (p6): Artificial stimulation by attaching scaffold complex to plasma membrane. For explanation, see text.

**Table 1.** Temporal characterisation of the signalling pathway and the feedback for selected cases. The characteristic time $\tau$ is a measure for the time at which a signal is active. The duration $\delta$ characterizes the period of active signal

|           | $G\beta\gamma$ | Fus3-P   | Sst2 active | Ste12 active | Bar1 active | Case |
|-----------|---------|----------|-------------|--------------|-------------|------|
| $\tau$ / min | 0.76    | 5.52     | 24.01       | 38.65        | 97.79       | Normal action |
| $\delta$ / min | 0.78    | 9.08     | 23.93       | 79.26        |             | Normal action |
| $\delta$ / min | 3.61    | 18.88    | 53.39       | 111.59       |             | No cleavage of $\alpha$ factor by Bar1 |
| $\delta$ / min | 349.82  | 1289.46  | 1656.12     | 5534.67      |             | No receptor degradation |
| $\delta$ / min | 711.74  | 2373.91  | 5628.49     | 3163.33      |             | No G$\alpha$GTP-hydrolysis by Sst2 |

$G\beta\gamma$ as well as Fus3-P are present at a much higher level for a much longer period (Fig. 2). Hence, this feedback loop seems to be very important for an appropriate cellular adaptation to pheromone.

### 3.1.4 Ste11 degradation

Ste11 mediates its phosphorylation via Ste7 and phosphorylated Fus3 performs feedback phosphorylation of the scaffold protein Ste5 and/or Ste11. Although we assumed that this stabilises the complex, stabilisation (in the sense that the complex is degraded more slowly) has not been introduced in the model. We simply assumed that the complex (Ste5P-Ste11PPP-Ste7PP) phosphorylates more than one Fus3 molecule. Computational analysis shows that if the scaffold complex cannot be reused for a second round of signalling then the pathway activity would be much lower. This is evident from an enhanced concentration of $G\beta\gamma$ accompanied by remarkably lower trajectories of Fus3-P and complex Far1-Cdc28 (Fig. 2).

Another effect of feedback phosphorylation may be an enhanced degradation of Ste11PPP, upon complex association, presumably preventing crosstalk to the filamentous growth or the HOG pathway (Wang et al. 2003). So far crosstalk has not been considered in the model. The effect on the performance or the downregulation of the pathway of enhanced Ste11PPP degradation was negligible in simulations (not shown).

The feedback effects discussed proceed on different time scales and the characteristic times and the duration of several signals have been calculated (Kofahl and Klipp 2004). Obviously, signal compounds can only exert an effect as soon and as long as they are present at a sufficient amount. In Table 1, the simulated characteristic times and signal durations for $G\beta\gamma$, Fus3-P, active Sst2, active Ste12, and active Bar1 for normal and disturbed function of feedback mechanisms, are shown. Missing feedback regulation lead to prolonged activity of signalling compounds,

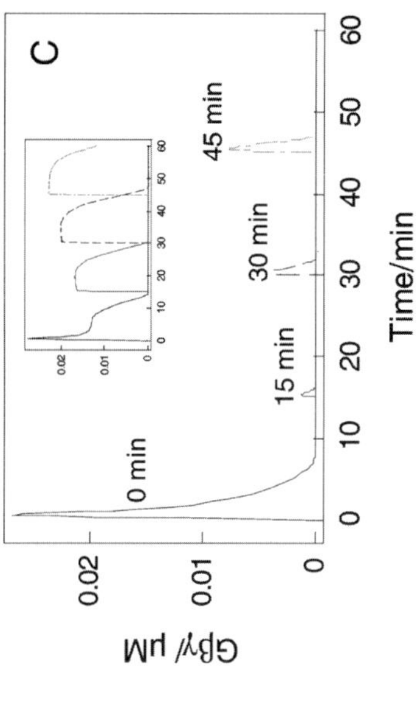


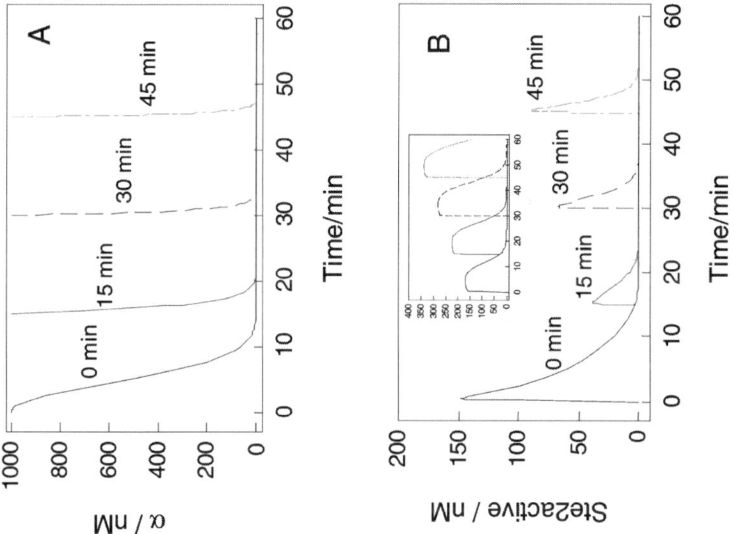


**Fig. 3.** Repeated stimulation of the Pheromone pathway. Shown are the concentration of α-factor (A), active receptor (B), and Gβγ (C). First stimulation with 1000nM α-factor, second stimulation with 1000nM α-factor after 15 min, after 30 min or after 45 min. Restimulation of the pathway is almost impossible after 15 min, but becomes stronger the later the second stress occurs. Insets in (B) and (C) show simulations for the case of missing receptor degradation (p2).

especially in the cases of missing receptor degradation and abolished $G\alpha$–GTP-hydrolysis by Sst2.

Pathway recovery can be studied by exerting a second stimulation with $\alpha$-factor (Fig. 3). At an early stage, the pheromone response can barely be re-stimulated due to active feedback regulation. In later stages, stronger responses are possible, indicating a release of the strong feedback actions. Without receptor degradation, the pathway reactivation would be much stronger and prolonged.

Taken together, adaptation mechanisms of the pheromone response aim at reducing the signal level itself (by degrading the pheromone, at least in the case of $\alpha$-factor), diminishing the capacity of the cell to perceive the signal (receptor internalisation), and to attenuate the transmission of the signal in different ways. Together, these mechanisms are crucial for desensitisation of the cell and, hence, recovery from pheromone-induced cell cycle arrest. This, in turn, is crucial to reset the cell cycle machinery after mating or in case mating was unsuccessful.

## 4 The high osmolarity glycerol response pathway

Yeast cells, being unicellular micro-organisms, are exposed to highly variable, often hostile, environments (Hohmann 2002; Tamás and Hohmann 2003). In particular, the water activity (availability of free water) in the environment varies, and yeast cells have to adjust their intracellular water activity accordingly. The ability to perform osmoadaptation and, hence, to control the availability of water, is a fundamental property of biological systems and by no means restricted to free living cells. Also, individual cells in the human body have mechanisms for osmoadaptation that are similar to those in yeast.

Cells suffer from dehydration when the external water activity drops. Such hyperosmotic shock activates the HOG pathway within one minute and a physiological response is initiated, which is best studied as a transcriptional reprogramming (de Nadal et al. 2002; Hohmann 2002; O'Rourke et al. 2002). At the same time, glycerol, a compatible solute, accumulates in the cytosol, allowing the cells to recover from the shock by regaining volume and turgor (Hohmann 2002; Tamás and Hohmann 2003). Glycerol accumulation is controlled by different mechanisms. (i) The glycerol facilitator, Fps1, closes rapidly after hyperosmotic shock, thereby, allowing accumulation of glycerol (Tamás et al. 2003); (ii) The MAPK Hog1 has recently been proposed to directly activate the enzyme 6-phosphofructo-2-kinase (PFK2), which catalyzes the synthesis of the glycolytic activator fructose 2,6-bisphosphate (Fru-2,6-$P_2$). Enhanced glycolytic activity may cause increased glycerol production (Dihazi et al. 2004); (iii) Active Hog1 promotes increased expression of the genes *GPD1* and *GPP2*, whose products are enzymes in glycerol production. Hence, this effect results in increased glycerol production capacity (Hohmann 2002; Tamás and Hohmann 2003).

The HOG pathway consists of two branches, the Sln1 branch and the Sho1 branch, which converge at the MAPKK Pbs2 (de Nadal et al. 2002; Hohmann 2002; O'Rourke et al. 2002). Each branch can apparently function independently and mediate activation of Hog1 upon osmotic shock, although it appears that the Sln1 branch is more sensitive to osmotic stimulation. In any case, it is necessary to block both branches to eliminate Hog1 activity. Sln1 is a sensor-histidine kinase of the eukaryotic type. These resemble bacterial two-component systems although commonly the kinase, the response regulator domain with the phosphorylated histidine residue and the receiver domain with an aspartic acid residue are located in one and the same protein (Wolanin et al. 2002). Under hypo-osmotic conditions, when the Sln1 histidine kinase is active, a phosphate group is transferred from the response regulator domain to the receiver domain of Sln1 and further to the response regulator domain of Ypd1 and finally to the receiver domain of Ssk1. Phospho-Ssk1 is unable to bind and activate the MAPKKK Ssk2 and Ssk22 and, hence, under hypo-osmotic conditions the MAPK cascade is not activated. Hyperosmotic shock blocks the histidine kinase activity of Sln1. Apparently this leads, within seconds, to a sufficient level of dephosphorylated Ssk1 for activation of the MAPK cascade. It is not known if a phosphatase is involved in this process. Dephosphorylated Ssk1 binds to and activates the MAPKKKs Ssk2 and Ssk22. Active Ssk2/22 phosphorylates and activates the MAPKK Pbs2 (de Nadal et al. 2002; Hohmann 2002; O'Rourke et al. 2002).

The Sho1 branch shares components with the pheromone response and the pseudohyphal development pathway. Two scaffold proteins control pathway activity. The first one is Sho1, which for some time has been thought to act as the sensor because it is a transmembrane protein. However, its only role appears to be tethering pathway components such as the scaffold/MAPKK Pbs2 to the cell surface (Raitt et al. 2000; Seet and Pawson 2004). Hence, the actual sensor in this system is not yet known. Sho1 binding to Pbs2 brings together a transient complex at the plasma membrane, in which Ste20 (activated and membrane-recruited by the G-protein Cdc42 – like in the pheromone response pathway) can phosphorylate and activate the MAPKKK Ste11, which in turn activates and phosphorylates Pbs2. Pbs2 kinase activity is necessary to break up the signalling complex, which, for this reason, can be detected at the plasma membrane only in such a mutant (Raitt et al. 2000; Reiser et al. 2000).

Pbs2, activated by Ssk2/22, Ste11, or both, phosphorylates the MAPK Hog1. Phosphorylation activates Hog1, which then accumulates in the nucleus. However, active Hog1 also has cytosolic targets and, hence, nuclear accumulation is incomplete. Such cytosolic Hog1 targets include the Ser/Thr protein kinase Rck2, which is necessary to attenuate protein synthesis after hypertonic shock (de Nadal et al. 2002; Hohmann 2002; O'Rourke et al. 2002). Recent data also show that Hog1 regulates cell cycle progression at the G1 phase by a dual mechanism that involves downregulation of cyclin expression and direct targeting of the CDK-inhibitor protein Sic1 (Escote et al. 2004).

Hog1 mediates transcriptional responses through several transcription factors, such as Sko1/Acr1, Hot1, Smp1, as well as the presumably redundant Msn2 and Msn4 (de Nadal et al. 2002; Hohmann 2002; O'Rourke et al. 2002). For Hot1, Sko1, and Msn2 it has been demonstrated that Hog1 does not only control the activity of these factors, but rather uses them to bind to promoter regions. It appears that Hog1 participates in transcriptional activation and probably is the actual transcriptional activator in this system (Alepuz et al. 2001, 2003; De Nadal et al. 2004).

Also for the HOG pathway, several feedback mechanisms that attenuate signalling have been reported. Expression of the genes *PTP2* and *PTP3*, which encode phosphotyrosine phosphatases, is stimulated in a Hog1 dependent manner, although this stimulation is moderate. Ptp2 and Ptp3 as well as the phosphoserine/threonine phosphatases Ptc1, Ptc2, and Ptc3 are negative regulators of the pathway and Hog1-dependent stimulation of phosphatase activity, or increased expression of genes encoding phosphatases, could provide a feedback loop (de Nadal et al. 2002; Hohmann 2002; O'Rourke et al. 2002). Another mechanism for negative regulation is degradation of unphosphorylated (active) Ssk1, although it has not been demonstrated to what extent this mechanism affects the activity profile of the pathway (Sato et al. 2003). As will be discussed in more detail below, signal cessation by successful glycerol accumulation and, hence, adaptation is another, and perhaps the most important mechanism, for pathway downregulation.

## 4.1 Feedback control of the HOG pathway

We refer here to a mathematical model of Klipp et al. (2004) which involves the following components (Fig. 4): (i) the dependency of the activity of two transmembrane proteins, Sln1 and Fps1, on the turgor pressure, (ii) the HOG signalling pathway including the phosphorelay system and the MAP kinase cascade, (iii) the nuclear import of activated Hog1, the change in transcription and translation level of Hog1-dependent genes, and (iv) a model for energy metabolism including glycerol production. In addition, the changes of volume, external and internal osmotic pressure, turgor pressure, and water flow over the cellular membrane are considered. This results in a system of 30 ordinary differential equations and 2 algebraic equations with in total of 57 parameters, which has been implemented in Mathematica. The qualitative and quantitative simulation results show strong correspondence with experimental results for five different mutants, various stress strengths, and different experimental scenarios, such as two subsequent stress treatments.

Based on experimental observations, four feedback loops, which are described below, are implemented in the model of osmostress response. The effect of the different feedback regulatory mechanisms on the cell performance under stress has been assessed by simulation of the network where these feedback loops have been eliminated individually.

### 4.1.1 Sensor activity

We first considered the feedback loop from turgor pressure change via sensor activation to turgor pressure adaptation (Klipp et al. 2004). Under ambient (hypo-osmotic) conditions the transmembrane receptor Sln1 is constantly phosphorylated while a decrease in turgor pressure (hyperosmotic shock) inhibits histidine kinase activity. Since hypophosphorylation of Sln1 and, eventually of Ssk1, leads to HOG pathway activation we refer to this event as activation. Subsequently, the

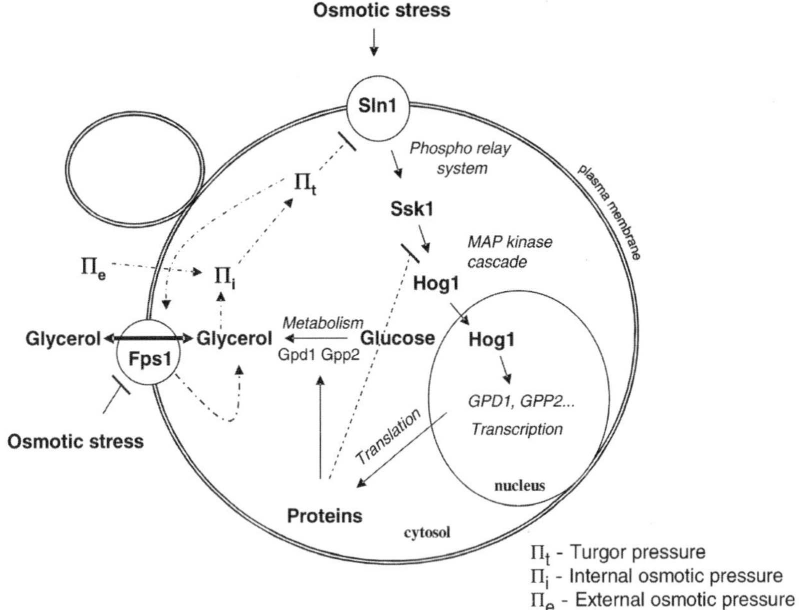



**Fig. 4.** Schematic representation of key events and regulatory interactions in the cellular response to osmostress. Osmostress activates the HOG signalling pathway including the receptor Sln1, Ssk1 and Hog1. Active (phosphorylated) Hog1 enhances the expression of several genes. Some of these genes code for enzymes (Gpd1, Gpp2), which accelerate the production of glycerol. Another gene codes for the phosphatase Ptp2, which has a negative effect on Hog1 activation. Glycerol is an osmotically active substance influencing the internal osmotic pressure and, thereby, turgor pressure and cell volume. Glycerol content of the cell is further regulated by the aquaglyceroporin Fps1, which in turn is sensitive to osmostress.

pathway encompasses phosphorylation of MAPK Hog1, alteration of gene expression, metabolic changes and increase of glycerol concentration. Increased glycerol levels lead to an increase in internal osmotic pressure and, hence, adaptation of turgor pressure. This adaptation then turns the signal around (from cell shrinking to cell swelling, i.e. from turgor drop to turgor increase), thereby, re-activating Sln1 phosphorylation.

The dependence of Sln1 phosphorylation on turgor pressure $\Pi_t(t)$ has been implemented using the relation

$$k_1^{TCS}(t) = k_1^{TCS,0} \cdot \left( \frac{\Pi_t(t)}{\Pi_t^0} \right)^2$$

where $k_1^{TCS}$ is the first order rate constant of Sln1 autophosphorylation. Superscript 0 denotes values at steady state before osmostress.

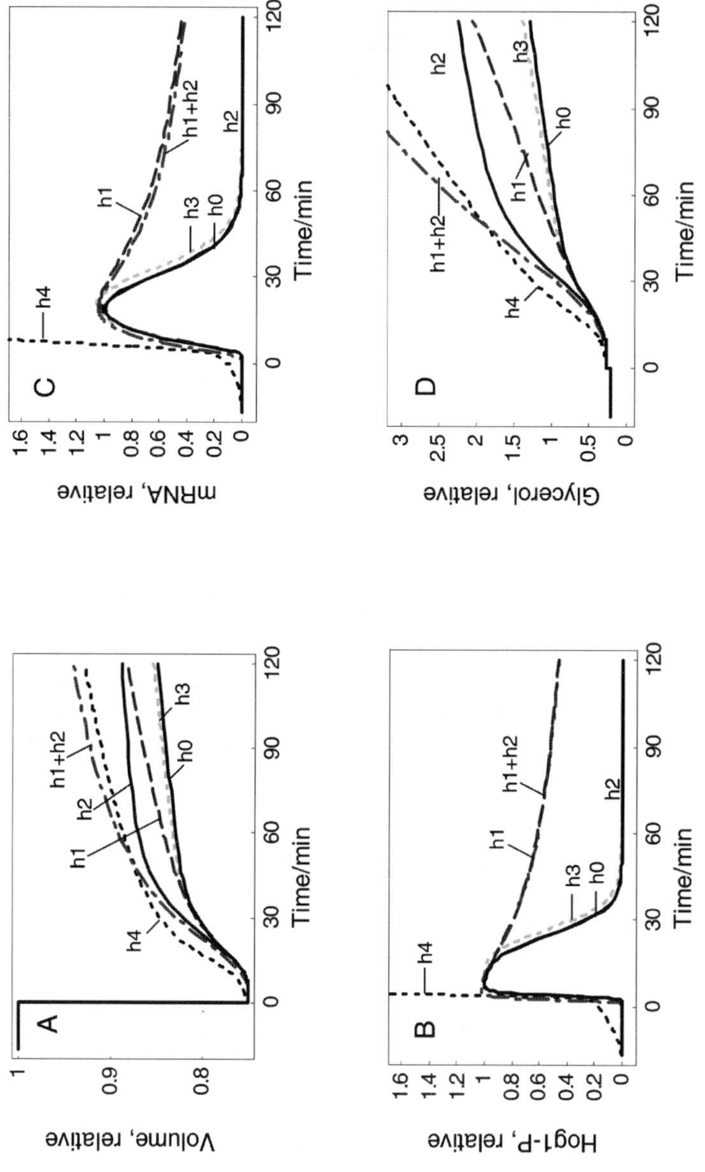

**Fig. 5.** Simulation of the temporal behaviour of key components of the Hog pathway and osmostress response in wild type cells, and in cells with artificially interrupted feedback loops. Panel A: cell volume, Panel B: phosphorylated Hog1; Panel C: mRNA, Panel D: internal glycerol. Curves in each panel: (h0): Wild type. (h1): Without feedback from turgor pressure to Sln1. (h2): Without feedback from turgor pressure to Fps1. (h1+h2): Without feedback from turgor pressure to Fps1 and Sln1. (h3): Without activation of Ptp2 expression. (h4): Abolished phosphatases. For explanation, see text.

This feedback loop is an example of signal cessation, where the cellular adaptation leads to a deactivation of the receptor, bringing it back to the initial state. If the cell is not able to adapt its turgor pressure (e.g. due to an inability of producing and accumulating glycerol) then the sensor histidine kinase system remains dephosphorylated.

The quick recovery of turgor pressure (and Sln1 phosphorylation) seems to be dependent on the closure of the aquaglyceroporin Fps1, which rapidly prevents glycerol efflux and leads to enhanced internal glycerol accumulation even without increased glycerol production. If, in simulations, turgor pressure changes can activate Sln1, but not deactivate it, then the activation of Hog1 and the expression of mRNA become amplified and prolonged (Fig. 5). At the same time, re-opening of Fps1 can prevent enhanced accumulation of glycerol. Therefore, in such a scenario volume and turgor pressure have a distinct temporal behaviour, but still show certain adaptation to osmostress. However, cells in which Sln1 signalling can not be deactivated by turgor increase during adaptation are not competent for an adequate response to a second shock (see further).

### 4.1.2 Turgor pressure and Fps1

Next we looked at regulatory interaction of turgor pressure and Fps1. Loss of turgor leads to a closure of the aquaglyceroporin (glycerol channel) Fps1. Therefore, if internal glycerol starts to accumulate, then turgor pressure increases and Fps1 re-opens. This feedback loop is short and also effects Sln1 phosphorylation. But regulation of the glycerol channel Fps1 alone is not sufficient for sustained glycerol accumulation and cellular stress adaptation. Instead, enhanced glycerol production is necessary for long-term adaptation.

If Fps1 cannot open after turgor pressure adjustment, then computational analysis predicts that the components of the signalling pathway (Hog1, mRNA) show the same temporal profile as in wild type, while glycerol accumulation is much stronger (Fig. 5). Although, at first glance, it seems advantageous to not to have a glycerol export channel, it is, however, needed when the cell faces hypo-osmotic stress.

### 4.1.3 Transcriptional activation of the phosphatases

It is generally assumed that transcriptional activation, and the production of the phosphatase Ptp2 is the most important feedback mechanism (de Nadal et al. 2002; Hohmann 2002; O'Rourke et al. 2002). Ptp2 catalyses the dephosphorylation of Hog1 in the nucleus. Although one should expect that the increase in Ptp2 accelerates Hog1 dephosphorylation and, therefore, pathway downregulation, the effect turns out to be negligible under the conditions studied. On the other hand, if the expression of *PTP2* cannot be stimulated, then the temporal profile of all considered quantities remains unchanged in the simulation (Fig. 5). It appears that activation of Ptp2 expression seems not to be important for the stress response or pathway downregulation under the modelling conditions currently studied. Conceivably, slightly enhanced Ptp2 expression may be necessary for proper adaptation under certain stress scenarios cells might face in their natural environment as well as for establishing a new steady-state in cells adapted to high osmolarity, ensuring appropriate thresholds for pathway activation/deactivation under these conditions.

### 4.1.4 Absent protein phosphatases

In addition to the "open" feedback loops, we have tested in simulations the effect of complete absence of phosphatases (Fig. 5). In this case, Hog1 would be completely phosphorylated and no pathway downregulation would occur. Volume restoration is stronger than with phosphatases, and a response to a second shock would not be possible. If, in addition, Fps1 could not re-open, then volume and internal osmotic pressure would dramatically increase.

An important conclusion can be drawn from these simulations: The interaction of all stress pathway elements ensures the optimal (or at least the experimentally supported) scenario of the cellular osmostress response. Elimination of only one single part of the adaptation mechanism does not abolish pathway activation or deactivation. Instead, combination of several defects is necessary to render the cell incompetent to respond properly. But only the wild type and the variant that cannot activate Ptp2 expression are able to respond to a second osmotic shock (see further).

Deactivation of the HOG pathway following initial stimulation is equally important as in the pheromone response pathway, because active Hog1 prevents cell division. However, the feedback mechanisms operate in such a way to allow for pathway reactivation by further osmotic stress stimulation, in essence monitoring the turgor pressure continuously, consistent with the variable natural environment of yeast.

# 5 Feedback control with and without pathway desensitisation

The feedback mechanisms of the pheromone response pathway and osmostress responsive HOG pathway initially have the same aim: to deactivate Fus3 and Hog1, respectively. This is necessary in order to overcome the cell cycle arrest mediated by active MAPK. Moreover, such downregulation of MAPK activity appears to be important to prevent a too strong response. It actually appears that a brief stimulation of MAPK activity is sufficient to stimulate gene expression over a longer period of time (Klipp et al. 2004).

Downregulation of MAPK activation is probably mediated in similar ways in both pathways. Protein phosphatases dephosphorylate and, thereby, inactivate the MAPK. As outlined here, stimulation of activity or level of the phosphatases probably only plays a minor role in adaptation, but the phosphatases as such are important to set appropriate thresholds and to inactivate the MAPK once other mechanisms turn off signalling from upstream components of the pathways. In addition to the phosphatases, there seem to be Fus3-activated feedback loops that attenuate signalling through the pheromone pathway. There is some indirect evidence that also active Hog1 mediates such feedback loops, at least in the Sho1 branch, whose feedback has not yet been studied in detail experimentally or theoretically.

In both pathways, the most important mechanism in downregulation appears to occur at the level of the sensors/receptors: the adaptation mechanisms seem to block sensor/receptor activation, thereby, preventing further activation of the pathway. However, the way this is achieved is very different in the two pathways. Sensor internalisation results in a desensitisation of the pathway, making it impotent for continued or subsequent stimulation until eventually the signal (pheromone) has disappeared from the environment. Such a desensitisation is necessary to allow the cell to initiate further rounds of proliferation in case mating was not successful. Indeed, in simulations re-addition of further pheromone leads to moderate if any reactivation of the pathway, as shown in Figure 3 at the level of G-protein activation, but the pathway gradually recovers its ability to respond.

Such desensitisation would, however, be fatal in case of the HOG pathway, because the cell needs to remain competent to respond to repeated osmotic shocks or continuous stimulation by increasing external osmolarity, for instance, on a grape drying in the sun. Hence, instead of degrading the sensor, it appears that the Sln1 osmosensor can readily toggle between an activated and inactive state. Presumably, hyperosmotic shock (turgor loss) turns the signalling on and it remains on until the cell starts to re-swell and regain turgor by accumulating glycerol. This scenario is well-supported by experimental evidence and simulations (Klipp et al. 2004). In other words, cellular adaptation by re-swelling removes the initial signal caused by shrinking even though the higher osmolyte concentration in the surrounding medium is still present. With such a toggling sensor system combined with constantly active phosphatases counteracting Hog1 activity the cell has a rather simple though effective system that allows rapid and repeated pathway activation and deactivation following osmotic changes.

Taken together, the pheromone response and HOG pathways are excellent and complementary model systems to study different mechanisms of MAPK pathway activation and deactivation, by experimental and theoretical approaches, because they receive different external signals and represent different mechanisms of adaptation to stimulation.

# 6 Data for modelling

Over the years, our understanding of pathway structures and organization has advanced significantly, mostly due to the generation of qualitative, commonly genetic, data. We clearly lack knowledge on the details of pathway dynamics including the mechanisms that maintain pathway specificity as well as quantitative aspects of the mechanisms that control pathway activity (thresholds, positive and negative feedback control, etc). At the same time, technological advances allow us to approach a point where it will be possible to monitor signal transmission in real time. This potential, which emerges from disciplines such as chemistry and physics, needs to be unlocked for the purpose of measuring parameters in the cell, rather than determining those by computational methods. We need to focus on establishing quantitative approaches that allow measuring different cellular components in (almost) real time and at absolute levels. Important are also proportions of, for instance, proteins in phosphorylated versus unphosphorylated states, or the fraction of a protein localised to certain compartments, and the fraction of proteins that are free or engaged in certain interactions. We also need to get a grip on, for example, binding constants and other parameters in the cellular environment. Measuring such parameters will make mathematical models far more realistic. Only then the combination of theoretical and experimental studies will lead to a true understanding of the complex operation of signalling pathways.

After the stimulus is sensed by the receptor/sensor, a signal is transferred via a cascade of proteins to the MAPK, which subsequently controls the activity of target genes and proteins. The transmission of the signal is a rapid event and involves several proteins that switch from inactive to active states. To date, signalling pathway activity is commonly monitored by specific antibodies, recognizing either tyrosine or dually phosphorylated MAPK. However, in order to determine as many parameters of pathway activity as possible, it will be necessary to measure phosphorylation states, activity and *in vivo* interaction of many individual components, both upstream and downstream of the MAPK. Such measurements could also provide useful insights into the regulation and the behaviour of the system. Presently, few specific antibodies are available that allow measuring activation states of individual kinases. With an increasing number of reagents, a large-scale approach using protein arrays becomes feasible, where, for example, specific antibodies are coupled to a solid support and bound proteins can be detected and quantified. Nielsen et al. describes the use of antibody arrays to monitor ErbB signal transduction in human cells (MacBeath 2002; Nielsen et al. 2003). A particular challenge is the determination of the fraction of a kinase that is phosphorylated. Presently, different laboratories work on mass spectrometry based approaches to address this important quantitative issue.

Inhibitors are useful tools to study the protein function. As *S. cerevisiae* is easy to manipulate genetically, mutants have been used to study the roles of proteins. While most powerful, there are several caveats to this approach: the cell could compensate for the missing component. Moreover, the organisation of protein complexes is likely disturbed when a component is missing and, hence, the results obtained may be misleading too. Therefore, the use of inhibitors is potentially a promising, complementary approach. Ideally, the inhibitors should be specific, readily taken up by the cell and rapidly acting on the target. For mammalian systems this approach has been used for many years, but the problem of specificity has remained. Specht and Shokat discuss the power of reverse chemical genetics, where a mutant allele of the target protein is sensitive to a specifically designed, small molecule, for example, a mutant kinase that only binds to a highly specific inhibitory compound (Specht and Shokat 2002). This approach could be extremely powerful in studying signal transduction and in a variation also to elucidate the targets of protein kinases, but so far very few studies have reported the successful use of the "Shokat approach".

The Green Fluorescent Protein (GFP) from the jellyfish *Aequorea victoria,* when tagged to a protein of interest, is a very powerful tool to study the dynamics of protein localization as well as *in vivo* protein interactions. However, the overall resolution and the possibilities for quantification are still limited, aspects that are central in technology development for quantitative biology. At present, there is a need for both, more stable and more short-lived variants of GFP. The GFP chromophore gets photo-bleached rather quickly, and following protein localisation over time will require more stable variants. On the other hand, the development of mutant forms of GFP less stable than wild type GFP will be useful when following, for instance, gene transcription of a GFP fusion protein as well as protein degradation. Spectral variants of GFP, such as CFP (Cyan) and YFP (Yellow), have been used to investigate interactions between proteins in living cells using the Fluorescence Resonance Energy Transfer (FRET) microscopy method. This technique relies on two different fluorophores, a so-called FRET pair, which has overlapping emission/excitation spectra. FRET occurs, when the emission spectrum of the donor significantly overlaps with the excitation spectrum of the acceptor (Gordon et al. 1998; Lippincott-Schwartz et al. 2001). The power of this approach has by far not been fully exploited and quantification is still difficult. Protein movements in the cell, either between different compartments or within a specific location in the cell, can be revealed by Fluorescence Recovery After Photo bleaching (FRAP). Fluorescent molecules lose their ability to emit fluorescence after the exposure to a high-powered laser beam, such as, photo bleaching. In a FRAP experiment, a protein of interest is fused to a fluorescent molecule, for example, GFP. A region of the cell is photo bleached, and fluorescence recovery, which is due to molecules moving from the non-bleached area to the photo bleached area until equilibrium is reached, is monitored (Lippincott-Schwartz et al. 2001). This technique is employed *in vivo* and measures the dynamics in real time, allowing quantifying the rate of protein movement. FRAP was used to investigate the MAPK dynamics in response to pheromones in *S. cerevisiae,* and the relocalization of components of the pheromone pathway was measured. For instance, the authors showed that Fus3 shuttles between the nucleus and the cytoplasm independently of pheromones, Fus3 phosphorylation and Ste5. FRAP could also reveal that after pheromone treatment and subsequent photo bleaching, Fus3-GFP recovered at the shmoo tip within a second while Ste5-GFP required 8-9 seconds (van Drogen et al. 2001). This is important information for dynamic modelling.

Activated MAPK usually affects gene transcription, either by modifying transcription factors or by direct binding to promoters. In the HOG pathway, it has been shown that activated Hog1 binds to chromatin, thereby, regulating transcription of stress-responsive genes. An important question is the extent to which promoter binding leads to altered gene transcription and in how much this eventually affects the production of proteins. The first aspect, promoter binding, can be monitored by ChIP (Chromatin Immuno Precipitation) analysis. Gene transcription can be monitored by several methods, such as the conventional Northern blot, real time RT-PCR and microarrays. Real time RT-PCR avoids the use of radiolabelled compounds and it provides the opportunity to quantitatively compare the transcription levels between genes (Bustin 2002; Vandesompele et al. 2002). To address the proportion of mRNA that is actually translated, polysomal RNA can be purified and used in microarray or RT-PCR experiments. Eventually, protein levels can be followed in time courses by conventional proteomics approaches. Taken together, a combination of quantitative ChIP, RT-PCR and proteomics could reveal the entire pathway from promoter binding to altered protein levels in a quantitative and time dependent manner.

For modelling purposes, the absolute levels of proteins and mRNA are extremely important, as they are used to estimate parameters correctly and assist deciding the appropriate modelling approach. Ghaemmaghami and co-workers determined the abundance of proteins by individually tagging each of the open reading frames in *S. cerevisiae* with the tandem affinity purification (TAP) tag, producing a fusion library. The quantification of each protein was conducted with quantitative Western blot where the detection of epitope-antibody interactions was compared to internal standards (Ghaemmaghami et al. 2003). Determination of absolute levels of mRNAs in the cell is approached by the group of Takashi Ito at Tokyo University.

In most experiments, a large population of cells is investigated and, hence, data obtained provide an average value of the quantitative measurements. While this approach provides intrinsically for statistical analysis, the time resolution of events might suffer. As a complement to large-population experiments, single-cell approaches could be useful for studying rapid, as well as slower, events. For instance, optical tweezers have been developed together with microfluidic systems, creating a versatile micro-laboratory. Cells are trapped by a laser beam and transferred from one micro-channel to another within seconds, and during this process, a fluorophore, such as GFP, can be visualised by microscopic techniques (Ericsson et al. 2000; Enger et al. 2004). This method is a powerful tool and provides the opportunity to follow protein movements and protein-protein interactions within a single cell at moderate throughput but at high resolution. The events can be visualised within seconds after the exposure to the stimulus and followed over several hours by video-microscopy. Flow cytometry is another single-cell analysis tool, which provides the opportunity to measure the distribution of properties within the population, overcoming the problem of treating cell population as "one cell". Most applications of flow cytometry are based on fluorescence monitoring, which is dependent on pre-treatment with reagents. Parameters, such as cell size, viability, DNA and total protein content, enzyme activity, and intracellular pH can be monitored, but also intracellular products, including GFP and its variants. Flow cytometry could also be employed to study protein-protein interactions, by using the FRET technology, or to measure the dynamics of gene expression by using reporter constructs (Winson and Davey 2000; Rieseberg et al. 2001). In contrast to optical tweezers combined with confocal microscopy (moderate throughout, high resolution), flow cytometry is a high throughout approach with low intracellular resolution.

# 7 Mathematical models

Mathematical modelling assists analysing the dynamics and behaviour of a cellular system, determining system parameters, and it has the potential of predicting the outcome of experiments. Mathematical models, based on experimental observations, can be set up to distinguish between possible scenarios and to identify the step of a system that is most sensitive to perturbations. For example, by *in silico* investigations, Swameye et al. showed that the parameters of nuclear shuttling of STAT5, in the JAK-STAT signalling pathway, were most sensitive to changes and this was verified by experiments (Swameye et al. 2003). Models can also be extremely useful to elucidate regulatory features of a signalling pathway. Brightman and Fell (2000) developed a mathematical model in order to distinguish between the quantitative differences in epidermal growth factor (EGF) and nerve growth factor (NGF) signalling in PC12 cells, where EGF causes transient activation of Ras, MEK and ERK (MAPK) while NGF causes sustained activation of the same components. In order to obtain comparable activation profiles between simulations and experimental results, the mathematical model required a negative feedback loop. This was mediated by active ERK on an upstream component, when the cells were exposed to EGF, and the signal was attenuated. In yeast, the heterotrimeric G protein cycle in the pheromone pathway has been characterised quantitatively. Yi et al. employed the FRET technique with different parts of the G protein cycle as FRET-pairs. By this method, the authors could compare the *in vivo* kinetics of G protein activation, after addition of α-pheromones, to transcriptional activation of pheromone responsive genes and cell-cycle arrest. Data from the G protein cycle were fitted to a mathematical model (Yi et al. 2003).

Although experimentally well-supported mathematical models are crucial steps to visualise actual processes, also purely computational models, which are set up with limited or without experimental values for the kinetic parameters, are useful to discuss a number of dynamic and regulatory features. For example, the frequently observed tendency of signalling pathways to convert gradual stimulation into an all-or-none decision was discussed with respect to the network structure, the kinetics of individual reactions and the choice of kinetic parameters. It could be shown that this switch-like behaviour may result from multilevel network structure of phosphorylation cascades (Huang and Ferrell 1996) or from the ultrasensitive stimulus-response curves of the individual reactions (Goldbeter and Koshland 1984). Furthermore, regulatory interactions in signalling cascades like feedback loops have been shown to be responsible for the occurrence of oscillations and bistability (e.g. Kholodenko 2000; Bhalla and Iyengar 2001; Ferrell 2002; reviewed by Tyson et al. 2003). Other important issues are the robustness of the functionality of signal transduction with respect to parameter changes (e.g. Barkai and Leibler 1997; Bluthgen and Herzel 2003), and the phenomenon of signal amplification or signal adaptation by means of dynamic regulation (Asthagiri and Lauffenburger 2000).

For the quantification of temporal changes, measures for characteristic times, signal duration or signal amplification have been introduced (Llorens et al. 1999;

Heinrich et al. 2002). These measures give a quantitative evaluation of signalling pathways in addition to qualitative characteristics, like the occurrence of bifurcations or oscillations.

## 8 Conclusions

Mathematical models have been employed to study different aspects of signalling pathways. In this chapter, we discussed the pheromone pathway and the HOG pathway in terms of feedback control, and the two mathematical models that have been set up were used to test scenarios of pathway downregulation. Deactivation of signalling is as important as activation, because inappropriate pathway activation causes cell cycle arrest, as in the cases studied here, or uncontrolled proliferation. Therefore, there must be rigorous feedback mechanisms, ensuring pathway shut-off. While the activation states of both the pheromone response and the HOG signalling pathways are attenuated by phosphatases, other mechanisms are different. For instance, the pheromone pathway is desensitised and cannot immediately respond to a second stimulus, due to receptor internalisation and degradation following stimulation. In contrast, the HOG pathway is readily reactivated by subsequent stimuli. Taken together, the main difference of regulation seems to lie at the level of the receptors/sensors, which is internalised in one instance (pheromone pathway) but remains signalling competent in the other (HOG pathway). This apparent difference fits well with the physiological function of each pathway: An osmosensing pathway needs to remain alert to further osmolarity changes, while a mating pathway should be able to recover from the pheromone-induced cell cycle arrest in order to resume proliferation when mating was either completed or unsuccessful. The two mathematical models used for these analyses suggested quantitative aspects not previously experimentally measured, but also suggested the steps most sensitive to perturbations. However, in order to advance our understanding of signalling, more quantitative data need to be generated that provide detailed information of how the different components behave in all four dimensions (X, Y, Z, and time). These data should be implemented in mathematical models, which could allow us to reach a system level understanding of signalling pathways.

## Acknowledgements

This work was supported by the European Commission (contract LSHG-CT2003-503230, the QUASI project), a collaboration grant by DAAD and STINT as well as positions to BN (Research School for Genomics and Bioinformatics) and SH (the Swedish Research Council). We also thank Marcus Krantz for critical reading of this review.

# References


Alepuz PM, de Nadal E, Zapater M, Ammerer G, Posas F (2003) Osmostress-induced transcription by Hot1 depends on a Hog1-mediated recruitment of the RNA Pol II. EMBO J 22:2433-2442

Alepuz PM, Jovanovic A, Reiser V, Ammerer G (2001) Stress-induced MAP kinase Hog1 is part of transcription activation complexes. Mol Cell 7:767-777

Asthagiri AR, Lauffenburger DA (2000) Bioengineering models of cell signaling. Annu Rev Biomed Eng 2:31-53

Bardwell L (2004) A walk-through of the yeast mating pheromone response pathway. Peptides 25:1465-1476

Barkai N, Leibler S (1997) Robustness in simple biochemical networks. Nature 387:913-917

Bhalla US, Iyengar R (2001) Robustness of the bistable behavior of a biological signaling feedback loop. Chaos 11:221-226

Bluthgen N, Herzel H (2003) How robust are switches in intracellular signaling cascades? J Theor Biol 225:293-300

Brightman FA, Fell DA (2000) Differential feedback regulation of the MAPK cascade underlies the quantitative differences in EGF and NGF signalling in PC12 cells. FEBS Lett 482:169-174

Bustin SA (2002) Quantification of mRNA using real-time reverse transcription PCR (RT-PCR): trends and problems. J Mol Endocrinol 29:23-39

de Nadal E, Alepuz PM, Posas F (2002) Dealing with osmostress through MAP kinase activation. EMBO Rep 3:735-740

De Nadal E, Zapater M, Alepuz PM, Sumoy L, Mas G, Posas F (2004) The MAPK Hog1 recruits Rpd3 histone deacetylase to activate osmoresponsive genes. Nature 427:370-374

Dihazi H, Kessler R, Eschrich K (2004) HOG-pathway induced phosphorylation and activation of 6-phosphofructo-2-kinase are essential for glycerol accumulation and yeast cell proliferation under hyperosmotic stress. J Biol Chem 279:23961-23968

Dohlman HG, Thorner JW (2001) Regulation of G protein-initiated signal transduction in yeast: paradigms and principles. Annu Rev Biochem 70:703-754

Elion EA (2000) Pheromone response, mating and cell biology. Curr Opin Microbiol 3:573-581

Elion EA (2001) The Ste5p scaffold. J Cell Sci 114:3967-3978

Enger J, Goksör M, Ramser K, Hagberg P, Hanstorp D (2004) Optical tweezers applied to a microfluidic system. Lab Chip 4:196-200

Ericsson M, Hanstorp D, Hagberg P, Enger J, Nystrom T (2000) Sorting out bacterial viability with optical tweezers. J Bacteriol 182:5551-5555

Escote X, Zapater M, Clotet J, Posas F (2004) Hog1 mediates cell-cycle arrest in G1 phase by the dual targeting of Sic1. Nat Cell Biol 6:997-1002

Ferrell JE Jr (2002) Self-perpetuating states in signal transduction: positive feedback, double-negative feedback and bistability. Curr Opin Cell Biol 14:140-148

Ghaemmaghami S, Huh WK, Bower K, Howson RW, Belle A, Dephoure N, O'Shea EK, Weissman JS (2003) Global analysis of protein expression in yeast. Nature 425:737-741

Goldbeter A, Koshland DE Jr (1984) Ultrasensitivity in biochemical systems controlled by covalent modification. Interplay between zero-order and multistep effects. J Biol Chem 259:14441-14447

Gordon GW, Berry G, Liang XH, Levine B, Herman B (1998) Quantitative fluorescence resonance energy transfer measurements using fluorescence microscopy. Biophys J 74:2702-2713

Gustin MC, Albertyn J, Alexander M, Davenport K (1998) MAP kinase pathways in the yeast *Saccharomyces cerevisiae*. Microbiol Mol Biol Rev 62:1264-1300

Heinrich R, Neel BG, Rapoport TA (2002) Mathematical models of protein kinase signal transduction. Mol Cell 9:957-970

Herskowitz I (1995) MAP kinase pathways in yeast: for mating and more. Cell 80:187-197

Herskowitz I (1997) Building organs and organisms: elements of morphogenesis exhibited by budding yeast. Cold Spring Harb Symp Quant Biol 62:57-63

Hohmann S (2002) Osmotic adaptation in yeast--control of the yeast osmolyte system. Int Rev Cytol 215:149-187

Hohmann S (2002) Osmotic stress signaling and osmoadaptation in yeasts. Microbiol Mol Biol Rev 66:300-372

Huang CY, Ferrell JE Jr (1996) Ultrasensitivity in the mitogen-activated protein kinase cascade. Proc Natl Acad Sci USA 93:10078-10083

Kholodenko BN (2000) Negative feedback and ultrasensitity can bring about oscillations in the mitogen-activated protein kinase cascades. Eur J Biochem 267:1583-1588

Klipp E, Nordlander B, Krüger R, Gennemark P, Hohmann S (2004) The dynamic response of yeast cells to osmotic shock - a systems biology approach. Nat Biotechnol under revision

Kofahl B, Klipp E (2004) Modelling the dynamics of the yeast pheromone pathway. Yeast 21:831-850

Krauss G (2001) Biochemistry of Signal Transduction and Regulation. Second Edition, Wiley-VCH, Weinheim ISBN 3-527-30378-2

Kusari AB, Molina DM, Sabbagh W Jr, Lau CS, Bardwell L (2004) A conserved protein interaction network involving the yeast MAP kinases Fus3 and Kss1. J Cell Biol 164:267-277

Lippincott-Schwartz J, Snapp E, Kenworthy A (2001) Studying protein dynamics in living cells. Nat Rev Mol Cell Biol 2:444-456

Llorens M, Nuno JC, Rodriguez Y, Melendez-Hevia E, Montero F (1999) Generalization of the theory of transition times in metabolic pathways: a geometrical approach. Biophys J 77:23-36

MacBeath G (2002) Protein microarrays and proteomics. Nat Genet 32 Suppl:526-532

Manney TR (1983) Expression of the BAR1 gene in *Saccharomyces cerevisiae*: induction by the alpha mating pheromone of an activity associated with a secreted protein. J Bacteriol 155:291-301

Marsh L, Neiman AM, Herskowitz I (1991) Signal transduction during pheromone response in yeast. Annu Rev Cell Biol 7:699-728

Nielsen UB, Cardone MH, Sinskey AJ, MacBeath G, Sorger PK (2003) Profiling receptor tyrosine kinase activation by using Ab microarrays. Proc Natl Acad Sci USA 100:9330-9335

O'Rourke SM, Herskowitz I, O'Shea EK (2002) Yeast go the whole HOG for the hyperosmotic response. Trends Genet 18:405-412

Raitt DC, Posas F, Saito H (2000) Yeast Cdc42 GTPase and Ste20 PAK-like kinase regulate Sho1-dependent activation of the Hog1 MAPK pathway. EMBO J 19:4623-4631

Reiser V, Salah SM, Ammerer G (2000) Polarized localization of yeast Pbs2 depends on osmostress, the membrane protein Sho1 and Cdc42. Nat Cell Biol 2:620-627

Rieseberg M, Kasper C, Reardon KF, Scheper T (2001) Flow cytometry in biotechnology. Appl Microbiol Biotechnol 56:350-360

Sato N, Kawahara H, Toh-e A, Maeda T (2003) Phosphorelay-regulated degradation of the yeast Ssk1p response regulator by the ubiquitin-proteasome system. Mol Cell Biol 23:6662-6671

Seet BT, Pawson T (2004) MAPK signaling: Sho business. Curr Biol 14:R708-710

Specht KM, Shokat KM (2002) The emerging power of chemical genetics. Curr Opin Cell Biol 14:155-159

Sprague GF, Cullen PJ, Goehring AS (2004) Yeast signal transduction: regulation and interface with cell biology. Adv Exp Med Biol 547:91-105

Sprague GFJ, Thorner JW (1992) Pheromone response and signal transduction during the mating process of *Saccharomyces cerevisiae*. In: Jones EW, Pringle JR, Broach JR (eds) The molecular and cellular biology of the yeast *Saccharomyces*. Gene expression. Cold Spring Harbor Laboratory Press, Cold Spring Harbor, pp 657-744

Swameye I, Muller TG, Timmer J, Sandra O, Klingmuller U (2003) Identification of nucleocytoplasmic cycling as a remote sensor in cellular signaling by databased modeling. Proc Natl Acad Sci USA 100:1028-1033

Támas MJ, Hohmann S (2003) The osmotic stress response of *Saccharomyces cerevisiae*. Topics Curr Genet 1:121-200

Támas MJ, Hohmann S (2003) The osmotic stress response of *Saccharomyces cerevisiae*. Topics Curr Genet 1:121-200

Tamás MJ, Karlgren S, Bill RM, Hedfalk K, Allegri L, Ferreira M, Thevelein JM, Rydstrom J, Mullins JG, Hohmann S (2003) A short regulatory domain restricts glycerol transport through yeast Fps1p. J Biol Chem 278:6337-6345

Tyson JJ, Chen KC, Novak B (2003) Sniffers, buzzers, toggles and blinkers: dynamics of regulatory and signaling pathways in the cell. Curr Opin Cell Biol 15:221-231

van Drogen F, Peter M (2001) MAP kinase dynamics in yeast. Biol Cell 93:63-70

van Drogen F, Peter M (2002) MAP kinase cascades: scaffolding signal specificity. Curr Biol 12:R53-55

van Drogen F, Stucke VM, Jorritsma G, Peter M (2001) MAP kinase dynamics in response to pheromones in budding yeast

van Drogen F, Stucke VM, Jorritsma G, Peter M (2001) MAP kinase dynamics in response to pheromones in budding yeast. Nat Cell Biol 3:1051-1059

Vandesompele J, De Preter K, Pattyn F, Poppe B, Van Roy N, De Paepe A, Speleman F (2002) Accurate normalization of real-time quantitative RT-PCR data by geometric averaging of multiple internal control genes. Genome Biol 3:RESEARCH0034

Wang Y, Ge Q, Houston D, Thorner J, Errede B, Dohlman HG (2003) Regulation of Ste7 ubiquitination by Ste11 phosphorylation and the Skp1-Cullin-F-box complex. J Biol Chem 278:22284-22289

Versele M, Lemaire K, Thevelein JM (2001) Sex and sugar in yeast: two distinct GPCR systems. EMBO Rep 2:574-579

Widmann C, Gibson S, Jarpe MB, Johnson GL (1999) Mitogen-activated protein kinase: conservation of a three-kinase module from yeast to human. Physiol Rev 79:143-180

Winson MK, Davey HM (2000) Flow cytometric analysis of microorganisms. Methods 21:231-240

Wolanin PM, Thomason PA, Stock JB (2002) Histidine protein kinases: key signal transducers outside the animal kingdom. Genome Biol 3:REVIEWS3013

Yi TM, Kitano H, Simon MI (2003) A quantitative characterization of the yeast heterotrimeric G protein cycle. Proc Natl Acad Sci USA 100:10764-10769

Zeitlinger J, Simon I, Harbison CT, Hannett NM, Volkert TL, Fink GR, Young RA (2003) Program-specific distribution of a transcription factor dependent on partner transcription factor and MAPK signaling. Cell 113:395-404

Hohmann, Stefan
Department of Cell and Molecular Biology, Göteborg University, Box 462, S-40530 Göteborg, Sweden
hohmann@gmm.gu.se

Klipp, Edda
Berlin Center for Genome Based Bioinformatics (BCB), Max-Planck Institute for Molecular Genetics, Dept. Vertebrate Genomics, Ihnestr. 73, 14195 Berlin, Germany

Kofahl, Bente
Berlin Center for Genome Based Bioinformatics (BCB), Max-Planck Institute for Molecular Genetics, Dept. Vertebrate Genomics, Ihnestr. 73, 14195 Berlin, Germany

Nordlander, Bodil
Department of Cell and Molecular Biology, Göteborg University, Box 462, S-40530 Göteborg, Sweden

# CELL LIFE AND DEATH

# Systems biology of the yeast cell cycle engine

Béla Novák, Katherine C. Chen, and John J. Tyson

## Abstract

One goal of systems biology is to obtain an integrated understanding of the physiological properties of cells from the detailed molecular machinery (the genes, proteins, and metabolites) that carry out these functions. Cell cycle regulation in yeast is an appropriate test case for this ambition, because the scientific community has amassed much information about the molecular components and functional properties of the control system. We propose a general mechanism for the regulation of cyclin-dependent kinases, the enzymes that control the major events of the cell cycle (DNA synthesis, mitosis and cell division). We translate the mechanism into differential-algebraic equations, and study solutions of these equations by numerical simulations and one-parameter bifurcation diagrams. We present results for wild type cells and a mutant that undergoes repeated rounds of DNA replication without intervening mitoses. We use bifurcation diagrams to reveal the general principles by which a cell controls its progression through the cell cycle.

## 1 Introduction

A living cell is a microscopic, chemical factory, using miniature protein machines to harvest energy and material for the purposes of continued existence and self-reproduction. The activities of the factory are coordinated by complex networks of interacting proteins, in response to internal and external cues (Bray 1995). The first step to understanding the machinery that underlies the regulated, goal-directed behaviour of a cell is to characterize its individual components (genes and proteins). This information has been provided in abundance by genome projects. The next level of description—how these parts fit together—is one of the principal goals of systems biology. In this paper, we illustrate the systems-biology approach to understanding molecular regulatory networks, using the eukaryotic cell division cycle as our example.

Cell reproduction requires, first and foremost, that the genetic information stored in the cell's DNA be copied and distributed between the two progeny cells. In eukaryotes, DNA copying and distributing are carried out in separate periods of time, called S phase (DNA synthesis) and M phase (mitosis) (The temporal gaps between S and M phases are called G1 and G2.). Successful completion of the cell

Topics in Current Genetics, Vol. 13
L. Alberghina, H.V. Westerhoff (Eds.): Systems Biology
DOI 10.1007/b137123 / Published online: 13 May 2005

**Table 1.** Cell cycle regulators in yeasts

| Regula-tors | *S. pombe* | *S. cerevisiae* | Functions and interactions |
|---|---|---|---|
| Cdk1 | Cdc2 | Cdc28 | cyclin-dependent kinases |
| CycB$^S$ | Cig2 | Clb5, Clb6 | S-phase cyclins |
| CycB$^M$ | Cdc13 | Clb1, Clb2 | M-phase cyclins |
| SPF | Cdc2/Cig2 | Cdc28/Clb5-6 | S phase promoting factor |
| MPF | Cdc2/Cdc13 | Cdc28/Clb1-2 | M phase promoting factor |
| MBF | Cdc10/Res1 | Mbp1/Swi6 | Transcription factor for CycB$^S$ Inhibited by SPF |
| CKI | Rum1 | Sic1 | Stoichiometric inhibitors of SPF and MPF Mutually antagonistic with SPF and MPF |
| Wee1 | Wee1 | Swe1 | Inhibitory kinase of Cdk1. Inhibited by MPF |
| Cdc25 | Cdc25 | Mih1 | Activatory phosphatase of Cdk1 Activated by MPF |
| APC$^M$ | APC/Slp1 | APC/Cdc20 | Degradation of MPF in M phase Activated by MPF |
| APC$^{G1}$ | APC/Ste9 | APC/Cdh1 | Degradation of MPF in G1 phase Inhibited by MPF |

cycle requires that the major events—DNA replication, mitosis, and cell division—occur in that specific order. If a cell cycle event cannot be completed (say, the cell's DNA is damaged or its replication is hindered), then the subsequent events of mitosis and cell division must be delayed. In addition, cell division must proceed in synchrony with the doubling in cytoplasmic mass called balanced growth and division (Fantes and Nurse 1981). If this requirement is not satisfied, then proliferating cells become progressively smaller or larger, which can happen for a limited period of time (as in oogenesis and embryogenesis) but not indefinitely. This 'size requirement' can be satisfied if the cell cycle engine is sensitive to the ratio of cell mass to DNA content, triggering cell cycle events with the same periodicity as the mass doubles. (For general review on cell cycle research, see Alberts et al. 1994; Nasmyth 1996b; Nurse 1999; Lodish et al. 2000.)

Satisfying these requirements is not a characteristic of any single component of the cell cycle machinery, but rather an emergent property of the way the components interact, as we intend to show.

## 2 Components of the cell cycle engine

The genes and proteins involved in cell cycle regulation, as well as their interactions, are well conserved throughout evolution from yeast to frogs to mammals (Nurse 1990). Table 1 shows a list of major cell cycle regulators in fission yeast and budding yeast. The three major cell cycle events—DNA replication, mitosis, and cell division—are controlled by cyclin-dependent protein kinases (Cdk's), which are active only when bound to a cyclin partner.

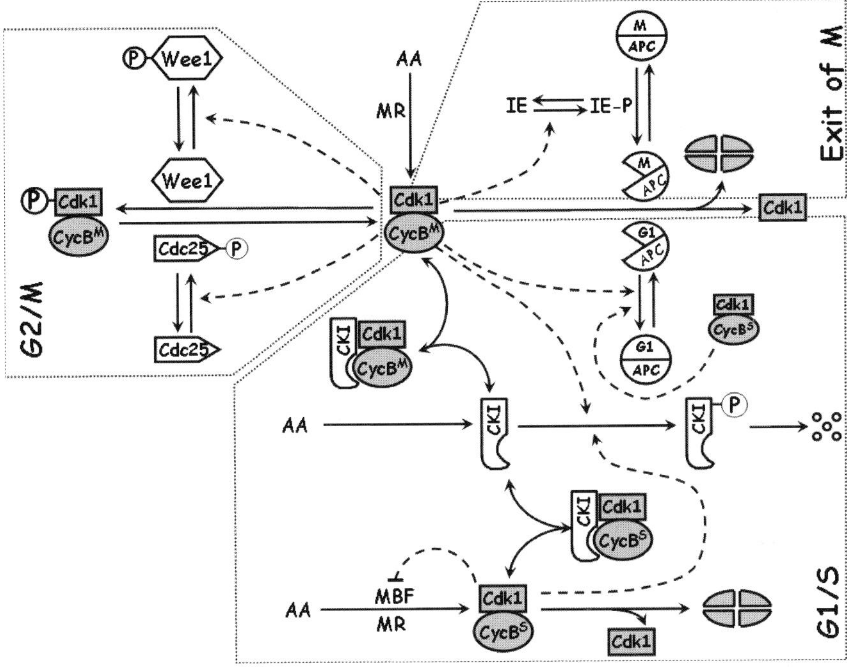

**Fig. 1.** The cell cycle engine in yeasts. The biochemical reactions controlling SPF (Cdk1/CycB$^S$) and MPF (Cdk1/CycB$^M$) activities are shown by solid lines with dashed lines representing enzymatic effects of Cdk1/CycB complexes on these reactions. The reactions are grouped into separate modules, which control different transitions of the cell cycle (G1/S, G2/M and exit from mitosis). The G1/S module is characterised by antagonistic relationships between Cdk1/CycB complexes and their enemies, CKI and APC$^{G1}$. The G2/M module is characterized by two positive feedback loops involving Wee1, Cdc25 and MPF. Mitotic exit is achieved by a time-delayed negative feedback loop, whereby MPF indirectly activates APC$^M$, which destroys CycB$^M$.

Yeast cells have only one essential Cdk subunit (referred as Cdk1), and they use different B-type cyclins in combination with the same Cdk1 subunit to initiate S phase and M phase (Fisher and Nurse 1995; Nasmyth 1996a; Mendenhall and Hodge 1998). These Cdk1/cyclin complexes are called S-phase Promoting Factor (SPF) and M-phase Promoting Factor (MPF). MPF can substitute for SPF in initiating S phase, but the reverse is not possible (Schwob and Nasmyth 1993; Fisher and Nurse 1996).

SPF and MPF not only promote cell cycle transitions but also inhibit them (Dahmann et al. 1995; Botchan 1996; Wuarin and Nurse 1996). For example, SPF and MPF initiate DNA synthesis by activating proteins at origins of replication on the DNA. After an origin has initiated, it loses a necessary component (called 'licensing factor'). Both SPF and MPF inhibit re-loading of licensing factor, thereby, preventing spent origins from being activated a second time until the cell has completed mitosis and destroyed all activity related to SPF and MPF. This dual signalling guarantees the alternation of S and M phases. Such dual control of Cdk activity on cell cycle events can be observed in case of mitosis and cell division as well. MPF activity promotes entry into mitosis and preparation for cell division, but it inhibits exit from mitosis and completion of cell division (Simanis 2003). Hence, cell division occurs only if the cell has first entered mitosis and then exited.

Cdk1/cyclin activities (SPF and MPF) are subjected to multiple levels of regulation (Fig. 1). Because Cdk1 subunits are usually present in excess, the total amounts of different cyclin subunits determine the relative levels of distinct Cdk1/cyclin complexes. Cyclin level can be controlled by synthesis (transcription) or by degradation (proteolysis). Cells can also regulate the activities of already assembled Cdk1/cyclin dimers: they can be inhibited by phosphorylation of the Cdk1 subunit or by formation of inactive Cdk1/cyclin/CKI trimers with a cyclin-dependent kinase inhibitor (CKI).

The activity of SPF (Cdk1/CycB$^S$) is regulated by the availability of S-phase cyclin (CycB$^S$), which depends on periodic activation of its transcription factor, MBF (Koch et al. 1993; Mondesert et al. 1996), and by binding to a CKI (Schwob et al. 1994; Correa-Bordes and Nurse 1995).

The regulation of MPF activity (Cdk1/CycB$^M$) is more complex: besides transcriptional control and CKI binding, the degradation of mitotic cyclins (CycB$^M$) is subject to complex controls, and the activity of the kinase subunit can be inhibited by phosphorylation. Budding yeast's two CycB$^M$ genes (*CLB1*, *CLB2*) are regulated transcriptionally (Amon et al. 1993), but the corresponding gene in fission yeast (*cdc13*) is transcribed constitutively (Fisher and Nurse 1996). Hence, for simplicity, we neglect transcriptional regulation of CycB$^M$.

The stability of mitotic cyclins changes drastically during the cell cycle in both yeasts: they are stable during S, G2, and early mitosis, but very unstable (half-life = 1– 2 min) in late mitosis and throughout G1 (Zachariae and Nasmyth 1999). CycB$^M$ is poly-ubiquitinated by the Anaphase Promoting Complex (APC), which targets it for proteolysis by proteasomes. Two different adaptors are used by APC to recognize mitotic cyclins for poly-ubiquitination. We call these two APC/adaptor complexes APC$^M$ and APC$^{G1}$ (Table 1). APC$^M$ is responsible for mitotic cyclin degradation at the end of mitosis, while APC$^{G1}$ takes over in G1 phase (Kitamura et al. 1998; Morgan 1999; Yamaguchi et al. 2000).

During G2 phase in fission yeast, MPF is inhibited by phosphorylation of a tyrosine residue of the Cdk1 subunit by a kinase called Wee1 (Russell and Nurse 1986). The action of Wee1 is opposed by a phosphatase, Cdc25, at the G2/M transition (Russell and Nurse 1987). Although corresponding enzymes are present in budding yeast, inhibitory tyrosine-phosphorylation is only significant if bud formation is blocked (Amon et al. 1992; Sorger and Murray 1992; Lew 2003).

# 3 Feedback loops and regulatory modules

The cell cycle engine cannot be defined simply by its components; the regulatory interactions among them are equally important. As described in the previous section, SPF and MPF are subject to both positive and negative controls. Interestingly, all of the transcriptional and post-translational controls that affect the activities of SPF and MPF are themselves subject to controls by SPF and MPF, creating feedback loops in the regulatory mechanism (Fig. 1). These feedback loops account for the emergent properties of the control system.

By 'positive feedback' we mean that a component promotes its own accumulation or activation. This can be achieved in many ways. A protein may activate its own transcription factor or inhibit its own proteolysis. It may activate its own activator, or somehow remove its own inhibitor. For example, phosphorylation of the stoichiometric inhibitor (CKI) renders it very unstable, and both SPF and MPF can phosphorylate the CKI (Verma et al. 1997; Benito et al. 1998). Hence, Cdk1/CycB and CKI are involved in a positive feedback loop (by mutual antagonism). By 'negative feedback' we mean that a component promotes its own removal or inactivation, which can likewise be achieved in diverse ways. For example, after S phase in both budding and fission yeast, the transcription factor MBF is inhibited (Amon et al. 1993; Ayte et al. 2001), and SPF appears to be responsible for this inhibition (Ayte et al. 2001), which creates a negative feedback loop.

Both forms of the APC ($APC^M$ and $APC^{G1}$) are controlled by MPF activity. $APC^M$, which is responsible for $CycB^M$ degradation as the cell exits from mitosis, is activated by MPF after a characteristic time-delay (Zachariae and Nasmyth 1999). (We model this delay by assuming that the activation is mediated through an intermediary enzyme IE.) Hence, MPF activation of $APC^M$, which subsequently degrades the cyclin component of MPF, constitutes a time-delayed negative feedback loop. On the other hand, $APC^{G1}$ is directly phosphorylated and inhibited by MPF (Zachariae et al. 1998; Jaspersen et al. 1999; Blanco et al. 2000; Peters 2002). Since $APC^{G1}$ destroys the CycB component of MPF and MPF inhibits $APC^{G1}$, these two complexes are involved in a positive feedback loop, via mutual antagonism (Novak et al. 1998).

Both Wee1 and Cdc25 are subject to MPF-mediated phosphorylation, which inhibits Wee1 and activates Cdc25 (Millar and Russell 1992; Aligue et al. 1997). Since Wee1 is an inhibitor of MPF, and Cdc25 is an activator, the phosphorylation of these two enzymes by MPF set up two positive feedback loops ($-\,-$ and $+\,+$) to enhance MPF activity.

The network shown in Figure 1 summarizes the interactions of SPF and MPF with their regulators. The network divides naturally into three modules. The G1/S module (controlling the transition from G1 into S phase) contains positive feedback loops (antagonisms of SPF and MPF with CKI and $APC^{G1}$) and a negative feedback loop (SPF inhibits MBF). The G2/M module contains only positive feedbacks (MPF inhibits Wee1 and activates Cdc25). The mitotic exit module

**Table 2.** Equations

$$\frac{\mathrm{d}}{\mathrm{d}t}[\mathrm{MR}] = k_{\mathrm{g}} \cdot [\mathrm{MR}], \text{ see footnote}[1]$$

$$\frac{\mathrm{d}[\mathrm{CycBM}]}{\mathrm{d}t} = k_{\mathrm{s,bm}} \cdot [\mathrm{MR}] - V_{\mathrm{d,bm}} \cdot [\mathrm{CycBM}]$$

$$\frac{\mathrm{d}[\mathrm{MPF}]}{\mathrm{d}t} = k_{\mathrm{s,bm}} \cdot [\mathrm{MR}] - (V_{\mathrm{d,bm}} + V_{\mathrm{wee}}) \cdot [\mathrm{MPF}] + V_{25} \cdot ([\mathrm{CycBM}] - [\mathrm{TrimM}] - [\mathrm{MPF}])$$
$$- k_{\mathrm{as,bm}} \cdot [\mathrm{MPF}] \cdot [\mathrm{CKI}] + (k_{\mathrm{di,bm}} + V_{\mathrm{d,cki}}) \cdot ([\mathrm{CycBM}] - [\mathrm{preMPF}] - [\mathrm{MPF}])$$

$$\frac{\mathrm{d}[\mathrm{preMPF}]}{\mathrm{d}t} = V_{\mathrm{wee}} \cdot ([\mathrm{CycBM}] - [\mathrm{preMPF}]) - (V_{25} + V_{\mathrm{d,bm}}) \cdot [\mathrm{preMPF}], \text{ see footnote}[1]$$

$$\frac{\mathrm{d}[\mathrm{TrimM}]}{\mathrm{d}t} = k_{\mathrm{as,bm}} \cdot ([\mathrm{CycBM}] - [\mathrm{TrimM}]) \cdot [\mathrm{CKI}] - (k_{\mathrm{di,bm}} + V_{\mathrm{d,bm}} + V_{\mathrm{d,cki}}) \cdot [\mathrm{TrimM}]$$

$$\frac{\mathrm{d}[\mathrm{MBF}]}{\mathrm{d}t} = \frac{k_{\mathrm{a,mbf}} \cdot (1 - [\mathrm{MBF}])}{J_{\mathrm{a,mbf}} + 1 - [\mathrm{MBF}]} - \frac{(k'_{\mathrm{i,mbf}} + k''_{\mathrm{i,mbf}} \cdot [\mathrm{SPF}]) \cdot [\mathrm{MBF}]}{J_{\mathrm{i,mbf}} + [\mathrm{MBF}]}$$

$$\frac{\mathrm{d}[\mathrm{CycBS}]}{\mathrm{d}t} = k_{\mathrm{s,bs}} \cdot [\mathrm{MBF}] \cdot [\mathrm{MR}] - k_{\mathrm{d,bs}} \cdot [\mathrm{CycBS}]$$

$$\frac{\mathrm{d}[\mathrm{SPF}]}{\mathrm{d}t} = k_{\mathrm{s,bs}} \cdot [\mathrm{MBF}] \cdot [\mathrm{MR}] - k_{\mathrm{d,bs}} \cdot [\mathrm{SPF}] - k_{\mathrm{as,bs}} \cdot [\mathrm{SPF}] \cdot [\mathrm{CKI}] + (k_{\mathrm{di,bs}} + V_{\mathrm{d,cki}}) \cdot [\mathrm{TrimS}]$$

$$\frac{\mathrm{d}[\mathrm{CKI_T}]}{\mathrm{d}t} = k_{\mathrm{s,cki}} - V_{\mathrm{d,cki}} \cdot [\mathrm{CKI_T}]$$

$$\frac{\mathrm{d}[\mathrm{APCG1}]}{\mathrm{d}t} = \frac{k_{\mathrm{a,cdh}} \cdot (1 - [\mathrm{APCG1}])}{J_{\mathrm{a,cdh}} + 1 - [\mathrm{APCG1}]} - \frac{(k''_{\mathrm{i,cdh,s}} \cdot [\mathrm{SPF}] + k''_{\mathrm{i,cdh,m}} \cdot [\mathrm{MPF}]) \cdot [\mathrm{APCG1}]}{J_{\mathrm{i,cdh}} + [\mathrm{APCG1}]}$$

$$\frac{\mathrm{d}[\mathrm{IEP}]}{\mathrm{d}t} = \frac{k_{\mathrm{a,iep}} \cdot [\mathrm{MPF}] \cdot (1 - [\mathrm{IEP}])}{J_{\mathrm{a,iep}} + 1 - [\mathrm{IEP}]} - \frac{k_{\mathrm{i,iep}} \cdot [\mathrm{IEP}]}{J_{\mathrm{i,iep}} + [\mathrm{IEP}]}$$

$$\frac{\mathrm{d}[\mathrm{APCM}]}{\mathrm{d}t} = \frac{(k_{\mathrm{a,c20}} \cdot [\mathrm{IEP}]) \cdot (1 - [\mathrm{APCM}])}{J_{\mathrm{a,c20}} + 1 - [\mathrm{APCM}]} - \frac{k_{\mathrm{i,c20}} \cdot [\mathrm{APCM}]}{J_{\mathrm{i,c20}} + [\mathrm{APCM}]}$$

$$[\mathrm{TrimS}] = [\mathrm{CycBS}] - [\mathrm{SPF}] \qquad\qquad [\mathrm{CKI}] = [\mathrm{CKI_T}] - [\mathrm{TrimM}] - [\mathrm{TrimS}]$$

$$GK(V_a, V_i, J_a, J_i) = \frac{2 J_i V_a}{V_i - V_a + J_a V_i + J_i V_a + \sqrt{(V_i - V_a + J_a V_i + J_i V_a)^2 - 4(V_i - V_a) J_i V_a}}$$

$$V_{\mathrm{d,bm}} = k'_{\mathrm{d,bm}} + k''_{\mathrm{d,bm,g1}} \cdot [\mathrm{APCG1}] + k''_{\mathrm{d,bm,m}} \cdot [\mathrm{APCM}]$$

$$V_{\mathrm{d,cki}} = k'_{\mathrm{d,cki}} + k''_{\mathrm{d,cki,s}} \cdot [\mathrm{SPF}] + k''_{\mathrm{d,cki,m}} \cdot [\mathrm{MPF}]$$

$$V_{\mathrm{wee}} = k'_{\mathrm{wee}} + (k''_{\mathrm{wee}} - k'_{\mathrm{wee}}) \cdot GK(k_{\mathrm{a,wee}}, k'_{\mathrm{i,wee}} + k''_{\mathrm{i,wee}} \cdot [\mathrm{MPF}], J_{\mathrm{a,wee}}, J_{\mathrm{i,wee}})$$

$$V_{25} = k'_{25} + (k''_{25} - k'_{25}) \cdot GK(k'_{\mathrm{a,25}} + k''_{\mathrm{a,25}} \cdot [\mathrm{MPF}], k_{\mathrm{i,25}}, J_{\mathrm{a,25}}, J_{\mathrm{i,25}})$$

Reset Rules: We assume that cells begin DNA replication when $[\mathrm{CKI_T}]$ drops below 0.25. At this time, we divide $[\mathrm{MR}]$ by 2. As cells exit mitosis, when $[\mathrm{MPF}]$ drops below 0.1, $[\mathrm{MR}]$ is unchanged.

Definition of variables for Table 2 (overleaf):

[MR] = mass-to-DNA ratio,

[PMPF] = the tyrosine-phosphorylated form of MPF,

[MC] = the MPF/CKI complex,

[PMC] = the PMPF/CKI complex,

$[CycBM] = [MPF] + [PMPF] + [MC] + [PMC]$,

$[preMPF] = [PMPF] + [PMC]$,

$[TrimM] = [MC] + [PMC]$.

Thus, $[PMPF] = [CycBM] - [TrimM] - [MPF]$, $[MC] = [CycBM] - [preMPF] - [MPF]$,

and $[PMC] = [MPF] + [preMPF] + [TrimM] - [CycBM]$.

(which controls downregulation of MPF activity at the end mitosis) is based on a time-delayed negative feedback loop (MPF activates IE activates $APC^M$ destroys $CycB^M$).


## 4 Mathematical formulation

The molecular network shown on Figure 1 can be described by a set of differential and algebraic equations, as proposed in Table 2. (For details, see the supplementary material.) It is important to realize that there is no one-to-one relationship between the wiring diagram (Fig. 1) and the equations (Table 2). The same network can be expressed in different types of equations, depending on assumptions made by the modeller. In addition, the equations cannot be solved until we have specified values for all the rate constants that appear therein. Because most of these rate constants have never been measured directly, it is up to the modeller to choose reasonable and efficacious values for these parameters. Parameter estimation is possible in our case, because we have a huge amount of quantitative and qualitative information about the control system. Yeast geneticists have mutated, deleted and overexpressed most of the genes involved in cell cycle control. Cells with such genetic changes can show either a delay or advancement in cell cycle transitions. If the effect is not severe, then the mutant cell can survive, although it likely differs from wild type (un-mutated) cells in average size and in the durations of G1-S-G2-M phases. If the delay or advancement of a cell cycle transition is too severe, then the mutant cell is inviable, blocking in a mutant-specific phase of the cell cycle. We assign values to the kinetic constants in the model so that the concentration profiles of cell cycle regulators are consistent with the phenotypes of wild type and mutant cells.


## 5 The role of the nucleocytoplasmic ratio

Size control is incorporated into the model by multiplying the rate of production of Cdk/cyclin complexes by the mass-to-DNA ratio of the cell. A recent experiment with budding yeast used the same argument in case of Cln3/Cdc28 to explain size control at G1/S (Alberghina et al. 2004). We imagine the following mechanism. Suppose Cdk1/CycB complexes (SPF and MPF) bind preferentially to a structure and get concentrated there. We assume that the binding is strong and the number of binding sites is not limiting. In this case, all Cdk1/CycB molecules will be bound to the structure, and the number bound will be proportional to cell mass, because the total number of Cdk/CycB dimers increases in proportional to cell mass. Furthermore, we assume that the molecules regulating or regulated by Cdk1/CycB complexes do not bind specifically to this structure but are uniformly distributed within the cell. In this manner, the rates of reactions between Cdk1/CycB and its regulators are assumed to be proportional to the number of Cdk1/CycB complexes per structure times the uniform concentration of the regulatory protein. Finally, we suppose that the structure doubles at DNA replication, so the number of Cdk1/CycB complexes per structure is proportional to the mass-to-DNA ratio of the cell. Our hypothetical nuclear structure could be chromosomes or the centrosome or something else; its identity is not important for our present purposes.

Next we have to specify how mass and DNA are changing during the cell cycle. We assume exponential growth in cell mass and an abrupt doubling of DNA content at S phase. We double the DNA content when SPF or MPF activity is high enough to keep total CKI level below a threshold value (0.25). At this point in the cycle, the mass-to-DNA ratio ([MR] in Table 2) of the cell is reduced by a factor of two. At mitosis, when MPF activity drops below a threshold value (0.1), we divide both the mass and the DNA content by two, leaving mass/DNA unchanged. These rules differ from our previous publications, where [MR] = mass per nucleus instead, which we reset at cell division. The benefit of this change in perspective is that we can explain nucleocytoplasmic ratio control over the endoreplication cycles observed in CycB$^{M}$ –deletion mutants in fission yeast, as will be described later.


## 6 Bifurcation diagrams and their biological significance

Bifurcation theory is a useful tool for understanding general properties of dynamic systems and how these properties depend on parameter values. (For an in-depth description of the theory, see Strogatz 1994; for an introduction to bifurcation theory, see Kaplan and Glass 1995; and for more details about bifurcation analysis of cell cycle controls, see Tyson et al. 2001.) To begin the analysis, we suppose that the cyber-cell, described by the equations in Table 2, is not growing ($k_{g} = 0$). In this case, we can look for steady state solutions of the Cdk control system, by setting the left hand sides of all the differential equations equal to zero. Of course, the number of steady states and their stability or instability depend on the values assigned to the parameters. As parameter values are changed, bifurcation theory tells us how steady state solutions appear and disappear, and how they change from attractors to repellers and vice versa. In particular, we want to know how the steady states of the control system change as the mass-to-DNA ratio changes.

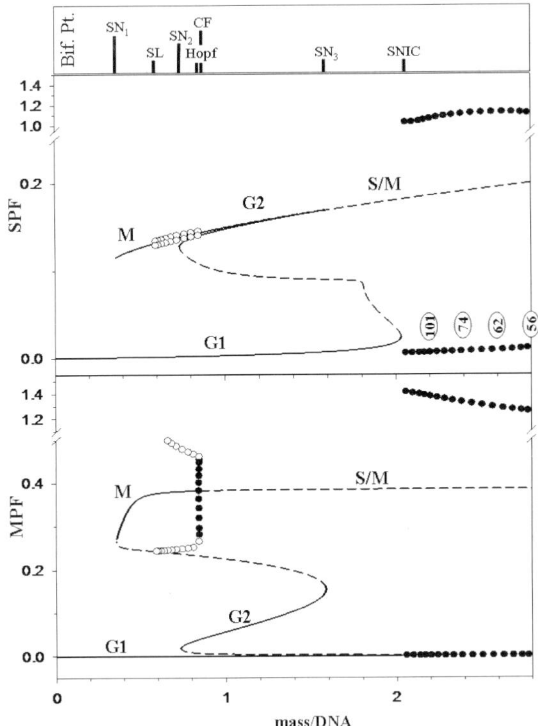

Fig. 2. Bifurcation diagram of the yeast cell cycle engine. Mass/DNA is the bifurcation parameter on the abscissa, and the control system is characterized by values of SPF and MPF. The steady state values for SPF in S, G2, and M are indistinguishable. Solid and dashed lines represent stable and unstable steady states, respectively. Filled and open circles indicate the amplitudes of stable and unstable limit cycles, respectively. Numbers in ovals give the periods of stable limit cycle oscillations (min). Bifurcation points are classified as SN (saddle-node), SL (saddle-loop), SNIC (saddle-node on infinite circle), CF (cyclic fold), and Hopf.

In Figure 2, we plot all steady states of the cell cycle engine as a function of the mass-to-DNA ratio. Two types of steady states are observed. Some steady states attract nearby trajectories from every direction. They are called stable nodes (or foci, if trajectories spiral into the steady state) and are indicated by solid lines. Saddle points attract trajectories from some directions and repel trajectories in other directions. They are unstable steady states and are indicated by dashed lines. Unstable steady states are sometimes surrounded by stable or unstable limit cycle oscillations (closed or open circles, respectively, in Fig. 2).

A bifurcation point is a specific set of parameter values where steady states of different stability properties merge with each other or inter-convert, or where limit cycles first appear or disappear. Five types of bifurcations are found in Figure 2.

The simplest is a 'saddle-node' (SN), where a saddle point and a node coalesce and disappear. At a 'Hopf' bifurcation, a steady state changes stability and a small amplitude limit cycle is born. At a 'cyclic fold' (CF), stable and unstable limit cycles coalesce and disappear. At SL and SNIC bifurcation points, a limit cycle merges with an unstable steady state, either a saddle point (called a 'saddle-loop' bifurcation, SL) or a saddle-node (called a 'saddle-node-invariant-circle' bifurcation, SNIC).

For mass/DNA between 0 and 2 in Figure 2, we find stable steady states with very small activities of SPF and MPF and significant activities (not shown) of the negative regulators (CKI, $APC^{G1}$ and Wee1). The transcription factor, MBF, which is inhibited by SPF, is fully active. $CycB^S$ is synthesised and $Cdk1/CycB^S$ level is appreciable, but SPF activity is small because CKI level is high. MPF activity is also very low: although $CycB^M$ is being synthesized, it is actively degraded by $APC^{G1}$, and the activity of any remaining dimer is inhibited by CKI. All of these characteristics lead us to identify this locus of steady states with G1 phase of the cell cycle. This locus of stable G1 steady states disappears at mass/DNA $\simeq 2$ by coalescing with a family of unstable saddle points. As the steady states merge and disappear, a limit cycle oscillation is born by a SNIC bifurcation.

A locus of stable steady states with medium SPF and MPF activities can be found for mass/DNA between 0.7 and 1.6. At these steady states, CKI level is low, $APC^{G1}$ is inactive, Wee1 is active, and MPF is mostly in the tyrosine phosphorylated form. Consequently MPF activity is neither very low nor very high. This intermediate activity of MPF is not enough to activate Cdc25 and $APC^M$. MBF is only partly active, due to inhibition by SPF. Based on these features, we associate this family of stable steady states with G2 phase of the cell cycle. These G2 states disappear at mass/DNA = 1.6 by a saddle-node bifurcation.

A third family of stable steady states, for mass/DNA = 0.35 − 0.8, is characterized by higher MPF activity (0.25 − 0.4) than that at the G2 state (0.02 − 0.18), with Wee1 inactive, and Cdc25 active. These properties suggest early M phase cells, but this region of the diagram, as we shall see, is never visited by normally cycling cells. Nonetheless, these stable mitotic states are of crucial importance for the spindle-assembly checkpoint, as we will note later.

The range of stable mitotic steady states is very limited, because they become unstable by a Hopf bifurcation at mass/DNA = 0.8. The time-delayed negative feedback loop is responsible for this bifurcation, which gives rise to small amplitude oscillations. Large amplitude limit cycle oscillations are not possible until mass/DNA exceeds 2.

In the oscillatory regime (mass/DNA > 2), SPF and MPF are oscillating out of phase. The maximal values attained by SPF and MPF are considerably larger than their values at the G1 and G2 steady states. Based on these characteristics, we associate these oscillations with S and M phases of the cell cycle. It is important to note that for mass/DNA between 1.6 and 2, the control system cannot do sustained limit cycle oscillations; rather it proceeds through part of the oscillation and is then captured by the stable G1 steady state.

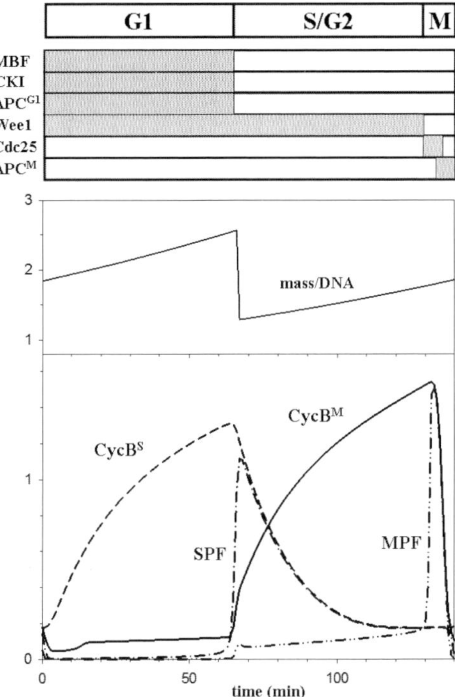



**Fig. 3.** Computer simulation of the yeast cell cycle, based on the model in Table 2. CycB$^S$ and CycB$^M$ represent all forms of S and M phase cyclins (including those in complex with the CKI and/or tyrosine-phosphorylated). The shaded areas at the top represent the period during the cycle when a particular regulator is present or active at more than 50% its maximal value.


# 7 Cell cycle progression on the bifurcation diagram

It is very informative to see how cycling cells move across this bifurcation diagram. To this end, we simulate cycling cells with the same equations and parameter values used for constructing the bifurcation diagram, with two differences. First, we now allow cell mass to increase exponentially with a specific growth rate, $k_g = 0.005$ min$^{-1}$. Second, we double the DNA value when SPF or MPF increases such that total CKI drops below a threshold value (0.25). The simulation of wild type cells (Fig. 3) shows relative concentrations of each cell cycle regulator as a function of time during the cell cycle. On Figure 4, we plot the SPF and MPF data from this simulation as a function of mass/DNA on top of the bifurcation diagram from Figure 2.

Cell division occurs at mass/DNA = 1.8, as the cell exits mitosis and MPF activity plummets. The newborn cell finds itself captured by the stable G1 steady state. However, the cell is growing and moving to the right on the bifurcation diagram. When mass/DNA exceeds the SNIC bifurcation point, the stable G1 steady state disappears, and the control system enters the oscillatory regime. This point marks the START transition of the yeast cell cycle, when a cell commits to a new round of DNA synthesis and cell division. The biochemical correlate of this event is that the amount of Cdk1/CycB$^S$ dimer surpasses the amount of CKI, and so the balance of power between them shifts in favour of SPF. Initially, the switch over (from low SPF, high CKI to high SPF, low CKI) is very slow close to the bifurcation point (a characteristic of SNIC bifurcations). Only when mass/DNA reaches 2.5 does SPF get abruptly activated, and MPF activity follows. (Activation of MPF is limited by tyrosine phosphorylation by Wee1.) Rising SPF triggers DNA replication, which causes mass/DNA to decrease by a factor of 2. Observe that SPF is at its maximum, while MPF is still low, when mass/DNA is halved.

The drop in mass/DNA (from 2.5 to 1.25) drives the system out of the oscillatory regime into the domain of attraction of the stable G2 steady state. During G2 phase, MPF activity increases only slightly with cell growth because of inhibition by Wee1. But when mass/DNA reaches the critical value (1.6), the stable G2 state disappears at an SN bifurcation. At this point MPF activity is high enough to inactivate Wee1 and activate Cdc25. The positive feedback loops are turned on, and MPF activity rises abruptly, triggering mitosis, and then begins to fall because APC$^M$ becomes active and destroys CycB$^M$. When MPF activity drops below a critical value, the cell divides and both mass and DNA are halved (mass/DNA unchanged). The cell cycle engine is drawn to the G1 steady state, which is the only attractor of the control system in this range of mass-to-DNA ratios.

## 8 Effects of cell cycle checkpoints on the bifurcation diagrams

The cell cycle engine must respond to checkpoint mechanisms which block (or delay) further progression if a cell cycle event cannot be completed properly. Such a problem initiates a signal transduction pathway that transmits an inhibitory signal to the cell cycle engine. The signal generally upregulates one of the negative regulators of Cdk1/CycB complexes, depending on which checkpoint pathway is activated.

In general, three major checkpoint mechanisms are operating in eukaryotes:

1. The G1 checkpoint blocks START (G1/S transition). This checkpoint is activated by DNA damage and many other signals.
2. The G2 checkpoint blocks entry into mitosis (G2/M transition), and it is activated, for example, if DNA replication blocked.
3. The spindle assembly checkpoint blocks exit from mitosis (metaphase to anaphase transition), and it is activated if spindle formation is impaired or chromosomes are not properly aligned on the metaphase plate.

During a normal division cycle (Fig. 4), a cell undergoes the G1/S transition at the SNIC bifurcation point and the G2/M transition at the SN3 bifurcation point.

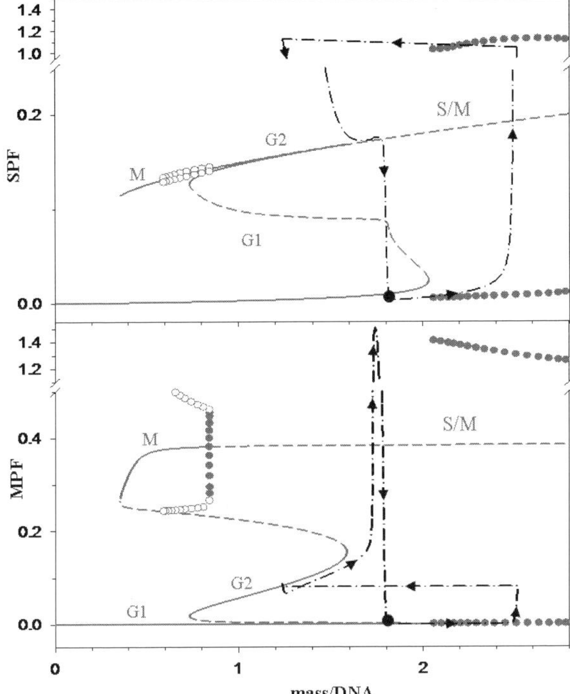


**Fig. 4.** Cell cycle progression on the bifurcation diagram. The SPF and MPF values from the computer simulation in Figure 3 are plotted as functions of mass/DNA on the one-parameter bifurcation diagrams in Figure 2. The • specifies the starting point for a newborn cell.

These bifurcation points function as size checkpoints during the normal cycle, because cells must reach certain critical mass-to-DNA ratios in order to pass to the next phase of the cell cycle. If the G1 or G2 checkpoint is activated, then the SNIC or SN3 bifurcation point is pushed to a much larger mass-to-DNA ratio, and the cell is delayed for a long time in G1 or G2 phase. It often happens, though, that a cell grows large enough to override the checkpoint (a phenomenon called 'adaptation').

The spindle assembly checkpoint also extends the range of a stable steady state. Activation of the checkpoint inhibits $APC^M$ and abolishes the negative feedback loop. Thus, the Hopf bifurcation point moves to a much higher value of mass/DNA, creating a stable mitotic state beyond the G2/M transition. After undergoing SPF and MPF activation, the cell becomes stuck in M phase, when the spindle checkpoint is active.

Examples of how the checkpoints affect the bifurcation diagrams in fission yeast are described in our previous publication (Tyson et al. 2002).

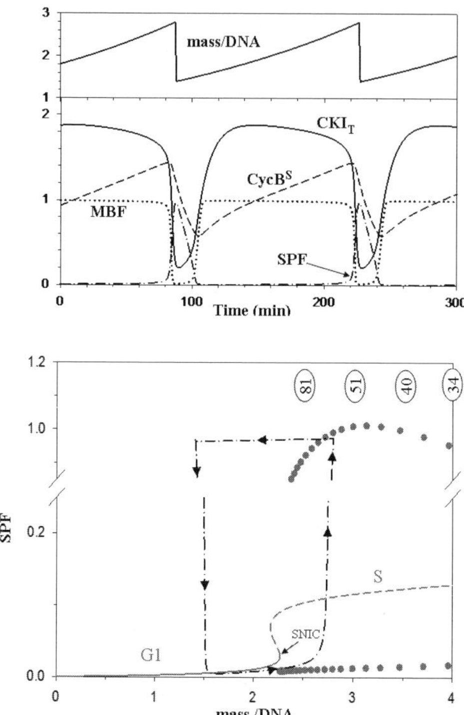


**Fig. 5.** Endoreplication cycles. Numerical simulation (A) and bifurcationdiagram (B) for endoreplication cycles, which are characterized by the lack of mitotic cyclins. SPF activity from the simulation is plotted on the bifurcation diagram, in order to show progression during the endoreplication cycle.

# 9 Endoreplication cycles

If the mitotic cyclin genes are deleted in yeasts, cells cannot synthesise $CycB^M$, MPF activity is not detectable, and cells cannot undergo mitosis. While budding yeast mutants of this sort arrest in G2 phase, fission yeast mutants undergo repeated rounds of DNA synthesis without intervening mitosis (G1, S, G1, S, etc.), a phenotype called endoreplication (Hayles et al. 1994).

According to the model, the difference in dynamic behaviour of the two yeasts can depend on quantitative differences in kinetic parameter values (e.g. how strong is the inhibition of MBF by SPF). Since the parameter values we use are much closer to the fission yeast situation, if we set the rate of $CycB^M$ synthesis to zero in the model, we get periodic activation of SPF (Fig. 5A). The period of oscillation is equal to the mass doubling time, because the endoreplication cycles are still driven by periodic resetting of the mass-to-DNA ratio. In this case, however, as the bifurcation diagram in Figure 5B shows, endoreplication cycles are organized around a SNIC bifurcation at the G1/S transition. The oscillations are created by the negative feedback loop whereby SPF inhibits its own transcription factor, MBF.

# 10 Conclusion

The eukaryotic cell cycle engine is regulated by many protein molecules, which are synthesized, degraded, and interact with each other in a very complicated and dynamic way. In this dynamic molecular control system, the influences of the components on each other are as important as components themselves. As a consequence, to understand how the eukaryotic cell cycle is controlled requires a systems biology approach, as outlined in this chapter.

Using cell cycle regulation in yeast as an example, we have shown that it is possible to create realistic, accurate kinetic models of molecular regulatory systems of considerable complexity. The wiring diagram of the control system and the numerical values of the kinetic rate constants can be estimated with some confidence, if we have sufficient information about the physiological properties of mutant cells having a deficiency or overabundance of crucial regulatory proteins. Each mutant can be thought of as an exploration of the parameter space of the control system in a certain direction.

A powerful mathematical tool for characterizing nonlinear dynamical control systems is bifurcation theory, which relates qualitative properties of the system to changing values of parameters. In particular, for cell cycle regulation, bifurcation diagrams illustrate how progress through the cell division cycle is coordinated to cell growth and DNA replication. The general approach and methods described in this review should be broadly applicable to many aspects of the molecular regulation of cell functions.

Does this systems-biology approach provide any new insights about cell cycle control? Using bifurcation theory for the cell cycle engine as a whole, we find an emergent property of the network that cannot be guessed by intuition alone. Namely, we find that the phases (G1, S/G2, and M) of the eukaryotic cell cycle are alternative stable steady states of the dynamical control system. At any particular ratio of cytoplasmic mass/DNA, the control system might be in one or another of these stable states, depending on the recent history of the transitions made by the control system. This behaviour, which is called multistability, cannot be attributed to any single molecules in the control system, but rather to the interactions (namely positive feedback and antagonism) among them. As a consequence, this subtle behaviour can be recognized only by taking the systems biology approach, as we have done here.

In our view, this type of integrative view is necessary to understand the details of cell cycle regulation. Although people have been modelling the Cdk-machinery of the cell cycle for more than 15 years, interactions between experimentalists and modellers have been few. The good news is that both communities have realized this problem, and fruitful collaborations are emerging (Cross et al. 2002, 2005;

Bai et al. 2003; Ciliberto et al. 2003; Pomerening et al. 2003; Sha et al. 2003; Thornton et al. 2004). To be sure, there are still many experimental questions about cell cycle regulation that do not require modelling at this time. However, as our understanding of the control network becomes ever more comprehensive and complicated, it becomes increasingly difficult to design and interpret experiments by intuitive reasoning alone, and mathematical modelling will become essential.

## Acknowledgements

This work has been supported by grants from the Defense Advanced Research Project Agency (BioSPICE: AFRL #F30602-02-0572), the James S. McDonnell Foundation (Complex Systems: 21002050), and the European Commission (COMBIO: LSHG-CT-2004-503568).

## References

Alberghina L, Rossi R, Querin L, Wanke V, Vanoni M (2004) A cell sizer network involving Cln3 and Far1 controls entrance into S phase in the mitotic cycle of budding yeast. J Cell Biol 167:433-443

Alberts B, Bray D, Lewis J, Raff M, Roberts K, Watson JD (1994) Molecular Biology of the Cell, 3rd edn. Garland Publishing, Inc., New York

Aligue R, Wu L, Russell P (1997) Regulation of *Schizosaccharomyces pombe* Wee1 tyrosine kinase. J Biol Chem 272:13320-13325

Amon A, Surana U, Muroff I, Nasmyth K (1992) Regulation of p34$^{CDC28}$ tyrosine phosphorylation is not required for entry into mitosis in *S. cerevisiae*. Nature 355:368-371

Amon A, Tyers M, Futcher B, Nasmyth K (1993) Mechanisms that help the yeast cell cycle clock tick: G2 cyclins transcriptionally activate G2 cyclins and repress G1 cyclins. Cell 74:993-1007

Ayte J, Schweitzer C, Zarzov P, Nurse P, DeCaprio JA (2001) Feedback regulation of the MBF transcription factor by cyclin Cig2. Nat Cell Biol 3:1043-1050

Bai S, Goodrich D, Thron CD, Tecarro E, Obeyesekere M (2003) Theoretical and experimental evidence for hysteresis in cell proliferation. Cell Cycle 2:46-52

Benito J, Martin-Castellanos C, Moreno S (1998) Regulation of the G1 phase of the cell cycle by periodic stabilization and degradation of the p25rum1 CDK inhibitor. EMBO J 17:482-497

Blanco MA, Sanchez-Diaz A, de Prada JM, Moreno S (2000) APC(ste9/srw1) promotes degradation of mitotic cyclins in G(1) and is inhibited by cdc2 phosphorylation. EMBO J 19:3945-3955

Botchan M (1996) Coordinating DNA replication with cell division: Current status of the licensing concept. Proc Natl Acad Sci USA 93:9997-10000

Bray D (1995) Protein molecules as computational elements in living cells. Nature 376:307-312

Ciliberto A, Petrus MJ, Tyson JJ, Sible JC (2003) A kinetic model of the cyclin E/Cdk2 developmental timer in *Xenopus laevis* embryos. Biophys Chem 104:573-589

Correa-Bordes J, Nurse P (1995) p25$^{rum1}$ orders S-phase and mitosis by acting as an inhibitor of the p34$^{cdc2}$ mitotic kinase. Cell 83:1001-1009

Cross FR, Archambault V, Miller M, Klovstad M (2002) Testing a mathematical model of the yeast cell cycle. Mol Biol Cell 13:52-70

Cross FR, Schroeder L, Kruse M, Chen KC (2005) Quantitative characterization of a mitotic cyclin threshold regulating exit from mitosis. Mol Biol Cell 16 (in press)

Dahmann C, Diffley JFX, Nasmyth K (1995) S-phase-promoting cyclin-dependent kinases prevent re-replication by inhibiting the transition of replication origins to a pre-replicative state. Curr Biol 5:1257-1269

Fantes PA, Nurse P (1981) Division timing: controls, models and mechanisms. In: John PCL (ed) The Cell Cycle. Cambridge Univ. Press, Cambridge UK, pp 11-33

Fisher D, Nurse P (1995) Cyclins of the fission yeast *Schizosaccharomyces pombe*. Semin Cell Biol 6:73-78

Fisher DL, Nurse P (1996) A single fission yeast mitotic cyclin B p34$^{cdc2}$ kinase promotes both S-phase and mitosis in the absence of G1 cyclins. EMBO J 15:850-860

Hayles J, Fisher D, Woollard A, Nurse P (1994) Temporal order of S phase and mitosis in fission yeast is determined by the state of the p34$^{cdc2}$ -mitotic B cyclin complex. Cell 78:813-822

Jaspersen SL, Charles JF, Morgan DO (1999) Inhibitory phosphorylation of the APC regulator Hct1 is controlled by the kinase Cdc28 and the phosphatase Cdc14. Curr Biol 11:227-236

Kaplan D, Glass L (1995) Understanding Nonlinear Dynamics. Springer-Verlag, New York

Kitamura K, Maekawa H, Shimoda C (1998) Fission yeast Ste9, a homolog of Hct1/Cdh1 and Fizzy-related, is a novel negative regulator of cell cycle progression during G1-phase. Mol Biol Cell 9:1065-1080

Koch C, Moll T, Neuberg M, Ahorn H, Nasmyth K (1993) A role for the transcription factors Mbp1 and Swi4 in progression from G1 to S phase. Science 261:1551-1557

Lew DJ (2003) The morphogenesis checkpoint: how yeast cells watch their figures. Curr Opin Cell Biol 15:648-653

Lodish H, Berk A, Zipursky SL, Matsudaira P, Baltimore D, Darnell J (2000) Molecular Cell Biology, 4th edn. W.H. Freeman, New York

Mendenhall MD, Hodge AE (1998) Regulation of Cdc28 cyclin-dependent protein kinase activity during the cell cycle of the yeast *Saccharomyces cerevisiae*. Microbiol Mol Biol Rev 62:1191-1243

Millar JBA, Russell P (1992) The cdc25 M-phase inducer: an unconventional protein phosphatase. Cell 68:407-410

Mondesert O, McGowan CH, Russell P (1996) Cig2, a B-type cyclin, promotes the onset of S in *Schizosaccharomyces pombe*. Mol Cell Biol 16:1527-1533

Morgan DO (1999) Regulation of the APC and the exit from mitosis. Nature Cell Biol 1:E47-E53

Nasmyth K (1996a) At the heart of the budding yeast cell cycle. Trends Genet 12:405-412

Nasmyth K (1996b) Viewpoint: putting the cell cycle in order. Science 274:1643-1645

Novak B, Csikasz-Nagy A, Gyorffy B, Nasmyth K, Tyson JJ (1998) Model scenarios for evolution of the eukaryotic cell cycle. Phil Trans Royal Soc London. Series B: Biol Sci 353:2063-2076

Nurse P (1990) Universal control mechanism regulating onset of M-phase. Nature 344:503-508

Nurse P (1999) Cyclin dependent kinases and regulation of the fission yeast cell cycle. Biol Chem 380:729-733

Peters JM (2002) The anaphase-promoting complex: proteolysis in mitosis and beyond. Mol Cell 9:931-943

Pomerening JR, Sontag ED, Ferrell JE Jr (2003) Building a cell cycle oscillator: hysteresis and bistability in the activation of Cdc2. Nat Cell Biol 5:346-351

Russell P, Nurse P (1986) cdc25$^+$ functions as an inducer in the mitotic control of fission yeast. Cell 45:145-153

Russell P, Nurse P (1987) Negative regulation of mitosis by wee1$^+$, a gene encoding a protein kinase homolog. Cell 49:559-567

Schwob E, Bohm T, Mendenhall MD, Nasmyth K (1994) The B-type cyclin kinase inhibitor p40$^{sic1}$ controls the G1 to S transition in *S. cerevisiae*. Cell 79:233-244

Schwob E, Nasmyth K (1993) *CLB5* and *CLB6*, a new pair of B cyclins involved in DNA replication in *Saccharomyces cerevisiae*. Genes Dev 7:1160-1175

Sha W, Moore J, Chen K, Lassaletta AD, Yi CS, Tyson JJ, Sible JC (2003) Hysteresis drives cell-cycle transitions in *Xenopus laevis* egg extracts. Proc Natl Acad Sci USA 100:975-980

Simanis V (2003) Events at the end of mitosis in the budding and fission yeasts. J Cell Sci 116:4263-4275

Sorger PK, Murray AW (1992) S-phase feedback control in budding yeast independent of tyrosine phosphorylation of p34$^{cdc28}$. Nature 355:365-368

Strogatz SH (1994) Nonlinear Dynamics and Chaos. Addison-Wesley Co., Reading, MA

Thornton BR, Chen KC, Cross FR, Tyson JJ, Toczyski DP (2004) Cycling without the cyclosome: modeling a yeast strain lacking the APC. Cell Cycle 3:629-633. Epub 2004 May 2003

Tyson JJ, Chen K, Novak B (2001) Network dynamics and cell physiology. Nature Rev Mol Cell Biol 2:908-916

Tyson JJ, Csikasz-Nagy A, Novak B (2002) The dynamics of cell cycle regulation. BioEssays 24:1095-1109

Verma R, Feldman RMR, Deshaies RJ (1997) Sic1 is ubiquitinated in vitro by a pathway that requires CDC4, CDC34 and Cyclin/CDK activities. Mol Biol Cell 8:1427-1437

Wuarin J, Nurse P (1996) Regulating S phase: CDKs, licensing and proteolysis. Cell 85:785-787

Yamaguchi S, Okayama H, Nurse P (2000) Fission yeast Fizzy-related protein srw1p is a G(1)-specific promoter of mitotic cyclin B degradation. EMBO J 19:3968-3977

Zachariae W, Nasmyth K (1999) Whose end is destruction: cell division and the anaphase-promoting complex. Genes Devel 13:2039-2058

Zachariae W, Schwab M, Nasmyth K, Seufert W (1998) Control of cyclin ubiquitination by CDK-regulated binding of Hct1 to the anaphase promoting complex. Science 282:1721-1724

## Abbreviations

Cdk: cyclin-dependent protein kinase
SPF: S-phase Promoting Factor
MPF: M-phase Promoting Factor

CKI: cyclin-dependent kinase inhibitor
APC: Anaphase Promoting Complex

## Supplement: balance equations

In general, the concentration of a protein, X(t), in a network can change due to synthesis, degradation, activation and inactivation, as summarized in the balance equation:

$$\frac{dX}{dt} = \text{synthesis - degradation + activation - inactivation}$$

Protein synthesis is a two-step process (transcription and translation). Assuming transcription to be the rate-limiting step, we equate protein 'synthesis' to the rate of transcription and we do not write differential equations for mRNA's.

In principle, we need a differential equation for each chemical species in the reaction mechanism, and so the number of equations can be very large. Fortunately, some simplifications are possible. For instance, if a protein changes back and forth between active ($X^*$) and inactive (X) forms without changing its total concentration ($X_T$), then the two differential equations can be replaced by one plus an algebraic equation:

$$\frac{dX^*}{dt} = \text{activation} - \text{inactivation}, \quad X = X_T - X^*$$

If the activation and inactivation steps are relatively fast compared to other processes in the system, then the active form of the protein ($X^*$) will quickly come to its pseudo-steady state level, given by 'activation = inactivation'. If both steps are governed by Michaelis-Menten kinetics, then the pseudo-steady state level of X* is given by the Goldbeter-Koshland equation:

$$X^* = \frac{2 \cdot \gamma}{\beta + \sqrt{\beta^2 - 4 \cdot \alpha \cdot \gamma}}$$

where $\alpha = V_{mi} - V_{ma}$, $\beta = V_{mi}X_T - V_{ma}X_T + V_{mi}K_{ma} + V_{ma}K_{mi}$ and $\gamma = V_{ma}K_{mi}X_T$. $V_{ma}$ and $K_{ma}$ are the maximum velocity and Michaelis constant for the activation reaction, and likewise for the inhibition reaction.

**Supplementary Table 1.** Parameter values for the model. All parameters that start with a lower case '$k$' are rate constants (min$^{-1}$). Parameters that start with '$J$' are dimensionless Michaelis constants.

| | | | |
|---|---|---|---|
| $k_g = 0.005$ | $k_{s,bm} = 0.05$ | $k'_{d,bm} = 0.05$ | $k''_{d,bm,g1} = 1$ |
| $k''_{d,bm,m} = 1$ | $k_{a,mbf} = 0.25$ | $k'_{i,mbf} = 0.01$ | $k''_{i,mbf} = 2$ |
| $k_{s,bs} = 0.03$ | $k_{d,bs} = 0.05$ | $k_{s,cki} = 0.2$ | $k'_{d,cki} = 0.1$ |
| $k''_{d,cki,s} = 1$ | $k''_{d,cki,m} = 10$ | $k_{as,bm} = 1000$ | $k_{di,bm} = 1$ |
| $k_{as,bs} = 200$ | $k_{di,bs} = 1$ | $k_{a,cdh} = 1$ | $k''_{i,cdh,s} = 10$ |
| $k''_{i,cdh,m} = 40$ | $k_{a,iep} = 0.1$ | $k_{i,iep} = 0.04$ | $k_{a,c20} = 1$ |
| $k_{i,c20} = 0.2$ | $k'_{wee} = 0.01$ | $k''_{wee} = 1$ | $k_{a,wee} = 0.25$ |
| $k'_{i,wee} = 0$ | $k''_{i,wee} = 1$ | $k'_{25} = 0.02$ | $k''_{25} = 3$ |
| $k'_{a,25} = 0$ | $k''_{a,25} = 1$ | $k_{i,25} = 0.25$ | |
| $J_{a,mbf} = 0.1$ | $J_{i,mbf} = 0.1$ | $J_{a,cdh} = 0.01$ | $J_{i,cdh} = 0.01$ |
| $J_{a,iep} = 0.01$ | $J_{i,iep} = 0.01$ | $J_{a,c20} = 0.01$ | $J_{i,c20} = 0.01$ |
| $J_{a,wee} = 0.01$ | $J_{i,wee} = 0.01$ | $J_{a,25} = 0.01$ | $J_{i,25} = 0.01$ |

**Supplementary Table 2.** Initial conditions

| | | | |
|---|---|---|---|
| $[MR] = 1.83925$ | $[CycBM] = 0.21038$ | $[MPF] = 0.08305$ | $[preMPF] = 0.08140$ |
| $[TrimM] = 0.07723$ | $[MBF] = 0.17138$ | $[CycBS] = 0.17651$ | $[SPF] = 0.15852$ |
| $[CKI_T] = 0.09748$ | $[APCG1] = 0.00251$ | $[IEP] = 0.46795$ | $[APCM] = 0.99265$ |

Chen, Katherine C.
   Department of Biology, Virginia Polytechnic Institute and State University, Blacksburg, Virginia, VA 24061, USA

Novák, Béla
   Molecular Network Dynamics Research Group, Budapest University of Technology and Economics and Hungarian Academy of Sciences,
   1111 Budapest, Gellért tér 4, Hungary
   bnovak@mail.bme.hu

Tyson, John J.
   Department of Biology, Virginia Polytechnic Institute and State University, Blacksburg, Virginia, VA 24061, USA

# A modular systems biology analysis of cell cycle entrance into S-phase

Lilia Alberghina, Riccardo L. Rossi, Danilo Porro, and Marco Vanoni

## Abstract

A modular systems biology approach to the study of the cell cycle of the budding yeast *Saccharomyces cerevisiae* is presented. Literature on the structure of yeast population and its relevance to the study of yeast cell cycle is reviewed. A model for the control of yeast cell cycle, with emphasis on a threshold mechanism controlling entrance into S-phase is presented. The simple model has been used as a framework to derive a molecular blow-up of the major upstream events controlling the G1 to S transition that involves two sequential thresholds cooperating in carbon source modulation of the critical cell size required to enter S-phase, a hallmark response of the cell cycle to changing growth conditions. The model is discussed as an aid to filter and give structure to post-genomic data. The iterative application of this approach allows to obtain more refined models capturing the major regulatory features and the molecular details of the circuits connecting cell growth to cell cycle.

## 1 Systems biology and complex cellular processes

The genetic and molecular studies of the last three or four decades and more recently the high-throughput technologies that measure the concentrations of different cell components on a global scale (for instance transcripts, proteins and metabolites) in various physiological conditions have made available a wealth of biological data that could previously only be dreamed of (Russell 2002; Willett 2002; Werner 2004). Starting from raw data, statistical methods allow us to obtain molecular profiles, i.e., snapshots of the presence or absence of each transcript, protein, or metabolite in a given condition giving indication on their modulation by genetic and/or environmental factors. But since we have only crude pictures of the regulatory circuits of most cellular processes (cell signaling, cell cycle, apoptosis, differentiation, transformation, etc.), it is often an unmet task to give logic structure to molecular profiles and to gain from them predictive ability.

The behavior of a cellular process, which may require the coordinated action of hundreds of different gene products on a specific time scale, is in fact the result of the action of regulatory circuits, which can be modulated by intra and extracellular signals. The molecular machines that take part in a given regulatory network are not fixed entities, but may change during time in quantity, subcellular localization,

Topics in Current Genetics, Vol. 13
L. Alberghina, H.V. Westerhoff (Eds.): Systems Biology
DOI 10.1007/b138746 / Published online: 25 May 2005

and molecular modification (for instance phosphorylation/dephosphorylation of enzymes). Positive and negative feedback loops, signal amplification, thresholds, switches are among the more common control functions that operate in biological systems, as well as in technological devices. Emerging properties such as adaptation, error correction, and robustness have thus originated in living systems through the interaction of these basic regulatory elements; thus, fitness constraints shaped living cells giving nonlinear, often intuitively unpredictable behavior. Models have been very useful for the study of technological systems and they are developed with the aim to offer a formal (mathematical) representation of the process under investigation, able to capture its essential features. Of course, the level of detail of the model varies by changing the precision of analysis and of predictions that are requested (Werner et al. 2003; Wiley et al. 2003; Yarmush and Banta 2003; Bugrim et al. 2004).

Systems biology aims to describe the structure of the system (that is the molecular components and their wiring in the regulatory network) and to predict its dynamics under a spectrum of different conditions (Kitano 2002b) with the goal to achieve a comprehensive body of knowledge of biological systems, solidly grounded at the molecular level (Kitano 2002a, Westerhoff and Palsson 2004).

For some relatively simple processes, like glycolysis, the structure of the pathway is generally sufficiently well defined and yet the systems biology approach has been able to identify a previously underestimated rate-limiting step (Rossell et al. 2002). For other processes, like cell cycle, we do not know sizable portions of their underlying molecular regulatory networks. For instance, it has been estimated (Spellman et al. 1998) that about 10-15% of the 6000 gene products of budding yeast are involved in execution or in control of cell cycle, but the representations of cell cycle events that we have available account only for a few tens of them.

In this review, we will collect findings indicating that to better understand complex bioprocesses it is useful to rely also on a modular systems biology approach, that dissects cellular processes into modules, subsystems of interacting molecules that perform a given function in a way largely independent from the context (Hartwell et al. 1999).


## 2 The modular systems biology approach

It is quite well established that cellular processes are carried out by distinct modules, which are subsystems of interacting molecules such as proteins, DNA, RNA, and metabolites that perform a given function in a way largely independent from the context. Such a concept has been made popular by Hartwell (1999) and was based upon previous work by different groups (Kahn and Westerhoff 1991; Schuster et al. 1993). Each module can, thus, be regarded as a functional entity that performs a given activity by processing material (i.e. metabolites or macromolecules) and/or information. Different instances of equivalent modules can be involved in different processes, for instance, as in the case with MAP kinase modules. Other

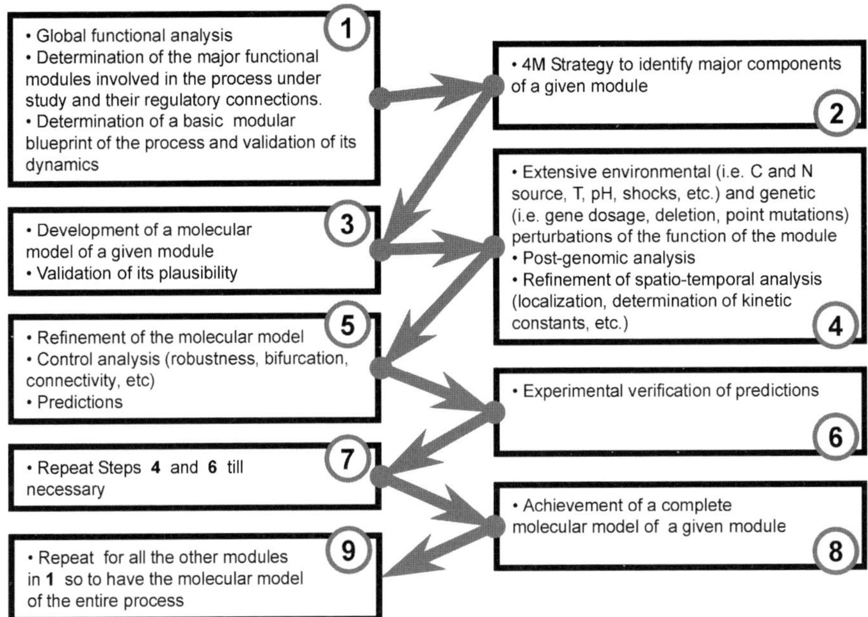


**Fig. 1.** The iterative roadmap of computational and experimental approaches to modular systems biology

modules, such as glycolysis, are instead unique inside a cell. The various modules are linked by governing interactions that follow general design principles well-known in engineering, such as switch, threshold control, positive and negative feedbacks, amplification, error correction, etc. (Nise 2004; Nurse 2003; Conzelmann et al. 2004). The modular systems biology approach to the analysis of a complex cellular process is summarized in Figure 1. The first step is to identify the biological functions of interest. A module may not simply correspond to a textbook pathway. Rather, its definition should include all those elements that allow the module to work as a functional unit in connection to other cell modules. At this stage, only those interactions that allow transfer of material or information among the different modules need to be considered, so to define inputs and outputs to and from each module. Once the basic modules have been identified, its blueprint can thus, describe a given process, which is a map in which basic modules and governing interactions of the process are identified. Simulation of the very basic blueprint map will allow evaluation as to whether it is able to capture the essential features of the processes (Fig. 1, Box 1). If so, the components of a module and their interactions can be identified following the **4M** Strategy (quoted by Henry 2003): **M**ining of literature and data banks; **M**anipulation by genetic means and by environmental changes of the module structure and function; **M**easurement of all putative regulatory components supposed to be active in the module (i.e. estimation of their concentration, localization, state of activation, time course changes, etc.); **M**odeling and simulation: construction of a model at higher resolution than the previous one, in which the putative regulatory components of the module are linked in a specific flow of events and then simulate the dynamics of the system under various conditions and compare the simulated behavior with experimental findings (Fig. 1, Box 2 and 3). The model can then be refined through molecular experiments that will include an extended set of metabolic and/or genetic perturbations as well as more extensive biological readouts, including post-genomic analyses and determination of relevant spatio-temporal parameters (Box 4). The new data will be fed into the module blueprint (Box 5). Based on control analysis and experimental verification of the predictions, steps in Box 4 and 5 will be repeated iteratively until the complete molecular model of the module is achieved (Box 7 and 8). The above steps will then be repeated for the other modules in order to obtain a molecular model of the whole process under study (Box 9).

Clearly, identification of the molecular structure of each module requires both hypothesis-driven experiments and computer simulations, whose predictions, in turn, will be tested experimentally. With an iterative repetition of modeling and experiments it should be possible to obtain satisfactory understanding and predictive ability of the process under investigation.

## 3 The control of cell cycle: an open question

Cell proliferation requires coordination of different processes: mass accumulation, DNA replication, and cell division. This tight coordination ensures maintenance of cell size and faithful partitioning of genetic material. The process is based on the cell ability to integrate external and metabolic signals with the activity of key cell cycle regulators. Such regulators are controlled by numerous mechanisms that reflect the diversity of signals, which they are able to integrate. In the budding yeast *Saccharomyces cerevisiae*, and possibly also in other eukaryotes (Wells 2002), coordination of mass accumulation with cell cycle progression relies on a sizer mechanism, so that DNA replication and/or cell division start only when cells have reached a critical cell size.

An appropriate model organism is invaluable in dissecting a regulatory network, since by allowing application of fine-tuned genetic and physiological perturbations, it allows to probe down to molecular detail the system under study. A model organism must have some distinctive characteristics: tools for forward (traditional) and reverse genetics, transgenic techniques must be well developed then it must be easily maintained and grown (and possibly with low cost). Finally, a completely sequenced genome and a strong research community are required. Budding yeast fulfills all these requirements and is probably one of the best and most used model organisms (Barr 2003; Castrillo and Oliver 2004). Therefore, it seems obvious that the most comprehensive cell cycle models have been developed for yeast (Tyson 1989; Chen et al. 2000, 2004: Novak et al. this volume), although no satisfactory molecular mechanism has been proposed for the more relevant physiological regulatory function, the cell sizer control over the entrance into

S-phase that is at the basis of the response of the cell cycle to changes in nutrients (Rupes 2002).

Even in a simple model organism, such as budding yeast, complete understanding of the cell cycle and its regulation is still an open question. Although tremendous efforts in recent years have allowed the elucidation of major components of the molecular machine driving the cell cycle, we still ignore its detailed mechanism and how intra and extracellular signals affect the entrance into and exit from the cell cycle in different experimental systems. Even if for a certain extent molecular mechanisms governing cell cycle progression have been computationally integrated in the models from Novak and Tyson groups, it is still not possible to link these mechanisms to the external environment through signaling pathways in a comprehensive way. But most importantly, we have not been able to integrate these mechanisms quantitatively with other broader regulatory signal transduction modules, such as metabolism, whose strongly influence on cell cycle is recognized but not completely described.


## 3.1 Cyclins, Cdks, and Cki are the evolutionary conserved molecular machines driving the cell cycle

Cyclin-dependent kinases (Cdks) play an essential regulatory role in cell cycle progression: it is in fact the sequential activation of Cdks by specific, unstable, regulatory subunits, named cyclins, that first triggers the onset of DNA replication and later initiates mitosis (Vermeulen et al. 2003; Murray 2004). Cdk activity is tightly regulated by different molecular mechanisms (Obaya and Sedivy 2002) that include regulatory phosphorylations, differential expression and/or localization and interaction with regulatory proteins, such as cyclin dependent kinase inhibitors (Cki) that inhibit Cdk activity by binding to Cyclin•Cdk complexes (Morgan 1995).

The evolutionary conservation of Cdks from yeast to mammalian cells is well established (Nurse 2000; Murray 2004), the cloning of the human Cdk-encoding *CDC2* gene by complementation of a *cdc2* mutant in *S. pombe* being one of the first pieces of evidence that cyclins are universal cell cycle regulators (Lee and Nurse 1987). In budding yeast, a single Cdk (Cdc28, now renamed Cdk1) is involved in the control of the cell cycle, while five Cdks active in the control of the cell cycle (out of a total of nine) have so far been identified in mammals. In the budding yeast cell cycle, there are cyclins associated with G1 (Cln1,2,3), S-phase (Clb5, 6), and mitosis (Clb1,2). These cyclins have a wide degree of redundancy and it is currently believed that their specificity in driving cell cycle progression depends mostly on the timing of expression and subcellular localization, rather than on a substrate specificity embedded in their molecule structure (reviewed in Murray 2004).

As the name implies, Ckis regulate cell cycle by inhibiting Cdk activity. In budding yeast, *SIC1* encodes a Cki that specifically inhibits complexes between Cdk1 and Clb-type cyclins (Mendenhall 1993). The second Cki, Far1 has been found to bind to and inhibit the activity of Cln•Cdk1 complexes in the response to

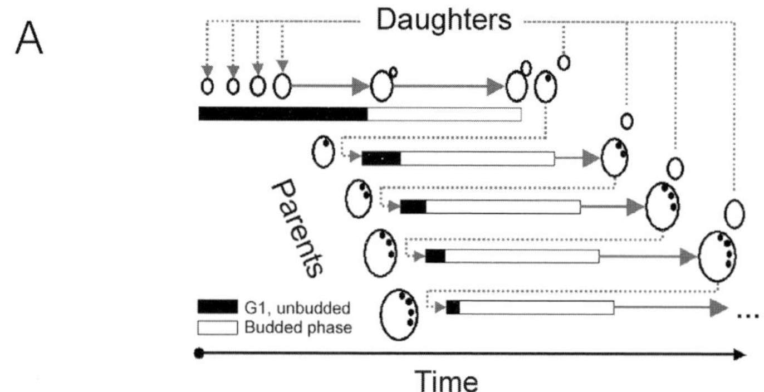
A
Daughters
Parents
G1, unbudded
Budded phase
Time

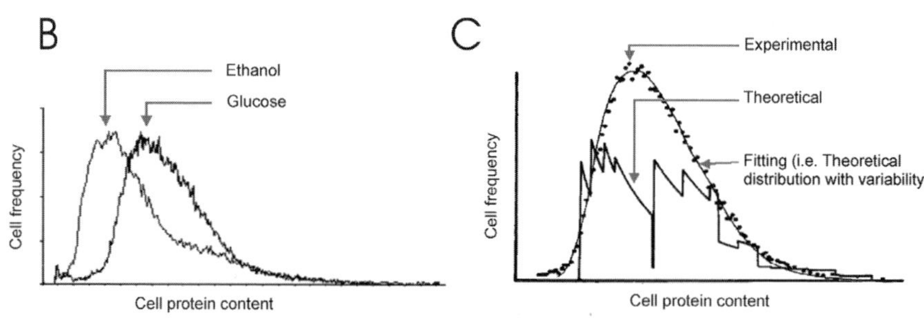
B
Ethanol
Glucose
Cell frequency
Cell protein content
C
Experimental
Theoretical
Fitting (i.e. Theoretical
distribution with variability)
Cell frequency
Cell protein content

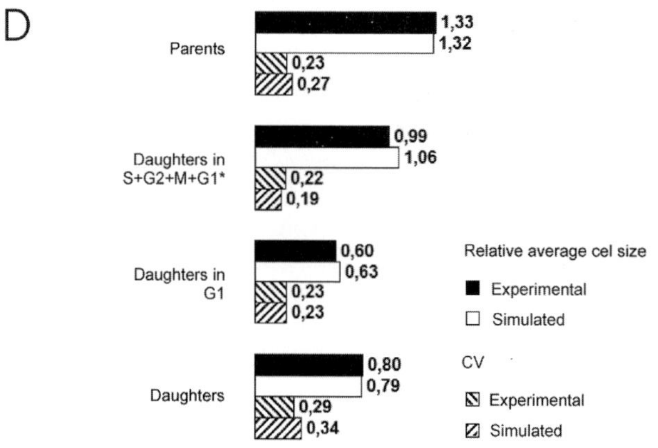
D
Parents
1,33
1,32
0,23
0,27
Daughters in
S+G2+M+G1*
0,99
1,06
0,22
0,19
Daughters in
G1
0,60
0,63
0,23
0,23
Daughters
0,80
0,79
0,29
0,34
Relative average cel size
Experimental
Simulated
CV
Experimental
Simulated

**Fig. 2. (overleaf)** A. A map of the yeast cell cycle incorporating the occurrence of genealogical age. It is assumed that: (1) each cell grows according to an exponential law; (2) all cells have to reach a protein content required to enter S-phase to start budding and DNA replication; (3) this protein content is the same for all daughter cells, regardless of their genealogical age; (4) it increases in parents of increasing genealogical age; (5) the duplication time of the daughter is longer that the duplication time of the parents; (6) all cells share the same budded period and (7) the duplication time of the parents is longer that the budded period (Carter and Jagadish 1978; Vanoni et al. 1983; Martegani et al. 1984; Woldringh et al. 1993). Only four populations of parent cells have been considered (Fig. 2A). B. Comparison of the cell protein content distributions for a yeast cell population exponentially growing on 2% v/v ethanol or 2% w/v glucose based media. Specific growth rates were 0.301 and 0.151 h-1, respectively. Cell protein contents have been obtained by flow cytometric analysis of FITC stained cells. Cell protein content (arbitrary unit). C. Theoretical protein density function and fitting of an experimental protein distribution (black points) for an exponentially growing *S. cerevisiae* population. Fitting was obtained by applying an appropriate variation cofficient (15-20%, accounting for both biological and instrumental variability) to the theoretical distribution. Modified from Martegani et al. (1984) D. Comparison of simulated cell size content distributions of the whole population, of the parent (P) and unbudded (U) and budded parent (B) subpopulations. The specific growth rate of the overall populations and the budding index values are 0.215 h-1 and 53.5%, respectively. The same values have been used for the simulations. Cell protein/size content (arbitrary unit). Unpublished data from our laboratory.

mating pheromones (Chang and Herskowitz 1990; Peter and Herskowitz 1994; Jeoung et al. 1998). Only recently, a role of Far1 in the control of the mitotic cell cycle has been proposed (Alberghina et al. 2004; Fu et al. 2003).

Many Ckis are multifunctional proteins (reviewed in Coqueret 2003). In budding yeast, Far1 affects cell cycle progression and cytoskeletal organization (Elion 2000) as p27$^{kip1}$ does in mammals. Kip1/Cip1 family members stably associate with cyclin D1-Cdk4 to assemble them into higher order enzymatically active complexes (Peter et al. 1993; Peter and Herskowitz 1994; Coqueret 2003; Besson et al. 2004). From a structural point of view, the Kip1/Cip1 inhibitors p27$^{Kip1}$ and p21$^{Cip1}$ share a homologous inhibitory domain (Russo et al. 1996). At the same time the Kip1/Cip1 proteins inhibit Cdk complexes containing cyclin D and E. Interestingly, as previously observed for Cdks and cyclins, also Cki structure and function is evolutionary conserved since a mammalian Kip1/Cip1 protein can substitute for Sic1 in yeast and, conversely, Sic1 can inhibit CyclinA•Cdk2 complex *in vitro* (Barberis et al. 2005). Whether the yeast Sic1 Cki has a scaffolding activity has not been directly addressed so far.

# 4 Global functional analysis of the G1/S transition in budding yeast

Cell cycle is an integrated cellular process well suited for a modular systems biology approach. Different kinds of models have been proposed based on chemical kinetics (Chen et al. 2004), on graph theory (Li et al. 2004), on block diagrams (Alberghina et al. 2001).

According to the above-described modular systems biology approach, a major aim in cell cycle research is to identify the molecular determinants of the regulatory network of the cycle (Ingolia and Murray 2004) and to quantitate emerging properties that arise from the interactions of the various components, such as robustness (Morohashi et al. 2002; Stelling et al. 2004; Kitano 2004a). It is likely that this new understanding will help to put on more rational ground the search for new drugs for proliferous diseases that are caused by a derangement of cell cycle controls (Kitano 2004b).

We have applied the modular systems biology approach to the study of a control mechanism of the G1 to S transition in budding yeast, thereby, obtaining indications of the fruitfulness of this approach.


## 4.1 Coordination between growth and the DNA division cycle: size distribution is a distinctive property of a yeast population

Cell growth (largely accounted for by increase in macromolecules, such as RNA and protein) and cell division have to be coordinated in order to maintain cell size homeostasis, thus, preventing cells from becoming too small or too large (Hartwell and Unger 1977). In *S. cerevisiae,* cell mass at division is unequally partitioned between a larger, parent cell (P) and a smaller, new daughter cell (D) that originate from an asymmetrical cell cycle. Figure 2A summarizes the current model for the asymmetrical yeast cell cycle. It is based on work by Hartwell and Unger (1977) and expanded by taking into account the heterogeneity in the cell cycles of cells of different genealogical age, i.e., the number of cell cycles a cell has passed through.

The method of choice for studying the coordination between growth and the DNA division cycle is the study of size distribution (volume distribution or distribution of the cellular protein content): it is generally accepted that the protein content is a good approximation of cell size (Alberghina and Porro 1993). In a given population, size distribution is a stable distinctive feature of each given balanced growth condition (Lord and Wheals 1980; Ranzi et al. 1986; Porro et al. 1995; Alberghina et al. 1998). As an example, Figure 2B shows the experimental protein distributions of cells growing on ethanol and glucose supplemented media. The average protein content is larger in glucose-growing cells than in ethanol-growing cells, the general trend both for batch cultures and chemostat-grown cells being that the average protein content and average protein content at Start are fairly constant at low and intermediate growth rates and increase at high growth rates (Vanoni et al. 1983; Ranzi et al. 1986; Alberghina and Porro 1993).

More detailed population information can be obtained by quantitative analysis of size distributions. Given the age density function of the population (the frequency of a cell of any age, i.e., the time elapsed from the previous cell division, in the population) and the exponential increase of protein content or cell mass, the theoretical cell size distribution can be derived, allowing to compute the expected frequency of cells of any given size (Martegani et al. 1984). By taking into account instrumental and biological variability (i.e. that individual cells differ because of their position within the cell division cycle, their genealogical age, and that cells of the same age and of the same cell cycle position not necessarily share the same size), satisfactory fitting of experimental protein distributions to theoretical size distributions (Fig. 2C) were obtained, once appropriate instrumental and biological variability was taken into account (Vanoni et al. 1983 and references therein; Martegani et al. 1984). Because of the tight interconnection between temporal parameters and size distribution, fitting of experimental distributions may be used to compute relevant parameters of a growing yeast population, as reported in Figure 2D.

## 4.2 Are metabolism and DNA division cycle coordinated?

In batch cultures, the average cell size remains at low and almost constant levels during growth on non-fermentable substrates (Fig. 3A, filled circles) while the average size of the cells linearly increases with the specific growth rate only during growth on fermentable substrates (Fig. 3A, open circles). Data from glucose-limited continuous cultures validate this observation, since the average cell size starts to increase after the critical dilution rate has been reached, i.e., after cells shift their metabolism from a fully oxidative to a respiro-fermentative one (Porro et al. 2003). The critical protein content at the G1/S transition, probably the more relevant parameter describing coordination between cell growth and the DNA division cycle, follows strictly that of the average protein content of the whole population. These results suggest a correlation between the rate of ethanol production and the increase of cellular protein content. By modulating the cellular metabolism through addition of formate to the culture medium of chemostat growing cells, increased production of NADH, which in turn enhances the rate of ethanol production, is observed. Significant increase in average protein content and protein content required to enter S-phase are observed even at very low growth rates (Porro et al. 2003), thus, linking the setting of cell protein contents to actual metabolism and not the specific growth rate.

## 4.3 Analysis of a shift-up

The complex effects of growth media on population growth can be seen during a nutritional ethanol/glucose shift-up. A rapid decrease in the fraction of budded cells has been observed; i.e., cells that were before Start at the moment of the shift delayed their entry into S-phase, whereas cells that had already executed Start delayed their exit from the cell cycle (Alberghina et al. 1998). The shift-up events can be dissected in different phases, all convergent to allow a rapid transition to cells growing faster, with an higher protein content at the G1 to S transition and a shorter budding time as compared to pre-shift cells. Immediately after the shift the

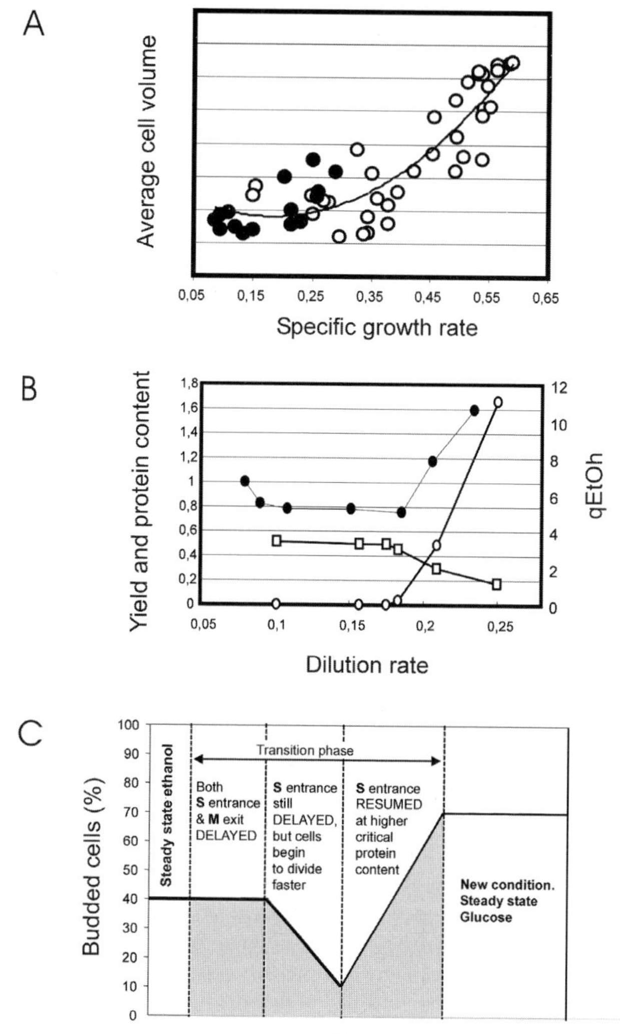



**Fig. 3.** A. *S. cerevisiae* yeast populations were grown in batch cultures on different media. (●) Average cell volume (arbitrary units) of the cells during growth on non-fermentable substrates. (○) Average cell volume (arbitrary units) of the cells during growth on fermentable substrates. Data redrawn from Tyson et al. (1979) B. Effect of specific growth rate on the average cell size content for S. cerevisiae populations growing in chemostat cultures. *S. cerevisiae* GRF18 cells were grown in glucose-limited continuous cultures at different dilution rates. (○) Specific ethanol production, $q_{eto}$, (mmol) (gram of dry weight biomass)$^{-1}$ (h)$^{-1}$ (●) average cell protein content (□) Yield ($Y_{SX}$), (g of dry biomass)(g of glucose consumed)$^{-1}$. The average protein content of the single cells, determined by flow cytometric analysis, has been normalized to 1 for the protein content of cells growing at D=0.08 h$^{-1}$. C. A schematic view of the effects of a nutritional ethanol/glucose shift-up on the composition of a yeast population. Unpublished data from our laboratory.

growth rate sets to the new value. During a first phase after the shift, both the entrance into S-phase and the exit from mitosis are severely delayed. Since the growth rate is enhanced, larger dividing cells are the consequence. In the second phase, the delay to enter S-phase is maintained, while cells are again able to divide, slowly shortening their budded period. After the shift, daughter cells are born at increasing sizes with the new setting of the protein content at the G1 to S transition so that they bud all together at about 4.3 hours after the shift. Pre-existing cells slowly recover their ability to enter S-phase at an increasing protein content, reaching the value characteristic of the new medium after 3.5 hours. At this same time also the budded period of the pre-existing cells has reached the value of the new medium (length of the budded phase = 1.4 hours). This analysis indicates that during the first and second phase of a shift-up, major resetting of the growth-sensitive cell size controlling the entrance into S-phase are taking place (Fig. 3C).

In conclusion, the molecular mechanism responsible for monitoring cell size is able to sense many internal (fermentative metabolism) and external (nutrients) signals, which allow a fast and efficient homeostatic response as exemplified by the complex resetting of a yeast population during the nutritional shift-up.

# 5 A new threshold control for the G1 to S transition in budding yeast

As outlined above, there is a general consensus that a critical cell size is required in yeast to enter S-phase (Rupes 2002; Wells 2002). This control is of outmost physiological relevance, since it allows the coordination of cell growth with cycle progression and it is responsible for cell size homeostasis (Tapon et al. 2001). Since the level of Cln3 is constant in G1 cells (Tyers et al. 1993), its amount per cell is proportional to cell mass. Increasing the level of Cln3 by overexpression or by a mutational stabilization of the protein decreases both cell size and protein content required to enter S-phase, while in cells in which the Cln3 gene has been disrupted, these two parametes increase (reviewed in Mendenhall and Hodge 1998). These data led to the commonly accepted hypothesis that Cln3 is involved in size regulation of the G1 to S transition (Futcher 1996). A popular model assumes that cell size is measured as nuclear Cln3 concentration that would occur because of steady Cln3 synthesis and constant nuclear volume during this cell cycle phase. Several data do not fit well with this hypothesis. First of all, it has been recognized that the G1 cyclin Cln3 is modulated by cell growth, its level being higher in fast growing cells than in slow growing ones (Hubler et al. 1993; Hall et al. 1998). Moreover, we have also recently shown that in cells undergoing a shift up from a poor to a rich medium, the level of Cln3 increases while the cells are unable to enter S-phase (Alberghina et al. 2004). These observations are consistent with the notion that Cln3 is not the only determinant of the cell sizer, thereby, supporting the hypothesis of a more complex mechanism. Moreover, recent data indicate that the nuclear/cytoplasmic ratio does not appreciably decrease prior of

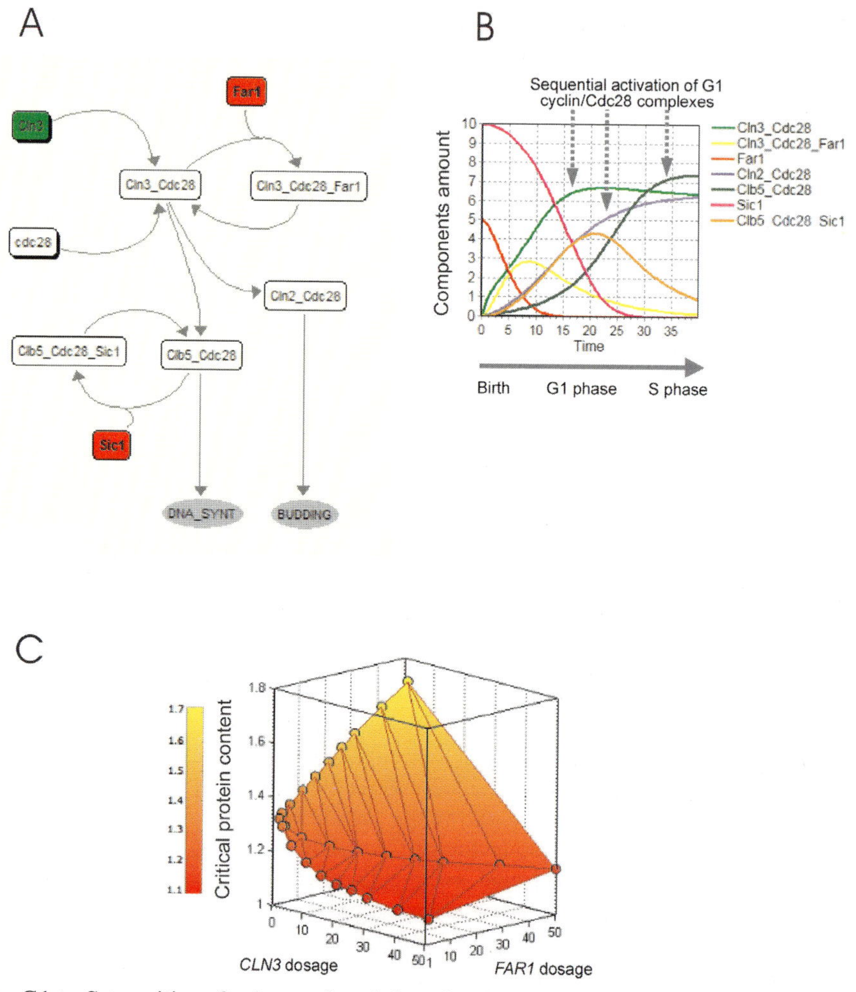



the G1 to S transition, further undermining simple models that connect nuclear accumulation of Cln3 to the G1 to S transition (Jorgensen and Tyers 2004 and references within).

A nutritionally modulated threshold given by an activator, which blocks an inhibitor, is a very simple and effective biochemical threshold mechanism (Alberghina et al. 1983). We have put forward the hypothesis that Far1, the Cki long recognized to inhibit the G1 to S transition in response to mating pheromones (Valtz et al. 1995), may have also a role in the standard mitotic cycle cooperating with Cln3 in a nutritionally modulated threshold controlling entrance into S-phase. Therefore, a basic blueprint of the cell cycle has been proposed based on a Cki•cyclin threshold acting as a Start function in which a cell sizer controls entrance into S-phase by activating waves of cyclins that set the timing for the onset of mitosis and cell division; computer simulations analyses based on this model

**Fig. 4. (overleaf)** A. The G1/S two threshold core model implemented with JDesigner (ver. 1.951, Systems Biology Workbench). A simple wiring diagram has been constructed, that consider cyclins, the Cdk Cdc28 and Cki Far1 and Sic1 and their interactions. Relative initial values of protein species are set consistent with experimental findings. Since dimerization or trimerization are not implementable in JDesigner complexes, formation is considered as a new component activation. The two thresholds acting at Start are designed as equilibria between active and inactive cyclin-Cdk complexes. B. Time from birth to S-phase is considered: such a system is able to reproduce the observable sequential activation of active G1 cyclin/Cdc28 complexes. The quantities "DNA synthesis" and "Budding" (not plotted in Panel B) burst in parallel after S-phase cyclin-Cdk complex activation (Clb5_Cdc28), which is consistent with experimental data. Start execution has been set putting equal to 10 the minimum level of *both* "DNA synthesis" and "Budding" quantities necessary to trigger its execution: this transition is passed at time 35 in wild type cells. C. The overexpression of *CLN3* and/or *FAR1* genes of an *n*-fold factor (simulated as *n*-fold higher initial amount of protein respect to wild type condition) alters the critical protein content required to enter S-phase. Made equal to 1 the size of the cell at the beginning of the cycle, the critical protein content (protein content when both "DNA synthesis" and "Budding" quantities reach value 10 in the simulation) is obtained and plotted as a function of *CLN3* and *FAR1* gene dosage. Critical protein content is calculated for wild type (*n*=1) and for *n*-fold overexpression of *CLN3* and/or *FAR1*, with *n* = 2, 5, 10, 15, 20, 25, 30, 40, 50. Obtained values are visualized as colored circles in the 3-dimensional graph.

are able to predict with accuracy the dynamics of growth and of budding in steady and in transitory states (Alberghina et al. 2001). A detailed analysis of the G1 to S transition has been performed in our laboratory following the 4M strategy. Deletion and overexpression of Far1 indicated a threshold role of this Cki in the control of the mitotic cycle. Furthermore, we have shown that the modulation of cell size given by carbon source requires the activity of both the Cln3/Far1 threshold and of the threshold Clb5,6/Sic1 (Alberghina et al. 2004).

Such thresholds are a typical systems-level property that is not present in each of the involved molecules alone, but rather emerges from the biochemical mechanism connecting the two entities, for instance the Cln3 activator and the Far1 inhibitor. This molecular mechanism sheds light on the modulatio of the size control operated by the cell. Both Cln3 and Far1 act as dynamical variables whose reversible interaction with each other and with Cdk1 brings to the formation of the Cln3-Cdk1 active kinase complex whose activity becomes the sensitive actuator regulating the first step in the G1 to S transition. The resulting critical size is then the outcome of several factors, like the initial amounts of Cln3 and Far1, in their free and complexed forms, and their specific dynamics (i.e. rate of synthesis and degradation). Such a mechanism, thus, allows plasticity of the threshold mechanism and its ability to respond to changing growth conditions in ways that could not be explained solely on the basis of the nuclear Cln3 concentration model. In fact by allowing independent regulation of the level - or functional state - of either the activator and/or the inhibitor, the size at the beginning of the cell cycle can be more easily fine tuned, as is shown by the results of simulations reported in Figure 4B and 4C. The existence of a molecule inhibiting the G1/S transition and whose properties strongly resemble those of Far1 (i.e. peak level expression in late M phase, Cln-primed phosphorylation) has been proposed in order to account for recent data relating growth rate and cell size modulation of G1 phase cyclins at Start (Schneider et al. 2004).

Starting from our systems level cell cycle model (Alberghina et al. 2001), we proceeded towards a molecular blow-up of the Start module regulating entrance into S-phase, concentrating on the molecular mechanism of the cell size by incorporating the experimental data reported above (Alberghina et al. 2004). The core of the G1/S transition is modeled as a simple network in JDesigner (System Biology Workbench; Sauro et al. 2003) taking into account the regulation of the Cdc28 Cdk by means of inactive Cki-cyclin-Cdk complexes formation, basic features of G1/S cyclin dynamics are conserved (Fig. 4A-B).

## 6 Post-genomic analysis of the G1/S transition

An increasing body of post-genomic data, notably transcriptome and interactome data is becoming available. Each genome-wide "omic" approach samples a "horizontal" slice (i.e. across all genes or gene products) of a multidimensional space encompassing a variety of relevant cellular properties connected with the function of any given gene or gene product (such as transcription, interaction, activity, metabolite(s) produced) (Vidal 2001). If we are interested in understanding the structure and regulation of complex cellular networks, then sampling a single dimension of a complex space will undoubtedly provide relevant information, but may not highlight the major regulatory features. As it has been pointed out by Nobel Laureate Paul Nurse, in order to explain the regulatory circuits of the cell cycle and of major cellular functions, we need to obtain and analyze experimental data in novel ways that allow to identify, characterize, and classify the components of the regulatory circuits that operate within cells (Nurse 2003). A precise description of the behavior of each of these components (see for instance Tyson et al. 2003) for a first attempt to such a classification may ultimately allow to build up a comprehensive catalogue of parts able to carry on a given regulatory function. It is to be expected that a given component of a regulatory circuit results from a specific, limited set of biochemical activities and molecular interactions. To identify and classify these unit operations might thus prove extremely useful in giving structure to genomic and post-genomic data, since these simple components may first be used as an aid to construct filters to extract relevant biological information from raw (conventional and post-genomic) data and later to assign molecular specific biochemical players to the identified component(s) of the regulatory circuit under investigation. Ultimately, each unit operation may become a sort of subroutine in systems biology simulation languages allowing to more easily assemble dynamic models of the biological processes.

One of the more reliable transcriptional data set available deals with cell cycle-regulated transcription (Spellman et al. 1998). By using DNA microarray and budding yeast cultures synchronized by different protocols (alpha factor arrest, elutriation, and arrest of a *cdc15* temperature-sensitive mutant), a comprehensive catalogue of yeast genes has been collected, the transcripts of which are characterized by a periodicity within the cell cycle. About 800 cell cycle regulated genes have thus been classified as G1, S, G2, M, and M/G1 specific. It is likely that second-generation experiments that combine hypothesis-driven (low or high throughput) experiments with genome-wide data will allow to extract information hidden in unbiased high throughput data set: for instance it has been possible to derive a model for the yeast cell cycle transcriptional regulatory network by crossing a genome-wide location analysis of yeast transcriptional activators with the Spellman data set (Lee et al. 2002).

Even this impressive achievement leaves open the question about the early events that precede and drive cell cycle dependent transcriptional remodeling. As briefly summarized before, cell cycle regulation is mediated also at the post-transcriptional level through formation and disruption of protein complexes that in turn are largely dependent on interconnected post-translational modifications, including phosphorylation/dephosphorylation, relocalization, regulated proteolysis.

Comprehensive efforts to characterize protein/protein interaction in budding yeast have been conducted using both genetic (two hybrid) and biochemical (affinity purification plus mass spectrometry) methods. These large-scale approaches yielded a large catalogue of interactions leading to a number of novel findings and the emergence of a huge network containing thousands of proteins. Comparison between two hybrid data sets (obtained independently from two groups) and mass spectrometry data (also obtained by two groups) showed a surprisingly small overlap: circa 10 and 14%, respectively (reviewed in Ito et al. 2002). Thus, interactome data require extensive efforts to assess their reliability, i.e., to filter false signals that have been estimated to be as high as 50% or more (Deane et al. 2002).

The presence of false negatives, i.e., failure of both large scale current genetic and biochemical methods to detect weak, transient interactions involving low abundance proteins (Ito et al. 2002) is also a relevant problem that can be exemplified with reference to the G1 to S regulatory network discussed in the previous paragraph. Within the network many protein/protein (as well as protein/DNA) interactions are present. Protein/protein interactions involve the formation of binary or higher order complexes as well as enzyme substrate interactions. Protein/protein and protein/DNA interactions predicted by the network are depicted as boxes with thick blue and red borders, respectively, in Figure 5A. Protein/protein, protein/DNA and synthetic lethal interactions detected in genome-wide analyses are reported in Figure 5B as colored boxes (see Figure Legend for further details). Interactions that are present in the YPD database, i.e., covering annotated literature are reported in Figure 5C. Panels D and E represent superposition of Panels A and B and of Panels A and C, respectively. It is quite clear that many relevant interactions detected in conventional molecular analysis are missing from the genome-wide data sets. Notably, in the genome-wide data set (Panels B and D), Sic1 appears to interact only with Cdk1 and no protein/protein interaction with either Clb5 or Clb6 is detected. Similarly, the well-known interactions of Far1 with Cln1,2,3 and Cdk1 are not detected in the experiments reported in the genome-wide data sets, but only in data sets compiled from literature data (compare Fig. 5D with Fig. 5E).

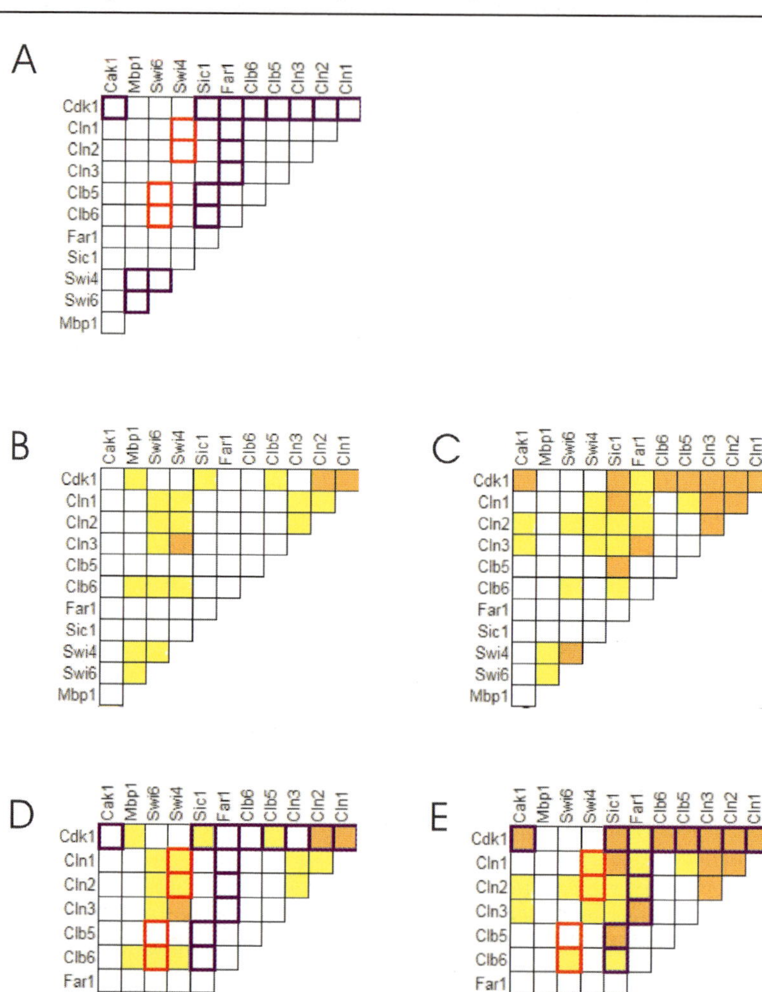

This preliminary comparison indicates that genome-wide data sets alone may miss important interactions and that they have to be considered with caution when aiming to identify regulatory networks. Thus, it is of primary importance that data from different "omics" approaches are integrated and analyzed together in order to identify hidden patterns and to allow recomposition of otherwise "missing" data (Begley et al. 2002). The availability of molecular blueprints derived from a systems biology approach could be very relevant to give structure and predictive ability to "omics" data.

**Fig. 5. (overleaf)** Mining of genome-wide and of literature data for interactions among cell cycle genes/proteins considered in the G1-S transition blueprint. Both physical and genetic interactions were considered. A. Gene product names of key regulators of G1-S transition are presented on the top and on the side of an "interaction board". Proteins from the model reported in Alberghina et al. (2004) are considered; interactions between two proteins are depicted with thick blue border at the crossing between the column and the row of those proteins. An analogue thick red border represents protein/DNA interactions (i.e. between a protein regulating expression of a gene and promoter sequences of the regulated gene). B. Interactions among cell cycle proteins derived from genome-wide datasets from two hybrid analysis (Uetz et al. 2000), search for complexes via TAP-MS (Gavin et al. 2002; Ho et al. 2002), synthetic genetic array (SGA) analysis on size mutants (Jorgensen et al. 2002), and identification of SBF and MBF target genes (Iyer et al. 2001). An empty box (white) means that no interaction is present between the two genes/proteins at the vertexes; yellow box interaction detected; orange box indicates an interaction detected in more than one dataset. C. Interactions among cell cycle proteins from literature obtained from mining of YPD database "Interaction report" (www.incyte.com; accessed May 31, 2004), and the *Interaction and Complex Viewer* of MIPS database (http://mips.gsf.de/proj/yeast/CYGD/interaction/; accessed May 31, 2004 and mined with both "physical" and "genetic" search criteria). Empty (white) box no interaction annotated; yellow box interaction annotated; orange box interaction annotated in both databases. D. Genome-wide interaction data as in Panel B overlaid with protein/protein and protein/DNA interactions (blue and red thick borders, respectively) considered in the G1-S transition blueprint of Panel A. E. Literature interaction data as in Panel C overlaid with protein/protein and protein/DNA interactions (blue and red thick borders, respectively) considered in the G1-S transition blueprint of Panel A.


# 7 What next?


The examples discussed in this paper indicate how the iterative integration of different genome-wide data sets can be fruitfully combined with "conventional" low throughput experiments in a systems biology framework. This approach may be the key to unravel the huge amount of hidden information present in genome-wide data sets, giving more refined models to be tested and validated with appropriately designed hypothesis-driven low throughput or second generation high throughput experiments (Fig. 1). Current frontiers may include development of computational tools to make cross-referencing of different data sets easier and (semi)automatic as well as developments of techniques allowing to measure with high precision and high throughput parameters that are of major relevance for cell cycle control such as intracellular localization of key proteins. Implementation of ever more accurate models may require measurement of actual affinity constants for selected protein/protein interactions *in vivo,* for instance by FRET technology (Yi et al. 2003). With current technology this is a typical low throughput task that limits its application to selected, key interactions in a given regulatory network.

The global functional analysis of the G1 to S transition has indicated a link between cell metabolism and the setting of protein content required to enter S-phase. The molecular mechanism so far unraveled does not account for this aspect, although the expression of several components of the cell cycle machinery is known to be affected by the carbon source (DeRisi et al. 1997; Gallego et al. 1997; Hall et al. 1998; Newcomb et al. 2003; unpublished results from our laboratory). It is of interest to recall that glucose-mediated *CLN3* induction does not require Tor-, Rgt2/Snf3- or Hxk2-mediated nutrient signaling, but requires transport and metabolism of glucose, being inhibited by the addition of iodoacetate, an inhibitor of glyceraldehyde-3-phosphate dehydrogenase (Newcomb et al. 2003). Recent data from our laboratory indicate that the Gpr1/Gpa2 system is involved in cell size control (Alberghina et al. 2004). Taken together, all these data suggest that both extracellular and intracellular glucose may cooperate in setting the cell size: the first through the Gpr1/Gpa2 signal transduction system and the latter through a direct or indirect link with glycolysis. The modulation of protein content required to enter S-phase by glucose metabolism appears to be conserved during evolution. In mammalian cells dependent on interleukin-3 for proliferation and survival, increased growth factor concentrations result in an increase of glycolytic rates and in an increase of cell size to more than a doubling (Vander Heiden et al. 2001).

Finally it should be remembered that a detailed knowledge of the structure of a yeast population under study is required in order to define experimental approaches able to enrich a specific sub-population(s), a mandatory prerequisite to study a given phenotypic trait in a target sub-population. At the same time, a detailed knowledge of the cell population is required to put back in context the acquired information, allowing to switch from phenotypic profiling of an unstructured non-existing 'average' cell to a more informative population-based profiling.

## Acknowledgements

This work has been partially supported by grants from AIRC and MIUR (Progetto Strategico Oncologia) to L.A. We would like to thank Uwe Sauer and Hans V. Westerhoff for critical reading of the manuscript and comments.

## References

Alberghina L, Martegani E, Mariani L, Bortolan G (1983) A bimolecular mechanism for the cell size control of the cell cycle. Biosystems 16:297-305

Alberghina L, Porro D (1993) Quantitative flow cytometry: analysis of protein distributions in budding yeast. A mini-review. Yeast 9:815-823

Alberghina L, Porro D, Cazzador L (2001) Towards a blueprint of the cell cycle. Oncogene 20:1128-1134

Alberghina L, Rossi RL, Querin L, Wanke V, Vanoni M (2004) A cell sizer network involving Cln3 and Far1 controls entrance into S-phase in the mitotic cycle of budding yeast. J Cell Biol 167:433-443

Alberghina L, Smeraldi C, Ranzi BM, Porro D (1998) Control by nutrients of growth and cell cycle progression in budding yeast, analyzed by double-tag flow cytometry. J Bacteriol 180:3864-3872

Barberis M, DeGioia L, Ruzzene M, Sarno S, Marin O, Coccetti P, Fantucci P, Vanoni M, Alberghina L (2005) Ck2 phosphorylation regulates inhibitory activity of the yeast cyclin dependent kinase inhibitor Sic1. Biochem J Immediate Publication, doi:10.1042/BJ20041299

Barr MM (2003) Super models. Physiol Genomics 13:15-24

Begley TJ, Rosenbach AS, Ideker T, Samson LD (2002) Damage recovery pathways in *Saccharomyces cerevisiae* revealed by genomic phenotyping and interactome mapping. Mol Cancer Res 1:103-112

Besson A, Gurian-West M, Schmidt A, Hall A, Roberts JM (2004) p27Kip1 modulates cell migration through the regulation of RhoA activation. Genes Dev 18:862-876

Bugrim A, Nikolskaya T, Nikolsky Y (2004) Early prediction of drug metabolism and toxicity: systems biology approach and modeling. Drug Discov Today 9:127-135

Carter BL, Jagadish MN (1978) The relationship between cell size and cell division in the yeast *Saccharomyces cerevisiae*. Exp Cell Res 112:15-24

Castrillo JI, Oliver SG (2004) Yeast as a touchstone in post-genomic research: strategies for integrative analysis in functional genomics. J Biochem Mol Biol 37:93-106

Chang F, Herskowitz I (1990) Identification of a gene necessary for cell cycle arrest by a negative growth factor of yeast: FAR1 is an inhibitor of a G1 cyclin, CLN2. Cell 63:999-1011

Chen KC, Csikasz-Nagy A, Gyorffy B, Val J, Novak B, Tyson JJ (2000) Kinetic analysis of a molecular model of the budding yeast cell cycle. Mol Biol Cell 11:369-391

Chen KC, Calzone L, Csikasz-Nagy A, Cross FR, Novak B, Tyson JJ (2004) Integrative analysis of cell cycle control in budding yeast. Mol Biol Cell 15:3841-3862

Conzelmann H, Saez-Rodriguez J, Sauter T, Bullinger E, Allgower F, Gilles ED (2004) Reduction of mathematical models of signal transduction networks: simulation-based approach applied to EGF receptor signalling. J Syst Biol 1:159-169

Coqueret O (2003) New roles for p21 and p27 cell-cycle inhibitors: a function for each cell compartment? Trends Cell Biol 13:65-70

Deane CM, Salwinski L, Xenarios I, Eisenberg D (2002) Protein interactions: two methods for assessment of the reliability of high throughput observations. Mol Cell Proteomics 1:349-356

DeRisi JL, Iyer VR, Brown PO (1997) Exploring the metabolic and genetic control of gene expression on a genomic scale. Science 278:680-686

Elion EA (2000) Pheromone response, mating and cell biology. Curr Opin Microbiol 3:573-581

Fu X, Ng C, Feng D, Liang C (2003) Cdc48p is required for the cell cycle commitment point at Start via degradation of the G1-CDK inhibitor Far1p. J Cell Biol 163:21-26

Futcher B (1996) Cyclins and the wiring of the yeast cell cycle. Yeast 12:1635-1646

Gallego C, Gari E, Colomina N, Herrero E, Aldea M (1997) The Cln3 cyclin is down-regulated by translational repression and degradation during the G1 arrest caused by nitrogen deprivation in budding yeast. EMBO J 16:7196-7206

Gavin AC, Bosche M, Krause R, Grandi P, Marzioch M, Bauer A, Schultz J, Rick JM, Michon AM, Cruciat CM, Remor M, Hofert C, Schelder M, Brajenovic M, Ruffner H, Merino A, Klein K, Hudak M, Dickson D, Rudi T, Gnau V, Bauch A, Bastuck S, Huhse B, Leutwein C, Heurtier MA, Copley RR, Edelmann A, Querfurth E, Rybin V,

Drewes G, Raida M, Bouwmeester T, Bork P, Seraphin B, Kuster B, Neubauer G, Superti-Furga G (2002) Functional organization of the yeast proteome by systematic analysis of protein complexes. Nature 415:141-147

Hall DD, Markwardt DD, Parviz F, Heideman W (1998) Regulation of the Cln3-Cdc28 kinase by cAMP in *Saccharomyces cerevisiae*. EMBO J 17:4370-4378

Hartwell LH, Hopfield JJ, Leibler S, Murray AW (1999) From molecular to modular cell biology. Nature 402:C47-C52

Hartwell LH, Unger MW (1977) Unequal division in *Saccharomyces cerevisiae* and its implications for the control of cell division. J Cell Biol 75:422-435

Henry CM (2003) Systems Biology. Chem Eng News 81:45-55

Ho Y, Gruhler A, Heilbut A, Bader GD, Moore L, Adams SL, Millar A, Taylor P, Bennett K, Boutilier K, Yang L, Wolting C, Donaldson I, Schandorff S, Shewnarane J, Vo M, Taggart J, Goudreault M, Muskat B, Alfarano C, Dewar D, Lin Z, Michalickova K, Willems AR, Sassi H, Nielsen PA, Rasmussen KJ, Andersen JR, Johansen LE, Hansen LH, Jespersen H, Podtelejnikov A, Nielsen E, Crawford J, Poulsen V, Sorensen BD, Matthiesen J, Hendrickson RC, Gleeson F, Pawson T, Moran MF, Durocher D, Mann M, Hogue CW, Figeys D, Tyers M (2002) Systematic identification of protein complexes in *Saccharomyces cerevisiae* by mass spectrometry. Nature 415:180-183

Hubler L, Bradshaw-Rouse J, Heideman W (1993) Connections between the Ras-cyclic AMP pathway and G1 cyclin expression in the budding yeast *Saccharomyces cerevisiae*. Mol Cell Biol 13:6274-6282

Ingolia NT  Murray AW(2004) The ups and downs of modeling the cell cycle. Curr Biol 14:R771-R777

Ito T, Ota K, Kubota H, Yamaguchi Y, Chiba T, Sakuraba K, Yoshida M (2002) Roles for the two-hybrid system in exploration of the yeast protein interactome. Mol Cell Proteomics 1:561-566

Iyer VR, Horak CE, Scafe CS, Botstein D, Snyder M, Brown PO (2001) Genomic binding sites of the yeast cell-cycle transcription factors SBF and MBF. Nature 409:533-538

Jeoung DI, Oehlen LJ, Cross FR (1998) Cln3-associated kinase activity in *Saccharomyces cerevisiae* is regulated by the mating factor pathway. Mol Cell Biol 18:433-441

Jorgensen P, Nishikawa JL, Breitkreutz BJ, Tyers M (2002) Systematic identification of pathways that couple cell growth and division in yeast. Science 297:395-400

Jorgensen P, Tyers M (2004) How cells coordinate growth and division. Curr Biol 14:R1014-1027

Kahn D, Westerhoff HV (1991) Control theory of regulatory cascades. J Theor Biol 153:255-85

Kitano H (2002a) Looking beyond the details: a rise in system-oriented approaches in genetics and molecular biology. Curr Genet 41:1-10

Kitano H (2002b) Systems biology: a brief overview. Science 295:1662-1664

Kitano H (2004a) Biological robustness. Nat Rev Genet 5: 826-837

Kitano H (2004b) Cancer as a robust system: implications for anticancer therapy. Nat Rev Cancer 4:227-235

Lee MG, Nurse P (1987) Cell cycle genes of the fission yeast. Sci Prog 71:1-14

Lee TI, Rinaldi NJ, Robert F, Odom DT, Bar-Joseph Z, Gerber GK, Hannett NM, Harbison CT, Thompson CM, Simon I, Zeitlinger J, Jennings EG, Murray HL, Gordon DB, Ren B, Wyrick JJ, Tagne JB, Volkert TL, Fraenkel E, Gifford DK, Young RA (2002) Transcriptional regulatory networks in *Saccharomyces cerevisiae*. Science 298:799-804

Li F, Long T, Lu Y, Ouyang Q, Tang C (2004) The yeast cell-cycle network is robustly designed. Proc Natl Acad Sci USA 101:4781-4786

Lord PG, Wheals AE (1980) Asymmetrical division of *Saccharomyces cerevisiae*. J Bacteriol 142:808-818

Martegani E, Vanoni M, Delia D (1984) A computer algorithm for the analysis of protein distribution in budding yeast. Cytometry 5:81-85

Mendenhall MD (1993) An inhibitor of p34CDC28 protein kinase activity from *Saccharomyces cerevisiae*. Science 259:216-219

Mendenhall MD, Hodge AE (1998) Regulation of Cdc28 cyclin-dependent protein kinase activity during the cell cycle of the yeast *Saccharomyces cerevisiae*. Microbiol Mol Biol Rev 62:1191-1243

Morgan DO (1995) Principles of CDK regulation. Nature 374:131-134

Morohashi M, Winn AE, Borisuk MT, Bolouri H, Doyle J, Kitano H (2002) Robustness as a measure of plausibility in models of biochemical networks. J Theor Biol 216:19-30

Murray AW (2004) Recycling the cell cycle: cyclins revisited. Cell 116:221-234

Newcomb LL, Diderich JA, Slattery MG, Heideman W (2003) Glucose regulation of *Saccharomyces cerevisiae* cell cycle genes. Eukaryot Cell 2:143-149

Nise NS (2004) Control Systems Engineering, 4th edn. John Wiley & Son, Inc.

Nurse P (2000) The incredible life and times of biological cells. Science 289:1711-1716

Nurse P (2003) Systems biology: understanding cells. Nature 424:883

Obaya AJ, Sedivy JM (2002) Regulation of cyclin-Cdk activity in mammalian cells. Cell Mol Life Sci 59:126-142

Peter M, Gartner A, Horecka J, Ammerer G, Herskowitz I (1993) FAR1 links the signal transduction pathway to the cell cycle machinery in yeast. Cell 73:747-760

Peter M, Herskowitz I (1994) Direct inhibition of the yeast cyclin-dependent kinase Cdc28-Cln by Far1. Science 265:1228-1231

Porro D, Brambilla L, Alberghina L (2003) Glucose metabolism and cell size in continuous cultures of *Saccharomyces cerevisiae*. FEMS Microbiol Lett 229:165-171

Porro D, Ranzi BM, Smeraldi C, Martegani E, Alberghina L (1995) A double flow cytometric tag allows tracking of the dynamics of cell cycle progression of newborn *Saccharomyces cerevisiae* cells during balanced exponential growth. Yeast 11:1157-1169

Ranzi BM, Compagno C, Martegani E (1986) Analysis of protein and cell volume distribution in glucose-limited continuous cultures of budding yeast. Biotechnol Bioeng 28:185-190

Rossell S, van der Weijden CC, Kruckeberg A, Bakker BM, Westerhoff HV (2002) Loss of fermentative capacity in baker's yeast can partly be explained by reduced glucose uptake capacity. Mol Biol Rep 29:255-257

Rupes I (2002) Checking cell size in yeast. Trends Genet 18:479-485

Russell RB (2002) Genomics, proteomics and bioinformatics: all in the same boat. Genome Biol 3:REPORTS4034

Russo AA, Jeffrey PD, Patten AK, Massague J, Pavletich NP (1996) Crystal structure of the p27Kip1 cyclin-dependent-kinase inhibitor bound to the cyclin A-Cdk2 complex. Nature 382:325-331

Sauro HM, Hucka M, Finney A, Wellock C, Bolouri H, Doyle J, Kitano H (2003) Next generation simulation tools: the Systems Biology Workbench and BioSPICE integration. Omics 7:355-372

Schneider BL, Zhang J, Markwardt J, Tokiwa G, Volpe T, Honey S, Futcher B (2004) Growth rate and cell size modulate the synthesis of, and requirement for, G1-phase cyclins at start. Mol Cell Biol 24:10802-10813

Schuster S, Kahn D, Westerhoff HV (1993) Modular analysis of the control of complex metabolic pathways. Biophys Chem 48:1-17

Spellman PT, Sherlock G, Zhang MQ, Iyer VR, Anders K, Eisen MB, Brown PO, Botstein D, Futcher B (1998) Comprehensive identification of cell cycle-regulated genes of the yeast *Saccharomyces cerevisiae* by microarray hybridization. Mol Biol Cell 9:3273-3297

Stelling J, Sauer U, Szallasi Z, Doyle FJ 3rd, Doyle J (2004) Robustness of cellular functions. Cell 118: 675-685

Tapon N, Moberg KH, Hariharan IK (2001) The coupling of cell growth to the cell cycle. Curr Opin Cell Biol 13:731-737

Tyson CB, Lord PG, Wheals AE (1979) Dependency of size of *Saccharomyces cerevisiae* cells on growth rate. J Bacteriol 138:92-98

Tyson JJ (1989) Effects of asymmetric division on a stochastic model of the cell division cycle. Math Biosci 96:165-184

Tyson JJ, Chen KC, Novak B (2003) Sniffers, buzzers, toggles and blinkers: dynamics of regulatory and signaling pathways in the cell. Curr Opin Cell Biol 15:221-231

Tyers M, Tokiwa G, Futcher B (1993) Comparison of the *Saccharomyces cerevisiae* G1 cyclins: Cln3 may be an upstream activator of Cln1, Cln2 and other cyclins. EMBO J 12:1955-1968

Uetz P, Giot L, Cagney G, Mansfield TA, Judson RS, Knight JR, Lockshon D, Narayan V, Srinivasan M, Pochart P, Qureshi-Emili A, Li Y, Godwin B, Conover D, Kalbfleisch T, Vijayadamodar G, Yang M, Johnston M, Fields S, Rothberg JM (2000) A comprehensive analysis of protein-protein interactions in *Saccharomyces cerevisiae*. Nature 403:623-627

Valtz N, Peter M, Herskowitz I (1995) FAR1 is required for oriented polarization of yeast cells in response to mating pheromones. J Cell Biol 131:863-873

Vander Heiden MG, Plas DR, Rathmell JC, Fox CJ, Harris MH, Thompson CB (2001) Growth factors can influence cell growth and survival through effects on glucose metabolism. Mol Cell Biol 21:5899-5912

Vanoni M, Vai M, Popolo L, Alberghina L (1983) Structural heterogeneity in populations of the budding yeast *Saccharomyces cerevisiae*. J Bacteriol 156:1282-1291

Vermeulen K, Van Bockstaele DR, Berneman ZN (2003) The cell cycle: a review of regulation, deregulation and therapeutic targets in cancer. Cell Prolif 36:131-149

Wells WA (2002) Does size matter? J Cell Biol 158:1156-1159

Werner T (2004) Proteomics and regulomics: the yin and yang of functional genomics. Mass Spectrom Rev 23:25-33

Werner T, Fessele S, Maier H, Nelson PJ (2003) Computer modeling of promoter organization as a tool to study transcriptional coregulation. FASEB J 17:1228-1237

Westerhoff HV  Palsson BO (2004) The evolution of molecular biology into systems biology. Nat Biotechnol 22:1249-1252

Vidal M (2001) A biological atlas of functional maps. Cell 104:333-339

Wiley HS, Shvartsman SY, Lauffenburger DA (2003) Computational modeling of the EGF-receptor system: a paradigm for systems biology. Trends Cell Biol 13:43-50

Willett JD (2002) Genomics, proteomics: What's next? Pharmacogenomics 3:727-728

Woldringh CL, Huls PG, Vischer NO (1993) Volume growth of daughter and parent cells during the cell cycle of *Saccharomyces cerevisiae* a/alpha as determined by image cytometry. J Bacteriol 175:3174-3181

Yarmush ML, Banta S (2003) Metabolic engineering: advances in modeling and intervention in health and disease. Annu Rev Biomed Eng 5:349-381

Yi TM, Kitano H, Simon MI (2003) A quantitative characterization of the yeast heterotrimeric G protein cycle. Proc Natl Acad Sci USA 100:10764-10769

## Abbreviations

Cdk: cyclin dependent kinase
Cki: cyclin dependent kinase inhibitor
NADH: nicotinamide adenine dinucleotide (reduced)
FITC: fluoresceine isothiocyanate

Alberghina, Lilia
University of Milano – Bicocca, Department of Biotechnology and Biosciences, Piazza della Scienza 2, 20126 Milano, Italy
lilia.alberghina@unimib.it

Porro, Danilo
University of Milano – Bicocca, Department of Biotechnology and Biosciences, Piazza della Scienza 2, 20126 Milano, Italy

Rossi, Riccardo L.
University of Milano – Bicocca, Department of Biotechnology and Biosciences, Piazza della Scienza 2, 20126 Milano, Italy

Vanoni, Marco
University of Milano – Bicocca, Department of Biotechnology and Biosciences, Piazza della Scienza 2, 20126 Milano, Italy

# Systems biology of apoptosis


Martin Bentele and Roland Eils


## Abstract


New approaches are required for the mathematical modelling and system identification of complex signal transduction networks, which are characterized by a large number of unknown parameters and partially poorly understood mechanisms. Here, a new quantitative system identification method is described, which applies the novel concept of 'Sensitivity of Sensitivities' revealing two important system properties: high robustness and modular structures of the dependency between state variables and parameters. This is the key to reduce the system's dimensionality and to estimate unknown parameters on the basis of experimental data. The approach is applied to CD95-induced apoptosis, also called programmed cell death. Defects in the regulation of apoptosis result in a number of serious diseases such as cancer. With the estimated parameters, it becomes possible to reproduce the observed system behaviour and to predict important system properties. Thereby, a novel regulatory mechanism was revealed, i.e. a threshold between cell death and cell survival.


## 1 Systems biology: paradigm shift from reductionism to holism in biology? The whole is greater than the sum of its parts

It is hard to believe that only 40 years ago Watson and Crick discovered the structure of DNA. Their work was based on a number of studies dating back to the work of Oswald Avery, Erwin Chagraff, and others that discovered the four letters A,C,G,T of the DNA and Chagraff's rule that the amount of nucleotides followed a simple relationship. Since the discovery of the DNA biology was mostly oriented towards a reductionist approach. Triggered by modern high-throughput sequencing technologies the code of life was deciphered letter-by-letter, word-by-word. It is now well known that the early hope of the genomics age could not be met that knowledge of the genetic code would help to comprehensively understand complex biological processes from the function of genes to the pathological mechanisms of genetic diseases such as cancer. Thus, the mistrust in the worldwide sequencing projects by Chagraff, who ironically paved the way for the later discovery of the DNA, has been fulfilled: "Niemand hat uns je gelesen, niemand wird uns je lesen" (Nobody has ever read us, nobody will ever read us). After almost two decades of genome research the leitmotif of reductionism that life is just


Topics in Current Genetics, Vol. 13
L. Alberghina, H.V. Westerhoff (Eds.): Systems Biology
DOI 10.1007/b137746 / Published online: 12 May 2005
© Springer-Verlag Berlin Heidelberg 2005

chemistry and physics and that once we have found the smallest components of life we will be able to understand the whole is widely rejected.

The obvious limitations of the reductionist's approach to biology has led to a revival of holistic approaches in biology. The approach to biology using the tools of systems engineering is not new (Bertalanffy 1973). In the first half of the last century, Jan C. Smuts laid the foundation for a holistic approach in natural sciences that has even inspired philosophers in their reflection of nature (Smuts 1926). However, success of a holistic approach has only become possible in the postgenomic era since the reductionists have provided information on the basic building blocks of many important cellular processes in a quantitative form. Accordingly, it is probably not correct to name the transition from reductionism to holism a paradigm shift in biology. Before the components of life were deciphered a holistic approach to biology was condemned to be restricted to a mystic holism that was prevalent at the beginning of the 20th century.

With the help of the reductionism, systems biology has now become an analytically rigorous discipline that aims at a description of complex biological processes by mathematical models. Systems biology can, thus, be considered a new discipline in biology that puts the theoretical foundations of system level dissection of living matter into the context of modern high-throughput quantitative experimental data, mathematics and *in silico* simulations in order to gain a holistic view on the complex workings of life. In this chapter, we will describe a mathematical model of programmed cell death called apoptosis. Motivated by this fundamental process prevalent in almost all higher organisms we developed a variety of modern methods for systems biology that are required to tackle complex biochemical pathways involved in cellular signalling.

## 2 Modelling signal transduction networks

A better understanding of signal transduction networks is one of the most challenging areas in systems biology. Cells show information processing by the biochemical interaction between molecules. Signals of external stimuli are, for example, passed into the nucleus to regulate gene expression, resulting in proliferation, mitosis (nuclear division), changes in metabolism or cell death (Alberts et al. 2002). Interactions like phosphorylation, exchange of smaller molecules, binding or cleavage, are the fundamental mechanisms, which form the signal transduction networks. Complexity arises from the huge number of different molecules and interactions between them. In eucaryotic cells, for example, the steadily growing number of known signalling molecule species is in the order of magnitude of $10^4$-$10^5$.

Different methods of transcribing signal transduction networks into the language of mathematics exist. Dynamic pathway models are constructed using a diversity of mathematical and computational methods. Petri Nets (Reisig 1985) are well suited to describe the state transition process of distributed systems. In agent-based approaches and cellular automata (Wolfram 1994), macroscopic system properties emerge from the individual properties of the single entities, interacting with each other. Other methods originate from the analysis of biochemical systems ranging from the examination of steady states and flux modes to a large variety of control theories (Kell and Westerhoff 1986; Heinrich and Schuster 1996; Schilling et al. 1999). More recently, theoretical models for describing the signalling behaviour on system levels have been developed, using modular approaches (Kitano 2002; Csete and Doyle 2002; Lauffenburger 2000). Loosely speaking, simulation of signal transduction networks is either based on discrete models describing signalling as information processing, or on continuous models, where the information flux is modelled by a biochemical control system. In the latter approach, which goes back to the pioneering work of Garfinkel and Hess in the mid 60s (Garfinkel and Hess 1964; Garfinkel 1968), the reaction network is translated into a system of ordinary differential equations (Bhalla and Iyengar 1999; Mendes 1997). Today, there is a variety of sophisticated simulation methods to analyze complex biochemical reaction systems (Mendes 1993; Tomita et al. 1999; Sauro and Fell 1991).

# 3 CD95-induced apoptosis

Programmed Cell Death, called *apoptosis*, is the natural and controlled death of cells, in which the cell and its nucleus shrinks, condenses, and fragments (Alberts et al. 2002; Evan and Littlewood 1998). Apoptosis is one of the most complex signalling pathways and an essential property of higher organisms. Defects in apoptosis result in a number of serious diseases such as cancer, autoimmunity, and neurodegeneration (Krammer 2000; Peter and Krammer 2003). To develop efficient therapies, fundamental questions about molecular mechanisms and regulation of apoptosis remain to be answered. Apoptosis is triggered by a number of factors, including UV-light, γ-radiation, chemotherapeutic drugs, growth factor withdrawal ('death by neglect'), and signalling from the death receptors (Nagata 1999; Ashkenazi and Dixit 1999). Apoptosis pathways can generally be divided into signalling via the death receptors at the membrane (extrinsic pathway) or the mitochondria (intrinsic pathway). Both pathways imply *caspases* as effector molecules (Salvesen 2002). Caspases belong to the family of proteases, which are enzymes that degrade proteins by hydrolyzing some of their peptide bonds (Thornberry and Lazebnik 1998). Caspases mostly exist in their inactive proforms (procaspases) and become active after getting cleaved. Various caspases are involved in both the initiation of the apoptotic process and the execution of the final apoptotic program. CD95-induced apoptosis is one of the best-studied apoptosis pathways. A detailed overview on this mechanism is for example given in Krammer (2000) and Danial and Korsmeyer (2004).

## 3.1 The CD95-receptor and the DISC

CD95 is a member of the death receptor family, a subfamily of the TNF-receptor superfamily (Nagata 1997). Crosslinking of the CD95-receptor either with its natural ligand CD95L or with agonistic antibodies such as anti-APO-1 induces apoptosis in sensitive cells. Upon CD95 stimulation with CD95L or anti-APO-1, the Death-Inducing Signalling Complex (DISC) is formed. The DISC consists of oligomerized CD95, the death domain-containing adaptor molecule FADD, procaspase-8, procaspase-10 and c-FLIP. The interactions between the molecules in the DISC are based on homophilic contacts. The death domain (DD) of CD95 interacts with the DD of FADD, while FADD interacts with procaspase-8 via the so-called death effector domain. Once the DISC is formed, procaspase-8 is autocatalytically cleaved: two procaspase-8 molecules bound at the DISC form the intermediate product p43/p41, followed by generation of an active caspase-8 complex p18/p10 (Lavrik et al. 2003). This process can be inhibited by c-FLIP, which binds to the DISC in various ways and blocks the latter mechanism (Krueger et al. 2001).

## 3.2 The caspase cascade

After formation of active caspase-8, the apoptotic signalling cascade starts. Caspase-8 cleaves and activates caspase-3 and -7; caspase-3 itself activates caspase-6, which again activates caspase-8, thereby, establishing a self-amplifying activation loop. Caspase-3, -6, and -7 are involved in the execution of the death process, for example, the chromosomal degradation of DNA and, therefore, called *executioner caspases*, whereas the others, responsible for transferring the death signal, are referred to as *initiator caspases*. The DNA degradation plays an important role in the cell death process. It is started after ICAD gets cut off the CAD-ICAD complex by caspase-3 and -7, thereby, terminating the inhibition of CAD, which directly fragments the DNA (Nagata 1999). In parallel, PARP, a molecule, which repairs broken DNA strands is cleaved by executioner caspases as well. Once the DNA fragmentation process is triggered, a complete degradation of the cell starts irreversibly.

## 3.3 Type I versus type II cells and the regulation of apoptosis

Two different CD95-signaling pathways are established in different cell types (Schmitz et al. 1999; Scaffidi et al. 1998). Type I cells are characterized by intensive DISC formation and mitochondria independent caspase-3 activation. In Type II cells, the formation of the DISC complex is reduced and the activation of caspase-3 occurs downstream of the mitochondria: the active form of caspase-8 cleaves Bid, followed by translocation of the cleavage product tBid to mitochondria, which results in the release of cytochrome-C. Subsequently, apoptosome, a complex consisting of Apaf-1, cytochrome-C and caspase-9, is formed (Zou et al.

2002), leading to the activation of caspase-9, which then activates caspase-3, triggering the subsequent apoptotic events. Here, a feedback loop is established by caspase-2: it is activated by caspase-3 downstream of mitochondria and it cleaves Bid, which in response leads to mitochondrial cytochrome-C release.

Furthermore, CD95-induced signalling is influenced and regulated by many other molecules, which mostly inhibit or amplify the apoptotic process, like XIAP, IAP1/2 and survivin (inhibitors of caspase-3,-7,-9) (Salvesen and Duckett 2002) or the BCL-2 family (Chao and Korsmeyer 1998) consisting of pro-apoptotic (e.g. Bak, Bax) and anti-apoptotic (e.g. BCL-XL) members, regulating the critical cytochrome-C release. An overview about the molecule families, which play an important role in this pathway, is given in (Westphal and Kalthoff 2003).

## 4 Mathematical models of apoptosis

Despite the steadily increasing number of biological papers on apoptosis, mathematical models of this complex process are very scarce. In a first attempt to theoretically describe apoptotic signalling, a mathematical model including more than 20 reactions was proposed (Fussenegger et al. 2000). However, this model was based on *ad hoc* fixed parameters and, thus, its potential for understanding the regulation of apoptosis remains very limited. More recently, the caspase cascade was translated into a reductionist model and analytical mathematical methods were applied to evaluate the system behaviour within a wide range of parameters (Eissing et al. 2004). System identification methods like parameter estimation (Deuflhard 1983) based on reliably measured time series of data, as it has been successfully applied for chemical reaction system (Bock 1981), are suggested (Swameye et al. 2003). However, system identification is severely impaired by the high number of unknown parameters and the *curse of dimensionality*. Curse of dimensionality refers to the problem that the space of possible parameter value sets grows exponentially with the number of unknown parameters impairing the search for the globally most probable parameter values. The high number of unknown parameters is mainly due to the complexity of signal transduction networks and the absence of reliable quantitative information about the underlying mechanisms. Consequently, data-based studies are typically restricted to small models in which the biochemical interactions are well understood.

Despite the ever-increasing number of studies on CD95-induced apoptosis, a systemic understanding of this complex signalling pathway is still missing. For this reason, we reconstructed the network topology of CD95-induced apoptosis by critically searching databases (Schacherer et al. 2001) and the literature. Molecules and reactions directly or indirectly interacting with the main components of this pathway were incorporated, leading to a network topology with more than 60 molecules and about 100 interactions. This complexity cannot be matched by experimental data at present.

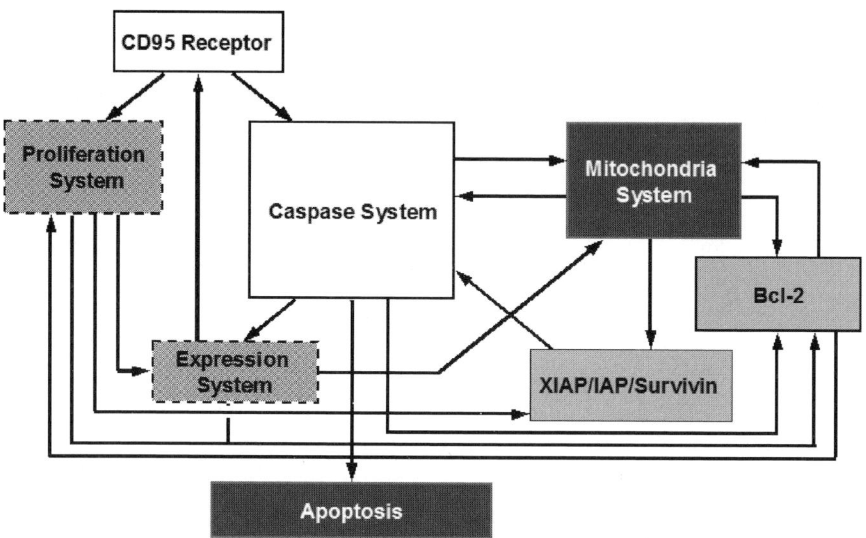



**Fig. 1.** Model of CD95-induced apoptosis combining subsystems of different information levels. The grey scale level of the boxes corresponds to the information quality (for details see text).

To tackle the high dimensionality of such systems, we developed an approach to large-scale modelling of signal transduction networks, which combines three methods (Bentele et al. 2004). Information on different levels of quality are combined in a unified form leading to the 'Structured Information Models'. Then, a global approach to parametric sensitivity analysis is introduced by which the dimensionality of the system can be significantly reduced. On this basis, a cluster-based and sensitivity-controlled parameter estimation method is set up.

## 5 Structured information models - The information problem

Information about signal transduction networks can be divided into different levels of information quality (Fig. 1). In most cases, interactions between two molecules are known on the semantic level only (e.g. A inhibits B or A activates B), thereby providing a network topology. For some well-investigated molecules and interactions the biochemical mechanism is also known (e.g. enzymatic process, formation of complexes) allowing quantitative modelling. However, information about the underlying biochemical parameters like reaction rate constants, Michaelis-Menten constants or dissociation rates, is mostly missing. Even if quantitative information is available, its usability is limited if it refers to different experimental settings, cell types or states of cells.

## 5.1 Network decomposition based on information quality

To reduce the complexity of the model without sacrificing essential components of the network, subsystems of different information qualities were identified and incorporated: subsystems mainly consisting of interactions with well-understood biochemical mechanisms are modelled as chemical reaction systems, whereas all others are modelled as 'black boxes', defined by their experimentally observed input-output behaviour. The subsystems are identified according to the following criteria:

- The input/output behaviour should be measurable.
- The number of input/output variables should be low.
- Subsystems should represent real functional systems (e.g. mitochondria).
- The information within one subsystem should be on the same level.

Notably, the black boxes do not assume knowledge of the exact underlying mechanisms. Instead, they reproduce the behaviour of the respective subsystems in a simplified way. Moreover, minimum sets of state variables and 'effective' parameters are introduced that do not necessarily correspond to molecule concentrations and biochemical parameters. As a consequence, the number of unknown parameters can be drastically reduced a priori.

Note that this concept is motivated by a typical situation in system identification of biochemical networks: due to missing information and limited experimental data, the models should be small and restricted to those network parts which are well understood. On the other hand, parts of biochemical networks can generally not be regarded independently of their environment. Thus, black boxes are introduced to reproduce the relevant effects of the surrounding network parts on a mechanistically well-understood subsystem, rather than for system identification of the surrounding network itself, for which the data basis would be missing anyway.

The degradation process of CD95-induced apoptosis is, for example, modelled as a decay function depending on a virtual state variable describing the 'apoptotic activity', which is influenced by executioner caspases. This function approximates the experimental observations, thereby, requiring a few parameters only.

The decomposition of the complete system into subsystems is an iterative and adaptive process. Based on new information, a subsystem might be split into further subsystems. A great advantage of the so-obtained 'Structured Information Models' is that it combines heterogeneous information in one model instead of dealing with isolated models.

## 5.2 Combined model definition

For the mathematical description of the mechanistic part of structured information models, interactions are modelled based on reaction rate equations. The state of a system is described by the concentration of l relevant signal transduction molecules ($x_1$, ..., $x_l$). The reaction rates depend on these concentrations and also on biochemical parameters ($\Phi_1$, ..., $\Phi_r$) like binding constants. To describe the temporal behaviour, a system of Ordinary Differential Equations is generated as linear combinations of the reaction rates $v_j$ :

$$dx_i\,/\,dt = \sum_{j=1}^{n} v_{ij} v_j(\vec{x}, \Phi_1, ..., \Phi_r),$$

where $v_{ij}$ denotes the stoichiometric matrix linking the reactions with the molecules affected. Given the initial concentrations, the time evolution of the reaction system can be propagated using an ODE Solver (Deuflhard and Bornemann 2002). Note that the initial concentrations $x_i(t=0)$ are often unknown and are considered unknown parameters as well.

Black boxes are defined by their experimentally observed input-output behaviour. Additional state variables $(x_{l+1}, ..., x_m)$ that do not necessarily correspond to molecule concentrations and additional parameters $(\Phi_{r+1}, ..., \Phi_s)$ can be introduced. The q-th black box is represented by the functions $f_i^q(x_1, ..., x_m, \Phi_1, ..., \Phi_s, t)$, $i=(1,...,m)$ that describe the changes of molecule concentrations and other state variables it affects. Boundary conditions like conservation laws have to be taken into account. Thus, the combination of subsystems modelled as chemical reaction systems and black boxes leads to an ODE system, which reads

$$dx_i\,/\,dt = \sum_{j=1}^{n} v_{ij} v_j(\vec{x}, \Phi_1, ..., \Phi_r) + \sum_{q=1}^{N} f_i^q(\vec{x}, \Phi_{r+1}, ..., \Phi_s) \ \text{ for } i=(1,...l),$$

$$dx_i\,/\,dt = \sum_{q=1}^{N} f_i^q(\vec{x}, \Phi_{r+1}, ..., \Phi_s) \ \text{ for } i=(l+1,...m),$$

where n denotes the number of reactions and N the number of black boxes.


## 5.3 The model of CD95-induced apoptosis

A quantitative model of CD95-induced apoptosis was derived from the network topology, thereby, taking into account the information on all underlying mechanisms (Fig. 2). A detailed reaction mechanism was established for the DISC- and the caspase-system. The mechanisms at the DISC are largely described by elementary reactions, whereas the caspase cleavage process is considered an enzymatic process (e.g. Stennicke and Salvesen 1999). In principle, these interactions could have been modelled in a more simplified way. The influence of caspase-3 on Bid, for example, could have been modelled directly thereby using 'effective' parameters without accounting for the intermediate caspase-2 cleavage. However, since time series about the concentration of caspase-2 have been available, this molecule was kept in the system in order to gain more information for system identification. In contrast, many molecules and interactions with equivalent properties were replaced by 'effective' molecules and interactions based on the analysis of parameter sensitivity correlations. The molecules XIAP, IAP1/2, survivin and their interactions with caspase-3,-7, and -9 are for example reduced to one 'effective' molecule called IAP and the interactions are described by effective binding parameters. Details about the model are given in Bentele et al. (2004).

## 5.4 Black boxes

Two subsystems with a significant influence on the signalling system were identified. The death process is described by the degradation of all molecules. It is modelled as an exponential decay-function dependent on the executioner caspase activity. The cytochrome-C release of the mitochondria is based on experimental observations (Goldstein et al. 2000), which describe a complete release within 5 minutes as soon as Bid reaches a certain level in comparison to Bcl-2/Bcl-XL.

To model the degradation process, a virtual state variable called $x_{apop}$ was introduced, which quantifies the 'apoptotic activity', by which the strength of the final death process is characterized. It is assumed that the velocity of cell degradation is directly influenced by this activity. It is also assumed that the activity itself is caused by active caspase-3, -6, and -7 and that the increase of the activity runs in parallel to the experimentally observable PARP cleavage. Thus, $x_{apop}$ represents the activity of the apoptotic processes triggered by executioner caspases. The degradation process is modelled by a decay function depending on $x_{apop}$, thus, decreasing the concentration of all molecules of the pathway.

## 5.5 Experimental data

A set of experiments to measure time series of concentrations of 14 different molecules and complexes (see framed molecules in Fig. 2) after activation of CD95-receptors was designed. Cells were stimulated with different concentrations of agonistic anti-APO-1 antibody, also referred to as ligands in the following, for various periods of time (from 5 minutes to 4 days). Each sample was evaluated by three independent approaches. See Bentele et al. (2004) for experimental details. In a first set of experiments, time series were measured for a 'fast' activation scenario with an oversaturated ligand concentration corresponding to more than one ligand per CD95-receptor. To gain additional information about the system's dynamic, several experiments with much lower ligand concentrations were performed resulting in a slower activation of apoptosis.

# 6 Model reduction by sensitivity analysis

The above-described model consists of 41 molecules and molecule complexes, 32 reactions, and 2 black boxes. It contains more than 50 missing parameters. Therefore, it is still too complex for reliable system identification and requires further reduction of complexity considering the limited number of data points. Here, we developed an approach for model reduction by sensitivity analysis.

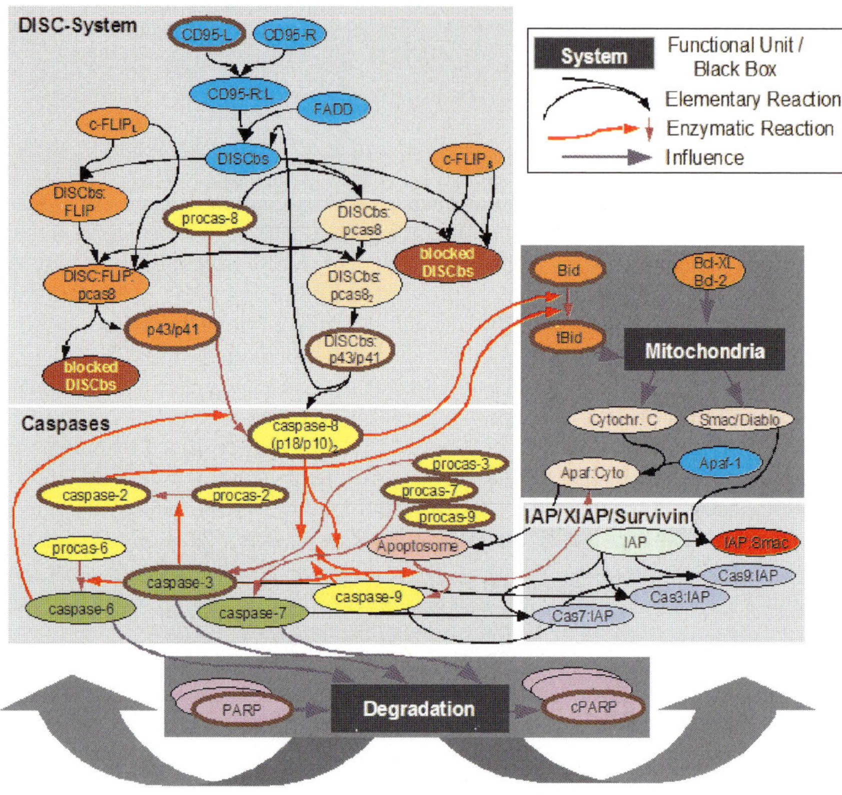


**Fig. 2.** Structured information model of CD95-induced apoptosis. In the mechanistic part (DISC, caspases, IAP), interactions are modelled as elementary reactions including competitive inhibitions and enzymatic reactions. Receptors are activated by ligands initiating the DISC formation. After binding to the DISC binding site (DISCbs), procaspase-8 is cleaved (initiator caspase), followed by the activation of executioner caspases (3, 6, 7). PARP cleavage was chosen as experimental end-point of the pathway. The mitochondria and the degradation process, which influences all molecules, are modelled as black boxes defined by their input-output behaviour. Experimental time series were measured for the molecules framed in red.

## 6.1 The sensitivity matrix

Parametric sensitivity analysis determines the changes of the system behaviour as a result of parameter variations (Varma et al. 1999). In a system with m state variables $(x_1, \ldots, x_m)$ and n parameters $(\Phi_1, \ldots, \Phi_n)$, the relative sensitivities $s_{ij} = ( \partial x_i / x_i ) / ( \partial \Phi_j / \Phi_j )$ describe the relative changes of the state variables as a result of changes of the parameters. In the signal transduction systems, the state variables mostly correspond to molecule concentrations. Note that the sensitivities are time-dependent $(s_{ij} = s_{ij}(t))$ and that the time points, for which sensitivities are

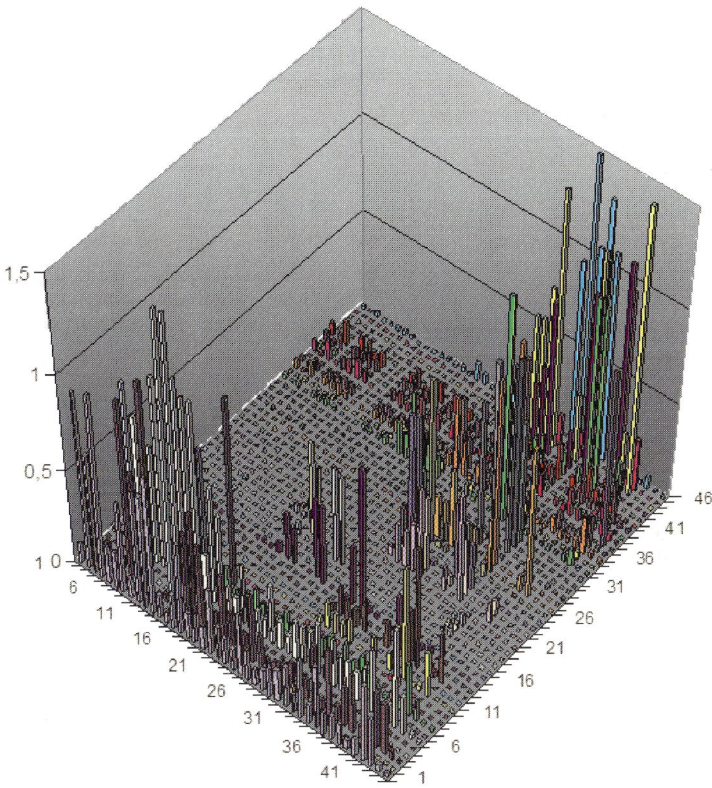

**Fig. 3.** Sensitivity Matrix. The sensitivity matrix elements $\hat{s}_{ij}$ show the relative changes of each state variable i (left to right), mostly referring to molecule concentrations, with respect to relative changes of each parameter j (front to back). The indices refer to the model definition of (Bentele et al. 2004).

computed, have to be chosen carefully. In *Metabolic Control Analysis* (Kell and Westerhoff 1986; Fell 1992) steady states depending on parameter variations are investigated, whereas in signal transduction systems, the transient behaviour is of high interest to analyse the regulation of a system. As a consequence, the complete time period, during which, for example, a signalling pathway is active and exhibits a dynamical behaviour, is relevant rather than a distinct time point:

$$\hat{s}_{ij} = \frac{1}{\Delta t} \int_{t_0}^{t_0 + \Delta t} s_{ij}(t)dt \, .$$

The time point $t_0$ corresponds to the start of the investigated scenario and $\Delta t$ to the period, in which the system shows reactions to parameter variations. For the apoptosis system, $t_0$ is the time point at which the pathway becomes activated by CD95-ligands and the time interval ends when the cell is completely degraded.

A sensitivity matrix with elements $\{ \hat{s}_{ij} \}$ is visualized in Fig. 3, showing two important facts:

- Sensitivities are low in general indicating high robustness and a lower 'effective' dimensionality of the parameter space since many parameters have only little impact on most molecules.
- Apparently, clusters can be identified that contain a subset of molecules, whose concentrations depend on a subset of parameters only. This inherent system property is an important feature for further modularization.

## 6.2 Local versus global sensitivity analysis

Usually, sensitivities of huge models can be computed numerically only and a general relation between sensitivities $\hat{s}_{ij}(\vec{\Phi})$ and the parameter set $(\Phi_1, ..., \Phi_n)$, at which the sensitivities are determined, cannot be deduced analytically. Instead, sensitivities are determined for specific points in parameter space only and are, therefore, called *local sensitivities*. In this study, however, sensitivity analysis is used as an essential tool for model reduction, which is required for system identification. As a consequence, it has to be performed in a virtual experiment prior to determination of parameters (see stochastic approach to sensitivity analysis below).

*Global sensitivity analysis*, which provides information about sensitivities for the complete space of possible parameter values, is impaired by the high dimensionality of the parameter space. Although this situation is ubiquitous for complex biological systems, a general solution to this problem does not exist.

## 6.3 Stochastic approach to global sensitivity analysis

In a virtual experiment, sensitivity analysis is performed for a large number of randomly chosen points in parameter space within specified ranges. The ranges are defined for each parameter type (e.g. bimolecular reaction rate constants, initial concentrations, Michaelis-Menten constants, etc.), unless more precise information was available. Thereby, the concept of 'Sensitivity of Sensitivities' is introduced, which examines the robustness of sensitivities with respect to parameter variations. This is motivated by two important facts:

- Biological systems often keep their system properties constant, although they are subject to high parameter fluctuations (Barkai and S 1997; Meir et al. 2002; Alon et al. 1999) suggesting that at least some sensitivities are insensitive with respect to parameter fluctuations (low Sensitivity of Sensitivities).
- Structure and connectivity of biochemical networks suggests that the influence of some parameters on distant network regions is limited and that the corresponding sensitivities are extremely low - independently of the parameter values.

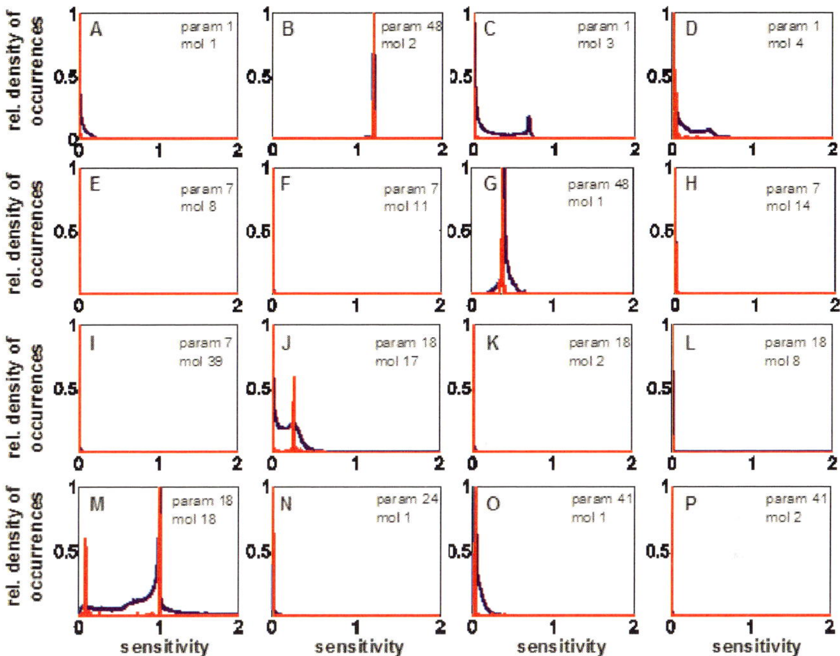

**Fig. 4.** Sensitivity Histograms: Sensitivity of Sensitivities. Each box shows a histogram for a specific sensitivity $\hat{s}_{ij}$, computed for a large number of randomly chosen points in parameter space. Parameter and molecule indices refer to the model definition of Bentele et al. (2004). The X-axis represents the relative sensitivity values from 0 to 2 and the Y-axis corresponds to the density of occurrences. The blue plot shows the uniformly weighted distribution of sensitivities, whereas for the red one, each contribution was weighted with the Boltzmann factor, resulting in sharper and sometimes slightly shifted peaks. The histograms are exemplary for all matrix elements $\{\hat{s}_{ij}\}$. Typically, histograms show clear peaks close to zero - an important property for further modularization. However, distributions like C, D, J, or M are not informative for modularization.

The distribution of the computed sensitivities $\hat{s}_{ij}(\vec{\Phi})$ for a large number of different points in the parameter space, $\{\vec{\Phi}^q\}$, are plotted in form of histograms for each sensitivity to show their distribution (Fig. 4). The histograms are generated in two different ways. In a first approach, all random parameter sets are equally weighted, independent of how much the resulting systems dynamics deviate from the real system. As an extension, information from experimental data was incorporated by introducing a Boltzmann factor: based on experimental time series of molecule concentrations, an objective function $E_q$ was calculated for each parameter set $\vec{\Phi}^q$ based on the differences between the experimental and simulation data:

$$E_q = \sum_{i,k} \frac{(x_{ik}^{\exp} - x_i^{\mathrm{model}}(t_k, \vec{\Phi}^q))^2}{\sigma_{ik}^2} \, ,$$

$x_{ik}^{\exp}$ : experimental values of the concentration of molecule i at time $t_k$,

$x_i^{\mathrm{model}}(t_k, \vec{\Phi}^q)$ : simulated values of the concentration at time tk for parameter set $\vec{\Phi}^q$,

$\sigma_{ik}$ : standard deviation according to experimental data,

$\{t_k\}$: time points of experimental data.

Then, it is assumed that the probability $p_q$ that a system with parameter set $\vec{\Phi}^q$ produces the experimental output $\{x_{ik}^{\exp}\}$ follows a Boltzmann distribution with the objective function as energy term:

$$p_q \propto \exp(-E_q / kT)$$

The assumption is motivated as follows: considering Gaussian random measurement errors, each characterized by $\sigma_{ij}$, the probability $p_q$ can be written as product (Gershenfeld 1999)

$$p_q \propto \prod_{i,k} \exp-\frac{1}{2} \frac{(x_{ik}^{\exp} - x_i^{\mathrm{model}}(t_k, \vec{\Phi}^q))^2}{\sigma_{ik}^2} \, ,$$

which is equivalent to the upper Boltzmann distribution using the definition of $E_q$. For generation of sensitivity histograms, the Boltzmann factor was used as weighting factor: instead of counting the number of parameter sets with sensitivities within a certain sensitivity interval, the contribution of each parameter set $\vec{\Phi}^q$ is given by $\exp(-E_q / kT)$. Thereby, the statistical impact of sensitivities for parameter sets that are more consistent with the experimental observations are amplified.

Additional information can be gained by varying the factor $kT$. If a sensitivity value is insensitive with respect to parameter variations, particularly within the subspace of possible solutions (areas in parameter space with a low objective function), 'cooling down' the system by decreasing $kT$ will result in histograms with a much sharper peak. Sensitivity histograms showing more than one distinct peak when kT is decreased indicate that the respective sensitivity strongly depends on the exact parameter set within the parameter subspace of probable solutions.

## 6.4 Sensitivity of sensitivities

The most crucial outcome of the 'global' sensitivity analysis approach presented here is the fact that sensitivities of sensitivities are extremely low in most cases, as shown in Fig. 4 for some exemplary sensitivity histograms. Obviously, most distributions show distinct and narrow peaks, indicating high robustness towards large variations of the parameter values. Whenever a sensitivity value $\hat{s}_{ij}$ is close to zero, an extremely sharp peak indicates that the state variable $x_i$ is likely not to be influenced by parameter $\Phi_j$, regardless of the exact parameter value set. By introduction of the Boltzmann weighting (Fig. 4, red line), most peaks become even sharper. Only in few cases, more than one peak remains or the distribution broadens as a consequence of the 'cooling', indicating that the system runs in different modes for certain areas of the parameters space.

As a consequence of the low sensitivity of sensitivities, subsets of parameters can be determined, which are unlikely to influence certain state variables. Considering the high number of sensitivities with $\hat{s}_{ij} \approx 0$, this step is crucial to reduce the system's dimensionality even without knowledge of the true parameter values. Thus, this method provides a basis for high-dimensional parameter estimation.

# 7 Sensitivity-controlled parameter estimation

Parameters are estimated based on experimental time series of measurable molecule concentrations using the maximum likelihood estimation (Gershenfeld 1999), which leads to the least-square problem:

$$E(\vec{\Phi}) = \sum_{i,k} \frac{(x_{ik}^{exp} - x_i^{model}(t_k, \vec{\Phi}))^2}{\sigma_{ik}^2} \rightarrow \min.$$

Thus, the objective function E, defined as the sum of squares of differences between experimentally measured and simulated molecule concentrations, divided by the standard deviation in order to lower the impact of experimentally less reliable values, has to be minimized. A review about methods commonly applied for biochemical reaction systems is given in Mendes and Kell (1998).

## 7.1 Cluster-based parameter estimation

This approach takes advantage of the fact that clusters of state variables and parameters can be identified in such a way as to have subsets of state variables whose temporal behaviour depends on a subset of parameters only. Considering the sensitivity matrix $(\hat{s}_{ij})$ of m state variables, n parameters, and an average sensitivity $\langle s \rangle$, it is assumed that sensitivities fulfilling $\hat{s}_{ij} < \Theta$ , $\Theta = \Theta_s \cdot \langle s \rangle$, where $\Theta_s$ is a low relative threshold (e.g. $\theta_s = 0.01$), indicate that the respective state variable and be considered to be independent of the respective parameter. Whenever a high percentage of sensitivities fulfil this property – and this is the typical case in large signal transduction systems – clusters of the remaining above-

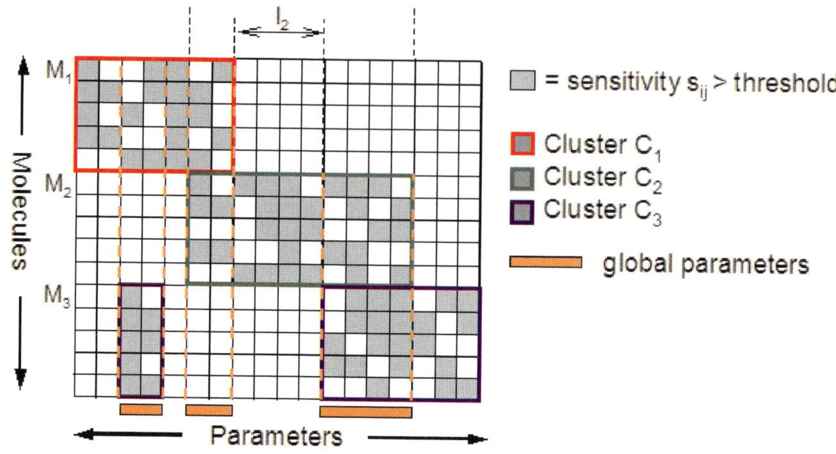



**Fig. 5.** The sensitivity matrix exemplifies the clustered dependence of state variables (vertical axis) and parameters (horizontal axis). Red bars indicate the global parameters on which state variables of more than one cluster are dependent.

threshold sensitivities can be established after reordering the rows of the sensitivity matrix. A cluster $C_q$ is defined as the sub-matrix of the sensitivity matrix (see Fig. 5):

$$C_q = (\hat{s}_{ij}), \ i \in \{M_q, \ldots, M_{q+1} - 1\},$$

$$j \in \{1, \ldots, n \mid \hat{s}_{ij} \geq \Theta \ \text{for at least one} \ i \in [M_q, \ldots, M_{q+1} - 1] \},$$

$M_q$ = index of first molecule of cluster q.

Let $C_q$ denote the q-th cluster and $l_q$ the number of 'local parameters' of $C_q$. Local parameters are defined as those parameters whose sensitivities are below threshold $\Theta$ for all state variables outside the cluster they belong to. If all parameters of one cluster influence state variables of the same cluster only, they can be estimated independently from all other clusters leading to a parameter estimation of much lower dimensionality. In general, this is, however, not the case and a cluster-wise estimation would determine the same parameters in the context of different clusters leading to inconsistencies. Therefore, parameters are split into local and global ones: a parameter, on which state variables of more than one cluster depend on, is called a 'global parameter'. Consequently, the total number of global parameters is given by $g = n - \sum l_q$.

Then, a hierarchical approach was designed, in which parameter estimation is performed on two levels. On the upper level, global parameters are estimated by optimising all clusters: for each cluster, parameter estimation is recursively called at the lower level. Thereby, the associated objective function is based on the estimated parameters of the single clusters (lower level). Thus, on the lower level, all remaining (local) parameters are estimated separately for each cluster, depending on the values of the global parameters proposed by the algorithm of the upper level, but independent of the parameters of all other clusters. This approach significantly reduces the dimensionality from the number of all parameters n to the sum of the number of global parameters g and the maximum number of local parameters $(g+max(l_q))$. If W(d) denotes the cost for a parameter estimation of dimension d, the total cost $W_{tot}$ of the algorithms can be compared by

Unclustered algorithm: $W_{tot} \sim W(n)$,

Clustered algorithm: $W_{tot} \sim W(g) \bullet \sum W(l_q)$.

Since W(d) strongly increases with dimension d - also known as curse of dimensionality, some methods even show an exponential relationship (Mendes and Kell 1998) - this relation reflects a drastic reduction of computational cost whenever the relative number of global parameters g/n is low, which is typically the case in signal transduction systems. To optimize the computation time for very large systems, it would be adequate to choose the clusters so that $g + max(l_q)$ is minimized.

Since the clustering is applied as a basis for efficient parameter estimation and thus before parameter values are given, there is no basis to compute a sensitivity matrix, which is, however, required for the clustering. A solution to this problem is given by the 'global' sensitivity analysis approach. Since the clustering method only requires information about which sensitivities are below threshold $\Theta$, the sensitivity histograms are evaluated. The clustering is then based on neglecting those sensitivities, whose histograms indicate that the property $\hat{s}_{ij} < \Theta$ is likely to be fulfilled within the complete space of possible parameters.


## 7.2 Parameter estimation algorithm

In order to further reduce the dimensionality of the parameter estimation problem sensitivities, which are below the threshold only locally in parameter space, are taken into account by integrating a sensitivity-control in the parameter estimation algorithm. As a robust algorithm for nonlinear least-square problems, the Levenberg-Marquardt method (Gershenfeld 1999) was chosen and combined with a multi-start algorithm. The ranges for the randomly initiated parameter values correspond to those used for sensitivity analysis. For adaptive sensitivity-control, local sensitivity analysis is performed after each iteration step. All those parameters, whose impact on the objective function is significantly below average are kept constant for the next step. This prevents the algorithm from being misguided by 'irrelevant' parameters. Since the distinction between relevant and irrelevant parameters is generally valid for a specific set of parameter values only, the sensitivities are recalculated after each iteration step.

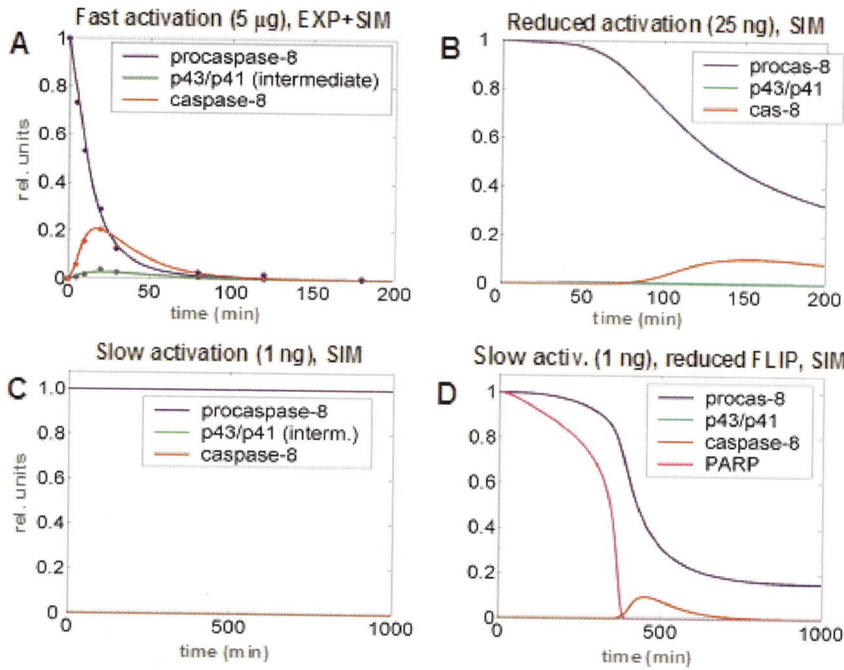

**Fig. 6.** A: Fast activation scenario: Parameter estimation led to a good fit between model simulations (solid lines) and experimental data (dots). Procaspase-8 is cleaved and active caspase-8 is formed resulting in the activation of executioner caspases, followed by cell death (data not shown). B: The simulation for much lower ligand concentration shows a delay, which is followed by a slow activation of caspase-8, also triggering the death process. C: For a below-threshold ligand concentration, the death process is completely stopped. According to simulation, active caspase-8 cannot be generated in a number sufficiently high to trigger apoptosis. D: Same activation scenario as in C with c-FLIP reduced by 75%. The simulation shows a slow and steady cleavage of procaspase-8, until executioner caspases are generated in a number sufficiently high to trigger the apoptotic process. Thus, the threshold can be shifted by varying the c-FLIP concentration, indicating the crucial role of c-FLIP in CD95-induced apoptosis.

# 8 Model simulation of apoptosis and experimental results

## 8.1 Parameter estimation based on multiple scenarios

In a first set of experiments, time series were measured for a 'fast' activation scenario achieved by 5 μg/ml anti-APO-1 corresponding to a ligand-receptor ratio of about 5:1. A good fit between model simulation and experimental data could be achieved reproducing the cleavage of procaspase-8 into its active form (Fig. 6A), followed by activation of the executioner caspases and cleavage of Bid and PARP.

To gain additional information about the system, activation scenarios with lower initial ligand concentrations were measured and the parameter estimation was based on these multiple conditions. It could be shown that the resulting parameters were capable of reproducing several activation scenarios (data not shown). To test the predictability of the model, simulations were then performed for scenarios with even lower initial ligand concentrations.

## 8.2 Threshold mechanism for CD95-induced apoptosis

The model predictions show that with decreasing ligand concentrations apoptosis is slowed down considerably, but cell death is still achieved (Fig. 6B). However, below a critical concentration corresponding to a ligand-receptor ratio of approx. $1:10^2$, apoptosis is completely stopped (Fig. 6C). This prediction was validated by experiments (Bentele et al. 2004).

As a next step, the model was used to reveal the underlying mechanism of the threshold. Even for a below-threshold scenario, the number of ligands per receptor is sufficient to cleave procaspase-8, thereby, triggering all subsequent caspases. In the model, the caspase-8 cleavage capacity at the DISC is assumed to be proportional to the number of active CD95 receptors since the DISCs are supposed to remain active after cleaving procaspase-8 molecules. Consequently, it could be assumed that even for a very low ligand concentration apoptosis should not be stopped entirely, but would only be slowed down. Instead, the model simulations show that c-FLIP, which binds to the DISC and which competes with activation of caspase-8 (Krueger et al. 2001), is responsible for completely stopping the death process. According to the parameter estimation, there are many more CD95 receptors and procaspase-8 molecules than c-FLIP molecules. The cleavage rate of procaspase-8 is assumed to be dependent on the number of active receptors. Whenever c-FLIP binds to a DISC, the respective binding site is blocked. Therefore, the simulated scenario with subthreshold concentrations of activating ligands shows a steady decrease of active DISCs until all of them are blocked by c-FLIP. As a consequence, the simulation shows no significant generation of active caspase-8 as a result of the early and complete DISC-blockage. Thus, the c-FLIP mechanism identified in the model can be considered a switch, which blocks the activation of caspase-8 for signals (ligand concentrations) below a critical quantity and passes on the activation signal above this level. As a result, it was predicted that the threshold mechanism is highly sensitive to the concentration of c-FLIP (Fig. 6D). This prediction was also confirmed by experiments (Bentele et al. 2004).

## 8.3 Delay of apoptosis and point of no return

Another important model prediction addresses the system behaviour above the threshold, where the combination of the c-FLIP mechanism with the amplification loop does not lead to a steadily decreased caspase cleavage rate upon a decreased ligand concentration. Instead, the simulation shows that the caspase cleavage, the amplification loop and the subsequent death process are delayed but not entirely stopped. As shown in Fig. 6B, there are no observable system changes for up to many hours after activation of the pathway. Then, the death process suddenly starts without any external stimulation of the system. This is due to an extremely slow increase of active caspase concentration, which reaches a critical level upon which the death process is triggered. Thus, for up to many hours, there maybe no phenotypic difference observed for cells, which are not (or insufficiently) stimulated by ligands. However, the death process is irreversibly triggered even for those cells and cannot be stopped anymore (point of no return). The predicted delays have also been verified by experiments.

# 9 Outlook

The investigation and analysis of complex biological networks and mechanisms in cells is probably one of the most challenging and fastest growing fields of science. The classic experimental approaches, which mostly focus on the investigation of molecule interactions in an isolated context and under specific experimental settings, cannot keep up with the steadily increasing number of potential interaction partners and the diversity and complexity of the real networks. The potential of entirely experimental approaches towards revealing network functionalities is, therefore, limited.

From a theoretical point of view, system identification of networks, characterized by an enormous complexity and a lack of information about the underlying mechanisms on both the qualitative and quantitative level, constitutes a new class of problems, which has not been sufficiently approached yet. Methods to describe complex networks in a qualitative way, for example, as scale free networks (Barabasi and Albert 1999), are promising for revealing general principles like robustness and fragility (Albert et al. 2000; Stelling et al. 2004). Their ability to describe the real system behaviour is, however, limited since single interactions are not quantitatively considered although the properties of real biological networks are often related to very concrete features of single mechanisms.

The emerging field of systems biology, which has been recently started with great enthusiasm, is an important step towards the investigation of biological processes on systems level. However, it has not been proved yet that it also provides qualitatively new methods addressing the new dimension of complexity. Instead, approaches from the field of engineering and numerical mathematics have been widely applied to biological systems, especially to subsystems, where they are expected to be well-suited and promising. Moreover, the majority of studies address specific applications with the goal of answering specific biological questions. Although this is a first important step to establish systematic and theoretical methods in the field of cell biology, it has to be followed by a second, much more challenging step concerning the development of new theoretical approaches for the description and system identification of complex and highly underdetermined biological systems.

It should be noted that numerical parameter identification methods originating from disciplines, where experimental data are generated in huge amounts and with high precision and where the number of unknown parameters is low, cannot be expected to be appropriate for the completely different situation in cell biology. Here, uncertain models and a large number of unknown parameters are facing a low number of experimental data points with high measurement errors. Current numerical methods are based on the assumption that the best parameter fit corresponds to the best solution. For biological systems, a huge solution space is typically obtained and a single parameter fit can be thus considered rather meaningless. One could argue that more experimental data should be generated and more quantitative information concerning the single mechanisms should be obtained to match the high dimensionality of parameter space. Considering the fact that quantitatively reliable *in vivo* measurements like time series of concentrations are still difficult to obtain on a large scale, this demand for a large amount of quantitative data cannot be met in the near future.

It would be fatal though not to attempt extracting information from available quantitative experimental data in the context of existing qualitative knowledge about, for example, network topologies. Therefore, alternative ways disregarding the restrictions of current numerical methods and by-passing the requirement of finding one distinct parameter set as suggested here have to be found. The methods presented here are well suited for theoretically tackling biological systems, as it is well-accepted that biological systems mostly keep their system properties constant although the real parameters are also subject to high variations. Thus, intrinsic biological properties like robustness and the fact that the function of biological systems does not require fine-tuned parameters indicate that the extraction of information is feasible even without exact knowledge of the true parameters. This is a new principle, which was approached here by evaluating randomly chosen parameter sets and by the generation of ensembles of estimated parameter fits based on randomly chosen initial values. The resulting histograms of parameter sensitivities provide a multitude of information, in particular by incorporation of a Boltzmann factor based on experimental data.

The currently developing methods such provided here or in the study of Brown and Sethna (2003) mark the beginning of a new methodology to investigate highly underdetermined systems. Such approaches have to be further refined, for example, by the 'High Dimensional Model Representation' (Li et al. 2001) for a systematic description of an 'effective' parameter space. Furthermore, it has to be extended by the concept of also considering alternative model choices instead of different parameter sets only, leading to model discrimination, which is currently examined by us (Vacheva et al. 2005). For a better understanding of the regulation of programmed cell death, the established loop between modelling, theoretical predictions and experiments has already proven to be highly efficient and has raised a lot of new detailed questions, for example, concerning the influence of spatial aspects on the network function or more detailed investigation of certain key regulatory mechanisms. The modular and hierarchical structure of the presented modelling framework provides a high degree of flexibility for future model extensions in various ways, either by adding additional pathways and systems like proliferation or gene expression, or by adding more detailed biochemical mechanisms with more information becoming available. Thus, our methods will be well-suited for tackling complex and highly underdetermined networks going far beyond the field of programmed cell death.

# References


Albert R, Jeong H, Barabasi A (2000) Error and attack tolerance of complex networks. Nature 406:378-382

Alberts B, Bray D, Lewis J, Raff M, Roberts K, Watson JD (2002) Molecular Biology of the Cell. Garland, New York

Alon U, Surette MG, Barkai N, Leibler S (1999) Robustness in bacterial chemotaxis. Nature 397:168-171

Ashkenazi A, Dixit V (1999) Apoptosis control by death and decoy receptors. Curr Opin Cell Biol 11:255-260

Barabasi A, Albert R (1999) Emergence of scaling in random networks. Science 286:509-512

Barkai N, Leibler S (1997) Robustness in simple biochemical networks. Nature 387:913-917

Bentele M, Lavrik I, Ulrich M, Stößer S, Heermann D, Kalthoff H, Krammer P, Eils R (2004) Mathematical modeling reveals threshold mechanism in CD95-induced apoptosis. J Cell Biol 166:839-851

Bertalanffy L (1973) General System Theory. Penguin Books, Harmondsworth.

Bhalla US, Iyengar R (1999) Emergent properties of networks of biological signaling pathways. Science 283:381-387

Bock H (1981) Numerical treatment of inverse problems in chemical reaction kinetics. Modelling of Chemical Reaction Systems. K Ebert, P DeuflhardW Jäger. New York, Springer. 8:102-125

Brown K, Sethna J (2003) Statistical mechanical approaches to models with many poorly known parameters. Phys Rev E 68:21904:1-9

Chao S, Korsmeyer S (1998) BCL-2 family: Regulators of cell death. Annu Rev Immunol 16:395-419

Csete ME, Doyle JC (2002) Reverse engineering of biological complexity. Science 295:1664-1669

Danial N, Korsmeyer S (2004) Cell death: Critical control points. Cell 116:205-219

Deuflhard P (1983) Numerical treatment of inverse problems in differential and integral equations. Birkhäuser, Basel

Deuflhard P, Bornemann F (2002) Scientific Computing with Ordinary Differential Equations. Applied Mathematics. New York, Springer

Eissing T, Conzelmann H, Gilles E, Allgower F, Bullinger E, Scheurich P (2004) Bistability analyses of a caspase activation model for receptor-induced apoptosis. J Biol Chem 279:36892-36897

Evan G, Littlewood T (1998) A matter of life and cell death. Science 281:1317-1322

Fell D (1992) Metabolic control analysis: a survey of its theoretical and experimental development. Biochem J 286:313-330

Fussenegger M, Bailey J, Varner J (2000) A mathematical model of caspase function in apoptosis. Nature Biotech 18:768-774

Garfinkel D (1968) The role of computer simulation in biochemistry. Comput Biomed Res 2:i-ii

Garfinkel D, Hess B (1964) Metabolic control mechanisms. J Biol Chem 239:971-983

Gershenfeld N (1999) The Nature of Mathematical Modeling. Cambridge University Press, Cambridge, UK

Goldstein JC, Waterhouse N, Juin P, Evan G, Green D (2000) The coordinate release of cytochrome c during apoptosis is rapid, complete and kinetically invariant. Nature Cell Biol 2:156-162

Heinrich R, Schuster S (1996) The regulation of cellular systems. Chapman & Hall, New York

Kell DB, Westerhoff HV (1986) Metabolic control theory: its role in microbiology and biotechnology. FEMS Microbiol Rev 39:305-320

Kitano H (2002) Systems biology: a brief overview. Science 295:1662-1664

Krammer P (2000) CD95's deadly mission in the immune system. Nature 407:789-795

Krueger A, Baumann S, Krammer P, Kirchhoff S (2001) FLICE-Inhibitory proteins: Regulators of death receptor-mediated apoptosis. Mol Cell Biol 21(24):8247–8254

Lauffenburger DA (2000) Cell signaling pathways as control modules: complexity for simplicity? PNAS 97(10):5031–5033

Lavrik I, Krueger A, Schmitz I, Baumann S, Weyd H, Krammer P, Kirchhoff S (2003) The active caspase-8 heterotetramer is formed at the CD95 DISC. Cell Death Differ 10:144-145

Li G, Rosenthal C, Rabitz H (2001) High dimensional model representations. J Phys Chem A 105:7765-7777

Meir E, von Dassow G, Munro E, Odell G (2002) Robustness, flexibility, and the role of lateral inhibition in the neurogenic network. Curr Biol 12:778-786

Mendes P (1993) GEPASI: A software package for modelling the dynamics, steady states and control of biochemical and other systems. Comput Applic Biosci 9:563-571

Mendes P (1997) Biochemistry by numbers: simulation of biochemical pathways with Gepasi 3. Trends Biochem Sciences 22(9):361-363

Mendes P, Kell D (1998) Non-linear optimization of biochemical pathways: applications to metabolic engineering and parameter estimation. Bioinformatics 14(10):869-883

Nagata S (1997) Apoptosis by death factor. Cell 88:355-365

Nagata S (1999) Fas ligand-induced apoptosis. Annu Rev Genet 33:29-55

Peter M, Krammer P (2003) The CD95(APO-1/Fas) DISC and beyond. Cell Death Differ 10:26-35

Reisig W (1985) Petri Nets, An Introduction. Springer_Verlag, Berlin

Salvesen GS (2002) Caspases: opening the boxes and interpreting the arrows. Cell Death Differ 9(1):3-5

Salvesen GS, Duckett CS (2002) IAP Proteins: blocking the road to death's door. Nature Rev Mol Cel Biol 3(6):401-410

Sauro HM, Fell DA (1991) SCAMP: A metabolic simulator and control analysis program. Math Comput Modelling 15:15-28

Scaffidi C, Fulda S, Srinivasan A, Friesen C, F L, Tomaselli K, Debatin K, Krammer P, ME P (1998) Two CD95 (APO-1/Fas) signaling pathways. EMBO J 17(6):1675-1687

Schacherer F, Choi C, Götze U, Krull M, Pistor S, Wingender E (2001) The TRANSPATH signal transduction database: a knowledge base on signal transduction networks. Bioinformatics 17(11):1053-1057

Schilling CH, Schuster S, Palsson BO, Heinrich R (1999) Metabolic pathway analysis: basic concepts and scientific applications in the post-genomic era. Biotechnol Prog 15(3):296-303

Schmitz I, Walczak H, Krammer P, Peter M (1999) Differences between CD95 type I and tpye II cells detected with the CD95 ligand. Cell Death Differ 6(9):821-822

Smuts J (1926) Holism and Evolution. Macmillan & Co Ldt., London

Stelling J, Sauer U, Szallasi Z, Doyle F, Doyle J (2004) Robustness of cellular functions. Cell 118:675-685

Stennicke H, Salvesen G (1999) Catalytic properties of the caspases. Cell Death Differ 6:1054-1059

Swameye I, Müller TG, Timmer J, Sandra O, Klingmüller U (2003) Identification of nucleocytoplasmic cycling as a remote sensor in cellular signaling by databased modeling. PNAS 100(3):1028-1033

Thornberry N, Lazebnik Y (1998) Caspases: Enemies within. Science 281: 1312-1316

Tomita M, Hashimoto K, Takahashi K, Shimizu T, Matsuzaki Y, Miyoshi F, Saito K, Tanida S, Yugi K, Venter J, Hutchison C (1999) E-CELL: Software environment for whole cell simulation". Bioinformatics 15:72-84

Vacheva I, Bentele M, Eils R (2005) Optimal experiment design for discriminating between competing signal transduction models. (in preparation)

Varma A, Morbidelli M, Wu H (1999) Parametric Sensitivity in Chemical Systems. Cambridge University Press, New York

Westphal S, Kalthoff H (2003) Apoptosis: targets in pancreatic cancer. Mol Cancer 2(1):6

Wolfram S (1994) Cellular Automata and Complexity: Collected Papers. Addison-Wesley, Reading, MA

Zou H, Yang R, Hao J, Wang J, Sun C, Fesik S, Wu J, Tomaselli K, Armstrong R (2002) Regulation of the Apaf-1/caspase-9 apoptosome by caspase-3 and XIAP. J Biol Chem 278(10):8091-8098

Bentele, Martin
Division Theoretical Bioinformatics, German Cancer Research Center (DKFZ), 69120 Heidelberg, Germany

Eils, Roland
Division Theoretical Bioinformatics, German Cancer Research Center (DKFZ), 69120 Heidelberg, Germany
r.eils@dkfz.de

# Scientific and technical challenges for systems biology

Hiroaki Kitano

## Abstract

Systems biology is an emergent discipline, yet can be rooted back almost a century when pioneering thoughts on system-oriented views were discussed. System-level understanding of life has consistently been a subject of the broad scientific community. With the progress of various molecular biology and genomics research, combined with advances in control theory, software, and computer science, we are now able to tackle this problem with renewed perspectives and powerful techniques. One of the significant questions is what is underlying principles of living systems. This paper argues that "robustness" is one of the fundamental properties of evolved biological systems and there are certain principles that govern biological systems at the system-level. Such a principle also provides us with insight into diseases and possible countermeasures.

## 1 Introduction

Systems biology aims at system-level understanding of biological systems (Kitano 2002a; Kitano 2002b). Investigations of biological systems at the system level are not a new concept; they can be traced back to homeostasis by Canon (1932), Cybernetics by Norbert Weiner (1948), and general systems theory by von Bertalanffy (1968). Various approaches in physiology have also taken a systemic view of biological subjects. The reason why systems biology is gaining renewed interest today is, in my view, due to emerging opportunities to solidly ground systems-level understanding on molecular level understanding, the possibility of establishing a well-founded theory at the system level, and increased recognition on the limitations of molecular approach in handling the complexity of biological systems. This is only possible today due to the progress of molecular biology, genomics, computer science, modern control theory, non-linear dynamics theory, and other relevant fields, which were not sufficiently mature at the time of early attempts.

However, "system-level understanding" is a rather vague notion and often hard to define, due to the fact that systems are not tangible objects. Genes and proteins are more tangible because they are identifiable matters. Although a system is composed of these matters, which are components of the system, the system itself cannot be made tangible. Often, diagrams of gene regulatory networks and protein

Topics in Current Genetics, Vol. 13
L. Alberghina, H.V. Westerhoff (Eds.): Systems Biology
DOI 10.1007/ b137124/ Published online: 12 May 2005
© Springer-Verlag Berlin Heidelberg 2005

interaction networks are shown as a representation of systems. It is certainly true that such diagrams capture one aspect of the structure of the system, but it is still only a static slice of the system. The heart of the system lies in the dynamics it creates and logic behind it. It is science on the dynamic state of affairs.

There are four distinct phases that lead us to system-level understanding at various levels. First, system structure identification enables us to understand the structure of the system. While this may be a static view of the system, it is an essential first step. Ultimately, the structure will be identified in both physical and interaction structures. Interaction structures are represented as gene regulatory networks and biochemical networks that identify how components interact within and among cells. Physical details of specific regions of the cell, overall structure of cells, and organisms are also important because such physical structures impose constraints on possible interactions, and the outcome of interactions impacts the formation of physical structures. The nature of interaction could be different if proteins involved in interaction were to move by simple diffusion or under specific guidance from the cytoskeleton.

Secondly, understanding dynamics of the system is essential and requires integrated efforts of experiments, measurement technology development, computational model development, and theoretical analysis. Numbers of methods, such as bifurcation analysis and wavelet, have been used, but further investigations are necessary to handle the dynamics of systems with very high dimensional space.

Third, methods to control the system need to be investigated. One of the goals is to find a therapeutic approach based on system-level understanding. Many drugs have been developed through extensive effect-oriented screening, but it is only recently that specific molecular targets have been identified and key compounds have been designed accordingly. Success in methods of controlling cellular dynamics may enable us to exploit the intrinsic dynamics of the cell, and thus precisely predict and control its effects.

Finally, we need to design a system for modifying and constructing biological systems with designed features. Bacteria and yeast may be redesigned to yield desired properties for drug production and alcohol production. Artificially created gene regulatory logic could be introduced and linked to innate genetic circuits to attain desired functions (Hasty et al. 2002).

Several different approaches can be taken within the systems biology field. One may decide to carry out large-scale high throughput experiments and try to find out an overall picture of the system at coarse-grain resolution (Ideker, Thorsson et al. 2001; Guelzim et al. 2002; Ideker et al. 2002; Ihmels et al. 2002). Alternatively, working on precise details of specific signal transduction (Bhalla and Iyengar 1999; Ferrell 2002), cell cycle (Tyson et al. 2001; Chen et al. 2004) and other biological issues to identify the logic behind them would be a viable research approach. Both approaches are essentially complementary and together reshape our understanding of biological systems.

# 2 Robustness as a fundamental organizational principle

Although systems biology is often characterized by the use of massive amounts of data and computational resources, there are significant theoretical elements that need to be addressed. After all, efforts to digest large data sets are intended to gain a deeper understanding of biological systems as well as to be applied for medical practices and other issues. In either case, there must be hypotheses to test the use of these data and computational practices.

I am particularly intrigued by the stunning diversity and robustness of biological systems, yet fundamental features are maintained across an astonishingly broad range of species. Robustness is the fundamental feature that enables diverse species to be generated and evolve. Such robustness is ubiquitous, and can be observed in virtually all species across different aspects of biological systems. Therefore, one of the central themes of my research is to understand robustness in biological systems (Kitano 2004a).

Why is robustness so important? First, it is a feature that is observed so ubiquitously in biological systems: from fundamental processes, such as phage fate decision switching (Little et al. 1999) and bacteria chemotaxis (Barkai and Leibler 1997; Alon et al. 1999; Yi et al. 2000) to developmental plasticity (von Dassow et al. 2000) and tumor resistance against therapies (Kitano 2003a, 2004b), which implies that it may be a basic principle that is universal in biological systems, as well as provide opportunities for finding cures for cancer and other complicated diseases. Second, robustness against environmental and genetic perturbations is essential for evolvability (Wagner and Altenberg 1996; de Visser et al. 2003; Rutherford 2003), which is an underlying basis of evolution. Third, it is one of the features that distinguish biological systems from man-made engineering systems. While some man-made systems, such as airplanes, are designed to be robust against a range of perturbations, most man-made systems are not as robust as biological systems. Fourth, robustness is distinctively a system-level property that cannot be observed by just looking at components.

Robustness is a property of the system to maintain a specific function against certain perturbations. The function may be maintained through a system's state returning to the original attractor, or transit to another attractor that ultimately maintains the function by other means. Robustness is not about whether the system is resistant to change and inflexible, but about the system's flexibly to handle perturbations to ultimately maintains a specific function. A specific aspect of the system, functions to be maintained, and the type of perturbations that the system is robust against must be well defined in order to make solid arguments. For example, a modern airplane (system) has a function to maintain its flight path (function) against atmospheric perturbations (perturbations). Across engineering and biological systems, there are common mechanisms that make systems robust against various perturbations.

First, extensive systems control is used, particularly negative feedback loops, to make the system dynamically stable around a specific state of the system. Integral feedback used in bacteria chemotaxis is a typical example (Barkai and Leibler

1997; Alon et al. 1999; Yi et al. 2000). Due to integral feedback, bacteria can sense changes of chemo-attractants and chemo-repellants independent of absolute concentration so that proper chemotaxis behavior is maintained over a wide range of ligand concentration. In addition, the same mechanism makes bacteria insensitive to changes in rate constants involved in the circuit. Positive feedback is often used to create bistability in signal transduction and cell cycle, so that the system is tolerant against minor perturbations in stimuli and rate constants (Tyson et al. 2001; Ferrell 2002; Chen et al. 2004).

Second, alternative (or fail-safe) mechanisms increase tolerance to component failure and environmental changes by providing alternative components or methods to ultimately maintain the functions of the system. Sometimes, there are multiple components that are similar to each other, thus, providing redundancy. In other cases, different means are used to cope with perturbations that cannot be handled by other means. This is often called phenotypic plasticity (Schlichting and Pigliucci 1998; Agrawal 2001) or diversity. Redundancy and phenotypic plasticity are often considered as opposite matters, but it is more consistent to view them as different ways to provide an alternative fail-safe mechanism.

Third, modularity isolates perturbations from the rest of the system. Cells are the most significant example, while more subtle and less obvious examples are modules of biochemical and gene regulatory networks. Modules also play an important role during developmental processes that buffer perturbations so that proper pattern formation can be accomplished (von Dassow et al. 2000; Eldar et al. 2002; Meir et al. 2002). The definition of modules and how to detect such modules are still controversial, but the general consensus is that modules do exist and play an important role (Schlosser and Wagner 2004).

Fourth, decoupling isolates low-level noise and fluctuations from functional level structures and dynamics. One example here is genetic buffering by Hsp90 in which misfolding of proteins due to environmental stresses are fixed, and so the effects of such perturbations are isolated from the functions of circuits. This mechanism applies also to genetic variations where genetic changes in the coding region that may affect protein structures are masked because protein folding is fixed by Hsp90 unless such masking is removed by extreme stress (Rutherford and Lindquist 1998; Queitsch et al. 2002; Rutherford 2003). Emergent behaviors of complex networks also exhibit such buffering property (Siegal and Bergman 2002). These effects may constitute canalization proposed by Waddington (1957). The recent discovery by Uri Alon's group on oscillatory expression of p53 upon DNA damage may exemplify decoupling at the signal encoding level (Lahav et al. 2004), because stimuli invoked pulses of p53 activation level, instead of gradual changes, effectively converting analog signals into digital signals. Digital pulse encoding may indicate robust information transmission, although further investigations are clearly warranted before any conclusions can be drawn.

# 3 Evolvability and trade-offs of robust systems

Robustness is based on evolvability. For a system to be evolvable, it must be able to produce a variety of non-lethal phenotypes (Kirschner and Gerhart 1998). At the same time, genetic variations need to be accumulated into a neutral network, so that pools of genetic variants are exposed when the environment changes suddenly. Systems that are robust against environmental perturbations entail mechanisms such as system control, alternativeness (fail-safe), modularity, and decoupling which also supports, by congruence, generation of non-lethal phenotypes and genetic buffering. In addition, the capability to generate flexible phenotypes and robustness requires the emergence of the bow-tie structure as an architectural motif (Csete and Doyle 2004). One of the reasons why robustness in biological systems is so ubiquitous is because it facilitates evolution, and evolution tends to select traits that are robust against environmental perturbations. This leads to successive addition of system controls.

Given that robustness is key in biological systems, it is essential to elucidate the trade-offs involved between robustness and fragility within a particular system. The "robustness index" interlocks with the complexity of a system and how the system performs in an environment with limited resources. Carlson and Doyle argued, using simple examples from physics and forest fires, that systems that are optimized for specific perturbations are extremely fragile against unexpected perturbations (Carlson and Doyle 1999, 2002). This means that when robustness is enhanced against a range of perturbations, then there must be corresponding fragility elsewhere as well as compromised performance and increased resource demands. Highly Optimized Tolerance (HOT) model systems are successively optimized/designed (although not necessarily globally optimized) against perturbations, whereas Self-Organized Criticality (SOC) (Bak et al. 1988) or Scale-Free Networks (Barabasi and Oltvai 2004) are unconstrained stochastic addition of components without design or optimizations involved. Such differences actually affect the failure patterns of systems and so have direct implications for understanding the nature of disease and therapy design.

Disease often reflects exposed fragility of the system. Some diseases are maintained to be robust against therapies because such states are maintained or even promoted through mechanisms that support robustness of the normal physiology of our body.

*Diabetes mellitus* is an excellent example of how systems that are optimized for near-starving, intermittent food supply, high-energy utilization lifestyle, and highly infectious conditions are fragile against unusual perturbations, in evolutionary time-scale, namely high-energy foods, and low-energy lifestyles (Kitano et al. 2004). Due to optimization to the near-starving condition, extensive control to maintain minimum blood glucose level has been acquired so that activities of the central nervous system and innate immunity are maintained. However, no effective regulatory loop has been developed against excessive energy intake, so blood glucose level is chronically maintained higher than the desired level, leading to cardiovascular complications.

Cancer is a typical example of robustness hijacking (Kitano 2003a, 2004b). Tumors are robust against a range of therapies due to genetic diversity, feedback loop for multi-drug resistance, and tumor-host interactions. Tumor-host interactions are, for example, involved in HIF-1 upregulation that then upregulates VEGF and uPAR and other genes that trigger angiogenesis and cell motility (Harris 2002). HIF-1 upregulation takes place because of hypoxia in tumor clusters and dysfunctional blood vessels due to tumor growth. This feedback regulation enables a tumor to grow further or cause metastasis. However, HIF-1 upregulation is important for normal physiology under oxygen deprived conditions such as high altitude and lung dysfunctions (Sharp and Bernaudin 2004). This indicates that mechanisms that provide protection for our body are effectively hijacked.

I would consider that there are three theoretically motivated countermeasures for such diseases. First, robustness of the epidemic state should be controlled by systematically perturbing biochemical and gene regulatory circuits using low-dose drugs. Second, a robust epidemic state implies that there is a point of fragility somewhere. Identification or active induction of such a point may lead to a novel therapeutic approach with dramatic effects. Third, one may wish to retake control of feedback loops that give rise to robustness in the epidemic state. One possible approach is to introduce a decoy that effectively disrupts feedback control or the invasive mechanisms of the epidemic. An example of such approach has been proposed for AIDS treatment that conditionally replicating HIV-1 (crHIV-1) vector with only has cis regions, but not trans, is introduced, so that a process to replicate HIV-1 virus is co-opted by crHIV-1 (Dropulic et al. 1996; Weinberger et al. 2003). Overall effect is that the epidemic state is maintained latent, but do not eliminate HIV-1 virus load.

It is not known how systematically we can identify such strategic therapies, and this will be the focus of major research in future. However, it is important to emphasize that the concept of viewing robustness as a fundamental principle of biological systems is a key aspect of this research program.

## 4 Computational tools in systems biology

For theoretical analysis to be effectively carried out, a range of tools and resources need to be made available. One of the issues is to create a standard for representing models. Systems Biology Mark-up Language (SBML: http://www.sbml.org/) was designed to enable the standardized representation and exchange of models among software tools that comply with the SBML standard (Hucka et al. 2003). The project was started in 1999, and has now grown into a major community effort. SBML Level-1 and Level-2 have been released and used in over 60 software packages (as of August 2004). Systems Biology Workbench (SBW) is an attempt to provide a framework where different software modules can be seamlessly integrated, so that researchers can create their own software environment (Hucka et al. 2002).

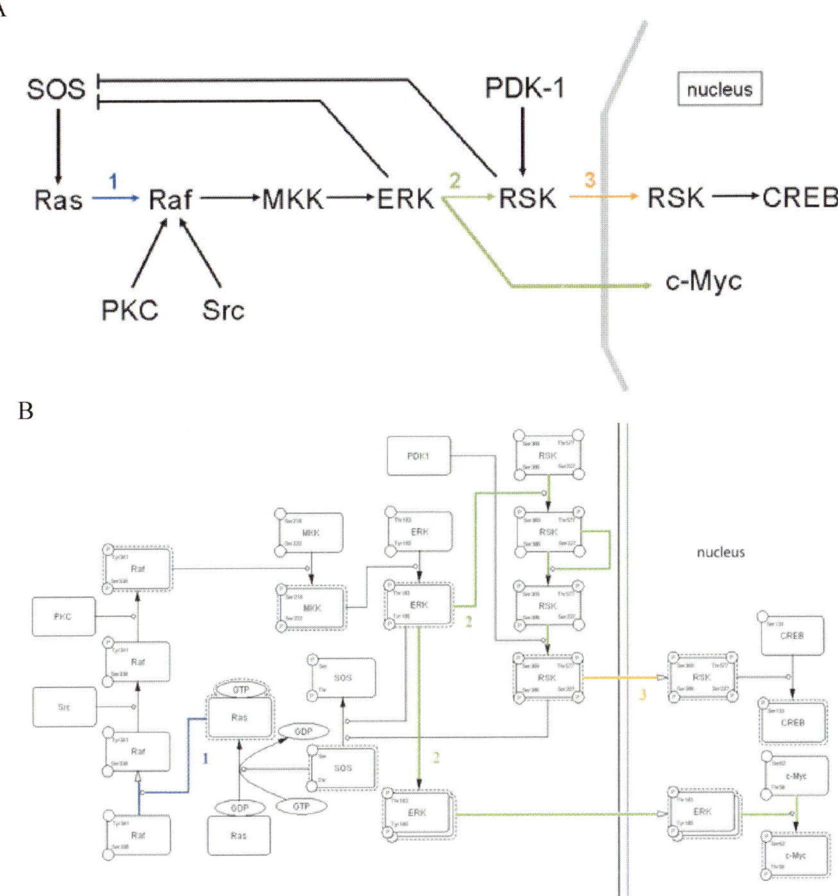


**Fig. 1.** (A) Conventional information diagram, (B) The process diagram.

Aside from these software standards and tool development efforts, one of the major missing pieces is a standard notation for visually representing complex gene regulation and biochemical interactions. The conventional arrow-and-circle notation is too informal to convey meaningful information, and is often used inconsistently. Development of a standard notation would provide unambiguous ways to convey knowledge of biochemical interaction networks with rich information content. An early attempt was made by Kurt Kohn with the famous "Kohn Map" (Kohn 1999). Unfortunately, it has not been widely used by the community, due to complexity of the notation and lack of software tools to support the notation. Several other proposals have been made since (Pirson et al. 2000; Cook et al. 2001; Maimon and Browning 2001), but not widely used to date. The author wishes to rectify this situation by providing a notation that remedies the problems of various previous proposals, and by providing software tools and resources (Funahashi and

Kitano 2003; Kitano 2003b). One of the problems of the Kohn Map is that the temporal order of interactions has not been explicit, so that the reader must untangle complex network diagrams to find out the order of reactions. However, the beauty of the Kohn Map is its consistency, whereby, one molecular species appears only one time in the diagram, and so all interactions are represented in a box that represents specific molecular species. In order to resolve this issue, the author proposes a new graphical notation, where two representations are used for visualization from one model. These two visualizations are: a process diagram that explicitly displays the temporal order of interactions, and a relation diagram that is close to the Kohn Map. In the process diagram, each node represents a specific state of the molecular species, so that one molecular species appears multiple times depending upon the number of different states it may take. This is more intuitive to readers without compromising the integrity of the diagram. CellDesigner has been developed to support the editing and viewing of such diagrams (Funahashi and Kitano 2003) (Fig. 2), and is freely available from the website http://www.systems-biology.org/. Figure 1 compares the conventional diagram and the process diagram for the Ras-Raf-MAPK cascade. A close look at these diagrams shows that arrows in conventional diagrams are used with at least three different meanings, causing substantial ambiguities in representation. Scalability of the process diagram is demonstrated by creating a 600 interaction scale diagram that represents most of the signal transduction pathways in macrophage (Oda et al. 2004) that are related to the Alliance for Cellular Signaling (http://www.afcs.org/), as well as four different cell types: adipocyte, skeletal muscle cell, beta cell, and hepatocyte (Kitano et al. 2004).

One of the major interests in computational aspects of systems biology is how numerical simulations can be used for gaining a deeper understanding of organisms and medical applications. There is no doubt that simulation, if properly used, can be a powerful tool for scientific and engineering research. Modern aircraft cannot be developed without the help of computational fluid dynamics (CFD). There are at least two issues that must be carefully examined in computational simulation. First, the purpose of simulation has to be well defined, and the model has to be constructed to maximize the purpose of the simulation. This affects the choice of modeling technique, levels of abstraction, scope of modeling, and parameters to be varied. Second, simulation needs to be well placed in the context of the whole analysis procedure. In most cases, simulation is not the only method of analysis, so it is important to coordinate what part of the analysis uses numerical simulation and what other parts use non-simulation methods in order to maximize overall analysis activity.

An example from racing car design illustrates these issues. CFD is extensively used in Formula-1 car design in order to obtain optimal aerodynamics of larger downward force and lower drag. Particular emphasis is placed on the effects of various aerodynamics components such as front wings, rear wings, and ground effects, but complicated interference between front wings, suspension members, wheels, and brake air intake ducts must also be investigated. Combustion in the engine is another example where simulation studies are often used, but is simulated separately from the CFD model. The success of CFD relies upon the fact that

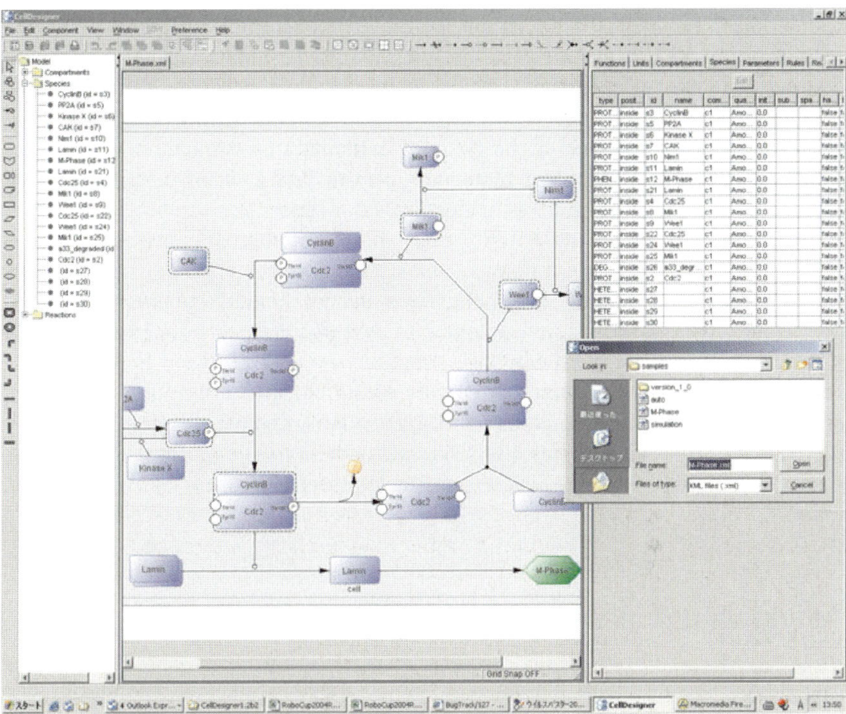

**Fig. 2.** A screen shot of CellDesigner

the basic principles of fluid dynamics are relatively well understood, although some issues remain to be resolved, and so simulation can be done with relative confidence. This exemplifies the practice of proper focus and abstraction. When receptor dynamics are being investigated, transcription machinery will not be modeled as it is only remotely related.

CFD is not the only tool for aerodynamics design. F-1 racing cars are initially designed using CFD (*in silico*), then further investigated using wind tunnels (*in physico*), followed by actual runs on a test course (*in vitro*) before being deployed in actual races (*in vivo*). CFD in this case is used for the initial search of candidate designs which will then undergo further investigation using wind tunnels. Computer simulation in biological systems perhaps plays a similar role, namely, upstream screening of candidate hypotheses, which will then undergo detailed experimental investigations. At the same time, this may imply that when specific aspects of modeling are sufficiently sophisticated, the use of computational models for predicting the effects of specific experiments and drugs could be widely accepted. In this case, government regulatory agencies such as the FDA would probably require computer simulations to be conducted before any drug could move into a phase-I study.

It should be noted, however, that CFD in racing car design has clear and explicit optimization goals: high downward force and low drag. The problem for simulation in biology is that what needs to be discovered by the simulation is not as straightforward as in racing car design. Here, it is important to remember the guiding principle of robustness. The guiding principle provides a view of what needs to be investigated and identified, thus, providing a starting point for a broad range of applications. One of the goals of computational simulation is to understand the nature and degree of robustness, and to find a set of perturbations that can compromise such robustness in a controlled manner.

In summary, I would like to emphasize the importance of research to identify fundamental system-level principles of biological systems, where numerous insights in both basic science and applications can come out. New opportunities are now emerging with the massive amounts of data that are being generated in large-scale experimental projects, but such data are best utilized when processed with certain underlying hypotheses that capture essential aspects of system-level properties. Robustness is a ubiquitous and fundamental principle. Investigation on the robustness of biological systems will provide us with guiding principles for understanding biological systems, diseases, as well as effective use of computational tools.

## Acknowledgements

The authors wish to thank members of Sony Computer Science Laboratories, Inc. and the ERATO Kitano Symbiotic Systems Project for valuable discussions. This research is supported by the Exploratory Research for Advanced Technology (ERATO) and the Solution-Oriented Research for Science and Technology (SORST) programs (Japan Science and Technology Organization); the NEDO Grant (New Energy and Industrial Technology Development Organization) of the Japanese Ministry of Economy, Trade and Industry (METI); the Special Coordination Funds for Promoting Science and Technology and the Center of Excellence Program for Keio University (Ministry of Education, Culture, Sports, Science, and Technology); the Rice Genome and Simulation Project (Ministry of Agriculture); and the Air Force Office of Scientific Research (AFOSR).

## References

Agrawal AA (2001) Phenotypic plasticity in the interactions and evolution of species. Science 294:321-326

Alon U, Surette MG, Barkai N, Leibler S (1999) Robustness in bacterial chemotaxis. Nature 397:168-171

Bak P, Tang C, Wiesenfeld K (1988) Self-organized criticality. Phys Rev A 38:364-374

Barabasi AL, Oltvai ZN (2004) Network biology: understanding the cell's functional organization. Nat Rev Genet 5:101-113

Barkai N, Leibler S (1997) Robustness in simple biochemical networks. Nature 387:913-917

Bertalanffy LV (1968) General System Theory. New York, George Braziller

Bhalla US, Iyengar R (1999) Emergent properties of networks of biological signaling pathways. Science 283:381-387

Cannon W (1932) The Wisdom of the Body. New York, Norton

Carlson JM, Doyle J (1999) Highly optimized tolerance: a mechanism for power laws in designed systems. Phys Rev E Stat Phys Plasmas Fluids Relat Interdiscip Topics 60:1412-1427

Carlson JM, Doyle J (2002) Complexity and robustness. Proc Natl Acad Sci USA 99 Suppl 1:2538-2545

Chen KC, Calzone L, Csikasz-Nagy A, Cross FR, Novak B, Tyson JJ (2004) Integrative analysis of cell cycle control in budding yeast. Mol Biol Cell 15:3841-62

Cook DL, Farley JF, Tapscott SJ (2001) A basis for a visual language for describing archiving and analyzing functional models of complex biological systems. Genome Biol 2:RESEARCH0012

Csete ME, Doyle J (2004) Bow ties metabolism and disease. Trends Biotechnol 22:446-50

de Visser JA, Hermisson J, Wagner GP, Ancel Meyers L, Bagheri-Chaichian H, Blanchard JL, Chao L, Cheverud JM, Elena SF, Fontana W, Gibson G, Hansen TF, Krakauer D, Lewontin RC, Ofria C, Rice SH, von Dassow G, Wagner A, Whitlock MC (2003) Evolution and detection of genetics robustness. Evolution 57:1959-1972

Dropulic B, Hermankova M, Pitha PM (1996) A conditionally replicating HIV-1 vector interferes with wild-type HIV-1 replication and spread. Proc Natl Acad Sci USA 93:11103-11108

Eldar A, Dorfman R, Weiss D, Ashe H, Shilo BZ, Barkai N (2002) Robustness of the BMP morphogen gradient in *Drosophila* embryonic patterning. Nature 419:304-308

Ferrell JE Jr (2002) Self-perpetuating states in signal transduction: positive feedback double-negative feedback and bistability. Curr Opin Cell Biol 14:140-148

Funahashi A, Kitano H (2003) Cell Designer: a process diagram editor for gene-regulatory and biochemical networks. Biosilico 1:159-162

Guelzim N, Bottani S, Bourgine P, Kepes F (2002) Topological and causal structure of the yeast transcriptional regulatory network. Nat Genet 31:60-63

Harris AL (2002) Hypoxia--a key regulatory factor in tumour growth. Nat Rev Cancer 2:38-47

Hasty J, McMillen D, Collins JJ (2002) Engineered gene circuits. Nature 420:224-230

Hucka M, Finney A, Sauro HM, Bolouri H, Doyle J, Kitano H (2002) The ERATO Systems Biology Workbench: enabling interaction and exchange between software tools for computational biology. Pac Symp Biocomput 450-461

Hucka M, Finney A, Sauro HM, Bolouri H, Doyle JC, Kitano H, Arkin AP, Bornstein BJ, Bray D, Cornish-Bowden A, Cuellar AA, Dronov S, Gilles ED, Ginkel M, Gor V, Goryanin II, Hedley WJ, Hodgman TC, Hofmeyr JH, Hunter PJ, Juty NS, Kasberger JL, Kremling A, Kummer U, Le Novere N, Loew LM, Lucio D, Mendes P, Minch E, Mjolsness ED, Nakayama Y, Nelson MR, Nielsen PF, Sakurada T, Schaff JC, Shapiro BE, Shimizu TS, Spence HD, Stelling J, Takahashi K, Tomita M, Wagner J, Wang J; SBML Forum (2003) The systems biology markup language (SBML): a medium for representation and exchange of biochemical network models. Bioinformatics 19:524-531

Ideker T, Ozier O, Schwikowski B, Siegel AF (2002) Discovering regulatory and signalling circuits in molecular interaction networks. Bioinformatics 18(Suppl 1):S233-S240

Ideker T, Thorsson V, Ranish JA, Christmas R, Buhler J, Eng JK, Bumgarner R, Goodlett DR, Aebersold R, Hood L (2001) Integrated genomic and proteomic analyses of a systematically perturbed metabolic network. Science 292:929-934

Ihmels J, Friedlander G, Bergmann S, Sarig O, Ziv Y, Barkai N (2002) Revealing modular organization in the yeast transcriptional network. Nat Genet 31:370-377

Kirschner M, Gerhart J (1998) Evolvability. Proc Natl Acad Sci USA 95:8420-8427

Kitano H (2002a) Computational systems biology. Nature 420:206-210

Kitano H (2002b) Systems biology: a brief overview. Science 295:1662-1664

Kitano H (2003a) Cancer robustness: tumour tactics. Nature 426:125

Kitano H (2003b) A graphical notation for biochemical networks. Biosilico 1:169-176

Kitano H (2004a) Biological robustness. Nat Rev Genet 5:826-837

Kitano H (2004b) Cancer as a robust system: implications for anticancer therapy. Nat Rev Cancer 4:227-235

Kitano H, Oda K, Kimura T, Matsuoka Y, Csete M, Doyle J, Muramatsu M (2004) Metabolic syndrome and robustness trade-offs. Diabetes 53(Suppl 3):S1-S10

Kohn KW (1999) Molecular interaction map of the mammalian cell cycle control and DNA repair systems. Mol Biol Cell 10:2703-2734

Lahav G, Rosenfeld N, Sigal A, Geva-Zatorsky N, Levine AJ, Elowitz MB, Alon U (2004) Dynamics of the p53-Mdm2 feedback loop in individual cells. Nat Genet 36:147-150

Little JW, Shepley DP, Wert DW (1999) Robustness of a gene regulatory circuit. EMBO J 18:4299-4307

Maimon R, Browning S (2001) Diagrammatic notation and computational structure of gene networks. Proceedings of the Second International Conference on Systems Biology. Pasadena CA

Meir E, von Dassow G, Munro E, Odell GM (2002) Robustness flexibility and the role of lateral inhibition in the neurogenic network. Curr Biol 12:778-786

Oda K. Kimura T, Matsuoka Y, Funahashi A, Muramatsu H, Kitano H (2004) Molecular interaction map of a macrophage. AfCS Research Reports 2:1-12

Pirson I, Fortemaison N, Jacobs C, Dremier S, Dumont JE, Maenhaut C (2000) The visual display of regulatory information and networks. Trends Cell Biol 10:404-408

Queitsch C, Sangster TA, Lindquist S (2002) Hsp90 as a capacitor of phenotypic variation. Nature 417:618-624

Rutherford SL (2003) Between genotype and phenotype: protein chaperones and evolvability. Nat Rev Genet 4:263-274

Rutherford SL, Lindquist S (1998) Hsp90 as a capacitor for morphological evolution. Nature 396:336-342

Schlichting C, Pigliucci M (1998) Phenotypic Evolution: A Reaction Norm Perspective. Sunderland, Sinauer Associates Inc

Schlosser G, Wagner G (2004) Modularity in Development and Evolution. Chicago, The University of Chicago Press

Sharp FR, Bernaudin M (2004) HIF1 and oxygen sensing in the brain. Nat Rev Neurosci 5:437-448

Siegal ML, Bergman A (2002) Waddington's canalization revisited: developmental stability and evolution. Proc Natl Acad Sci USA 99:10528-10532

Tyson JJ, Chen K, Novak B (2001) Network dynamics and cell physiology. Nat Rev Mol Cell Biol 2:908-916

von Dassow G, Meir E, Munro EM, Odell GM (2000) The segment polarity network is a robust developmental module. Nature 406:188-192

Waddington CH (1957) The Strategy of the Genes: a Discussion of some Aspects of Theoretical Biology. New York, Macmillan

Wagner GP, Altenberg L (1996) Complex adaptations and the evolution of evolvability. Evolution 50:967-976

Weinberger LS, Schaffer DV, Arkin AP (2003) Theoretical design of a gene therapy to prevent AIDS but not human immunodeficiency virus type 1 infection. J Virol 77:10028-10036

Wiener N (1948) Cybernetics: or Control and Communication in the Animal and the Machine. Cambridge, The MIT Press

Yi TM, Huang Y, Simon MI, Doyle J (2000) Robust perfect adaptation in bacterial chemotaxis through integral feedback control. Proc Natl Acad Sci USA 97:4649-4653

Kitano, Hiroaki

Sony Computer Science Laboratories, Inc, 3-14-13 Higashi-Gotanda, Shinagawa, Tokyo 141-0022 Japan, and, The Systems Biology Institute, Suite 6A, M31 6-31-15 Jingumae, Shibuya, Tokyo 150-0001 Japan

kitano@symbio.jst.go.jp

# AND NOW...

# Systems Biology: necessary developments and trends

Lilia Alberghina, Stefan Hohmann, and Hans V. Westerhoff

## Abstract

At the end of this definition of Systems Biology through exampling, we discuss ambitions, goals, and challenges relating to this new discipline. We estimate the impact that Systems Biology may have on health management, both in terms of drug discovery and in terms of enabling healthier lifestyles. In this context, we indicate what aspects of Systems Biology need to be stimulated most. We also touch on its effects on competitiveness of high-sophistication industries. Finally, we suggest special requirements that Systems Biology imposes on the organization and funding of Life Sciences research in the 21$^{st}$ century.

## 1 Various facets of Systems Biology

Two of us (LA and HW) began this book with an iconoclastic analysis of what our own field, biochemistry, purports for science and society. We gave a short definition of the Systems Biology that might serve as the new generation biochemistry/molecular biology, and which should serve science and society better. We also argued that the definition of Systems Biology should not be vague, yet, it should also be dynamic and heterogeneous. But who are we to say? What weight does our opinion carry? We, therefore, decided to ask some of the most active scientists in Systems Biology to give their vision of this new field and to prove their points by example. The result – this book - may now serve as a definition of Systems Biology based on scientific evidence. That definition is very close to that given by others recently (e.g. Kirschner 2005, Aderen 2005).

The authors of this book have accomplished their task excellently. With this book in hand, we can now decide for any funding agency whether a proposal is or is not about Systems Biology.

But of course, even more exciting was reading the spectrum of rapidly developing Systems Biology. Clearly Systems Biology extends far beyond massive genome-wide transcriptomics. We now think we understand, a bit better, how flux patterns change with conditions in some microorganisms, why the yeast cell cycle is not by itself a limit-cycle oscillation, how the cell size may be set by metabolism, what affects cycling, how developmental biology depends on a variety of new types of processes, how regulation can be traced through signal transduction networks, why *E. coli* is subtle in catabolite repression and inducer exclusion, why

Topics in Current Genetics, Vol. 13
L. Alberghina, H.V. Westerhoff (Eds.): Systems Biology
DOI 10.1007/4735_87 / Published online: 21 June 2005
© Springer-Verlag Berlin Heidelberg 2005

fluctuations can be much larger in cell biology than suggested by the Poisson distribution, how one can make replica of cell biology in computers and web-connect them to all scientists, how exemplary Metabolic Control Analysis has been for Systems Biology, how apoptosis involves a dynamic interplay of various signal transduction systems, how yeast uses Systems Biology to deal with stress and external dynamics, and how metabolism and signal transduction can be integrated. In short, we begin to understand what networks do and how they work in living cells.

Perhaps we should have been satisfied with this definition of Systems Biology by the system biologists themselves. However, we should also like to know what Systems Biology will become. After all, more important than the definition of Systems Biology are its aims and ultimately its accomplishments.

## 2 Long and medium-term goals of Systems Biology

The ambition of establishing a systemic understanding of cells and organisms (Von Bertalanffy 1962) is cited more and more, if only because it seems to have become more realistic. The success achieved in the past decade by the high-throughput methods for measuring the global expression of genes in terms of mRNA, proteins, and metabolites suggests that we should be close to meeting this ambition. This aspiration compels research into finding new ways to structure the resulting wealth of information. That structuring should not be flat bioinformatics, but synchronized with the various physiological and pathological states and perturbations, if we want to move from description to understanding and perhaps to improvement of the living state.

Databased mathematical modeling is required to give unambiguous dynamic structure to the networks and pathways that emerge from the interplay of genes, proteins, small molecules between them, and with their dynamic environment. Most interactions of the various molecular constituents of pathways and networks are nonlinear. Their dynamics, which cannot be appreciated by mere intuition, lead to the emergent properties of biological systems. Thereby, Systems Biology is a necessary complement to the successful reductionist approach that dominated biological research in the second half of the 20th century.

Systems Biology is often associated with bioinformatics. Although bioinformatics is an important contributor to Systems Biology, it differs from the latter, both in its approaches and goals. Explicating these differences may help identify the aims of Systems Biology. Bioinformatics aims to extract information from biological findings without itself engaging in experimentation: prediction of 3D structure from sequence data, clustering of mRNA expression data, and molecular evolution analysis, constitute examples. Both disciplines utilize biological data and computer methods, but Systems Biology should go a step further. It should also engage in quantitative wet-lab experimentation, in recursion with the theoretical approaches.

Most bioinformatics' results are correlations. Systems Biology aims at cause-effect relations. Systems Biology should aim to understand and to discover new principles and mechanisms of biological function. It, therefore, needs to integrate and structure "omics" data of many levels in order better to define and understand control circuits and executive steps of the many complex processes that govern cell function. The latter include signaling, gene regulation, cell cycle, differentiation, apoptosis, aging, *and* transformation, not just the primary sequence. Systems Biology, therefore, needs to do invasive experiments, where it changes molecular properties, measures the systemic effects, and determines if the latter concord with proposed mechanisms. The phenomenon in biology that much of causality is spiraling if not circular (*cf.* Rosen 1991 and the chapter by Westerhoff and Hofmeyr) sets Systems Biology further apart from bioinformatics, and makes it transcend most of the Natural Sciences. The long-term goal is to understand cells and eventually tissues and organs. The long-term 10-mile stone is the computer replicas of the same, which should serve as an ultimate substrate for proving that understanding (*cf.* the chapter by Snoep and Westerhoff). It should be an ultimate aim of Systems Biology to convert biology and medicine from being descriptive (even at the molecular level) to being precise and explicit, from being correlative to fostering understanding and even prediction of how genetic, metabolic, and chemical perturbations affect the functions of living organisms in health and disease.

## 2.1 Quantitative measurements on single cells?

Quantitative measurements of gene expression, protein levels, and metabolites are needed to *describe* any cellular process fully. Because of the nonlinearities involved, the measurements often need to be *quantitative* to describe the processes or their regulation *in their essence* (Westerhoff and Van Dam 1987; Rosen 1991; the chapter by Westerhoff and Hofmeyr). In many cases, also the spatial organization and the temporal dynamics of each molecular species is required to be able to extract the putative structure of the hidden regulatory circuits and to test these then *in vivo* and *in silico*. These kinds of data are rarely available in the literature, holding back the Systems Biology approach. Besides, the minimum detection levels of most measurement technologies necessitate average determinations on large numbers of cells, thereby, hiding the cell-to-cell variation of the phenomenon under investigation. Full-fledged Systems Biology requires new technologies enabling measurements of the various cell components non-destructively and in real time, possibly on single cells. In the meantime, it should standardize measurements and set quality assurance and quality control procedures for collecting data in Systems Biology data banks.

## 2.2 Systems Biology data and model-bases

The huge ambition of Systems Biology requires that each scientist can build upon prior knowledge. Findings, starting from previous publications on conventional biochemistry, genetics and physiology, and proceeding to genome-wide high-throughput analyses, should be made accessible automatically, with enough information to make different experiments comparable.

At present, experimental findings are accumulated and readily forgotten by the scientific community, thereby, strongly reducing the cost-effectiveness of its research activities. Systems Biology should induce a change of paradigm by developing ways (models) to structure experimental data and to store them in a way useful for future investigations. The creation and maintenance of a database, possibly divided in sections dedicated to specific organisms and/or cellular processes, is expected to be both a condition for the development of Systems Biology and a measure of its success. The many problems posed by database creation are discussed in the report from the American Academy of Microbiology "Systems Microbiology: beyond microbial genomics", 2004.

The data management problem of Systems Biology is different and perhaps less absolute than that for bioinformatics. For, sadly enough, most data that is collected is irrelevant for Systems Biology, as it has been collected for molecules under unphysiological conditions, or because the type of data is inconsequential for understanding functional behavior of the molecules (*e.g.* mass). Systems Biology offers an approach to data storage that is radically different from that used by bioinformatics: it stores the data in computer replica of the actual system. Only data that are relevant for the replica are stored there and the relevance of the data is represented by the sensitivity of the mathematical model predictions to the data values (*cf.* the chapter by Snoep and Westerhoff).

## 2.3 Standard notation and visualization

As discussed very clearly in Kitano's chapter of this book, one of the major missing pieces for Systems Biology development is a standard and unambiguous notation for pathways and regulatory circuits. The notation should be intuitive and display the temporal order of events, even highly complex ones. A new graphical notation has been proposed by Kitano and corresponding computer software has been developed. Time will tell if this notation will be widely accepted by the scientific community.

This development reminds us of the SBML standard, developed for the exchange of information between mathematical modeling programs, which has been highly successful (*cf.* the chapter by Hucka *et al.*), as well as of the standard of making models accessible through the World Wide Web, as set by the Silicon Cell program (*cf.* the chapter by Snoep and Westerhoff).

## 3 The challenges of Systems Biology

Even our 'test-tube' microbial cells, both prokaryotes and lower eukaryotes, are extremely complex entities. Both in *Escherichia coli* and in *Saccharomyces cerevisiae*, thousands of components interact. Presumable, this is necessary in order to perform the functions required for living, in a strict and fine-tuned interplay, and with proper responsiveness to the environment. The goal of obtaining a computer replica of a whole microbial cell might, therefore, seem impossible to achieve. Indeed, if we have 6 000 gene products, there are 18 million possible binary interactions. Or, if we consider a pathway with 14 enzymes, there are thousands of protein-protein interactions to worry about. Indeed, interactome studies in yeast of a few years ago, spell trouble (see the chapter by Alberghina *et al.* and Han *et al.* 2004), even though the number of actual interactions that were observed was much, much smaller than the potential number estimated above.

Some mathematical models *have* been made. They *were* based on empirical characterization of all the individual enzymes in a pathway. However, these pathways were limited in length or complexity, or not all their component enzymes had been characterized in full detail. Out of the multitude of mathematical models that exist in biology, this is then the very limited subset of precise models, or silicon cells (*cf.* www.siliconcell.net). In these 'silicon cells' (which should perhaps rather be called 'silicon pathways') the behavior of pathways could be reproduced fairly well, sometimes after some adjustment. Glycolysis in yeast and *Trypanosoma brucei* are among the examples where this approach has been quite successful (see the chapter by Snoep and Westerhoff). How is this then possible in view of the vast number of possible interactions?

Well, reality is not as bad as it may seem from the above computation. Even though intracellular biochemistry abounds in interactions, the number of interactions any individual enzyme has with its environment is highly limited. In practice, a protein in yeast cannot interact with all 6 000 other proteins and with all 800 or so metabolites. The reason is that for interactions to mean something they need to be specific. In 55 M of water, it is difficult for any substance to compete with water for interactions. Therefore, only interactions that are hard-coded in protein structure, hence, gene sequence are effective (even though many more interactions may be measured experimentally with techniques that are not optimizing for specificity). For a metabolite to interact effectively with a protein with any relevant affinity that protein needs to have a domain with a specific structure. There are only so many domains that befit any protein, because of the limited surface area that is accessible in a stable protein. A glycolytic protein in glycolysis, therefore, may be expected to interact with at most some 12 metabolites, if not much less. That such a protein may bind non-specifically to membranes or polymers, such as microtubules, merely means that its properties are different from those determined *in vitro*, not more complex.

For protein-protein interactions to be important, the interacting protein must be expressed at the same or higher level. In growing organisms such as *S. cerevisiae* and *T. brucei* when glycolysis is effectively the only free-energy source of ATP, the concentrations of the glycolytic enzymes are so high that few other proteins could bind stoicheiometrically. Moreover, the carbon flux through glycolysis is so dominant in magnitude that this alone reduces the *metabolic* control a process outside this main carbon and energy metabolism can have on the process (Westerhoff and Van Dam 1987) (with the exception of course of ATP consumption which is a connected high flux system (*cf.* Koebmann *et al.* 2002)). By choosing intracellular pathways that dominate mass flow, and by focusing on *metabolic* regulation, some Systems Biology has been possible (see the chapter by Snoep and Westerhoff for a further review). For some signal-transduction and gene-expression systems, the situation is less clear. The abundance of many important factors is low and they may be influenced by binding to other proteins present at much higher concentrations. Perhaps then Systems Biology should start studying networks where abundances are high. On the long run, however, such a limitation will not be satisfactory.

## 3.1 The modular approach

Therefore, for Systems Biology to flourish, it should find methods to deal with the complexity of the living cell. And, there is hope. A breakthrough in Systems Biology has come from the recognition that cellular functions may be dissected into modules, subsystems of interacting molecules (proteins, DNA, RNA, and small molecules), which perform a given task in a way largely independent from the context of the other modules (Westerhoff *et al.* 1990a, 1990b; Kahn and Westerhoff 1991; Hartwell *et al.* 1999). Not only spatially defined macromolecular complexes, such as mitochondria and ribosomes are examples of modules, but also 'levels' in signal transduction cascades (Kahn and Westerhoff 1991) and parts of pathways in metabolic networks (Westerhoff and Van Dam 1987; Westerhoff *et al.* 1990; Schuster *et al.* 1993; Ravasz *et al.* 2002). Statistical analysis of gene expression data can identify regulatory modules (Segal *et al.* 2003). A different type of modularity has been that of elementary flux modes, corresponding to various possible trains with different destinations running over the same rail network (Schuster *et al.* 2002). Similarly, extreme modes follow routes that have been optimized for some performance (extreme pathways) (Price *et al.* 2003).

Yet another appreciation of the organization plan of modules may be obtained by focusing on the proteome that furnishes the active components in the modular structure. By analyzing protein-protein interaction networks it has been confirmed (see above) that most proteins interact with few partners, while a small number of proteins, called "hubs", interact with many different proteins. Such "hubs" belong to either of two categories, *i.e.*, "party hubs", which interact with most of their partners at the same time, or "data hubs", which bind their different partners at different times and locations (Han *et al.* 2004). Accordingly, one type of modularity is therefore organized by "party hubs" that give structure to each module, and by "data hubs" that function as global connectors among modules in a higher hierarchical role than "party hubs". Another type of modularity derives from the networking of the data hubs. In appraising the networks that are reported on the basis of yeast two-hybrid screens, or co-precipitation, one should take into account that structural interaction and catalytic interaction (or interaction with chemical consequences) are not necessarily related. A set of proteins may all associate to a microtubule network, but this does not mean that they modify the microtubules or that the microtubules have much effect on the catalytic performance of those proteins, or that the proteins affect each others' function. On the other hand, one protein may phosphorylate another one, but the contact between the two proteins can be hit-and-run (data-like) rather than continuous (party-like).

Clearly, venues are open towards the dissection of the entire cell or complex processes therein into modules, followed by global functional analyses identifying the important interactions among modules. A given process can then be described by a blueprint in which its basic modules and their regulatory interactions (positive and negative feedback, thresholds control, amplification, error correction, etc.) are represented (Alberghina *et al.* 2004). Mathematical methods have been developed to analyze the signal transfer through an entire network considering only the interactions between modules (Kholodenko *et al.* 2002; Brazhnik *et al.* 2002). The reactive strengths of such regulation routes can be asserted, particularly when a silicon cell is available (Bruggeman *et al.* 2005).

It is for Metabolic and Hierarchical Control Analysis that the modular methods have been elaborated furthest perhaps. This is because the frameworks of these Analyses enable precise results and mathematical proof. For Metabolic Control Analysis, Schuster and colleagues (Rohwer *et al.* 1996) have given a precise definition of a monofunctional unit, which is a part of metabolism that can be replaced mathematically by a single enzyme. Modules of this type can be connected by mass fluxes, but for monofunctionality this should be a single flux (or a number of, but then strictly coupled, fluxes), and processes outside the modules should not be regulated by concentrations inside the module, nor share moieties. Multifunctional units of metabolism can be defined analogously. Such modules can be characterized in terms of their sensitivities to changes in concentrations of metabolites that are external to them ('overall elasticities'; Westerhoff *et al.* 1990; Westerhoff and Van Dam 1987), and as a unit exert control on fluxes and processes in the rest of the system (Global Control). The methodology, certainly when applied in approximate modes, should greatly rationalize the control analysis of large networks. Approaches that are standard to Control Engineering may further help in this respect.

Molecular biology has championed an important simplifying paradigm according to which DNA determines mRNA, which determines protein, which determines metabolism and function. This inspired Hierarchical Control Analysis to present all cellular processes in terms of cascades of various levels that communicate only through influences, *i.e.*, not through mass transfer (Westerhoff *et al.* 1990). At each level of such a cascade a multitude of processes engages in mass transfer (such as metabolic pathways at the level of metabolism). This concept of cell function enables quite a reduction in the complexity of describing the control of cell function. Kahn and Westerhoff (1991) elaborated the method for signal transduction, where similar cascades may be discerned. Recent genetic network methodologies, tacitly or sometimes explicitly, assume the prerogatives of the cascade scheme (Brazhnik *et al.* 2002; Kholodenko *et al.* 2002).

Living cells are too complex, certainly when looked at in random ways. In order to be successful, Systems Biology needs to engage in optimum modularization of the intracellular networks.

## 3.2 Models at different levels

Since we have presented evidence of the existence of modular Systems Biology approaches, let us now see the two kinds of modeling that are feasible. For processes like metabolism or cell signaling the activity of each subsystem can be simulated from its known components in a "bottom-up" modeling approach. The chapters contributed by Heinrich, Nordlander *et al.,* Sauer, and Snoep and Westerhoff, exemplify this kind of approach. This approach can also be called synthetic Systems Biology, as it synthesizes the system from its components.

For processes such as the cell cycle and apoptosis many components are not yet known. Then it is better to use "top-down" approaches in which measures of the global activity are used to structure the blueprint of the process. Modeling algorithms will allow to test whether the very basic blueprint is able to capture the essential features of the process. If so, the components of each module will be identified by the 4M strategy discussed in the chapter by Alberghina *et al.* Several contributions to this book elucidate various aspects of the "top-down" approach: Kremling *et al.,* Bentele and Eils, Kholodenko *et al.,* Müller *et al.,* and Alberghina *et al.* The top-down approach can also be called Analytic Systems Biology. Of course, the ultimate aim is to combine the two approaches into Integrative Systems Biology (*cf.* Westerhoff and Palsson 2004).

Another aspect of Systems Biology modeling, which has not been covered in this book, is that of low resolution models of organs and of entire organisms. These models try to capture the essential features of a given process to gain better understanding and predictive ability of a body function. A very interesting example is given by the pioneering work by Noble and colleagues on heart modeling (Noble 2004).

# 4 Potential applications of Systems Biology

The most significant potential for applied Systems Biology usually cited in discussions is that of improving human health. Quite a few diseases are due to dysregulation of cellular systems and since Systems Biology aims at understanding cell regulation by experimentation and computer simulation, Systems Biology could contribute to a better understanding of such disease processes. Deregulation of cellular systems implies the involvement of a multitude of molecular factors, encoded by a multitude of gene functions. Accordingly, the diseases we are writing about here are observed as multifactorial diseases. Their impact on a patient is affected by many polymorphisms. Because other diseases have been decimated through successful therapies, most diseases that require treatment in the Western world are multifactorial, involving pleiotropic dysregulation. Examples include cancer, neurological, and cardiovascular diseases. The same is true for the ageing process. In order to improve our understanding of such complex diseases, it is essential to develop new strategies, based on the understanding of how the functional system is controlled simultaneously by many factors, *i.e.,* by Systems Biology approaches.

At some stage this requires the application of high-throughput techniques from functional genomics to acquire information on most or all genes and gene products involved in the disease process, as well as on the response of the entire organism to any possible treatment. At a second stage, however, simplification may be useful in trying to understand the essence of the disease in terms of the interactions of a large but limited number of molecules. In a third phase then, the proposed system mechanisms should be analyzed *vis-à-vis* the very large quantity of information combining clinical, experimental, and computational inputs. This may again require the use of whole genome models that translate the information into predictions of the effects of different therapeutic schedules.

Systems Biology has the potential to become part of the drug development process. Network-based drug design makes use of systems properties, for instances by trying to identify the "weak points" in an otherwise robust parasite or tumor cell. Whole organism models could then be used to simulate the effects, and side-effects, of drug treatments. In the foreseeable future, this concept is likely to become a requirement of the Food and Drug Administration for any new drug to be accepted. Because they may well substitute for animal testing and for part of the clinical testing, computer simulations could reduce that testing period as well as eliminate false drug leads at a much earlier stage. In this manner Systems Biology could do something about three of the most costly parts of the drug development process. Systems Biology should also open the possibility for development of drugs and drug combinations that are directed to specific genetic and physiological backgrounds. The concept of personalized medicine should become more realistic if the parameters of a generic model of the human body (or parts of it) are adapted to an individual and their values are obtained from personalized functional genomics. A better understanding of diseases based on Systems Biology could lead to alternative treatment strategies, for instance, based on special diets or on tailor-made nutraceuticals. Part of Systems Biology should be developed towards these venues, as its applications are likely to increase public health quite significantly and to reduce public health cost.

Further along these lines, Systems Biology could play a major role in preventing diseases and developing an improved, healthier life style. While it is not difficult to the common human sense to foresee the consequences of massive calorie or alcohol intake, the consequences of many other human activities are more difficult to assess in advance. In principle, it may be possible in the future that everyone who so wishes runs its own personalized simulation to assess the consequences or intake of certain foods, physical exercise, traveling, sleeping, etc. Popular computer games actually already go into this direction. That this scenario has significant ethical aspects, and even holds potential for Systems Biology being rejected by the public, is easy to imagine. This issue needs careful consideration.

Systems Biology will certainly be developed and applied in various areas of biotechnology, such as the engineering or breeding of industrially important microorganisms as well as plants and animals. Today, 25% of all medicines are plant derived, and the spectrum of medicinal feedstocks and efficiency of production can still be greatly enhanced. Systems Biology also holds great potential to foster sustainable development, by accelerating and rationalizing the production of plant-derived biofuels and chemical feedstocks in preparation for the inevitable depletion of fossil carbon. Probably closest to actual application is the use of simulations in the design of physiologically adapting microorganisms for biotechnological processes, such as yeasts for bio-ethanol production.

We envisage that biotechnological production methods will become highly rationalized. Whenever organisms will be used to produce food, the processes in which that food is produced, the organism by which it is produced, as well as the spectrum of compounds that end up in that food, will have to be defined and controlled with the help of Systems Biology. From a scientific point of view, the composition of contemporary food, including that produced in fairly high-technology biological processes is presently ill-defined. The specifications that a new chemical needs to meet are far more stringent than those that need to be met by traditional biologicals. Of course, the traditional biologicals, such as bread and wine, have undergone a long selection for not obviously threatening human health. On the other hand, marginal negative effects on human health from certain wine stocks, certain beer brands, or certain types of marihuana cannot be excluded, and might well turn up once the analytic methodology is put in place. Cleary, our food industry also has to become one of high sophistication. Systems Biology may help in such a development. It is no surprise, therefore, that most major food, pharmaceutical, and cosmetic industries show a profound interest in Systems Biology. Also in view of the geo-economical role the West will want to play 20 years from now, they need to turn into high-sophistication industries. For their management the situation is difficult however; Systems Biology is too big to be dealt with by an individual company. On the other hand, most companies are not very keen on pre-competitive work with other companies. EU regulations are so focused on competition that they do not stimulate such essential endeavor either.

## 5 Systems Biology: towards new ways of organizing research?

Worldwide, Systems Biology has stimulated the organization of new research alliances and even institutes. American and Japanese research institutes have already attracted wide visibility and private and public funds to invest into Quantitative/Systems Biology. For example, Kitano, a driving force in Systems Biology, runs well established activities both in Japan and the USA. Also with the E-cell program around Tomita and colleagues, Japan has become a leader in the Systems Biology field. Hood has established the Institute for Systems Biology in Seattle. This institute will soon have ample revenues from patented systems-biology diagnostic tests. The Molecular Science Institute, founded by Brenner, explicitly uses Quantitative/Systems Biology approaches. Palsson heads a rather prolific Systems Biology setup with commercial Software spin-offs. Harvard has recently established a Department of Systems Biology with Kirschner as director. Examples of interdisciplinary organizations of Quantitative Biology include Stanford University and the San Francisco Bay region. These investments are made because Quantitative/Systems Biology holds significant future potential. The demand for scientists trained in this field will increase drastically in academia and industry. The best recognized European Systems Biology group is in Amsterdam, but major other constellations and institutes are being put in place, for instance in Germany (Bioquant, Heidelberg), Switzerland (SystemsX, Basel), and the UK (MIB, Manchester). Transnational initiatives such as between Göteborg, Berlin, and Amsterdam are accelerating.

In a number of respects Systems Biology is an entirely new way of doing biology. Faced with the abhorrent complexity of biology, biologists have hitherto either reduced their object of study to a single molecule and then engaged in hypothesis driven research, or looked at large systems but then in a rather more descriptive fashion. The entire systems of Life were simply too complex for hypothesis driven research, as always an as yet unknown molecule could turn the corner and explain the inexplicable within a tested hypothesis. With functional genomics, this has come to an end: the most relevant entire living systems, *i.e.*, autonomously living cells have become knowable in terms of the identity and concentrations of all their molecules. Consequently, whole systems of Life can be analyzed quantitatively in all their components. Properties that emerge from the interactions should be understandable using nonlinear mathematics. Hypothesis driven, synthetic Systems Biology should be possible, in principle. As we discussed above, the practice is less forgiving; after all the minimum genome size of 300 is not directly a system that lends itself to simple hypothesis-driven research on that whole system; it simply is too complex. So much data is accumulating, that this is now often seen as a disadvantage, because comprehending the data in terms of preexisting hypotheses is arduous. Data-driven hypothesis generation, or Analytical Systems Biology may here come to the rescue. The form hypotheses take, may have to change. Importantly, Systems Biology mechanisms are bound to be multifactorial, and it may be better to have the data generate the more Systems Biology type of hypotheses, than to propose traditional single factor mechanisms. Ultimately, data-driven and hypothesis-driven Systems Biology should find an optimum way of mingling. Part of the procedure may be robotized (King *et al.* 2004).

Four aspects pose an enormous challenge to the organization of this new science. First, Systems Biology is BIG SCIENCE. Studying one part of the living cell now has to be done in connection with the studies of all other parts, and has to use many different methods at the same time, ranging from functional genomics, to molecular biophysics and well tuned physiology. There is little chance for a gene sequencing group at a standard size university to find the most excellent molecular biophysics groups working on the same set of macromolecules in the cellular context at the same university. Second, whereas most biologists have looked down upon mathematics as being something dull and uninteresting, and mathematicians have looked down upon biology as being much too impure and applied, these two extremes should now learn to cooperate. Students of mathematics should now be stimulated by their professors to take biology as their second topic, and biology students should be credited when taking mathematics as their second course. Although we here address the issue in terms of biology versus mathematics, Systems Biology also requires much contribution from biophysics and analytical chemistry. Naturally, Systems Biology is interdisciplinary. Collecting the quantitative data needed to build the models requires interdisciplinary research, also involving chemists and physicists. Large alliances need to be made, therefore, combining wide areas of expertise. A new group of scientists has to be grown in this new discipline of Systems Biology, which requires new types of teaching. Advanced courses dedicated to Systems Biology may be part of the answer (*cf.* www.febssysbio.net). And, in the third place, the funding and refereeing systems need to be changed. Up to this moment, funds are concentrated on the topmost excellent groups and grant applications are judged by experts on individual topics. When one needs to fund a consortium of groups that should study a topic that integrates a range of molecular topics, then it can be proven by simple mathematics that they cannot all be equally excellent, nor can the evaluation committee ever judge all parts of the grant application. And then a fourth aspect is the size of the funding units, which is linked to the size of the countries that fund the science. Systems Biology grants need to be large to be comprehensive. Moreover, because diverse expertise is needed, it is relatively unlikely that the best needed expertise is all found in a country like Luxemburg. It is more likely that the one group is in Luxemburg, a second in Lithuania, and the third group in Spain. Transnational research programs are needed in which countries that have decided to fund Systems Biology put their minds and budgets together. As of 2005, the Germans and Dutch are among the ones engaging in this for Microbial Systems Biology, with many other countries eager but perhaps not quite ready to join.

Hence, in Systems Biology, groups of scientists from different disciplines jointly work on a common biological (and often also technological) objective. Notwithstanding the lack of funding, much of this is happening. The International *E. coli* Alliance (IEcA), the Yeast Systems Biology network (YSBN), and the Receptor Tyrosine Kinase network are among the examples (for updated information see www.systembiology.net). To foster the interaction between disciplines a dedicated interdisciplinary research institute may be a suitable setup. Such a central institute could also serve to put the necessary standardization in place and to offer some of the most expensive instrumentation and expertise to all other research institutes connected. On the other hand, each of these different groups also needs to flourish within its own environment and develop its strength within its own scientific discipline. Therefore, a suitable way is efficient networks of scientists, perhaps performing a virtual research institute (*cf.* www.esbl.org). Such a setup indeed then calls for novel ways to fund and manage research.

The European Commission has had funding instruments in place that could have supported such interdisciplinary networks with significant funding. However, when the corresponding initiatives were taken, the foreseeable happened: the committees judging the proposals found them too wide, too ill focused, *i.e.*, not focused on a single molecule or a single cell type. They found Systems Biology not yet ripe for a major initiative. The field has had to develop essentially without major pan-European funding, hence, largely from Japan and the USA as we described above. National governments had to act first, including the Finnish, German and British. Especially the Germans have started to invest in networks as well as in virtual centers. The UK is now investing in actual Systems Biology centres.

Of course, here too, the management aspect remains an issue to be addressed. Often success of an interdisciplinary network depends on a coordinator that functions as a driving force for cooperation of different groups on a daily basis. Funding of interdisciplinary networks in Systems Biology not only requires the resources for doing the research and development but also for employing skilled and dedicated research managers with a strategic vision.


# References


Aderem A (2005) Systems Biology: Its practice and challenges. Cell 121:511-513

Alberghina L, Chiaradonna F, Vanoni M (2004) Systems biology and the molecular circuits of cancer. Chembiochem 5:1322-1333

Brazhnik P, de la Fuente A, Mendes P (2002) Gene networks: how to put the function in genomics. Trends Biotechnol 20:467-472

Bruggeman FJ, Boogerd FC, Westerhoff HV (2005) The multifarious short-term regulation of ammonium assimilation of *Escherichia coli*: dissection using an *in silico* replica. FEBS J 272:1965-1985

Han JD, Bertin N, Hao T, Goldberg DS, Berriz GF, Zhang LV, Dupuy D, Walhout AJ, Cusick ME, Roth FP, Vidal M (2004) Evidence for dynamically organized modularity in the yeast protein-protein interaction network. Nature 430:88-93

Hartwell LH, Hopfield JJ, Leibler S, Murray AW (1999) From molecular to modular cell biology. Nature 402:C47-C52

Kahn D, Westerhoff HV (1991) Control theory of regulatory cascades. J Theor Biol 153:255-285

Kholodenko BN, Kiyatkin A, Bruggeman FJ, Sontag E, Westerhoff HV, Hoek JB (2002) Untangling the wires: a strategy to trace functional interactions in signaling and gene networks. Proc Natl Acad Sci USA 99:12841-12846

King RD, Whelan KE, Jones FM, Reiser PG, Bryant CH, Muggleton SH, Kell DB, Oliver SG (2004) Functional genomic hypothesis generation and experimentation by a robot scientist. Nature 427:247-252

Kirschner MW (2005) The meaning of Systems Biology. Cell 121:503-504

Koebmann BJ, Westerhoff HV, Snoep JL, Nilsson D, Jensen PR (2002) The glycolytic flux in *Escherichia coli* is controlled by the demand for ATP. J Bacteriol 184:3909-3916

Noble D (2004) Modeling the heart. Physiology 19:191-197

Price ND, Reed JL, Papin JA, Famili I, Palsson BO (2003) Analysis of metabolic capabilities using singular value decomposition of extreme pathway matrices. Biophys J 84:794-804

Ravasz E, Somera AL, Mongru DA, Oltvai ZN, Barabasi AL (2002) Hierarchical organization of modularity in metabolic networks. Science 297:1551-1555

Rosen R (1991) Life itself. Columbia University Press, New York

Rohwer JM, Schuster S, Westerhoff HV (1996) How to recognize monofunctional units in a metabolic system. J Theor Biol179:213-228

Schuster S, Fell DA, Dandekar T (2002) A general definition of metabolic pathways useful for systematic organization and analysis of complex metabolic networks. Nat Biotechnol 18:326-332

Schuster S, Kahn D, Westerhoff HV (1993) Modular analysis of the control of complex metabolic pathways. Biophys Chem 48:1-17

Segal MR, Dahlquist KD, Conklin BR (2003) Regression approaches for microarray data analysis. J Comput Biol 10:961-80

Von Bertalanffy L (1962) General System Theory - A critical review. Gen Syst 7:1-20

Westerhoff HV, Van Dam K (1987) Thermodynamics and control of biological free energy transduction. Elsevier, Amsterdam

Westerhoff HV, Palsson BO (2004) The evolution of molecular biology into systems biology. Nat Biotechnol 22:1249-1252

Westerhoff HV, Aon MA, Van Dam K, Cortassa S, Kahn D, Van Workum M (1990) Dynamical and hierarchical coupling. Biochim Biophys Acta 1018:42-146

Westerhoff HV, Koster JG, Van Workum M, Rudd KE (1990) On the control of gene expression. In Control of Metabolic Processes (Cornish-Bowden A, ed), pp 399-412, Plenum, New York

Alberghina, Lilia
Department of Biotechnology and Biosciences, University of Milano-Bicocca, Piazza della Scienza 2, 20126 Milano Italy
lilia.alberghina@unimib.it

Hohmann, Stefan
Department of Cell and Molecular Biology/Microbiology, Göteborg University, Box 462, S-405 30 Göteborg, Sweden
hohmann@gmm.gu.se

Westerhoff, Hans V.
Systems Biology, Manchester Interdisciplinary Biocentre, and Molecular Cell Physiology, BioCentrum Amsterdam, Faculty of Earth and Life Sciences, BioCentrum Amsterdam, Free University, De Boelelaan 1087, NL-1081 HV Amsterdam, The Netherlands
hans.westerhoff@falw.vu.nl

Printing: Krips bv, Meppel
Binding: Stürtz, Würzburg